PLANETARIUM
THE UNIVERSITY OF TEXAS AT ARLINGTON
www.utaplanetarium.com
817-272-1183

NUCLEAR FORCES

NUCLEAR FORCES

The Making of the Physicist Hans Bethe

Silvan S. Schweber

HARVARD UNIVERSITY PRESS
Cambridge, Massachusetts
London, England
2012

Printed in the United States of America

Library of Congress Cataloging-in-Publication Data
Schweber, S. S. (Silvan S.)
Nuclear forces : the making of the physicist Hans Bethe / Silvan S. Schweber.
p. cm.
Includes bibliographical references and index.
ISBN 978-0-674-06587-1 (alk. paper)
1. Bethe, Hans A. (Hans Albrecht), 1906–2005. 2. Bethe, Hans A. (Hans Albrecht), 1906–2005—Political and social views. 3. Nuclear physicists—United States—Biography. 4. Manhattan Project (U.S.)—Biography. 5. Atomic bomb—Moral and ethical aspects—United States. 6. Nuclear weapons—Moral and ethical aspects—United States. 7. Nuclear warfare—Moral and ethical aspects—United States. I. Title
QC774.B4S39 2012
530.092—dc23
[B] 2012005094

For Rose Bethe

with deep affection and admiration

Contents

NUCLEAR FORCES

Introduction

> Hands off: neither the whole of truth nor the whole of good is revealed to any single observer, although each observer gains a partial superiority from the peculiar position in which he stands. Even prisons and sick-rooms have their special revelations. It is enough to ask of each of us that he should be faithful to his own opportunities and make the most of his own blessings, without presuming to regulate the rest of the vast field.
>
> William James, "On a Certain Blindness in Human Beings," 1899

In her review of Charles Thorpe's 2006 biography of Robert Oppenheimer, the distinguished scholar Sheila Jasanoff recalled that on a cool day in the fall of 1995, after having had lunch at the Terrace Café in Cornell University's Statler Hall, she went to retrieve her trench coat before returning to her office. Her coat was not where she had left it, but hanging nearby she found a similar-looking one and inferred that someone had taken hers by mistake. The mystery was solved when Hans Bethe, "the eminent theoretical physicist, then close to ninety in age, came shambling into the cloak room and sheepishly confessed that he was the one who had accidentally taken her coat off instead of his." They quickly swapped garments, and when writing her review in 2008 she still remembered "the slight frisson" she felt from thinking that her coat had been worn for a few minutes by a man who had seen the explosion of the first plutonium bomb at the Trinity Site in 1945—"the moment that separated the twentieth century into a time before and a time after humanity had learned self-consciously to destroy itself." Moreover, she indicated that she was not alone in reacting to Bethe in this way. Hans Bethe in his nineties still projected intense charisma and drew huge audiences to his public lectures. He was a compelling force on the Cornell campus, legendary for his continued scientific productivity into an advanced age, and hugely respected for his moral stance against nuclear proliferation.[1]

With Bethe's death on March 6, 2005, an era came to an end. He was the last of the young physicists who, from 1925 to the early 1930s, established quantum mechanics, the theory that made possible the intellectual mastery of the microscopic world. The initial formulation of quantum mechanics was the work of Max Born, Louis de Broglie, Paul Dirac, Werner Heisenberg, Pascual Jordan, Wolfgang Pauli, and Erwin Schrödinger in 1925–1926. Many of its wide-ranging applications and extensions thereafter were made by the young physicists who came of age as it was being created: Bethe, Felix Bloch, Edward Condon, George Gamow, Walter Heitler, Lev Landau, Fritz London, Neville Mott, Robert Oppenheimer, Linus Pauling, Rudolf Peierls, John Slater, Igor Tamm, Edward Teller, John van Vleck, John von Neumann, Victor Weisskopf, Eugene Wigner, and others. Many of them were later involved in the design of nuclear bombs, weapons that shaped a good deal of the history of the twentieth century. Bethe was at the center of these developments, first as head of the Theoretical Division of the Los Alamos Laboratory, then by helping design H-bombs, and subsequently when serving on influential governmental committees trying to restrain the further expansion of the atomic arsenal and preventing the proliferation of atomic weapons. And whenever possible he informed the public about the threats that nuclear weapons pose to mankind.

Bethe was an outstanding physicist. He made seminal contributions to virtually all fields of physics and was awarded the Nobel Prize for his explanation of energy generation in stars. He was a superb teacher, the much-admired and respected mentor of several generations of Ph.D.s, post docs, and research associates, and the person who molded the Cornell department of physics and the Newman Laboratory of Nuclear Studies into outstanding research centers.

It should be clear even from these brief remarks that Bethe's life should be narrated. Bethe asked me to write his scientific biography in the late 1980s. Initially I conceived of writing his biography as a study in parallel lives, contrasting him to some of his close friends and thereby allowing me to point out his strengths and his weaknesses. Although part of this undertaking appeared in an earlier publication dealing with Bethe and Oppenheimer, it was only fairly recently that it became clear to me how to narrate his life.[2] There are many reasons for this.

Biography has always been considered "weak" history. Claude Lévi-

Strauss in *La Pensée Sauvage* noted that "Biographical and anecdotal history, which is at the bottom of the ladder, is weak history that does . . . not contain its own intelligibility; which it gets only when it is transferred *en bloc* into a history stronger than itself; and the latter stands in the same relation to a class above it."[3] Most historians would agree with him. Historians of science who have written biographies have constantly felt the need to justify the enterprise, and the role of biography in the scholarship of the discipline has been the focus of constant discussions. Thus in 2006 the editor of *Isis,* the lead periodical of the American history of science community, asked four distinguished historians of science—Mary Jo Nye, Joan Richards, Theodore Porter, and Mary Terrall—who each had recently written an important and well-received biography of a scientist, to address the issue, "Biography in the History of Science."[4] Not surprisingly, given that biography has been one of the principal narrative modes of the history of science and that each one of them had invested considerable time and effort in writing a biography, their conclusion—for different reasons—was that there is still considerable merit in scientific biography as a vehicle for narrating the history of science.

In 2007 Mott Greene, another distinguished historian of science and the biographer of the German geophysicist, meteorologist, and climatologist Alfred Wegener, wrote a sensitive, insightful, and critical essay on the role and function of scientific biographies. He came to the conclusion that "biography, however useful, exerts a powerfully distorting influence on the image of how most science gets done."[5] Given his criticisms, it is important to review them. Doing so will also allow me to indicate what I have tried to do that is different.

Greene gives convincing arguments that in the past the writing of biography has had many similarities to the writing of historical novels. Greene contends that both are controlled by the extensive genre conventions of the *Bildungsroman* and by the conceptual frame of Max Weber's theory of ideal types. These in turn derive from the folkloric conventions of the tales of the "hero's quest." It is Greene's claim that these genre conventions, which are "independent of historiographic stance and even of the choice of a historical master narrative, . . . exert a powerful selection pressure that determines whether or not a scientist's life gets written, sometimes irrespective of the importance of that scientist's work."[6]

Additionally, Greene noted that when science is taken to be "a cumulative and progressive enterprise," scientific biographies provide the origin stories for the developments in science that the community at large considers important. They also aim to explain how the inner experience of those individuals deemed responsible for the progress in a particular science becomes public knowledge. And often, they serve "the function of providing exemplary lives for the edification of apprentice scientists, as algorithms of 'scientific lives well lived.'"[7]

But "for all the good that biography does" and however useful, Greene concludes that biography addresses principally the lives of extraordinary people, and thus cannot "by its very nature, portray most scientific careers and cannot therefore instantiate the general development or movement of the sciences . . . [and] works against the attempt to portray science as a collaborative social enterprise" developing cumulatively and incrementally.[8]

I believe that Greene has focused too much on the scientific biographies of the "great men" of science before the twentieth century—Newton, Faraday, Kekulé, Darwin, Kelvin, and others. And although he notes the important transformation of the sciences during the past half-century—as reflected for example in the multiple authorships of most scientific papers—the different sciences have their own development histories and cannot be treated collectively. The history of theoretical physics—which as a discipline is primarily a twentieth-century creation—is a history dominated by "off-scale" people. The post–World War II intellectual agenda of the subdiscipline that used to be called elementary particle physics was set for various periods of times by "off-scale" individuals: Julian Schwinger, Richard Feynman, Murray Gell-Mann, Gerard 't Hooft, Steven Weinberg, Kenneth Wilson, and others. The issue is to understand why this was so in this particular field and much less so in what was called "solid state physics."[9]

Bethe was an "off-scale" physicist. My biography does have some of the features that Greene attributes to scientific biographies. But it is as much an attempt to understand the self-development of Bethe as an individual and as a physicist as it is an attempt to understand the development of theoretical physics during his lifetime. Recall Plutarch's observation: "What we achieve inwardly will change outer reality." Having been a Darwin scholar, I considered it self-evident that I must situate Bethe in the institutions and networks that made his career possible. But equally important for me was to understand and elucidate how local differences—among Frankfurt University, Sommerfeld's seminar in Munich, Ewald's in Stuttgart, the Cavendish in Cambridge, Enrico Fermi's institute in Rome, Niels Bohr's in Copenhagen, and, most important, Cornell—were reflected in the physics produced in them, in Bethe's persona, in his presentation of self, *and* in his physics.

Making the biography verifiable presented no difficulties. In fact, I was overwhelmed by the available documentation. Cornell University has over 200 boxes of Bethe materials in its archives. They contain his scientific and professional correspondence; extensive notes for many of the papers he wrote; notes for the courses he taught and for the speeches he made; materials on nuclear shielding, test bans, antiballistic missiles, energy and energy policy, the Union of Concerned Scientists, the Strategic Defense Initiative, and much else. There are also many technical reports at MIT written during the year Bethe spent in the wartime Radiation Laboratory; a large number of reports at Los Alamos dating from World War II up until the 1960s; and technical reports Bethe wrote for the GE Knolls Laboratory, for Detroit Edison, General Atomics, and Avco Corporation. Extensive materials can be found at the Institute of Physics in Maryland relating to his presidency of the American Physical Society; transcriptions of extended interviews with Charles Weiner and Jagdish Mehra, and with Judith Goldstein; transcriptions of other interviews and speeches he made; and newspaper and periodical articles that are not in the Cornell Archives. And of course there are the books and near 300 scientific articles he published, some of them of book length. There are also my recorded interviews with his colleagues and the extensive tape-recorded sessions I had with him in which he related his memories; as well as the numerous conversations with his wife and children.

To manage this overabundance of available data, of information of what Bethe did and what others had done that impacted on him and his work, the criterion that I adopted to narrate Bethe's life is relevance. I took Bethe to be representative of the role played by physicists during the twentieth century, and my biography is an attempt to understand, interpret, and communicate the meaning of and the reasons for Bethe's involvement, and that of other physicists, in the shaping of the history of the twentieth century. What guided me in my selection was to make possible a narration of Bethe's life that would aid understanding of the field at present and explain how we came to be where we are.

Bethe was born in 1906 in Strassburg (now Strasbourg), then part of Germany. At the time, his father, Albrecht Bethe, was an *Assistent* to the physiologist Richard Ewald. His mother was the daughter of a professor of medicine at Strassburg University. Hans was the only offspring of the marriage.

In 1915 Hans's father accepted the directorship of the Institute for Animal Physiology at the newly established Frankfurt University, and Hans attended *Gymnasium* in Frankfurt. He was so superior to his fellow students in physics at Frankfurt University that one of his teachers there urged him to go to Munich to study with Arnold Sommerfeld. In the spring of 1926, just when Erwin Schrödinger's seminal papers on wave mechanics were being published in the *Annalen der Physik,* Sommerfeld accepted him to his seminar, then the outstanding school of theoretical physics in Europe. By 1931 Bethe had published ten papers, three of which became classics.[10]

Bethe's off-scale ability to absorb and analyze huge amounts of complex, difficult data and synthesize the materials into clear, coherent, useful knowledge with the help of transparent, readily understandable models was first demonstrated in the two encyclopedic *Handbuch der Physik* articles he published in 1933. In them he gave a masterly exposition of the application of quantum mechanics to atomic, molecular, and solid state physics. Each was written in less than a year and reflected his uncommon energy, his extraordinary powers of concentration, and his great ambition. These articles set standards for subsequent contributions to these fields and they have remained classics to this day.

In 1933, after Hitler came to power, Bethe lost his position as an assistant professor in Tübingen. His mother had been Jewish before she converted to Protestantism. Sommerfeld helped him obtain a fellowship in England. In February 1935 he joined the physics department of Cornell University, at which institution he stayed until the end of his life.

During the 1930s the frontier of physics shifted to nuclear physics, and Bethe became an acknowledged leader in this field, co-authoring with Stanley Livingston and Robert Bacher three lengthy articles in the *Reviews of Modern Physics* that became known as the "Bethe Bible" of nuclear physics.[11] His mastery of nuclear physics made it possible for him to put forward in 1938 his explanation of energy generation in stars, for which he won the Nobel Prize in 1967.

This biography is concerned with the molding of the young Hans Bethe. It focuses on the factors that endowed him with an identity. It concentrates on his parents; on his upbringing; on the cultural environment in which he grew up; on his studies and the institutions in which they took place; on his friends; and on the contingencies that made possible his becoming the person he was when World War II broke out: a uniquely gifted personality able to seize the opportunities before him and successfully accomplish all he did thereafter.

It was Bethe's father who stimulated and nurtured his interest in science. Chapter 1 analyzes his influence in shaping Bethe's scientific outlook. Bethe's first exposure to the practice of science was in his father's laboratory. He there became aware of the amazing diversity of animal life and learned that individual behavior can never be considered by itself but must always be seen through interactions with the environment and all the entities that make up that environment. He there also first experienced science as a social activity in which his father, his *Assistenten, Doktoranden,* and laboratory assistants were in constant interaction in a shared physical and intellectual environment. His father taught in the medical school, and as Hans grew up, he made him aware of the dual significance of his professorial activities. On the one hand, his research on the structure of the nervous system of invertebrates helped elucidate their behavior, helped give direction to research on the nervous systems of higher organisms, and helped recognize a common feature of all nervous systems: their plasticity. But these researches also proved to be of great practical value, and the insights gleaned from them proved very helpful during World War I and thereafter in rehabilitating soldiers who had suffered brain injuries.

Bethe's introduction to science in his father's laboratory was as a practice in which knowledge is created by experiments that use instruments to measure with limited accuracy, and produce data that have to be analyzed statistically and interpreted with mathematical models that idealize the context in which the interactions take place.[12] The aim of the knowledge produced was to understand the complexity and diversity of the biological world. There was no attempt to find an ultimate theory that would explain all biological phenomena.

Like his eminent predecessor Hermann von Helmholtz, Albrecht Bethe saw the mid-nineteenth-century developments in physics—in particular the establishment of the first and second law of thermodynamics—as giving physics a universal character. He understood the drive in physics toward atomic and microscopic explanations as a drive to free it as much as possible from context dependence. The possibility of explaining macroscopic phenomena in terms of microscopic, ahistoric components is what stimulated Albrecht and other physiologists to investigate those chemical and physical processes in living systems whose descriptions were local, had a seemingly universal character, and could be mathematized.

In the first scientific paper on which the name H. A. Bethe appeared, Hans was a co-author with his father and one of his father's *Assistenten.* He supplied the solutions of the diffusion equation applicable to an investigation of dialysis that his father was carrying out with an assistant. For the paper Hans tested whether the data fit the predictions of the diffusion equation when the parameter in that equation—the diffusion constant—was suitably adjusted to the given environment. In addition Hans tried to understand when the data did so and when they did not—all when he was seventeen.

It was Arnold Sommerfeld who molded the young Bethe into a researcher whose investigations of physical problems were careful, detailed, and characterized by mathematical virtuosity. Chapter 2 analyzes Sommerfeld's background, his association with Felix Klein and the Göttingen circle of mathematicians, his metaphysics, his demeanor, and how his views of mathematics, science, applied science, and technology influenced his students—Bethe in particular. Like Bethe's father, Sommerfeld indicated to Bethe that *how* science was practiced was a significant factor in giving meaning to being a scientist. The communicative aspect of Sommerfeld's seminar was one such indication. In his lectures and in his interactions with students, *Doktoranden,* and *Assistenten,* Sommerfeld also demonstrated how to be a responsible teacher, *Doktorvater,* and mentor. Sommerfeld was there for Bethe when he needed his help after he had been dismissed from Tübingen in 1933. In addition, Sommerfeld buttressed one of Bethe's most important traits: his self-confidence. Sommerfeld early on told Bethe that he was one of his best students, among whom were Peter Debye, Paul Ewald, Gregor Wentzel, Wolfgang Pauli, and Werner Heisenberg, and therefore that he could aspire to a professorial career.

The following incident bears telling in this connection. In the fall of 1988, the American Academy of Arts and Sciences celebrated the eightieth birthday of Bethe's close friend Victor (Viki) Weisskopf. Bethe was one of the speakers at the gathering in Cambridge. Kurt Gottfried, a distinguished physicist and a close friend of both Bethe and Weisskopf, was the master of ceremonies. When he introduced Bethe, Gottfried told the following story: in 1934, Weisskopf was Pauli's assistant. They had formulated the Pauli-Weisskopf quantum field theory of spin zero particles, and Pauli asked Weisskopf to compute the cross section for the pair production by photons of such particles. Bethe had previously carried out the more difficult calculation for spin 1/2 particles, in other words, for electrons and positrons. Bethe and Weisskopf both happened to be in Copenhagen at the time, so Weisskopf went to see Bethe to ask him how long it would take to do the calculation. Hans answered: "Me it would take three days. You it will take three weeks!" Before starting his lecture, Bethe indicated that the story was true and admitted that "I was very conceited at that time. I still am. But now I can hide it better!"

In 1934 Bethe published a paper in *Zeitschrift für Physik* entitled "Critique of Schachenmeier's theory of superconductivity." It was a detailed, highly critical answer to Richard Schachenmeier's lengthy response to a brief criticism that Bethe had previously offered of Schachenmeier's theory of superconductivity in Sommerfeld and Bethe's *Handbuch der Physik* article, *Elektronentheorie der Metalle.* Bethe's response to Schachenmeier was all set in small print, as Bethe had requested. This was because Bethe had told the editor of the *Zeitschrift für Physik* that Schachenmeier's theory didn't deserve any more space. And the concluding sentence of Bethe's article was: "This enumeration of the mistakes in Schachenmeier's papers makes no claim to completeness." The sentence was not meant to be funny.[13] If "conceited" was a correct designation in 1934, it is important to understand the cultural factors that made such conceit and behavior necessary, and by extrapolation why "self-confident" would be the more accurate description for the older Bethe. Chapter 4 takes up these matters.

If one facet of Bethe's character was the self-confidence he acquired as a young man, another was the modesty he displayed in his mature years. Starting in the early 1990s I recorded long interviews with him regarding his life. In large measure they formed an autobiographical study. What struck me at the time, and what struck me even more when I studied the transcriptions, is how unpretentious he was. He would always insist on giving credit to others when collaborative efforts were involved and he would always indicate the sources that gave rise to his inquiries.[14] Another characteristic of these interviews is the acuity of his amazing memory and the honesty with which he admitted any lapses. Whenever he did not remember a particular event or some detail of one, he would say forthrightly "I don't remember." A third characteristic of these interviews is the objectivity with which he related his personal interactions. He did not shy away from acknowledging the pain he had caused others. He was well aware that his success had exacted a heavy price on his family—on his wife and on his children. He could talk about this because he had come to terms with it—and because his wife, Rose, and his children, Henry and Monica, had helped him come to be at peace with it.

In 1931 and 1932 Bethe went to Fermi's Institute of Physics in Rome as a Rockefeller Fellow. Bethe's way of doing theoretical physics became an amalgam of Sommerfeld's and Fermi's. It combined the thoroughness and the mathematical rigor of Sommerfeld with the "lightness," simplicity, and back-of-envelope calculational insightfulness of Fermi.

There was another mentor who affected Bethe deeply, but whose influence Bethe did not acknowledge in his oral interviews: Paul Ewald. In his doctoral dissertation Bethe applied Ewald's theory of X-ray scattering by crystals to electron diffraction by crystals. In 1929 he became Ewald's *Assistent* for a semester in Stuttgart and there learned of the solid state problems that had been intractable in the old quantum theory but that the new quantum mechanics could solve. In Ewald's Institute of Theoretical Physics he also became aware that physics research could be carried out in cooperation rather than in competition with others.

Bethe's experiences before coming to Cornell in 1935 made him aware of what Karin Knorr Cetina has succinctly stated about knowledge practices: "Epistemic cultures [are] those amalgams of arrangements and mechanisms—bonded through affinity, necessity, and historical coincidence—which, in a given field, make up *how we know what we know.* Epistemic cultures are cultures that create and warrant knowledge."[15] Theoretical physics was indeed practiced differently in Munich, Stuttgart, Cambridge, and Rome, and scientific knowledge was produced and warranted differently in each of these places—though a consensus on the value of the work done in the different institutions was eventually arrived at by the community at large, this because experiments could be repeated and improved upon, calculations checked, assumptions analyzed, and results compared with experimental data. For the physics community "Nature" was the final arbiter.

Just as Bethe's interactions with his father, with Sommerfeld, Ewald, Douglas Hartree, Fermi, Patrick Blackett, Neville Mott, and others shaped him as a theoretical physicist, similarly, the social milieu in which he grew up shaped his personality and cultural outlook.

Albrecht Bethe's first marriage was to Hans's mother, Anna, a woman of Jewish descent and the daughter of an assimilated professor of medicine.[16] His second wife was also of Jewish descent. After Hans's parents moved to Frankfurt in 1915, most of Hans's friends—male and female—came from assimilated Jewish families. This affinity continued when he went to Munich, where he met Rudolf Peierls, Edward Teller, Margarete Willstätter, and others; and again later when he spent time in Stuttgart as Paul Ewald's *Assistent* and got to know Ella Philippson Ewald, Paul's wife and the mother of Rose Ewald, who later would become Hans's wife. Bethe found the warmth, liberality, and cosmopolitanism of the homes of his assimilated Jewish friends gratifying. In the case of the Ewalds, it was Ella, a distant relative of an influential Reform rabbi, Ludwig Philippson, who made the household so. A part of Chapter 2 attempts to formulate some insights to explain Bethe's affinity to these individuals of Jewish descent and to address the remarkable creativity of assimilated Jews in Germany. I want to stress that like Ivan Strenski in his *Durkheim and the Jews of France* I do not believe in essentialist strategies to explain Jewishness. Strenski put the matter succinctly: "Jewishness is something that can be altered and transformed, then maximized or minimized in response to circumstances and either brought out or obscured by other identifications. The extent to which Jewishness becomes a significant marker of identity for an individual depends, then, on the reading of historical and social circumstances both by the individual and by the surrounding society."[17] Like Strenski, I have focused on the context in which Bethe's interactions took place—whether these were social or intellectual interactions. This is why I have described at some length the families of Bethe's friends in Frankfurt, and some of Albrecht's Jewish students.

Given the centrality of physics and of science in Bethe's life and this biography's focus on the history of physics in the twentieth century through Bethe, my narrative includes a presentation of some of the physics produced by Bethe, and more generally of some aspects of the theoretical physics created in the first half of the twentieth century. It is, however, not a chronicle of what Bethe did in physics. I do try to convey the significance and importance of his scientific endeavors, and do so by including some of the technical details. But my focus is on when, how, where, and why he undertook solving particular problems, and why his solutions proved—or did not prove—to be generative. Some of the more technical expositions have been relegated to appendixes. What has been included in the text proper can be skipped by readers if desired.

I devote more space to an exposition of Bethe's research for his doctorate and the research submitted by him to give him the right to teach at a university than to what he later did in nuclear physics. His earlier works reflect his activities when working by himself: they highlight his assessment of the conceptual insights and technical tools that the new quantum mechanics provided. His later research during the 1930s reflects his intimate relationship to the nuclear physics group at Cornell and his close connection to the wider nuclear physics community. His early works indicate that there is much greater continuity in the models and the mathematical methods used when "paradigms" are replaced than suggested by the thesis presented by Thomas Kuhn in *The Structure of Scientific Revolutions.* The concepts in the paradigms that governed the physics community in the twentieth century before and after a Kuhnian conceptual revolution had occurred are less "incommensurable" than Kuhn postulated. Furthermore, the problems that Bethe solved at that stage give proof of the progress that the new quantum mechanics made possible and, contrary to Kuhn's assertion, imply that there is a coherent direction of ontological development in the history of physics: there is an accretion of stable knowledge concerning what exists and occurs in the world. Its stability stems from the recognition that the knowledge and its representations are delimited to specific domains. And it is the community at large that warrants the correctness of the attribution of the entities and representations that are to explain the possible phenomena in that domain. But this warrant has a historicity attached to it. It is in this sense that the knowledge can be said to be objective.

The exposition of the physics Bethe did reflects, on the one hand, his ever-greater integration into the physics communities and, on the other, the evolution, growth, and maintenance of these very same physics communities. At a given time, the scientific knowledge he produced and that produced by the community in which he was immersed—whether Sommerfeld's institute in Munich, the Cavendish in Cambridge, Bragg's institute in Manchester, or Cornell University—demonstrate that such knowledge is problem-specific, institution-specific, and culture-specific, only later to be assessed, warranted, and amalgamated by the wider community.

My biography of the young Bethe concludes with his marriage to Rose Ewald in September 1939. The year was a turning point for Bethe in many respects. After his marriage, he cut the umbilical cord to his mother, got ever more deeply involved in war work, and was recognized as one of the outstanding physicists of his generation.

In January 1939 Niels Bohr brought the news of Otto Hahn and Fritz Strassman's discovery of fission to the United States. By the summer of that year the possibility of making an atomic bomb became apparent. In September 1939, war broke out when Germany attacked Poland. Bethe's involvement with weaponry began after the outbreak of hostilities. Thereafter, until he returned to Cornell in February 1946, all his energies were devoted to the solution of problems with military applications: armor penetration, radar, shock waves, atomic bombs.

During the summer of 1942 Bethe took part in extensive discussions in Berkeley regarding the feasibility of atomic weapons. Although he had been skeptical of the possibility of building an atomic bomb in time to affect the course of the war, Edward Teller convinced him otherwise. He joined the Manhattan Project in March 1943 and became the director of the Theoretical Physics Division at Los Alamos.

His marriage was an unambiguous decision for Hans, but not so for Rose. She knew what it meant to be the wife of a "Herr Professor." Her father, Paul Ewald, had been one and she was well aware that Hans was so inclined. She eventually overcame her apprehensions and felt that she could indeed be a good partner for him. Yet the first year of their marriage was very difficult. Bethe's mother lived with them and almost wrecked the marriage.

The nature of their partnership is indicated by the fact that Hans told Rose the implication of the discovery of fission for weaponry and informed her about what was going to be discussed in Berkeley in the summer of 1942. Except for Fermi, all of the European scientists who came to Los Alamos shared this information with their partners, but none of the Americans did. Rose became deeply involved in discussions with Hans on the morality of building atomic bombs whose scale of destruction was orders of magnitude greater than any conceivable chemical bomb. The belief that Great Britain and the United States were in a desperate race with Nazi Germany, which presumably had started developing such bombs much earlier, carried the day. Going to Los Alamos was a joint decision. The morality of *using* the bomb came up only after the bombings of Hiroshima and Nagasaki. Bethe believed that the Hiroshima bomb was justified: it had prevented the deaths of hundreds of thousands of Americans and Japanese by ending the need for an invasion of the main islands of Japan.

In the fall of 1949 the issue of whether to build an H-bomb arose—a response to the Soviets having detonated an atomic bomb. Rose argued emphatically that it would be wrong to do so on moral grounds. Two close friends, Victor Weisskopf and George Placzek, were of the same opinion, and the three of them convinced Bethe. Bethe initially declined Teller's invitation to work with him on the project. Bethe next wrote an article for *Scientific American* in which he eloquently and effectively argued against the development of the hydrogen bomb, even after Truman had ordered a crash program to build one. The tone of his article reflected Rose's moral passion and the vehemence of her moral objections. She had made clear to Hans the inhumanity of the project, even though pretenses of humanity were being invoked to justify it. Sensitized by her upbringing and trained as a social worker, Rose was well aware of the brutality and heartlessness that can exist beneath pretenses of humanity and civility.

In this *Scientific American* article Bethe asserted that the most important question to answer in proceeding with the development of an H-bomb was the moral one: "Can we who have always insisted on morality and human decency between nations as well as inside our own country, introduce this weapon of total annihilation into the world?" After indicating the genocidal consequences of a war in which hydrogen bombs were used, Bethe adduced the one and only reason that could justify the building of an H-bomb by the United States, namely, "to deter the Russians from using it against us, if only for fear of our retaliation." And if this were the reason for "us" building it, Bethe wanted the United States to declare that "it will never be the first to use an H-bomb." Rose played an important role in Bethe's moral evolution. The intensity of his moral concerns increased with her presence in his life. Chapter 9 is devoted to Rose's life before she married Hans.

After Stanislaw Ulam and Teller found a mechanism to build a fusion bomb, Bethe agreed to work on implementing its design. The reason was deterrence. If the United States could build one, so could the Soviets, and the balance of terror would prevent their use. Bethe came to subscribe to the doctrine of mutual assured destruction (MAD). Thus the moral questions were addressed and answered by separating the issues of *building* a thermonuclear bomb and *using* it; believing all the while that its existence would prevent its use.

Bethe's experiences during World War II transformed his life. Bethe is the paradigmatic example of why theoretical physicists proved to be so valuable in the war effort. It was his ability to translate his technical mastery of the microscopic world—the world of nuclei, electrons, atoms, and molecules—into an understanding of the macroscopic properties of materials and into the design of macroscopic devices such as radar junctions and atomic bombs that rendered his services so valuable at the Radiation Lab at MIT and at Los Alamos.[18]

Physicists had conceived of atomic bombs, and only they had the expertise to make them a reality. During the war they believed that because of this they would have a say in their use. The Trinity test in July 1945 indicated the destructive power of such bombs and confirmed that they—the scientists—had transformed the world by the creation of these weapons. Hiroshima and Nagasaki made them aware of the scale of human suffering they had wrought and also made clear to them the political realities concerning who decided their use.

The creation of nuclear weapons owed much to Oppenheimer's directorship of Los Alamos. Very few people at Los Alamos made technical and managerial contributions as important and consequential as Bethe, however. Norris Bradbury, who became the director of Los Alamos after Oppenheimer resigned in December 1945, when interviewed by the FBI in 1947 in connection with the renewal of Bethe's security clearance, stated, "Dr. Bethe did as much toward the actual development of the bomb as any one individual connected with the program."[19] It was the Theoretical Division under Bethe's leadership that designed the uranium bomb and estimated the damage that would result from its use. It was never tested before being dropped on Hiroshima. Similarly the Theoretical Division played a crucial role in the design of the plutonium bomb, with Bethe himself designing its neutron initiator, his design being superior to Fermi's.

It is ironic that many of the physicists who participated in the creation of nuclear weapons looked back on wartime Los Alamos as almost a utopia.[20] When they returned to their universities after the war, they tried to recreate the spirit of cooperation and commitment that had existed at Los Alamos and the illusory feeling of wholeness that had permeated the place. The Newman Laboratory for Nuclear Studies at Cornell was Bethe's attempt to do this. It was to be devoted to "pure" research. And in contrast to other laboratories of nuclear studies—such as the ones at MIT, Chicago, and Iowa—at Cornell chemists, "nuclear" engineers, and metallurgists were excluded. It was Bethe's assertion of what Pierre Bourdieu has called the "autonomy" of a scientific field. With the physicists gathered around him—Richard Feynman, Phillip Morrison, Dale Corson, Boyce McDaniel, Robert Wilson, and others—Bethe created not only one of the outstanding centers of high energy physics but also a research institute imbued with a intense sense of community.

It was part of Bethe's greatness that he was able to endow the Newman

Laboratory for Nuclear Studies and the physics department at Cornell with the qualities and norms to which he was so deeply committed. Bethe believed that communities devoted to the growth of knowledge maintain themselves by virtue of a commitment to civility, cooperation, trust, and truthfulness; and that they must be free to set their goals and their intellectual agenda. These communities were for him the guarantors that one of the most exalted of human aspirations, namely, "to be a member of a society which is free but not anarchical"—as Isidor Rabi later put it—could indeed be realized. Additionally, Bethe believed that such communities were models for how larger democratic societies could function.

Science was the passion of Bethe's life. Beginning in 1928 and continuing for over seventy years until he was well into his nineties, Bethe published over 300 scientific papers on an astonishing array of subjects: atomic and solid state physics, nuclear physics, astrophysics, quantum field theory, high energy physics, shock waves, neutron stars, neutrinos, and supernovae, among others. And many of these articles—as in the case of his 1939 article on energy generation in stars—were the starting point for subsequent developments in the field. He probably wrote an equal number of technical reports after 1939 as a consultant to industry and the government. Many of them are still classified.

I have stressed Bethe's physics, his creativity and generative productivity in physics, because doing science—and physics in particular—was not only his passion but also his anchor in integrity. But Bethe was aware that his field was also the locus of a competitive struggle, and he was conscious of the fact that he was very successful in that competition.[21] The reason for his success was that he was conservative and cautious in choosing research topics and did not devote his off-scale intellectual powers to somewhat speculative endeavors. He would only tackle problems about which he could say to himself, "I can do that," and only address those he believed his technical capacities and stored knowledge would allow him to solve. He was very much aware that his successes in that struggle—his achievements—gave him scientific authority, status, recognition, and fame. He also understood that his continuing achievements were responsible for his status after World War II as the much-sought-after expert in the corridors of power in Washington. And he was very conscious of the fact that it was not only his technical competence but also his integrity that had made him influential both in the scientific community and in Washington, and that these combined qualities had given him a unique moral authority.

There is another trait that Bethe acquired as he matured: the ability to gauge whom to trust among those who were responsible for the institutions he was associated with—and when to do so. Thus he came to trust Sommerfeld; to trust Clifton Gibbs, the chair of the department of physics at Cornell when he came there in 1935; Livingston Farrand, the president of Cornell at that time; and Edmund Ezra Day, who succeeded him. Similarly during the war he came to trust Oppenheimer, Rabi, and Robert Bacher; and after the war Robert Wilson and Dale Corson, his colleagues in the nuclear lab. And later when he was a member of the President's Science Advisory Committee (PSAC), he came to trust Dwight D. Eisenhower, James Killian, Robert McNamara, and McGeorge Bundy. Trust meant that he would accept the judgment of these people in matters he considered himself not an expert in or had insufficient knowledge of. Trust meant that he had confidence in the way they ran the institution they headed and that he could leave its running in their hands without apprehension.

After World War II Bethe had to decide how best to amalgamate his expertise and his concerns concerning nuclear weapons and in what capacity he would most effectively help control their development and curb their proliferation. For awhile he was very active in the Federation of American Scientists (FAS), giving lectures and radio talks to educate the public about the new world they were living in. But as he kept going to Los Alamos every summer—thereby maintaining his expertise—and as he became extensively involved in advising the armed forces, de facto he gradually became an insider rather than an outsider in his advice and criticisms of governmental policies regarding nuclear power—both civilian and military. He maintained his Q clearance—the highest level of security clearance—until the late 1960s.[22]

One of the consequences of the secrecy that the weapons work entailed was that it eroded some of the bonds between Hans and Rose. Secrecy "cut off" a vital part of Bethe from Rose, namely, his work, and "separated" them.[23] After the war, upon his return to Cornell, Bethe devoted more and more time to doing pure physics and discharging his responsibilities as the leader in the Laboratory for Nuclear Studies and the department of physics at Cornell. At the same time, however, he became ever more involved in matters relating to atomic bombs as an adviser to the armed forces and the Department of Defense, and in his role as a consultant to Los Alamos and to industry. Rose therefore had to assume ever greater responsibilities in raising their two children. A rift developed between them that was not easily overcome, and was bridged and healed only after painful experiences on the part of their children.

The high esteem in which Bethe was held as an outstanding scientist, as an expert in weaponry and nuclear matters, and as an upright citizen resulted in his being appointed to some of the most influential committees in the Pentagon and the White House. In 1956 he was invited to join the committee that advised the president on scientific matters, the committee that later became PSAC, after Sputnik.

In 1985 Bethe reminisced about PSAC and indicated that one of the most memorable hours of his life was the time when PSAC had its meeting with President Eisenhower after the launching of Sputnik.[24] "Eisenhower was truly amazing in the speed of his understanding of our suggestions. After each suggestion, he might ask for a few details, but then he quickly agreed, and asked his adjutant, Brigadier General Goodpaster, to make a note and see that our suggestion was implemented. I have never before been present at a session where so much was decided in such a short time. Eisenhower was most impressive."[25] Eisenhower appointed James Killian as his special assistant, serving as his science adviser. Killian was elected chair of PSAC, and shortly thereafter PSAC established several panels. One of these dealt with disarmament and another one with strategic military problems. Bethe served on both these panels.[26] As a member of the disarmament panel he proposed that the United States might start with a ban on nuclear weapons tests. The panel accepted his suggestion. Killian then took the recommendation to Eisenhower, who was very much in favor of investigating this possibility. As a consequence of this favorable reaction by Eisenhower, an interagency committee was set up to investigate (a) whether test cessation could be verified, and by what means, and (b) whether test cessation would be to the net advantage of the United States. Bethe was made chair of this committee.[27] Though excluded from the official delegation, Bethe's participation in the deliberations at Geneva played a key role in the signing by the United States and the Soviet Union in 1963 of a treaty that forbade atmospheric and underwater nuclear tests and limited underground testing to low-yield weapons. For all its limitations, the treaty was for Bethe a first step in nuclear arms limitation, and also a justification for his insider status.[28]

In retrospect, Bethe felt that perhaps PSAC's most important accomplishment was offering resistance to the technological imperative that drove the military industrial complex. Eisenhower had warned the country about such an imperative in his farewell address to the nation. PSAC gave Eisenhower the technical resources and thereby the means to counter the most extreme policies that the armed forces and some members of Congress—not to mention the clique of Ernest Lawrence, Teller, and Lewis Strauss—wanted to implement.

Lawrence Radiation Lab at Livermore was the second weapons laboratory, established in 1952 at the insistent and obsessive urging of Edward Teller and his allies and backers in the Air Force and Congress, to pursue more intensely the development of thermonuclear weapons. Livermore was the paradigmatic incarnation of the technological imperative.[29] The lab's working philosophy called for always pushing at the technological extremes. Everyone working there accepted it. Its staff's commitment to thermonuclear weapons was summarized by Herbert York, its first director: "We set out from the start to construct nuclear explosive devices that had the smallest diameter, the lightest weight, the least investment in rare materials, or the highest yield-to-weight ratio or that carried the state of the art beyond the currently explored frontiers. We were completely confident that the military would find a use for our product after we proved it and that did indeed usually turn out to be the case." Its leadership was continuously engaged in efforts to sell its ideas, to anticipate military requirements, and to suggest to the military ways in which its new designs could be used to support the United States' nuclear strategy. It operated under a doctrine of technology first, requirements after the fact.[30]

It was this operating philosophy that PSAC was able to oppose with partial success during Eisenhower's and Kennedy's presidencies. Although it was deeply involved in making recommendations regarding missile and antiballistic missile (ABM) systems, and many other weapons systems, it was also instrumental in stopping the development of a nuclear-powered airplane, in reducing the duplication of missile development, in founding NASA as a civilian establishment, in making recommendations regarding high energy accelerators, and much else. Since reliance on technology was crucial in obtaining information about Soviet nuclear weaponry, ABM systems, and more generally about Soviet military preparedness and knowledge of Soviet industrial production, PSAC also became involved with the development of the U2 plane. The information gathered by the U2's high-altitude flights over the Soviet Union was a key factor in making Eisenhower skeptical about the information and advice he was receiving from the Central Intelligence Agency and the Air Force.

But because PSAC had not been created by Congress, it became easy for a president to change its role. The technological demands and possibilities of the Vietnam War led to its demise under Richard Nixon. Although Bethe had initially supported Lyndon B. Johnson's expansion of the American involvement in the Vietnam War, by 1968 he had become disillusioned with Johnson. His escalation of the war in Vietnam, his decision to erect a "soft" ABM system to safeguard the nation against possible Chinese missiles, his wanting to have PSAC suggest technologies to aid in the Vietnam conflict—all led Bethe to join George Kistiakowsky and Weisskopf to support Eugene McCarthy in his bid to get the Democratic Party's nomination for the presidency.[31]

Already under Johnson's imperial presidency the litmus test for an appointment to PSAC became whether the candidate would support Johnson in the construction of an ABM defense system. Nixon further politicized PSAC appointments after his re-election in 1972: no Democrat would even be considered, and all appointees were expected to support and defend all of Nixon's policies. Nixon disbanded PSAC in early 1973 after Richard Garwin, a member of PSAC, testified before a congressional committee, stating his opposition on technical grounds to the development of a supersonic transport aircraft that Nixon had decided to have built for political reasons.

Nixon's re-election in 1972 marked a turning point in Bethe's life. The fact that he was a Democrat meant that his advice and expertise were no longer sought by the government. This, however, freed him to resume an extremely productive career as an astrophysicist, after having made major contributions to the theory of nuclear matter.[32] With Gordon Baym, Christopher Pethick, and others he worked on neutron stars; with Gerald Brown on the formation and death throes of supernovae and binary star systems; and with John Bahcall on the unraveling of the solar neutrino problem, the fact that there are far fewer neutrinos being emitted than the solar models predicted.[33]

What is remarkable is that in addition to these very fruitful but very demanding scientific activities, Bethe got deeply involved with energy policies after the energy crisis of 1973.[34] He widely and forcefully argued for the reconsideration of nuclear power in the aftermath of the oil embargo the Arab states had imposed following the Yom Kippur War. There are over a dozen boxes in his archive filled with post-1973 energy-related materials. And during that same period, as a consultant at Avco Corporation, he was actively investigating the possibility of separating the uranium isotopes using lasers and designing high-power "chirping" lasers. Similarly in 1983 and thereafter he collaborated with Garwin, Kurt Gottfried, Henry Kendall, and other Union of Concerned Scientists (UCS) members to challenge the claims President Reagan and Edward Teller were making for their Strategic Defense Initiative (SDI) scenario.

Bethe's life ended on a somber note. Throughout his life he had felt a heavy responsibility for his contributions to the creation of nuclear weapons and had invested huge efforts to constrain their development and to seek peaceful applications for nuclear energy. He found justification and consolation for his participation in the development of fission and fusion bombs in the fact that the Soviet Union and the United States did not come to blows in the face of many provocations.

Bethe's anxieties concerning nuclear weapons culminated in 1995 on the occasion of the fiftieth anniversary of the leveling of Hiroshima. He made an appeal to scientists to take an oath like the Hippocratic Oath—not to work on weapons of mass destruction:

> As the Director of the Theoretical Division of Los Alamos, I participated at the most senior level in the World War II Manhattan Project that produced the first atomic weapons.
>
> Now, at age 88, I am one of the few remaining such senior persons alive. Looking back at the half century since that time, I feel the most intense relief that these weapons have not been used since World War II, mixed with the horror that tens of thousands of such weapons have been built since that time—one hundred times more than any of us at Los Alamos could ever have imagined.
>
> Today we are rightly in an era of disarmament and dismantlement of nuclear weapons. But in some countries nuclear weapons development still continues. Whether and when the various Nations of the World can agree to stop this is uncertain. But individual scientists can still influence this process by withholding their skills.
>
> Accordingly, I call on all scientists in all countries to cease and desist from work creating, developing, improving and manufacturing further nuclear weapons—and, for that matter, other weapons of potential mass destruction such as chemical and biological weapons.

After George W. Bush became president in 2001, Bethe became deeply concerned by the fact that independent scientific and technological advice was playing an ever-smaller role in governmental policies. More particularly, he became very disturbed by the actions of the Bush administration in disbanding many governmental scientific advisory bodies and replacing a large fraction of the members of the still-existing ones with people who were either drawn from the industrial scientific community and were less independent than scientists in the academy or were ideologically committed to the Bush policies irrespective of the scientific facts. It was with profound anguish that he observed the paths taken by the Bush administration when addressing issues relating to nuclear weaponry, test ban treaties, the environment, and the dramatic increase in the information it stamped secret. Secrecy prevented people from knowing. Only if they had knowledge could they act rationally—and rationality was essential to Bethe. He decried the Bush administration's involvement in Iraq and the secrecy involved in the justification for the military actions taken. And he lamented the fact that the Bush administration was giving political and military considerations priority over all other factors, including scientific realities, and this at a time when science and technology were of paramount importance in making possible the United States' economic and social well-being. He came to regret the role he had played in making some of this possible. His despair stemmed from the fact that he had a drastically different vision of the aims and responsibilities of the United States in the world and of the role that science would play in its growth and evolution than the one projected by the George W. Bush administration. And perhaps most painful was that his faith in reason and rationality—which had given him hope, resilience, and buoyancy all his life—had been deeply shaken and undermined.

He did not live to see Barack Obama elected president. He surely would have been elated. And given his father's closeness to Max Weber after World War I, he might have recalled the conclusion of Weber's "Politics as a Vocation":

> Politics is a strong and slow boring of hard boards. It requires passion as well as perspective. Certainly all historical experience confirms the truth that man would not have achieved the possible unless time and again he had reached out for the impossible. But to do that a man must be a leader, and more than a leader, he must be a hero as well, in a very sober sense of the word. And even those who are neither leaders nor heroes must arm themselves with that steadfastness of heart which can brave even the crumbling of all hopes. This is necessary right now, otherwise we shall not be

able to attain that which it is possible to achieve today. Only he who is certain not to crumble in the process should hear the call of politics; he must endure even though he finds the world too stupid or too petty for that which he would offer. Only he who in the face of all has the resolve to say, "In spite of it all!"—has he the calling for politics.[35]

I

Growing Up

> What we achieve inwardly will change outer reality.
>
> Plutarch

> If you were not to find satisfaction *in the search for knowledge,* you would despairingly put your hands in your lap and say: It is too difficult for us humans.
>
> Albrecht Bethe, 1899

This chapter recounts Hans Bethe's childhood, until he entered university. It tells of his parents, of his home, of his *Gymnasium* experiences, and of his bout of tuberculosis. I have devoted a great deal of attention to the role played by his father, Albrecht Bethe, for a number of reasons. Albrecht was deeply committed to Darwin's theory of evolution, and Hans grew up in an environment where describing how things came to be was a natural form of explanation and understanding. His first exposure to doing science was through an introduction to the extraordinary variety and complexity of living entities, entities that according to his ethologist father could only be understood in relation to their environment.

It is interesting to note that after he had won the Nobel Prize the mature Bethe had a second career as an astrophysicist. Astronomical observations have revealed an astounding variety of objects and phenomena populating the universe, many of which are yet to be analyzed and explained. Astrophysics has been described as "the realm of the many-body problem," in which "individual behavior can never be considered in itself and is always seen through interactions with many particles."[1] It may well be that Bethe's interest in astrophysics and stellar evolution has its origin in Albrecht's evolutionism. It is also the case that Bethe proved himself to be an outstanding astrophysicist.

Strassburg, Kiel, Frankfurt

Hans Bethe was born in Strassburg on July 2, 1906, and was christened in an Evangelical Lutheran church shortly thereafter. Strassburg, now Strasbourg, the capital of Alsace, was then part of Germany. Alsace and Lorraine had been ceded to Germany after France's defeat in 1871. Hans's father, Albrecht, was a physiologist who, at the time of Hans's birth, held an appointment at the University of Strassburg as *Assistent* to J. Richard Ewald, a distinguished physiologist,[2] under whom he had obtained his *Habilitation*.[3] Already Albrecht Bethe was recognized as an outstanding scientist then charting new directions in neurobiology.[4]

Albrecht Bethe was a tall, handsome, and quite remarkable man. He was born in 1872 in Stettin, the capital of Pomerania, the city where the Bethe family had lived since the sixteenth century. The Bethes who originally settled in Pomerania were Scottish Protestants who had migrated when Scotland was still Catholic. Albrecht was the youngest of the five children in the family.[5] His father was a very successful doctor, a passionate beetle collector, but also a very strict disciplinarian. As a young boy Albrecht was sent to a humanistic *Gymnasium* and thus forced to learn Latin and Greek, subjects in which he had no interest and came to hate. At age fifteen he ran away from home and joined a circus. But juggling was not his forte, and so after six weeks he returned home to the deafening silence of his father, who didn't speak to him for a month. Already as a teenager in *Gymnasium* he was a liberal in his political views, and he told Hans that "he and his friends went sometimes to political meetings, speaking against anti-Semites."[6] Failures in Latin and Greek kept him from being promoted to the next grade on four occasions. Thus he was twenty when he graduated from the *Gymnasium.* His love of animals led him to study zoology, first in Freiburg and then in Munich. His studies there made it clear to him that he was really interested in physiology, and in particular, in the neural mechanisms that determine and regulate animal behavior. He went to Berlin to study neuroanatomy with Wilhelm von Waldeyer and then returned to Munich, where he earned his doctorate in 1895 working with Richard Hertwig,[7] carrying out extensive histological work on the central nervous system of the crab *Carcinus maenas.* He thereafter spent some time at the Naples Zoological Station.

The Naples experience was very important. The Stazione Zoologica had been founded in 1872 by the German zoologist Anton Dohrn, who had been a student of Ernst Haeckel. Haeckel was the remarkable and inspiring biologist who had disseminated Charles Darwin's work in Germany.[8] Dohrn became "an unwavering defender of evolutionary theory" and made the Naples station a mecca for evolutionists, the hub of a worldwide network of research facilities for studies in marine biology and an international model for laboratory research in morphology and physiology.[9]

During his stay in Naples Albrecht became a committed evolutionist. Understanding something meant being able to account for how that something came to be. Dohrn influenced Albrecht in another important way. He believed that the science of biology could not be confined to the laboratory or the museum but must include the study of organisms in their natural environments. He stressed the importance of the study of the habits and conditions of the lives of animals *(Lebensweise der Tiere).* This became Albrecht's perspective, and he implemented this vision of biology when he became the editor of *Pflügers Archiv.*

After his stay in Naples and one in Heidelberg, Albrecht accepted the position of Richard Ewald's *Assistent* in Strassburg.[10] He there continued making anatomical and physiological investigations of the nervous system of invertebrates. Albrecht had an enormous capacity for work and was very productive in his histological research. He earned a medical degree (Dr. med.) in 1898 and one year later submitted his *Habilitationsschrift* (dissertation), becoming a *Privatdozent.* This gave him the right to teach physiology at a university.[11]

Joseph Fruton in his book on research groups in the chemical and biochemical sciences in Germany at the end of the nineteenth century noted that although Albrecht could not be considered a biochemist, nonetheless his influence on that community was deep "because of his leading role in bringing together the younger members of various departments in the Strassburg medical faculty in a journal club that met frequently."[12]

Hans's mother, Anna, was born in Strassburg in 1876. She was the daughter of Abraham Kuhn, a professor of medicine at the Strassburg University Hospital. His specialty was diseases of the ears, nose, and throat. Neither of Anna's parents was Alsatian. Abraham was born in Worms, Germany, where his quite well-off family owned vineyards and his father was a wine merchant. Abraham was the first member of the family to attend university. He went to Paris to study medicine, wrote a book in French on diseases of the ear, and thereafter settled in Alsatian-speaking Strasbourg, staying there after its annexation by Germany in 1871.[13] He clearly was an unusually able and well-informed physician, for it was very difficult for a Jew to be appointed to a professorial position in a German university. Anna's mother, Amalie Seligmann, was born in Karlsruhe into a very prosperous Jewish family of cloth merchants and bankers. The Kuhns were well on the road to total assimilation. Thus in 1893 they allowed Anna's older sister, Bertha, to marry Carl Fitting, a German artillery officer. To do so she had to convert to Protestantism. At the conversion ceremony Anna also converted.[14]

As a child Anna was considered frail, so her parents decided not to send her to public school. Instead she was instructed with a half-dozen other girls by a private teacher, who taught them reading, writing, arithmetic, and some French. In her teens she went to the Strassburg Conservatory to study violin and singing and "became quite good at both." As a young woman she began writing little sketches for the *Strassburger Post* that were well received by its readership. As she was quite pretty, very lively, and witty, she was courted by many of the young lieutenants stationed in the town. Her deportment concealed a highly intelligent, very talented, strong-willed and very capable woman.

Hans's parents met at a social gathering of the faculty of medicine sometime after Albrecht had come to Strassburg. They married in October 1900 on the heels of a tragedy that had befallen Anna's family: a month earlier, on September 14, Anna's father had died of scarlet fever, and a week later her sister Bertha succumbed to the same illness. It was a traumatic experience for Anna.[15] The first ten years of Albrecht and Anna's marriage were seemingly happy. They had more than adequate financial resources, a modest part of which came from Albrecht's salary as *Assistent* and from the students' fee in the courses he taught as a *Privatdozent.*[16] The major part of their income derived from Anna's considerable dowry of 200,000 marks, which had been invested in bonds that paid regular interest until the end of World War I and the subsequent inflation.[17]

Hans recalled that "Mama loved music, sang and played the violin many evenings. She also began writing fairy tales for children. Papa was not musical but liked to hear music." Albrecht excelled in the use of his hands—a necessary requirement to be a good neuroanatomist—and was quite gifted as a wood carver. He would often spend the evening carving pieces of furniture while listening to Anna sing or play the violin. He was physically active, and passionate about mountain climbing. But the foremost and central passion of his life was science—his laboratory and his scientific research.

In 1911 Albrecht had an operation to repair a hernia, which resulted in a blood clot in his leg. Medical practice at the time required patients with a thrombosis to stay in bed for an extended period; in Albrecht's case this turned out to be over five months in a hospital. Hans marveled at his father's patience, but also remembered that his father became somewhat religious during the episode. It took a long time to get the leg to function normally again. Only in the summer of 1915 after the family had moved to Frankfurt could Albrecht go for long walks in the nearby Taunus Mountains. But by the early 1920s he was again climbing moderate mountains.

A measure of Albrecht's professional accomplishments is indicated by the fact that by 1900 Albrecht had published a monograph on the nervous system of the crab *Carcinus maenas,* subtitled "An anatomical-physiological experiment";[18] written a long review that presented a comparative study of the functions of the nervous system of arthropods; investigated the role of certain chemicals in ant colonies and bee hives that gave these insects the ability to find their way back to their homes;[19] and written a very important, lengthy paper on whether one should attribute psychological qualities to ants and bees.[20] The work on arthropods had been carried out at the Naples Marine Biological Station. While there in the mid-1890s he had met Hans Driesch and had been introduced to his views on the plasticity of organisms. He also got to know Theodor Beer and Jakob von Uexküll there.[21] The three of them wrote an influential paper—an elaboration of Albrecht's paper on the inappropriate attribution of psychological traits to ants and bees—that recommended the use of a new, "objective" terminology in the sensory physiology of lower animals to replace anthropomorphic descriptions, and thus not prejudge the mental capabilities of the animals in question. They urged replacing "seeing" with "photoreception," "hearing" with "audioreception," "smelling" with "stibireception," and more generally replacing "sense organs" with "reception organs" and "receptors." Their paper played an important role in the history of behaviorism. Many psychologists were influenced by the paper and considered it "fundamental" in the development of "objective" psychological research in the period from 1900 to 1925.[22] Albrecht's professional standing can be gauged by the fact that already in 1901 he was identified in Baldwin's *Dictionary of Philosophy and Psychology* as one of the eight authors of "important recent books on comparative psychology."[23] The publication in 1903 of his *General Anatomy and Physiology of the Nervous System* made him an internationally known, widely respected physiologist, and its publication led to an invitation to be on the editorial board of the *Journal of Comparative Neurol-*

ogy. Soon thereafter he was asked to join the editorial board of *Pflügers Archiv für die gesamte Physiologie des Menschen und der Tiere,* then the most prestigious and widely read physiology journal. In 1918, Albrecht became its co-editor and discharged that responsibility until his death in 1954.

Albrecht and Anna stayed in Strassburg until 1912 and had an apartment in Anna's mother's house. It had an enormous garden in which Hans remembered playing very happily. The house was located at Ruprechthauer Allee 59, off a boulevard a few blocks away from the Orangerie. The Orangerie is a beautiful park in the western part of the city, to which Hans was taken almost daily by his mother.

Hans was an only child. A bout of syphilis Albrecht had contracted before he and Anna were married affected their sexual relations, and Hans was only born six years after they were married.[24] Anna thereafter was very concerned about Hans's health, and she remained very apprehensive as he grew up. She became overprotective and allowed him but few contacts with the outside world. Hans wore dresses until he was old enough to go to the toilet by himself. Anna also kept his hair long and curled until he was three and a half years old. Albrecht then decided "that he has to look like a boy" and took him to a barber. Anna shed bitter tears that day.[25]

On the surface, relations between the parents were amiable, but Hans's father would later tell him that there had been difficulties from the beginning of the marriage. Albrecht sublimated his marital relations in his work, and his professional successes must have made him content. After Hans was born, Hans's obvious precocity and mathematical abilities gave Albrecht great joy. But, in part due to Albrecht's wartime duties, it was only when Hans became a teenager that Albrecht began spending time with him, nurturing his potentialities. Anna, on the other hand, must have felt isolated and frustrated. Though she could sublimate her desires somewhat in her music and in her writing, she became deeply attached to Hans, and he became the focus and the center of her life.

One of Bethe's earliest memories was being interested in numbers and playing with numbers. His numerical and mathematical abilities manifested themselves very early. Richard Ewald, the professor of physiology in whose laboratory Hans's father was working, became a close friend of the family and Hans's godfather.[26] Ewald took great interest in the little Hans and was fascinated by his love of numbers. One day, when Hans was four, Ewald asked him, "What is 0.5 divided by 2?" Hans's answer, believing that Ewald needed the information, was "Dear Uncle Ewald, that I do not know myself!" A few days later, upon seeing Ewald, Hans ran to him across the wide boulevard through the thick of traffic and informed him triumphantly, "0.25!" Hans's father told of Hans at age five sitting on the stoop of their house, a piece of chalk in each hand, taking square roots of numbers.[27] By the age of five Hans fully understood fractions and could add, subtract, multiply, and divide any two of them. In an early (undated) letter to his mother (probably when he was five years old and had just learned how to write in Gothic letters) Hans wrote:

> Dear Mama,
>
> . . . Today I have figured out—it took me about a half an hour—how many seconds there are in a year! There are 300015000 . . . I told Clara [their cook] She always forgets about the calendar, and for every day that she has not torn it off she has to pay me 10 Pfennig, and she already owes me 80 Pfennig . . .
>
> 100 000 000 greetings and
> 1 000 000 000 kisses

By the age of nine Hans was finding ever larger prime numbers and had made a table of the powers of two and three up to 2^{20} and 3^{10} and had memorized them. When he was ten, he was introduced to algebra and later in life commented: "I became fascinated with algebra as I learned about it. I never was much interested in geometry."[28]

Hans started reading at the age of four and began writing in capital letters at about that same age. He developed a distinct mode of writing: he was left-handed and would write the first line from left to right, the next line from right to left, and continue in this way! Many years later while visiting Crete, Bethe was pleased to learn that that was how the Greeks wrote on their tablets in 700 BCE: They too wrote in capital letters, the first line left to right, the next one right to left, and so on.[29]

In the spring of 1912 Albrecht accepted the professorship and directorship of the Institute of Physiology of the University of Kiel. Undoubtedly, the marine zoological station in nearby Helgoland was a factor in his acceptance of the directorship. The director's apartment was on the top floor of the institute. Hans's father later told him that when Hans came for the first time to the director's office on the lower floor, he had looked at the ceiling and stated, "There are 72 squares there." So Albrecht asked him how he had counted them so quickly. The six-year-old Hans told him, "It's very easy. There are nine rows and each has eight squares." The clarity of Hans's reasoning at such a young age had impressed Albrecht very much and was the reason for his telling Hans the story many years later.

Albrecht's institute was located at the bottom of a small hill, whose paved sidewalk ran into a flat area in front of the institute. Hans had a steerable little wagon in which he could sit and in which he would ride down that sidewalk. He recalled that being "wonderful." The institute also had an enormous garden attached to it in which vegetables were grown in one part and flowers in the other. Weather permitting, Hans would spend a great deal of time there. Life in Kiel was very comfortable. The Bethe household employed two maids and a cook. Clara, the cook, stayed with them "for many, many years." She was very deaf and somewhat surly, and Hans didn't like her when he was a child.

As Hans had turned six when his parents moved to Kiel, the question of what to do about his schooling had to be addressed. Anna had kept Hans socially isolated, and his inability to interact with children of his age was forcefully brought home when they first came to Kiel. Anna had invited a few young boys from the neighborhood to come to their house and play with Hans. During the entire time they were there Hans stood in the corner of his room, not saying a word as the other children played with his toys. After they had left, surveying the disarray that the children had created and conscious of the inner turmoil their presence had generated, Hans forcefully told his mother, "Let us not do this again."[30]

The schooling issue was resolved by Hans joining a small group of some eight children who were taught privately by a very nice teacher. They went to his house three days a week to have their two-hour lessons there. He was clearly a very fine teacher, for in a year and a half he taught his wards the equivalent of three years of *Vorschule* and got them ready for entering *Gymnasium.*[31] Hans enjoyed these lessons. The students, an equal number of boys and girls, were supposed to do a little homework, principally writing stories. Hans's mother, who was then writing children's books, "very much" encouraged this activity. Hans would tell her the stories before he wrote them down, and this made both of them very happy. Interestingly, the stories he wrote down were not changed for having been told to his mother. They usually were about one of the animals in his large collection of stuffed animals he called "his children" and "loved dearly."[32]

Bethe recalled seeing his mother "constantly" in Kiel. She walked him to the teacher's house and spent much time with him in the garden. "Occasionally while there, I would read Grimm and Andersen fairy tales, and similar things." His recollections of a typical day during this period, when not going for lessons to the teacher's house, were of being by himself: "writing numbers and stories into little books. I liked to invent stories at that time. I invented a country, . . . and so I described that country and its organization in those stories."

Much of these materials were recently rediscovered.[33] Most remarkable among the young Hans's extensive writings and drawings from age six to age ten are the hundred pages or so that describe the world he had created. He had invented a large country in the shape of Italy, and to describe its geography he drew numerous colored maps that indicated its neighboring countries, its topography, and the road and railway connections between its cities and towns. To specify the political and social structure of the country he provided information about its cities and towns: their size, their governmental institutions, their political setup, the demography of their population, the schedules of the trains connecting them, and their past history, including information revealed by archeological excavations. That the seven- or eight-year-old Hans would include detailed archeological information surely reflects the influence of his father's evolutionary views. Also included in the description of the country are numerous tables containing information about its currency and that of its neighbors, and tables of the exchange rates of the various currencies and how they have changed over time . . . and much more. All this was invented, crafted, and made consistent as Hans grew up, from age six or seven until he was ten years old. When he turned ten he created similar things for the various planets, which he assumed had become settled by humans. Until he attended *Gymnasium,* when left by himself Hans evidently inhabited the world of numbers, the rich imaginary worlds he had created and those of the fairy tales he read.

Also in Hans's mother's trunk were the diaries he kept between the ages of thirteen and fifteen in which he outlined his beliefs. In them are to be found the reasons why he gave up Christianity and religion after he had been confirmed at age thirteen.[34]

All these materials give proof that already as a young boy and as a teenager Hans was "off-scale." Had his talents been in music, we would be hearing little symphonies composed by him as a youngster. Hans had no musical education, however. He was not very musical, but as a child he probably was not as "unmusical" as he was led to believe. "My mother, being very musical, thought that my talents were zero, so she put no emphasis on this at all. I couldn't sing, I couldn't keep a tune, so I was not to be musically educated. I probably could have been under more normal circumstances."[35]

Except for vacation time during the summer, Hans didn't see much of his father in Kiel as he was very busy, but what he saw of him "was always pleasant." Even during vacations interactions between Hans and his father were somewhat limited. His parents would rent a house an hour away by train from Kiel. Hans and his mother would stay there during the week and be joined by his father on weekends. But Albrecht would also bring his work there.

Bethe vividly remembered one of the medical episodes that took place while in Kiel. Being located at a fairly northern latitude the climate in Kiel was not pleasant: short summers and raw, cold, wet winters during which both Hans and his mother had "many, many colds." In the winter of 1913–14, one of Anna's colds developed into a severe flu and a streptococcal infection of the ears that left her almost deaf. Thereafter, she couldn't hear music well, and when she sang, it was off key. This was extremely hard on her, for she had been a talented singer and an accomplished violinist and loved music. The illness and its aftereffects magnified the inner tensions she had been living with, stresses that she had been able to manage until then. It affected her psychologically and she became "very difficult" to live with.

Two years after the Bethes moved to Kiel, World War I broke out. That was Hans's first "detailed" memory. He recalled being a very patriotic young boy, but also being very conscious that people were getting killed. He read the headlines in the newspapers and was aware of the course of the war. Hans's father, because of his age, was not required to serve in the army; but because he had a medical degree, soon after the outbreak of the war he joined a hospital train and took care of wounded soldiers for over two months. As the war progressed he devoted more and more time to this activity. He became very interested in prostheses for arms and legs, and designed an artificial hand that was much better than what existed at the time: its fingers could be moved individually by the nerves and muscles of the upper arm.

Hans's mother, on the other hand, was very much against the war. She thought it was "awful that people were killing each other." Early in the war she remarked to Hans, "How can God exist if he has to listen to the French and British at the same time as us—with both sides claiming that he was on their side?" Her statement made the eight-year-old Hans skeptical about religion. But Hans did not feel torn between the differing views of his parents. "I agreed with my father in patriotism, and I did not, at that time, listen to the anti-war feelings of my mother, but only remember this remark about God." The differing political viewpoints and moral outlooks of father and mother would come to represent the tension between the rationalist and the artist. Anna had strong views regarding morality, justice, and righteousness, and her opposition to the war reflected these. Despite all the difficulties associated with Kiel, for Hans "Kiel was a happy time."

In the spring of 1915 Hans's father received an offer from the Königliche University of Frankfurt-am-Main to head the new Institute of Physiology.[36] The university had been founded the year before with considerable financial support from the Jewish community of Frankfurt.[37] It thereby acquired a unique position among German universities by being under local rather than state control and by the fact that its bylaws prohibited religious discrimination in the appointment of faculty members. Many members of the initial professorial staff—for example, Richard Wachsmuth—were individuals who had held appointments at academic or research institutions in Frankfurt. In particular, the entire clinical faculty of the medical school was recruited from the local municipal hospital, and the whole medical school, including the basic medical sciences, such as anatomy, physiology, biochemistry, bacteriology, and immunology, became located in that hospital, or next to it.

The large, new building the Institute of Physiology occupied had been made possible by the benefaction of Theodor Stern, a wealthy Frankfurt Jew. It housed Albrecht's Institute of Animal Physiology, which concentrated on biophysical processes, and an Institute of Vegetative Physiology, which focused on biochemistry, headed by Gustav Embden.[38] But in contradistinction to Kiel, the professor of physiology in Frankfurt did not have living quarters in his institute. The Bethes found an apartment quite close to Frankfurt University, at Kettenhofweg 126, but very far from the hospital where Albrecht worked. The apartment was sizable—"some seven rooms, but somewhat dark."[39] As they lived on the first floor, they had the use of the small gardens attached to the house. During the war Hans's father grew vegetables in the garden in the front of the house. Although he never took to growing vegetables, Hans did take to eating them.

Hans's mother's querulous disposition, the result of her illness in Kiel, was exacerbated by the Bethes' move to Frankfurt. During her first few years there she "hated" Frankfurt. Although it was a nicer and healthier environment for her—Frankfurt had a much milder climate than Kiel—she thought Frankfurt was a very unpleasant, "terrible" city. Her feelings about the city changed when, in 1917, she started writing plays for children, and her outlook improved. Except for an occasional breakdown "she got cheerful." Bethe remembered that "some of the more exciting times [in the house] were when her composer, Hans Hermann, came and played the tunes on the piano and discussed the play with my mother."[40] One of her plays for children, *Das Neugierige Sternlein* (The Curious Little Star), proved to be a great success.[41] It opened in Mannheim shortly after the Armistice was signed in early 1919 and continued to be staged in many cities for quite a number of years thereafter. "Mama felt very famous, made friends with some of the actors, with a few of the directors, and with the theater critic of the *Frankfurter Zeitung.* She had a great time and was really happy for several years. Papa and I were very happy with Mama's career in the theater." However, none of the other plays for children that she wrote was a success.

The daily routine in Frankfurt was much the same as it had been in Kiel. The main meal was lunch, which took place very late. Hans's father had the habit of going to his laboratory in the morning, and coming back home at around 2:00 or 2:30 for lunch. He would then have his nap, and afterward would work at home until dinner. Dinner was late, served at around 8:00, and usually consisted of cold meat and bread.

Father, mother, and Hans would sit down together for lunch and dinner. Hans remembered those occasions as "very pleasant." There would be conversation among the three of them, and sometimes between just his mother and his father. As Hans grew older most of the table talk would be between Hans and his father, discussing science. Albrecht was very interested in Hans's mental development. He would tell him of his work, and already as a young teenager Hans would go to his laboratory and help with various chores. Although Hans wasn't interested in physiology, he did become interested in the physical experiments that his father was doing, "which had lots to do with electricity, in particular in his experiments on the legs of dead frogs which could be made to twitch when stimulated by an electrical impulse, and their motion recorded as a line on a rotating drum of paper." But he disliked the numerous dogs used in various experiments "that barked a lot."

Hans's parents had very little social life in either Kiel or Frankfurt. Hans remembered that he was "tremendously impressed" by his mother's dress on one occasion in Kiel when his parents went out to a ball, and had said, "Why, Mama, you look like a fairy." But he could not recall any evening parties in his parents' house in Kiel, and except for the dinner party Albrecht hosted when he was *Rektor* of the University of Frankfurt in 1917 Hans could not remember any other in Frankfurt. "That was kind of sad." But Hans did get to know his aunts, uncles, and cousins on both sides of his family and recalled them warmly in the memoirs he wrote for his children and grandchildren.

The Bethe household was nominally Christian, but one in which religion and rituals did not play an important role. There would be a tree at Christmas, and gifts were given out—but no social gathering, neither for Christmas nor for New Year's Day. In his interview with Charles Weiner, Bethe put it thus: "My father was, I think, slightly religious. I was taught to pray in the evening before going to bed, and I attended the Protestant religious instruction, which was given in the schools in Germany. I was also confirmed, and the instruction which I got in this connection got religion out of my system completely. It was never very strong before, and the confirmation had the consequence that I just didn't believe."[42]

As the war progressed Albrecht started discussing with Hans the implications of its development. After the entry of the United States into the war in 1917, Albrecht grew skeptical of a German victory. His views reflected those expressed by Max Weber's occasional articles in the *Frankfurter Zeitung,* the leading liberal newspaper in Germany.[43] When in the summer of 1918 he received an invitation from the University of Strassburg to become its professor of physiology, although tempted, he foresaw the defeat of Germany and declined the offer. Anna was greatly disappointed. Strassburg was her home, and she had looked forward to returning there.

Gymnasium

When the Bethes moved to Frankfurt, the nine-year-old Hans started regular school in the Goethe Gymnasium. The Goethe Gymnasium was a "modernized" *humanistisches Gymnasium* in which great emphasis was placed on Latin and Greek, Roman and Greek history and literature (Caesar, Cicero, Ovid, Virgil, Homer, Plato, Aristotle, and others), and the "classic" German literature of the eighteenth and first half of the nineteenth century (Johann Wolfgang von Goethe and Friedrich Schiller in particular). It was "modern" in that the first language taught was French and not Latin, and in the upper grades English was part of the curriculum.[44]

At the time it was normal to start *Gymnasium* in fourth grade. Hans encountered two difficult situations at this school to which he had to adjust. First, he was left-handed, and until he entered *Gymnasium,* he had done all his writing with his left hand. But this was unacceptable in the *Gymnasium,* and the school forced him to write with his right hand only. This is the origin of his very striking and distinctive handwriting. Second, although Hans was well-prepared intellectually, he was "so terribly naive, so terribly unprepared socially" that it was a great shock for him to be in a class with forty other boys, many of whom had been together in the *Vorschule* for three years. They made fun of him, and one of them was particularly "nasty," and at times threw things at him, though he didn't try to hit Hans. But three boys in the *Gymnasium* who were half a year older than Hans, Paul Berlitzheimer, Werner Sachs, and Heinrich zür Strassen, liked him from the beginning and "sort of adopted" him. Paul was the son of a very successful physician who became the Bethes' family doctor. The other two were sons of Frankfurt University professors: Werner's father was the professor of immunology, and Heinrich's the professor of zoology. Heinrich was very bright and always the top student of the class, with Hans second.

Hans, Werner, and Paul became close friends. Hans often played with Paul and Werner at their homes. Less frequently, Hans would play with Heinrich zür Strassen, "but that was quite separate." Hans would usually go to Werner's or Paul's house to play, but Heinrich always came to Hans's house. Heinrich's mother had multiple sclerosis, and boys' play was not encouraged because it had to be very quiet in the house. On the other hand, both Werner and Paul lived in big apartments, much bigger than the Bethes', and both sets of parents were very welcoming. So it seemed natural for Hans to go to their apartments.

Bethe's recollection of his first year at *Gymnasium* was that he "felt uncomfortable in school, except for these three friends." The curriculum included German, history, some mathematics, "which was quite elementary," religion, handwriting, and some general science, which was primarily botany, and in which he "was not interested at all." Botany was "just classification, observation, and not plant physiology." Had it been physiology he would have been interested. He had two hours a week of gymnastics, at which he was "very bad." He also learned a good deal of French, as French was taught intensively by a very good teacher one hour a day, even though Germany and France were at war. At the time Hans could recite by heart with a good French accent fables by La Fontaine and poems by Victor Hugo.

Hans went to school six days a week. The school day began at eight in the morning, and he would return home at around one in the afternoon for lunch. The rest of the day would be spent doing homework, which he did "quite religiously." "Quite religiously" can be translated into acquiring the discipline to do all his homework at that fixed time, avoiding other interests or activities that might take precedence over this task, and concentrating totally on that endeavor. After he had finished his homework he would go with his mother to the *Palmengarten,* the large nearby park, where "interesting and beautiful plants could be seen throughout the year."

This being 1916, the students were made aware that there was a war going on. Once every month they went off for an excursion to the nearby woods to collect beech nuts to supplement the available food. Also "every time there was a big battle, there was an assembly at which the director would talk to us about the significance of this battle. They were of course all German victories in those days, and so we cheered, and if the battle was really big, then we got off, and we got a holiday."

Summer vacations were one part of the year that were very important for Hans. Bethe had very fond memories of vacation time when growing up in Frankfurt, for his parents always vacationed in lovely locations. The first place they went to after coming to Frankfurt was Niedernhausen, a village in the foothills of the low-lying Taunus Mountains west of Frankfurt. At that time the village was becoming a *Luftkurort,* a health resort, and was the terminal station of one of the suburban trains from Frankfurt. As had been the case while they lived in Kiel, Hans's father would come only on weekends. His mother generally didn't want to walk very much, but she let the nine-year-old Hans walk wherever he wanted. He learned all the markings on the trails and every day went for extended walks that took several hours. He commented, "It is to the credit of my mother that she just let me go by myself." Hans loved this first communion with nature. "It was in the Taunus, which is all wooded, and so you could walk in the forest for hours, mostly up and then down. And that was a great experience, and since then I have always liked to walk in the mountains . . . I never felt lonely. I don't remember what I thought about, probably I did think about something. Maybe not. I don't know what a nine year old . . . But anyway I enjoyed it."

Hans's schooling in the Frankfurt Gymnasium was interrupted for two years. Like his mother, Hans had repeatedly been ill with bronchitis and high fevers in Kiel. When he came to Frankfurt he was sickly and lacked the energy expected of a boy his age. His condition became aggravated by the fact that as the war went on, food became scarce and rationed, and civilian life more difficult. In the summer of 1916, while on vacation in the Black Forest, Hans was diagnosed as having tuberculosis, with the infection located in the bronchi.[45] On the advice of Dr. Berlitzheimer, their house physician and the father of Hans's friend Paul, his parents decided that Hans should go to a *Kinderheim* (children's home) in Bad-Kreuznach. And so early in the fall of 1916 Hans was admitted to the *Haus Bartenstein—Erholungsheim für Kinder* (children's recuperation home), which was directed by the pediatrician Dr. Bartenstein in Kreuznach.

Bad-Kreuznach is a beautiful town some sixty kilometers southwest of Frankfurt on the Nahe River, a tributary of the Rhine. It is well known for its sanatoria and its salt inhalations for people with respiratory diseases. Hans was one of ten children in the *Erholungsheim.* He had to share a room with one of the other boys, and that turned out to be very difficult at first, because not only did it mean sharing the same space but also sharing the toys and stuffed animals Hans had brought with him. He eventually adjusted to the new realities and Kreuznach proved to be an important event in Hans's socialization.

The first autumn in Kreuznach was spent resting a good deal of the time and having weekly salt inhalations. Kreuznach is located in a wine-growing area, and Hans also took walks in the surrounding vineyards. In addition, he received private instruction in writing and in French. By the spring of 1917 he was strong enough to attend the local *Realschule,* where he felt much better educated than the local students. His mother visited him several times during his stay in Kreuznach, and a few times he took the train by himself and came home. Bethe remembered Kreuznach as a "pleasant place" and having "a good time there."

In early 1918 Hans was considered sufficiently recovered to be discharged from the *Erholungsheim.*[46] His parents sent him to a recently founded, "fancy," progressive, co-educational boarding school located not far south of Frankfurt. The school, Odenwaldschule, offered instruction in all the grades of *Gymnasium.* It was committed to a Pestalozzian educational philosophy that emphasized not only academic studies but also practical skills such as carpentry, and in particular, the building of bodily strength. The boys always wore short pants and sandals and went bare-legged. The teachers and the director dressed the same way. There were some 350 students in attendance, with about fifty of them the same age as Hans. Odenwaldschule was very "permissive" as to the subject matter of the courses one took, as long as one took the required number of these quite intensive courses; and every two months students were allowed to rearrange their academic program. The instruction was outstanding, and Hans got "a quite good course in algebra, a good course in history, and a good German course." But what Hans also remembered of the school was "that we had the most abominable food, which was not all the fault of the Odenwaldschule, but of food in Germany in the last year of the war." Initially, Hans disliked the other children, and in particular, his roommate. But after overcoming some of his shyness, he began to like the company of other children, especially that of some of the slightly older girls, who encouraged him to be more self-assertive. There was another reason for his initial displeasure with the school. Hans had acquired "a whole menagerie of stuffed play animals—elephants, chickens, horses" and had insisted on taking many of them to Kreuznach during his recovery. Odenwaldschule "didn't tolerate that." In addition, the physical demands at the school "to toughen him" were too strenuous. For example, to reach the train that took him to his mother during summer vacation in 1918, the twelve-year-old Hans had to hike three hours carrying a very heavy rucksack on a very hot day. He probably got badly dehydrated on the march, and Anna was frightened when she saw him. She said, "You have to leave that school." Hans was very happy to do so.

Kreuznach and Odenwaldschule were critical events in Hans's life. His stay at these places pulled him out of his social isolation.[47] Odenwaldschule had severed Hans's ties to his stuffed animals. When he came back home he asked for, and got, a Meccano set, which "gradually got bigger and bigger."[48] With it he built large bridges, cranes, and buildings, some a meter high. He was very proud of these constructions. Hans returned to Frankfurt just before the war ended. The instruction Hans had gotten at Odenwaldschule, together with some private lessons in Latin, allowed him to rejoin his friends Werner, Paul, and Heinrich in the fourth year of *Gymnasium,* which made him "very happy." But it was also

> a very hungry time, [because of the British naval blockade] and I remember that very vividly, and I never got enough to eat, and what we got was pretty miserable stuff. There was a ration of a hundred grams of meat a week. Often we didn't get that, and often it was not very good meat. I think there

> was a ration of two hundred grams of bread a week, which is very little, and my father had a letter scale, and weighed the bread out for us. You could have a thin slice of thirty grams, or a medium slice of 50 grams. Also the potatoes we ate were mostly frozen, which made them taste awful. Mostly we got turnips, which is awful food if you get it regularly. And then they had burnt turnips, from which you were supposed to make soup. It was pretty awful. And it got progressively worse as the war dragged on. It got better only about the middle of 1919, when there came a lot of food from America. And we praised Woodrow Wilson for that. Actually it probably was Herbert Hoover who was responsible. Towards the end of the war I read the newspapers quite diligently. I was very much aware that there were no more German victories, and that when the location of the German army was mentioned it was always a little further back from the previous mention. And then came the Armistice. I remember vividly my father's reaction to the Armistice. The Armistice conditions were really very harsh. Germany had to surrender its fleet, planes, tanks, . . . Well, that was more or less obvious, but also the larger part of our railway stock—hundreds of locomotives, thousands of freight and passenger cars—and that really hurt. It hurt the country because the transportation of food depended on it, and so my father was quite despondent on reading these Armistice conditions. He knew perfectly well that Germany would lose the war. But I don't think he anticipated that it would lose the war that badly. Nor, that it would then be treated so harshly. The peace conditions seemed even harsher, especially the territorial changes. Most Germans did not accept this dismemberment though my father accepted the conditions gracefully. They especially resented that the peace treaty had been drawn up entirely by the Allies, with the Germans just asked to sign it, and it was called the Dictate of Versailles, or the disgraceful treaty of Versailles. I remember many debates [on the issue] in *Gymnasium* in liberal Frankfurt, with the students about evenly divided. And there were student meetings in Munich in 1926–29, long before Hitler, to protest the treaty—with nearly everyone disagreeing with me. Only there did I really encounter strong nationalist feeling.

With the Armistice the impact of the war was felt mostly in terms of food and the general economic condition and, although better by mid-1919, these remained very bad until 1920. In fact, times were very tough until the end of Hans's *Gymnasium* in 1924.

After the Armistice the twelve-year-old Hans grew more interested in politics because his father became very active in the Frankfurt branch of the newly founded *Deutsche Demokratische Partei* (German Democratic Party).

He followed closely the election to the National Assembly in January 1919, and diligently studied the constitution that had been written in Weimar thereafter. Already at age thirteen Hans identified himself as a social democrat. His father had told him of the differences among the political views of social democrats, those he held as a member of the German Democratic Party, and those of the communists. He explained to him that both kinds of democrats were committed to religious and cultural tolerance, but social democrats believed in the virtue of using the state as an instrument for achieving the collective good. Hans thereafter decided that he was in favor of taxing everyone so that the state would provide health insurance and other social goods to everyone. As a result of his discussions with his father, he adopted his father's views that democracy and social justice were needed in Germany, but that radicalism and revolution were to be shunned, and he was glad that the republic was resisting the communists. His father had told him about the socialist republic that had been set up by Kurt Eisner in Munich just before the Armistice. He knew about Eisner's assassination by a right-wing extremist while on his way to tender his resignation to parliament after his Independent Social Democratic Party of Germany had lost the Bavarian elections of February 1919. Hans was also aware of the Soviet-style *Räterepublik* that had been set up when the communists seized power during the unrest that Eisner's assassination had generated, and of the terrible White Terror after the *Räterepublik* had been overthrown by the military forces sent to Munich by the national government.

In retrospect, Bethe thought that the Weimar Republic had been "a brave attempt" at democracy but that the German people were not "sufficiently disgusted" with the kaiser to make it last. The harshness of the Treaty of Versailles and the explicit policy on the part of the Allies, especially France, to punish Germany made many Germans believe that there indeed had been a *Dolchstoss von Hinten*—that Germany had been "stabbed in the back" in 1918. Many believed that she not been defeated militarily on the battlefields but had been betrayed by corrupt, leftist civilians in the government, cowardly soldiers in the rear, and mutinous sailors.[49] In addition, the many-party system that came into being after the war resulted in an irremediable confrontation between the two extreme right-wing parties, which represented the former régime and the Communist Party.

It is interesting to note that the eighty-year-old Bethe, who throughout his life constantly read a good deal of history, nonetheless characterized the

Versailles treaty as harsh. The reparations demanded of Germany were not intended as punishment. They were to cover the cost of repairing the extensive damage done to northern France and to Belgium by the four years of trench warfare there. Furthermore, the German forces, when retreating in 1918, had carried out a scorched-earth policy, destroying bridges, flooding mines, plundering factories, and burning farms and farmlands. The reparations were also meant to help France, Italy, and England repay the debts they had incurred to the United States in order to finance the war. At the Paris talk in 1919 Woodrow Wilson had insisted that there were to be no punitive fines on Germany and its allies, and Lloyd George and Georges Clemenceau, the prime ministers of Great Britain and France, reluctantly acquiesced and signed the treaty.[50] The cost of the repairs was not known in 1919, and the United States was unwilling to forgive the Allies' debts, so the amount of reparations to be levied was not included in the Treaty of Versailles. This was exploited in Germany by all those opposed to the Weimar government and what it stood for. To many it indicated that the Allies would exploit Germany indefinitely. The total amount of reparations to be paid by Germany was not set until 1921 and came to about $250 billion in 1990 dollars, a sum less than what France had paid to Germany as a result of its defeat in 1870. Germany, for political reasons, never paid.

Reparations were blamed for all the woes Germany suffered after the war, and the Weimar government was blamed for signing the treaty. Moreover, many in Great Britain came to concur that the reparations were unjust, imposed on the Germans by vengeful Frenchmen. The sense of guilt this generated was in part responsible for the appeasement of Hitler by the British during the 1930s. The views Bethe expressed in the 1990s indicate how effective had been the propaganda disseminated within Germany—by both the Weimar government and its foes—while Bethe was growing up.

From 1920 on, Germany experienced an accelerating spiral of inflation.[51] The deterioration of the German currency had begun during the war. Since it had anticipated a rapid victory the government did not raise taxes to cover the cost of the war. This in turn forced it to increase the supply of money from 12.5 billion marks in 1914 to 63.5 billion marks in 1918. The Versailles peace treaty and its demands of reparations aggravated Germany's precarious economic condition. Germany's refusal to pay any reparations led to the French occupation of the Saar in January 1923. The severity of the crisis intensified thereafter with the mark collapsing totally.[52]

Since France withheld coal and steel shipments from the Saar in order to obtain the payments that Germany owed France in reparations, there wasn't enough coal to heat Hans's *Gymnasium* building. Hans's school was relocated to the Viktoria Schule, an *Oberreal Gymnasium* for girls. The girls had their classes in the morning so that Hans's *Gymnasium* could hold its classes there in the afternoon. The Viktoria Schule was a four-minute walk from the Bethe home.

In 1923 Albrecht would get his salary—adjusted for inflation—twice a week. On the day that the salary was paid Hans would go the bursar's office to collect it. He would arrive home three minutes later and take his bicycle to the food stores with a list that Clara, the Bethes' cook, had given him. He had to finish buying all the provisions for the next four days by 1 PM because the stores closed at that time. When they reopened at 2 PM, the new rate for the dollar would be applied, and the new prices were usually twice those of the morning! To help the family's finances Hans started tutoring and gave private lessons in mathematics. He was paid in dollar scrip issued by the German government at a rate of twenty-five cents per hour. Hans found the hyperinflation that had started in 1923 "terribly interesting" and kept a record of its progress by making a chart on graph paper of the value of the dollar.[53] The plot was on semi-logarithmic paper. In 1922 keeping records on a single sheet was still possible, but by 1923 Hans had to attach new sheets fairly frequently. By October-November 1923 a factor of two per day was common.

After the war Hans started collecting postage stamps, and stamp collecting became a lifetime passion. The central post office in Frankfurt was on the way from the *Gymnasium* to his home, and it always had all the new stamps that were being issued, which Hans would buy. Some of them became quite rare, especially when stamped, and quite valuable.[54] He vividly remembered that a particular stamp, which when he had bought it cost a few marks, cost 1 million marks in September 1923. By November of that year that same stamp cost 1 trillion marks!

Hans's dealings with hyperinflation left scars. A feeling of financial instability and insecurity never left him, even after he had obtained tenure at Cornell. His worries over money stayed with him in later life as well, even though he had invested shrewdly and successfully in the stock market. The scars manifested themselves in a certain reluctance to spend money and a measure of stinginess. Thus the furnishings in the house he and Rose bought after returning from Los Alamos to Cornell remained the same throughout their forty-year stay in that house, until their move to the Kendal retirement home in Cayuga Heights.

But despite the rampant inflation that gripped Germany, Hans's final years of *Gymnasium*—from 1922 to 1924—were a "happy time," a period during which he grew more self-confident. When he was sixteen years old Hans was giving financial advice to his parents. A letter, dated July 26, 1922, written while Hans was visiting his friend Werner Sachs in Heidelberg, is filled with two pages of analysis of the economy. At its conclusion Hans made the following specific recommendations: "I would sell Raab-Graaz . . . it will yield barely more than 2000M and its price is excellent. Things will unfortunately continue to go up and it will go down . . . I also want to say that it seems *urgently* advisable to sign a compulsory loan *now*." Hans's growing self-confidence is also apparent in his handwriting. In 1919, before going to the *Gymnasium*, Hans's handwriting was very neat and small in size. By 1922 it had become much larger, bolder, and more assertive.

During his last three years in *Gymnasium* Hans read a good deal of mathematics. When he was fifteen he had gotten hold of his father's copy of Nernst and Schönflies's book on calculus, published in 1918. He read it but didn't tell his father that he had taken the book. When he finished studying it, he put it back on the shelf where he had found it. He didn't tell his father he had read it because he thought his parents would disapprove. In his interview with Thomas Kuhn Bethe commented, "my parents did not like my spending so much time on these matters [reading mathematics]. They said I should play and I should play with other boys and if I was always alone and not doing schoolwork they would much rather have me build structures with my Meccano set than go into mathematics. However, mathematics was the thing that interested me and they thought this was much too difficult for me, but I did enjoy it . . . They thought it would make me one sided."[55] Hans's mother, in particular, felt that he was too focused on mathematics and the sciences, and that this prevented his development in the fine arts and literature. But Hans found calculus "fascinating," and from Nernst and Schönflies he obtained a good introduction to the subject.

Since in their last two years in *Gymnasium* students could take extra mathematics courses as electives, Hans did so. He was fortunate in having an outstanding teacher, Mr. Wirtz, for these courses and got extremely good training in applying calculus. Moreover, Mr. Wirtz recognized Hans's outstanding abilities, nurtured his talents, and encouraged him to continue studies in the quantitative sciences.

Hans was also very lucky that in his last two years of the *Gymnasium* his primary teacher, the one responsible for the instruction of Latin, Greek, German, and history, was an exceptional scholar and "a very fine person" by the name of Dr. Bruhn. Dr. Bruhn was also the principal of the school, and for his doctoral dissertation he had studied the writings of Hans's uncle, the Greek scholar Erich Bethe. Bruhn made the *Odyssey* and *Iliad* "very lively." He was able to make ancient Greece come to life by lecturing on all aspects of its social, cultural, intellectual, political, and military history. Bethe remembered those last two years of Greek "very, very favorably." However, it should be mentioned that after ninth grade, Hans was "very disgusted with learning Greek and Latin vocabulary," and was considering the possibility of quitting school like his friend Paul Berlitzheimer had done. "My father, without great difficulty, persuaded me that after all I did not want to do that, because I probably would want to go to the university, and that would be very difficult if now I quit."

One of Hans's classmates, Karl Guggenheim, who emigrated to Palestine in the 1930s, late in life in an interview reminisced about the Goethe Gymnasium and his impressions of Hans. Guggenheim had been in the same class as Hans from the time that Hans came back from Odenwaldschule in 1918 until they both graduated in 1924. Karl came from an observant, traditional Jewish family. His father was a merchant in Frankfurt. He remembered the *Gymnasium* classes as being quite large—some thirty-five boys, of whom "1/3 were Jewish, 1/3 Protestant and 1/3 Catholic, and most of the students belonged to upper middle class families. Our fathers were lawyers, physicians, merchants, university professors, bankers, high school teachers. There was no anti-Semitism, . . . and all the teachers were men and not Jewish."[56] One of Karl's closest friends in the class was Felix Goldschmidt, "one of the brightest boys in our class," who emigrated to Palestine in 1933 and changed his name to Benyamin Ben Yosef. As an indication of the influence Dr. Bruhn had on his students in his Greek and Latin classes, Guggenheim related that he once went visiting Ben Yosef when he was employed in the phosphate works on the southern tip of the Dead Sea and found him at work reciting loudly and with great pleasure Homer in Greek and Horace and Ovid in Latin.

Guggenheim remembered Hans as having been among the most gifted in the class in every subject and to have particularly distinguished himself in mathematics and physics;[57] also, that Hans had been very helpful to the other pupils in Mr. Wirtz's mathematics classes. He remembered Hans as slightly heavy and quite poor in gymnastics. What had struck Guggenheim was the incongruity of Hans being so good in the classroom and so uncoordinated in gymnastics.

One of Karl Guggenheim's most vivid memories of his years at school with Hans and Felix Goldschmidt was the following: "The 16 or 17 years old Hans and Felix played chess by memory and without a chess board. I still see them before my eyes walking together in the yard of the *Gymnasium* during intermission, playing chess by heart. They were interrupted by an hour of instruction and continued their play in the next intermission!"[58]

Friendships

Hans's friendships with Heinrich, Paul, and Werner were deep and meaningful: they enabled him to achieve a measure of independence from family ties. That Werner and Paul were Jewish and Heinrich Christian made no difference. When he grew up Hans "never felt any division between being Jewish and Christian."

After Hans's return from Odenwaldschule, Hans, Paul, and Werner would often go to the Palmengarten and row on the lake or take long walks there. They called one another by their inverted names, Snah, Luap, and Ren (Ren for Werner) and called their group the Talkeb Bund, partially inverting Kleeblatt (plant leaf). Hans also remembered the three of them together with two or three other boys for awhile going regularly to the Palmengarten to build a dam to divert the little stream that fed the park's lake, "surely a destructive activity." Their subversive labors made them fear the gardeners.

When the Talkeb Bund was not getting together outdoors, it usually met in Werner's apartment, and Bethe recalled that they kept a typed report of their activities. Both Paul and Werner admired Hans's abilities in mathematics, and some of the afternoons were spent having Hans tutor Paul and Werner in algebra and other math subjects. In other matters it might be Paul or Werner who would lead the group. Thus Hans, Paul, and Werner would go to the Palmengarten to read the plays of Lessing, Schiller, and Goethe, with each of them taking various roles.[59] In this activity, it was Werner Sachs, the most mature of the three, who picked the plays and assigned the roles.

Hans's friendship with Werner Sachs deepened as time went on, and they remained friends throughout their lives. Werner's parents were quite welcoming and their house very friendly. Hans went there often and was present at many of their dinner parties. He looked forward to these occasions and characterized the people at them as having "wide horizons." When in 1920 Werner's father accepted the professorship of immunology at Heidelberg University and the family moved to Heidelberg, Hans would frequently travel there for weekend visits and stay at their house.[60] Hans "loved their easy hospitality. Both adults and teenagers came and went, and I was among them." He continued making these visits until he left Germany in 1933.[61] Hans and the other young people who were visiting would go for walks in the surrounding hills, have extended discussions, and dance and party in the evening. On one of his early visits to the Sachses in Heidelberg Hans met Fips Boehm, a friend of Ilse Sachs, Werner's sister. Fips was two years younger than Hans. The seventeen-year-old Hans "fell in love with her" and often visited her in Frankfurt. "But we never kissed."

Paul Berlitzheimer didn't finish *Gymnasium* and quit school after the ninth grade—he evidently wasn't a very good student—and joined a bank. Hans continued seeing him on weekends, when they would play roulette with paper money. Paul's father was quite successful as a family doctor: he owned his house and had a chauffeur-driven car. Paul's uncle was likewise quite well off: he sat on the board of several flourishing commercial enterprises. Paul's father and his uncle introduced Hans "to owning stock, rather than only bonds." Hans concluded at that time "that one should put a large part of one's money into stocks and not bonds. The bonds in Germany were terribly depreciated, and I think that was a good thing to learn, and also that you shouldn't trust any kind of money too much." Eventually Paul likewise became quite well to do. Contact between the two of them became less frequent after Hans went to Munich in 1926. After the advent of Hitler Paul went to the United States and settled in New York. Hans and Paul renewed their friendship when Hans was on a sabbatical at Columbia University in 1941. He then would borrow Paul's car to visit his mother, who was being taken care of in Long Island.[62]

Hans and Heinrich zür Strassen's somewhat more reserved friendship lasted through the 1920s. They both went to Frankfurt University after they graduated from *Gymnasium* and both went to Munich in 1926. Heinrich had decided to become a chemist and chemistry in Munich was "absolutely excellent." They saw much of one another in Munich. Heinrich remained in Germany during the Nazi period, and they would see one another only during the summer when Hans visited his mother. After she emigrated to the United States in 1939 there was no contact between Hans and Heinrich until after World War II. Hans did visit him shortly after the war, but thereafter Heinrich became mentally ill and their ties were severed.

Mama and Papa

Anna's illness in Kiel in the winter of 1913 proved to be deeply consequential. Not only did it affect her hearing and her music making, it also changed her personality, and "nothing was right for her" thereafter. The management of her bodily ills became a major preoccupation. She would take a daily nap after lunch, "a very long nap of about two hours" during which she couldn't be disturbed; and "she had to be approached with caution" at other times as well.

Already in Kiel the seven-year-old Hans found her "very difficult" and so did Albrecht, in fact, even more so. Hans's memory was that his parents slept in separate bedrooms from that time on. "For many years, my father slept in the same, rather big, room with me."

The difficulties between Hans's parents antedated Anna getting ill in 1913. There was a marked difference in the sexual temperament of Albrecht and Anna; Albrecht alluded to these differences and consequences in conversations with Hans in 1926. Anne Fischer, whose husband studied with Albrecht Bethe in the mid-1920s and later became his *Assistent,* in an interview in the early 1990s stressed the "Victorian, puritanical atmosphere that permeated the Bethe, the Levi, the Strauss, and the Sachs households."[63]

Albrecht had an affair with one of the secretaries in the Kiel institute. Such liaisons became more frequent after the Bethes moved to Frankfurt as Albrecht cultivated an active social life without Anna. "My father liked to go to fancy dress balls, my mother did not. In the mid 1920s he had a mid-life crisis, and fell in love with another woman." Anna found this out in the spring of 1926 while she was taking care of Albrecht, who was once again bedridden with a bout of phlebitis. She became intensely jealous, and "things became very acute" with the household spiraling into a hotbed of recriminations, angry arguments, and very unpleasant exchanges. This was just before Hans went to Munich to study with Sommerfeld. "After I left, I wrote many letters to my mother. My mother was very much let down by this of course. I wrote her trying to console her." By the beginning of the summer of 1926 the situation became intolerable, and Albrecht asked for a divorce. Anna felt rejected but "did not want to separate, although she was the aggrieved party. She wanted to continue, and my father wanted to separate and marry that other woman, whom he finally didn't marry. I wrote my mother many letters trying to make her accept the separation, which by that time was, in my opinion, unavoidable. The divorce finally took place on 2 July 1927, my 21st birthday, a date chosen very carefully by the judge, because he then didn't need to assign me to either parent."

Anna broke down during all this, became severely depressed, and in the summer of 1926 had to go to a psychiatric hospital in Badenweiler, near Basel. Almost every year thereafter Anna had bouts of depression that lasted for several months and required her to stay in a sanatorium.

Hans's mother kept all the letters that Hans ever wrote her or wrote to her and Albrecht jointly when they were married.[64] The letters written to both parents have a different tone from the ones addressed to Anna alone. The latter are extremely affectionate and intimate, and some of them have a lyric quality to them. They give proof of Hans's emotional maturation, but also reveal the exceptionally strong ties between them—bonds from which Hans could not free himself until much later. They also provide evidence of his insights into his mother's condition. Thus in most of his letters to his mother while she was in Badenweiler feeling abandoned and utterly alone, Hans addressed her as "My dear best friend" *(Freundin).*

In a letter to his mother not long after she had gone to Badenweiler, Hans wrote that he had found her "improving from day to day" during his extended visit and that he had left "at peace." Evidently, there had been an earlier incident while she was still in Frankfurt that had been extremely difficult because Hans had thought that she had made an attempt on her life by taking an overdose of barbiturates and he had "sermonized" her, and she had said "sadistic hate things against Papa." In that same letter he made a point of telling her that from that moment on he had enormous sympathy for both of them, and "could not properly imagine how things would develop without [him]." As Hans felt that she was improving, he indicated that he thought that his constant interventions were no longer necessary, and that the situation between her and Papa would improve as a result of their mutual efforts. Hans concluded his letter by stressing that

> If you again feel oppressed by anything, please write to me, and do not think you need to go easy on me for any reason. It has almost become a need for me to comfort people—not because of any reasons of religious duty, but simply because it gives me contentment to give something to people who are close to me, and I don't want to deny that and generally not make myself out better than I am . . . Beyond that, I am very, very glad that I have come so close to both of you . . . I only wish that it will have for both of you as good consequences as it has for me.

The statement "It has almost become a need for me to comfort people . . . simply because it gives me contentment to give something to people who are close to me" was not idle or momentary talk. Hans's letters to his friend Rudolf Peierls in 1928 and 1929 record several occasions he went visiting recent widows. On one occasion he made a special trip from Munich to Frankfurt to visit a friend whose mother had just died because "he wanted to help a little bit."[65]

In a letter to his mother dated November 27, 1926, written after she had been sent home, Hans advised her to socialize: "The acquaintances that you do not value because they desert one in rough times, can be helpful in many respects *just because* one does not let them see one's soul. They help by providing diversion." In a subsequent letter he advised her not to obsess about the divorce and not to be upset by every bit of news from Frankfurt. In fact, Albrecht was trying to be helpful and had begun writing to Anna, and it seemed to Hans that at least superficially the situation between them was better.

And in a letter to his mother after seeing a performance of Henrik Ibsen's *Ghosts* and having been deeply moved by the play because he saw many parallels in it to the tragedy they had gone through, Hans made the following observations:

> The piece even offers something beautiful for life: For the mother the many years of ruined marriage are not only a sorrow, but afterwards also good fortune, fortunate in her child and also perhaps in the sorrow itself. That sounds perhaps like exaggerated Christianity, but I feel exactly so. In every

> misfortune lies, if one suffers it, simultaneously the strength to take it, which at least gives me a certain pride. There is nothing noble in that, but purely the feeling of being better, of being able to endure more, and of having a more mature standpoint than other people.

What is also striking about these handwritten letters is the fact that there are but a handful of corrections of single words in all of them. The few times this happens the word is struck through, and a word written underneath it. Otherwise there is a continuous narrative from beginning to end without any corrections. Every sentence is perfectly clear and the letter as a whole coherent and informative.

After the divorce, relations between Hans's parents remained thorny and demanding.[66] In many ways, Albrecht had as difficult a time as Anna coming to terms with what he had done and with the divorce. Hans tried to help him as much as he could and in whatever way he could. He worked hard after they separated at getting them to re-establish some sort of civil, even friendly, relationship, and to some extent he succeeded. Occasionally, Hans's father would visit Anna for a day or two in Baden-Baden. "My mother still loved him and hated him, and many times the visits were quite peaceful, and many times they were the opposite. I don't believe he went to visit anymore after he remarried in 1929." Though divorced, she continued calling herself *Frau Geheimrat Professor* Bethe.

From the time that Albrecht asked for a divorce the responsibility of looking after Anna fell on Hans. He was there for her throughout her stay in Badenweiler, visiting her often and writing her a long letter almost every week. It was Hans who found her a place to live near Baden-Baden, and in late 1927 she moved to her new dwelling.[67] She was quite happy there except for her recurrent depressions. She lived in quite spacious quarters in a house with a big garden "where she spent a lot of her time." From the house she had a beautiful view of meadows bounded by a forest. A ten-minute walk up the road from her house brought her to a small farming village, and a little beyond the village were woods with lots of trails. "Mama walked often in these woods." After Anna moved to Baden-Baden Hans wrote her often and kept her informed about the progress he was making on his dissertation. Later on, he would write her about the papers he was publishing, briefly outlining their content, and send her reprints of them. And he would spend part of every summer with her.

A deep sense of filial responsibility bound Hans to his mother. But there was also a deep love between them. On May 7, 1930, three days after successfully defending his *Habilitation* and obtaining his *Venia legendi,* the right to lecture at a university, Hans sent a copy of his *Habilitationsschrift* to his mother with the following inscription: "The three day old 'famous head' to his Mama, to whom he is always bound by an invisible umbilical cord."[68] It took a long time to cut that umbilical cord.

A more complete portrait of Hans's mother must include the recollections of her grandniece. Charlotte Levi Litt, the granddaughter of Bertha Kuhn Fitting, the older sister of Anna, was born in Leipzig in 1920. She visited Tante Anna several times when she lived in Baden-Baden and remembers her as a kind, loving, and caring person. Before coming to the United States in 1936, Charlotte went to Baden-Baden to say goodbye to her. "I loved Tante Anna."[69] Charlotte looked after Anna in Ithaca in the fall of 1940. Later when Charlotte got married, Anna gave the couple valuable presents. Charlotte and her husband, Mortimer Litt, visited Anna when she was in a nursing home on Long Island and later in one in Ithaca. Mortimer gave the following description of the trumpet-using and bedridden Anna when in Ithaca: "She was the matriarch personified. When you entered a room with her present she was the dominant figure."

Papa

If when Hans was growing up Anna was the doting, possessive mother who at times was very demanding, Albrecht was the person who took great pride and felt great fulfillment in his son's precocity and unusual abilities. He had recognized these quite early and nurtured Hans's intellectual growth. Through his studies of Kant in *Gymnasium* Albrecht had taken to heart Kant's admonition: "The human being and every human being gifted with reason exists as an end in himself, not merely as a means for arbitrary use by this or that will. Rather, he must be regarded as an end, in his dealings both with himself and with other reasoning beings."[70]

Albrecht always interacted with Hans with this dictum in mind. It is corroborated by the way Albrecht reacted to Hans having been diagnosed with tuberculosis when the family was vacationing in the Black Forest in the summer of 1916. Albrecht told the ten-year-old Hans what the disease was. He then took Hans to a sanatorium in the Black Forest near where they were staying to show him people whose tuberculosis had progressed to the point that their bones had been infected and told him, "You don't want to get in one of those places." Although not explicitly stated, the trip was meant to tell Hans that although he did not feel sick, there was a purpose to his being sent to Kreuznach and that he had a responsibility "to take care of himself."

The closeness between father and son in personal matters and also an indication of Hans's reaction to girls as a young teenager is revealed by a comment his younger half-sister Doris made.[71] Albrecht told her that when Hans was a teenager he would sometimes ask him, "Can you tell me if this girl is a pretty one?"[72]

It was his father who made Hans socially conscious. When he was ten Hans witnessed his father "helping the helpless." He recalled that on several occasions in Kiel "when seeing a man trying to push a heavy cart up a steep hill Papa would rush to help him." And it was his father who indicated to Hans the meaning and responsibilities of familial bonds: Albrecht helped his older sister Lisbeth, Hans's favorite aunt, survive financially during the 1920s.

Hans became aware of the respect accorded his father by his colleagues at the university when in 1917, two years after he had joined Frankfurt University, the faculty elected him *Rektor.* And in late 1918, after the Armistice had been signed, the faculty once again asked him to serve for another year as *Rektor.*[73] And Hans was very conscious of the fact that the only time that there was any social function in the Bethe household was when Papa was *Rektor!*

As noted earlier, it was also his father who made Hans politically conscious as a teenager by discussing with him the implications of the stands taken by the various political parties on the issues of the day. Shortly after the Armistice Albrecht joined the newly formed *Deutsche Demokratische Partei* and was very active in it until 1925, giving many speeches in support of its candidates.[74] In the early 1920s he himself became a candidate for the Frankfurt City Council. He was politically savvy and courageous. In the spring of 1933, shortly after the national socialists had won the national elections, Albrecht at a meeting of the medical faculty had asserted: "We now have the Nazis, we have to be very careful, but we have to resist them. They will take over and do very bad things to the universities." Most of the faculty disagreed: "No, no, they are all right, they are just conservative." A few months later, in the summer of 1933, one of the most conservative professors came to Albrecht and said, "Bethe, you were right. We should have resisted at that time, and now it is too late."

Albrecht had recognized that only collective action might have had some political impact. Individual action could not. Although he admired James Franck for having resigned his directorship of the experimental physics institute in Göttingen in the belief that this would change the policy of dismissing non-Aryans from civil service jobs, it was clear to Albrecht that such individual action was futile.

As he matured Hans found his political views to be somewhat to the left of those of his father, and he came to identify himself as a social democrat. When he became eligible to vote he cast his ballot with the *Sozialdemokratische Partei Deutschlands,* which was committed to reforming capitalism democratically and to introducing legislation that would remove the social injustices they believed capitalism fostered. He supported their resolute opposition to the authoritarian, uncompromising Marxist revolutionary socialism that the communists were advocating.

Once his Institute of Animal Physiology in Frankfurt was set up, Albrecht spent a good deal of time with Hans, as much as his professional and wartime duties would allow. After the Armistice nearly every Sunday the two of them would go on long four- or five-hour walks, either to the *Stadtwald* south of Frankfurt or to the Taunus forests. During summer vacations they would go on even longer hikes, sometimes staying overnight for one or two nights in the huts along the trail. Later, when Hans was older, they would travel to the Bavarian and Austrian Alps. Already as a young boy Hans was overwhelmed by the beauty of some of the mountain scenery. Mountain climbing became one of the great loves in Bethe's life. Mountaineering had been one of his father's passions. While in *Gymnasium* Albrecht had done some serious mountain climbing with a classmate. Later when he was an *Assistent* to Ewald in Strassburg, Albrecht, this *Gymnasium* friend, and another mutual friend would hire a guide in Switzerland for one or two weeks and would climb challenging mountains. They succeeded in climbing to the summit of the Dom and of Monte Rosa in the Pennine Alps near Zermatt, the two highest peaks in Switzerland—both demanding mountains over 4,500 meters high.[75] Hans acquired his father's passion for mountain climbing. For the rest of his life the communion with nature and the "sense of liberation" that stemmed from hiking in the mountains became almost a physical need for Hans.

After their hikes Albrecht and Hans would talk about politics, and Hans would tell his father about his studies. But most of the conversation would be taken up by Albrecht telling Hans about science, and in particular about his scientific investigations, his approaches to the solutions of problems, and how he had come to posing the questions that had led to them. He told him of his evolutionary views, of his research on arthropods, of his writing extensive review articles, of their value and rewards, and of the growth of his scientific toolkit as he tackled new problems. They discussed Kant and his philosophy, and Albrecht stressed to Hans the fact that Kant had founded his philosophy on contemporary scientific knowledge. He felt strongly that future philosophies must do the same, but that they must also recognize the limits of scientific discoveries and knowledge. It was the recognition of these limits that allowed him to be somewhat religious, admit to metaphysical preconceptions, reject Comtian positivism, and to be a good scientist, in fact, a very good scientist.

Albrecht Bethe is well known not only for his physiological and neurohistological research but also for his important contributions to understanding animal behavior. Bethe recalled his father telling him that when he first started working on the nervous system of arthropods, bees, and ants, he thought of them as reflex machines. But later, when studying the regulation of motor coordination in dogs, he came to see animals as much more than machines that responded passively when reacting to stimuli. Animals could differentiate themselves from their environment, had memories they could recall, and could by virtue of these memories behave differently under identical circumstances. Living entities had histories!

In his research Albrecht was committed to the psycho-physical view of nature that Wilhelm Wundt had defined in his *Physiological Psychology* as "that view which starts from the empirically well-established thesis, that nothing takes place in our consciousness which does not find its foundation in definite physical processes."[76] It would take us too far afield to elaborate on Albrecht Bethe's changing views of living creatures as "machines" to creatures that learned, had volition, and whose nervous systems had plasticity.[77] Albrecht was aware of the philosophical, cultural, and political connotations that the notion of "machine" evoked. Anne Harrington has given an informative and insightful account of the "holistic" reactions to the mid-nineteenth-century attempts by Ernst Brücke, Karl Ludwig, Emil Du Bois–Reymond, Hermann von Helmholtz, and others to account for biological phenomena in a reductionist, mechanistic, causal fashion by emulating and adopting where possible the approaches of the quantitative physical sciences, such as physics and chemistry.[78] Driesch and von Uexküll were at the center of the end-century "wholeness" movement that rejected this approach, and Albrecht had been in close contact with them at the Naples Biological Station.[79] Although he was influenced by them and adopted some of their viewpoints, he remained a physiologist committed to Wundtian views. As Harrington explains, in Germany at the beginning of the twentieth century, "there were two approaches to understanding life processes: one the province of what [Uexküll] called *biology*, and one the province of *physiology*. Physiologists concerned themselves with the material, causal substances, and forces operating within the organism. Biologists, on the other hand, . . . were interested in accounting for the activities of a particular animal in terms of its functional logic and underlying plan."[80]

The democrat Albrecht Bethe was also aware of the political connotations of "machine." The amalgamation of the various German states under Prussian rule with Bismarck at the helm had initially made people—particularly, the more liberal component of the population, the middle classes that formed the *Bürgertum*—believe that the unification of 1871 had fulfilled a national yearning for political unity and cultural "wholeness." But as the century wore on, Bismarck's "iron fist" policies and his political maneuvering reminded the *Bürgertum* more and more of the "machine." The "machine" became identified with the efficient, impersonal, and ruthless Prussian Army; with the iron mines and steel plants and factories of the Ruhr valley and what they had done to the environment; and with the atomization, bleakness, and impersonality of the rapidly growing cities and their squalid and seedy tenements.

With Wilhelm II's dismissal of Bismarck in 1890, the cultural, political, and intellectual settings once again resonated with talk of "wholeness"—with its various and varied meanings.[81] In biology, Driesch and Uexküll expressed one of its manifestations; Kurt Koffka, Wolfgang Köhler, Max Wertheimer, and Kurt Goldstein articulated another in psychology and psychiatry with their holistic, Gestalt views. Bethe remembered that during the war his father had told him about the research he had conducted with Kurt Goldstein concerning the recovery of wounded soldiers who had suffered severe brain injuries.[82] They had found that these soldiers could relearn many functions and activities, "not only speech, but, for example, also moving a hand when the corresponding nerve center in the brain had been injured beyond recovery." Stimulated by his earlier contacts and interactions with Driesch, Albrecht Bethe concluded that the nervous system was plastic, "that one part of the brain could learn the functions of another part if the latter was damaged." Whereas Goldstein went on to consider the brain's plasticity in humans, Albrecht Bethe confirmed his findings with experiments on animals.[83] Hans remembered his father telling of his experiments with spiders and daddy longlegs, in which he would "take off one, or two, or three legs, and then observe how they would change their walking." Since they were able to adapt to the new conditions, Albrecht had concluded that there must be something in the nervous system that made this possible. "He emphasized this very much. He was quantitative and mechanistic and all that," but at the same time, he saw that organisms were self-organized entities endowed with histories and plasticity, entities that reproduced themselves with great fidelity, in which structure and function, anatomy and morphology, were closely connected.

For the post–World War I generation of neurophysiologists Albrecht Bethe is best known as the originator and formulator of important ideas concerning the plasticity of the central nervous system regarding the functions it performs. In his publications on this topic he used evidence derived from the mammalian and human brain—the loci in which he and Goldstein had first encountered the phenomenon—but he himself indicated that his early and later experiments on arthropods played an important role in formulating and making precise the concept of plasticity of the central nervous system.[84]

Albrecht embodied the ideals of the Enlightenment in many ways and he inculcated in Hans a commitment to a rational approach to the world and a belief in progress. In scientific matters, he instilled in Hans the faith that reproducible experiments are the determinants of acceptable answers to questions; and that it was the scientific community that decided the validity of the answers. The experiments that Albrecht designed dealt with biological phenomena, which made reproducibility more difficult. It meant focusing on observations that could be duplicated and reproduced. When quantitative measurements could be made, it meant that they would exhibit stable results with little fluctuation about the measured values. These requirements made it possible to attempt mathematical analysis and a mathematically based model of the system. Albrecht recognized the special role of mathematical models: they allow one to specifically formulate assumptions and rigorously deduce from them consequences or expectations. This was the reason that, in contrast to most biologists of the day, Albrecht had learned calculus, had studied the elements of chemical thermodynamics, and tried to keep abreast of developments in atomic physics and its mathematical models.[85] In this Albrecht reflected the great influence that Helmholtz had exerted on the development of the sciences in the second half of the nineteenth century. David Cahan, a leading Helmholtz scholar, put it thus: "Helmholtz . . . profoundly altered his principal scientific disciplines of physiology and physics, and influenced the related disciplines of medicine, mathematics, physical chemistry, psychology and meteorology . . . His views on science and society were listened to and solicited by ministers of state, and late in his career he participated in the interactions of science and industry."[86] As was the case for Helmholtz, empiricism was a fundamental and essential component of Albrecht Bethe's philosophical outlook and commitment. And Albrecht followed in Helmholtz's footsteps in the employment of quantification and mathematical analysis in physiology.[87] And like Helmholtz, Albrecht "consistently pursued scientific problems that stood at the boundary of two or more sciences using the methods or techniques of one science to work on problems in another."[88] There was also a marked affinity between Albrecht and Helmholtz in their quest for intellectual synthesis. Helmholtz's work "attempted syntheses both within given disciplines and between parts of different disciplines with one another," and so did Albrecht's works and writings.[89]

Hans Bethe would do the same. In fact, if one were to speak of Hans Bethe's "genius," one would characterize it as an ability to synthesize that overwhelms. And like Helmholtz, Albrecht became deeply committed to Darwin's theory of evolution. He undoubtedly had read Helmholtz's *Popular and Philosophical Essays.* It would be difficult to find a more insightful presentation of Darwin's theory than the one given by Helmholtz in his 1869 lecture on "The Aim and Progress of Physical Science."[90] Albrecht's stay in Naples had given him an even wider perspective best summarized by the great American astronomer George Ellery Hale in 1908: "We are now in a position to regard the study of evolution as that of a single great problem, beginning with the origin of the stars in the nebulae and culminating in those difficult and complex sciences that endeavor to account, not merely for the phenomenon of life, but for the laws which control a society composed of human beings."[91] Albrecht would certainly have agreed, but might have refrained from including human social life. Hans grew up with this understanding of the history of the cosmos. More specifically, Hans got from his father an overview of the history of biology during the nineteenth century. Albrecht conveyed to him that this history was not the story of a steady march toward Truth, but rather the narration of the introduction of new instruments, new experimental techniques, and new methodologies. It also constituted the description of a process of creating and recreating classifications, formulating and reformulating concepts, constructing and reconstructing models, and learning from the more mature sciences, such as chemistry and physics, what it means to be a science. In the fields Albrecht carried out research—histology, neuroanatomy, and neurophysiology—developments for the most part had been cumulative. But Bethe recalled that his father had made him aware of the conceptual ruptures that had occurred with the introduction of the notion of the cell by Schwann, of natural selection as the principal mechanism of evolution by Darwin, and by Mendel's research and by his introduction of the idea of a character as an element of heredity. Albrecht made it a point to bring home the impact of Mendel's use of mathematical modeling.[92] In their garden in Frankfurt Albrecht replicated for the thirteen-year-old Hans Mendel's pea experiments! And Hans remembered becoming very interested when his father did these experiments and planted beans of various colors to verify Mendel's findings. Hans commented, "the beans did indeed perform most beautifully."

Another deeply influential lesson that Hans learned from his father was the special role that mathematics played in the sciences. As I mentioned in the Introduction, the first paper that Hans wrote was a scientific publication with his father in which he supplied the solutions of the diffusion equation applicable to an investigation of dialysis that Albrecht was carrying out with his assistant. For the paper Hans tested whether the data fit the predictions of the diffusion equation when the parameter in that equation—the diffusion constant—was suitably determined. In addition Hans tried to understand when it did so and when it did not. And all this when Hans was 17!

This was Bethe's memory of it in 1996:

> My father was very interested in dialysis. How do various substances go through membranes, and so he had a very good Japanese research associate, by name of Terada, who did the experiment. They got certain results; they had observed the concentration as a function of time it took a few days usually. They asked me whether I could do any mathematics on that, and just having learned calculus, I eagerly grasped that problem, and reinvented the elementary law of diffusion, that it goes exponentially, that's in the paper. Well, it didn't go exponentially, or rather it did so for some substances, but

> not for others, and I don't remember which it did, and which it didn't. I think for electrolytes it did not, and, but I don't know whether my father and Terada made anything of the discrepancies. In some cases it worked, and followed my exponential formula beautifully. That's that paper.[93]

The fact that Albrecht's institute was part of the medical school was also of importance in conveying to Hans the character of scientific progress exhibited in the biological and medical sciences. Because of his responsibilities for the training of physicians Albrecht was very attentive to and cognizant of developments in medicine. Albrecht, both in Kiel and in Frankfurt, was very much part of the effort to make physiology a key component of "scientific medicine."[94] Since a large fraction of the students he taught would become physicians, his courses emphasized the factors that had made physiology quantitative, amenable to precision, and objective: namely, its experimental approach and its reliance on instruments and measuring devices.[95] Through his own training as a physiologist and as a physician he was very much aware of the role that instruments such as Helmholtz's ophthalmoscope had come to play in medicine. Similarly, he was aware how conceptual and technical advances in chemistry had been amalgamated into biology and medicine to yield new practices and subdisciplines such as biochemistry and histology.[96] If not explicitly stated, nonetheless the message was that progress in medicine and biology up to that point had for the most part been driven by technical and instrumental advances and by conceptual innovations, not by theory. But like his eminent predecessors, Ernst Brücke, Emil Du Bois–Reymond, and in particular Helmholtz, Albrecht saw the mid-nineteenth-century developments in physics, in particular the establishment of the first and second laws of thermodynamics, as giving physics a universal character. He understood the drive in physics toward atomic and microscopic explanations as freeing it as much as possible from dependence on context and history. The possibility of explaining macroscopic phenomena in terms of microscopic ahistoric components is what stimulated Albrecht and other physiologists to investigate those chemical and physical processes in living systems whose descriptions had a seemingly constant character among many species, and could be mathematized.

Bethe's first exposure to science was in his father's laboratory, which meant that he first experienced science as a communal practice in which knowledge is produced by experiments, with reliable instruments that measure with some precision, and with mathematical models that attempt to interpret the results;

that the knowledge thus gained is valid to a greater or lesser degree of probability and always subject to revision. His father's view of animals as reflex machines was a memorable illustration of this. Bethe also became aware that his father's science was not "disinterested": In addition to meeting the highest standards set by the best "pure" science that was undertaken to understand physiological processes, Albrecht's research was also meant to be of service in bettering medical care.[97]

A judicious attitude toward theory—stemming from his awareness of the role of technical and instrumental innovations in the advancement of scientific knowledge and a recognition that theory deals with idealized situations—became part of Bethe's approach to theoretical physics. He never believed that any given theory was an ultimate one, he always tested the limits of the applicability of a given theory, and he always sought confirmation of the experimental "reality" of the entities to which the theory referred and the limits of the domain in which they retained their validity and integrity. His first scientific investigation also introduced him to a subject that would concern him the rest of his life: the relationship between macroscopic descriptions—in the case at hand, the diffusion equation with its experimentally determined diffusion constant—and microscopic ones that referred to the dynamics of the atomic and molecular entities involved.

In his collaboration with the young Hans on diffusion through membranes, Albrecht also demonstrated to him that it was straightforward to publish good scientific papers. To some extent the joint paper had made them peers, and in recognition of this Albrecht subsequently took Hans to various scientific meetings. Hans recalled attending the 1924, 1925, and 1926 meetings of the German Society of Scientists and Physicians. One of the things that he must have observed at these gatherings was the deference accorded to his father, and more generally to professors and to those who had carried out important research.

But, perhaps, the most valuable lesson that Albrecht Bethe taught his son concerning doing research was that the enjoyment of the search and the satisfaction and gratification from the search are to be valued more than the knowledge gained. Already in 1899, when twenty-seven years old, he had stated, "If one were not to find satisfaction in the search for knowledge, one would despairingly put one's hands in one's lap and say: It is too difficult for us humans."[98] Albrecht placed greater value on the process than on the goal of learning, exploring, and creating new knowledge.[99] And since all scientific knowledge is always only probable to a greater or lesser degree and never absolutely certain, and its validity limited in time, the accepted knowledge reflects the values of the scientific community that gives it its approval and determines its acceptance. There therefore cannot be an absolute separation between facts and values. Values—what *ought* to be—belong to the realm of metaphysics, and in practice in the realm of politics, and therefore politics will always intrude—to a greater or lesser degree depending on the degree of certainty of the scientific knowledge—even when answering questions of what is. As a young man Albrecht had concluded what Weber was to state later: "*All* ultimate questions without exception are touched by political events, even if the latter appear to be superficial."[100] Since this was so, Albrecht had recognized the need to formulate his own coherent political outlook, to live by its tenets, and to implement his political views by political action. And the same was true in his professional endeavors. For Albrecht practice became the answer both in his science and in politics. In politics this meant being actively engaged in political action; formulating relevant, responsible, and feasible policies for the benefit of all citizens. It meant possibly running for political office and, if elected, serving with integrity.

Albrecht found contentment in his search for knowledge by organizing his laboratory as a communal, collective enterprise—a place where men and women were treated equally, irrespective of status or religious affiliation, or of social or ethnic background. He did so within the limits imposed by the culture of German educational institutions: he, being the director, controlled both the finances as well as the intellectual agenda of the institute. Organizing his laboratory and institute in this way was in part the legacy of his years working for his doctorate with Richard Hertwig. Richard Goldschmidt, who was Hertwig's *Assistent* for six years after obtaining his doctorate with Otto Bütschli in 1903, identified the secret of Hertwig's success in the laboratory as the result of

> his immense human interest in every one of his students, to whom he was like a father and friend. There was no part of their work, including the menial tasks, into which he would not enter and give a helping hand. He was always ready to suggest new lines of approach and even try them out first himself. There was a constant collaboration between the professor and the advanced student, who thus got all the benefits of his experience, knowledge, skill, and enthusiasm, while the whole group worked like a family, each member helping the other.[101]

Albrecht Bethe did the same, and in addition made it become an expression of his political views, especially with regard to the treatment of Jews and women in German society. He came to Frankfurt as an eminent and well-known physiologist specializing in the structure and evolution of the nervous system. The Institute of Animal Physiology he built there became a highly respected training center for young physiologists. One of the first students who came to the institute was a young Jewish woman named Trude Neugarten. Trude was born in Mainz in 1894, where her father was a wine merchant and her mother taught at the conservatory.[102] Trude went to *Mädchen* school and then to a *Realgymnasium* from which she graduated in 1913. She then went to Freiburg University but after the war broke out she enrolled in the newly founded Frankfurt University to continue her medical studies. In 1915 she took Albrecht Bethe's physiology course. There were between fifteen and twenty students attending the class, and Trude was the only woman. Two years later there were eight women taking the course.

Trude remembered Albrecht Bethe as "a very good lecturer and a very good teacher, very patient, who in his lectures only used brief notes."[103] It was always clear to her that he had a thorough mastery of the materials he presented. She did quite well in his course, and he invited her to become an assistant in his laboratory when she finished it. When he offered her the job she warned him that she didn't know "anything." The physiology course had consisted of lectures with demonstrations, but no laboratory. The first thing Albrecht asked her to do was an experiment using some phosphorus. When she weighed the phosphorus without using any paper on the pan she recalled that "Prof. Bethe was dismayed." When she reminded him that she had warned him that she didn't know anything, he calmly indicated, "You will learn." She did learn and proceeded not only to help him in the preparations for his lecture demonstrations but also to do some fine research. Most of her work with him dealt with nerve regeneration, and in particular, the results of transplanting nerves from one dog to another. This work was carried out during the war and was intended to help soldiers wounded in battle. She remembered "Prof. Bethe" often being away on military duty. Some of the work she did led to a publication that "Professor Bethe" thought was good enough to earn her a doctorate in physiology.[104]

Trude met Hans for the first time in 1917 when he came to the laboratory and asked her what she was doing with her instruments. She also remembered the later father-son collaboration that brought Hans to the lab for more extensive stays and led to their co-authored publication.[105] Trude was never invited to the Bethe home. She did meet Hans's mother in the mid-1920s at the performance of one of her children's plays in Mainz, where she then lived with her husband and young son.

Another student of Albrecht Bethe was one of Hans's classmates, Karl Guggenheim. Karl's friends were Jewish and Zionists. Karl's closest friend was Nahum Glatzer, who introduced him to the circle around Franz Rosenzweig and Martin Buber.[106] Glatzer advised him to go into medicine—which he did. In the fall of 1924 he enrolled in Frankfurt University. Albrecht Bethe taught the animal physiology course he took in 1925 and 1926, and Guggenheim remembered him as a tall, attractive person: "A fascinating lecturer, always interesting and lively."[107]

Ernst Fischer, who completed his *Habilitation* with Albrecht Bethe and was his *Assistent* until 1933, became an outstanding physiologist. In his eulogy for Albrecht, Fischer noted that *Assistenten* in Bethe's institute were given freedom to undertake their own projects and financial support to do so; and that the same was true for younger co-workers. This is probably the reason that one does not speak of the Albrecht Bethe school of physiology: too many diverse projects were undertaken and carried out. Fischer observed:

> Under Albrecht Bethe's influence all of us working at his institute felt ourselves as members of a close-knit working community, in which everyone, whether secretary, researcher, clean-up woman or director, gave his or her best to serve the total effort . . . With his kindly nature he could only see the good in nearly every person, and—mostly through small rebuffs, often by good natured mocking comments—he knew how to squelch every possible dissension between two hotheads. He took an active and warmhearted part in our personal fates, steadily supportive in advice and action, for which his fine capacity for human empathy and his generally optimistic temperament served him well.[108]

Hans Bethe's love and admiration for his father were movingly conveyed in the letter he wrote after his father's death to his uncle Martin, Albrecht's only surviving brother.

> Ithaca, 8.2.55
> Dear Uncle Martin,
> . . . I thank you for your loving letter . . . It was a great joy for me to be able, once more, to be with him [Albrecht] a great deal this past summer, even

> though he no longer was wholly his former self. He tired easily and slept thereafter on the couch. He could no longer follow difficult conversations. But he was still interested in many things, and still well informed, and it was nice to see how he could still perk up during an interesting conversation and forget about all physical hardships.
>
> He was particularly fond of remembering past times . . . [and] in this connection there was also a "reconciliation" ["Versöhnung"] with my mother—or perhaps this was more in order for me to write to my mother, who knew that he was very ill (though not the exact nature of his illness until after his death) and therefore that as earlier, she might write him more affectionately and more often.
>
> Yes, he was really a very good man . . . He had a warm heart for everyone, and I remember how 30 (or 25) years ago everyone would come to him for help and encouragement, or to be the peacemaker in a dispute. He was concerned with the well being of everyone in his institute and the father-confessor of all his colleagues. I did not inherit this from him—I believe, I have some of his goodness, but not the patience, to listen as long to the woes of my fellow men.
>
> As you say: many will long mourn him. It often occurs that I think "Oh, I must tell this to Papa"—and then I remember that he is no longer here. The science [animal ethology and physiology] mourns him; there is no question that till the end he was totally up to date and active. A quite famous zoologist, Konrad Lorenz from Westfalen, recently visited our university and gave some lectures here. We invited him to the house, and he told how two years ago at a conference he was attacked by everyone, until my father stood up and clarified how Lorenz's observations agreed with his own of 30 years ago, and how all this is consistent with many other facts. It is clear that he, the 80 year old, understood the new theory immediately![109]

In 1972, on the occasion of the hundredth anniversary of his father's birthday, Bethe received a letter from the dean of the faculty of medicine of Frankfurt University informing him that "We shall remember this great teacher and researcher on the occasion of his birthday with a lecture" and that "your father is not forgotten here." He wrote Vera, his stepmother and "the other person his father was closest to," to inform her of the memorial lecture and expressed the hope that the dean had also written her. After telling Vera that he recently had a discussion with a young graduate student at Cornell who told him that Albrecht's research on pheromones and on the plasticity of the brain was still referred to, Bethe took the occasion to reminisce about his father.

> [You and I] knew Papa not so much as a scientist, but as a human being. His understanding of everything human was enormous. I have scarcely known anyone else who had this understanding in such full measure . . .
>
> We both had the good luck to live with him for many years. I constantly realize how much he gave to me: humaneness *(Menschlichkeit),* a liberal outlook, a love of science and of nature. You know all that, and moreover, you helped him so much. Without you he would not have survived the difficult times of the war and after the war. I therefore thank you once again on this occasion for having been *there,* and thereby having made it possible that I could see Papa again three times—in 1948, 51 and 54.

Siddhartha

In his interview with Charles Weiner, Bethe recalled, "There were many books in the Bethe home. There was the usual German literature—Goethe and Schiller and novels of the 19th century and the 20th century. I read quite eagerly the German novels in the 1920s."[110] Partly through the influence of Werner Sachs and partly because of his own interests Hans read a great deal: all the Joseph novels of Thomas Mann and his *Magic Mountain,* Erich Remarque, many of Dostoyevsky's novels, and Hermann Hesse. Hesse was a friend of his mother, and she had admired his courage in publishing in November 1914, shortly after the outbreak of war, an appeal to German intellectuals not to be blinded by patriotism.

Probably during his last year in *Gymnasium,* Hans read Hesse's *Demian: The Story of Emil Sinclair's Youth.* He perhaps identified with Emil Sinclair, the protagonist of the novel, who is in search of meaning in his life. He may have identified Werner Sachs with Max Demian, who in the novel becomes Emil's friend and mentor; identified Franz Kromer with the bullies he had encountered in *Gymnasium*; and Frau Eva, Demian's mother, with the ideal woman, the life companion he sought. Emil Sinclair's recognition that "Schicksal und Gemüt sind Namen eines Begriffs" (Destiny and temperament are two words for one and the same concept) struck a chord in Hans.

Sometime during the summer of 1927, while his mother was in Badenweiler recovering from the trauma her separation from Albrecht had caused, Hans read Hesse's *Siddhartha.* The book affected him deeply. After reading it Hans wrote his mother a long letter about his reaction to it. In a subsequent letter he commented, "You found my *Siddhartha* letter so good. It was written entirely under the impression of the book, so much so, that today, I no longer know what was in it. The book, just as *Demian* earlier, had transported me directly into a trance lasting one, two days. I would like to be able to write down such ideas more often, especially in my journal, but I never get around to it and so my last entry—6 February 1926—is about dancing lessons."[111]

The story of Siddhartha's journey of self-discovery resonated with Hans. Like Siddhartha, he was somewhat of a child prodigy; like Siddhartha he experienced multiple departures—from Kiel, Kreuznach, Odenwald, Frankfurt—and, though very successful in his academic studies in Frankfurt, like Siddhartha, the twenty-year-old Hans was unsettled, somewhat of a loner socially, unable to open up to anyone, and in "search for the source of the self within the self."

The ninety-year-old Hans vividly remembered Hesse's description early in the book of Siddhartha's encounter with the Buddha and Hesse's depiction of the Buddha as a very special human being who had found peace. The Buddha could not communicate to Siddhartha, either in words or in teaching, the secret of what had happened to him in his hour of enlightenment, which is what Siddhartha wanted to know. Therefore Siddhartha decided to go on his way "not to seek another and better doctrine, for I know there is none, but to leave all doctrines and all teachers and to reach my goal alone—or die."[112]

Siddhartha's reflections after his encounter with the Buddha left their mark on Hans. He thought, "I, also, would like to look and smile, sit and walk like that, so free, so worthy, so restrained, so candid, so childlike . . . A man only looks and walks like that when he has conquered his Self. I also will conquer my Self."[113]

Mary Oliver perhaps puts it best in her poem "The Buddha's Last Instruction," published in 1990, in which the Buddha counsels the poem's subject to make of the self a light. And Hans did. In the process he achieved much inwardly. As Plutarch observed: "What we achieve inwardly will change outer reality." The love his mother had showered on him had made him feel very special. His father had corroborated his brilliance as far as his scientific talents were concerned: at age seventeen Hans could do things in science that his father—the director of an institute—couldn't. His teachers in the *Gymnasium* had made it clear to him that he was outstanding in mathematics. He graduated from *Gymnasium* confident in his abilities. Frankfurt University reinforced that feeling.

If the reading of *Siddhartha* resonated with Hans's search for an authentic, individual self and with the romantic component of his self-shaping that had been nurtured by his mother, its individualistic message needed to be reconciled with the ideals of the collective demanded by his molding as a scientist, and more specifically as a theoretical physicist. His identity as a member of this new disciplinary collective was the part of the self that had been nurtured by his father. His father had resolved the tension between the individualistic and collective aspects of membership in a scientific community by making his laboratory a communal enterprise wherein each member could be creative, thrive, and grow. But he did so within the context of the German "plutocratic" university: the physiology institute was *his* institute, *he* controlled its budget, and the problems people worked on were determined by *him*.[114]

The theoretical physics community resolved the tension between creative individualism and collective resources by creating hierarchies with the most creative theorists—as determined and recognized by the community—at the top, and the others ranked by the importance of their scientific productions, their educational backgrounds, and their current institutional affiliations. The hierarchies had similarities with the English class system. Theoretical physicists were conscious of their standing, and social interactions were partly determined by one's standing within the hierarchies. How Bethe resolved this tension will be part of this study of his life. Let me here only state that Bethe came to recognize very early in his career that physics was also the locus of a competitive struggle and that he would be successful in that competition only if he did what his off-scale technical capacities and his impressive powers of synthesis would allow him to do successfully.[115] Before tackling any problem he would always ask himself whether he had the tools and intellectual capabilities to successfully find a solution. Only if the answer was "I can do that" would he undertake the project.

2

Maturing

This chapter describes Hans's social and intellectual world while attending Frankfurt University. Although in his interviews with Thomas Kuhn and Charles Weiner Bethe minimized the value of his studies at Frankfurt and emphasized that he "came alive" in Sommerfeld's seminar in Munich, a careful examination of his stay in Frankfurt disagrees with his assessment. The chapter also introduces the social circle of the teenage Hans, the *Tanzstunde*, which played an important role in his emotional maturation. In addition I discuss the "Jewishness" of Hans's friends, and that of his cohort in theoretical physics, and their place in the broader history of the German middle-class intelligentsia.

Frankfurt University

In the summer of 1924, upon passing with flying colors his *Abitur,* the final *Gymnasium* examinations that allowed him to matriculate at any German university, Hans enrolled in Frankfurt University. There was never any question of leaving home and going to a university other than the local one. The ten-year-old Frankfurt University had a very good reputation. Professors at

Frankfurt used their university experience there as a stepping stone to a professorship at other leading German universities. Thus, Max von Laue had been its first professor of theoretical physics until he went to Berlin in 1919. He was succeeded by Max Born. Born in turn left Frankfurt in the fall of 1921, when he accepted the theoretical physics professorship in Göttingen, at the time, together with Munich, the most prestigious universities in Germany after Berlin. And in 1922 Otto Stern and Walther Gerlach carried out an experiment in Frankfurt that won wide acclaim: it established the validity of the quantization of angular momentum in the Bohr model of the atom. They showed that in a magnetic field the component of the magnetic moment of an atom in the direction of the field was quantized, as predicted by Arnold Sommerfeld.

Hans's friend Heinrich zür Strassen was also going to attend Frankfurt University. They both decided to study physics, mathematics, and chemistry, since in the first year it did not matter which of these disciplines one was going to specialize in. Hans's mother, wanting him to have a profession in which he would readily find employment, had recommended he become an engineer.

In the course of their first year at the university Heinrich decided to concentrate in chemistry rather than in physics. After having taken the laboratory courses in general chemistry and in quantitative analysis as well as a "very good course in organic chemistry," Hans decided not to major in chemistry because, not being good with his hands, he "always made a mess of the lab experiments." "I was clumsy, and did most of the experiments on my lab coat, including work with sulfuric acid which disintegrated it."[1]

Although Hans had outstanding young mathematicians as instructors in his mathematics courses, he didn't "like mathematics." Carl Ludwig Siegel's lectures on number theory were "very good." The course had essentially no prerequisites, yet Siegel presented to the class "very interesting facts." However, his lectures didn't satisfy Hans. Evidently, he had become pragmatically oriented by his father, and wondered "What do I do, by knowing so much about prime numbers?" The way mathematics was being taught at the university seemed to him to have no connection with the real world, and no connection with the sciences—and the latter is what Hans wanted. This feeling of a lack of connection with the sciences was reinforced by the fact that in the advanced calculus course Hans took during the first semester, taught by another very fine mathematician, Otto Szász, the emphasis was on "epsilons and deltas" and "seemed to prove things that are obvious." Hans found it "boring." Both Siegel and Szász had studied in Göttingen and had been deeply influenced by David Hilbert, who had made mathematics undergo what Jeremy Gray has called its "modernist transformation." Gray, in his magisterial review of mathematics from 1890 to 1930, defined modernism in mathematics as "an autonomous body of ideas, having little or no outward reference, placing considerable emphasis on formal aspects of the work and maintaining a complicated—indeed, anxious—rather than a naïve relationship with the day-to-day world."[2] Both Siegel and Szász presented mathematics as an autonomous discipline with its own aims and foundations. These courses convinced Hans that he didn't want to major in mathematics. But in them Hans acquired confidence in his use of mathematics, and no mathematics—however abstract—would frighten him. Mathematics, and its special connection to physics, would acquire a new meaning under the influence of Arnold Sommerfeld.

Richard Wachsmuth, the professor of experimental physics and director of the Physikalisches Universitäts Institut, gave the introductory course in physics.[3] He was not a very inspiring lecturer, and Hans often skipped Wachsmuth's classes. Wachsmuth knew Hans's father and met him once a month when they administered exams to medical students, where he often would tell him, "Your son comes very seldom to my lectures." When that happened Albrecht would come home and tell Hans, "Please go to your physics lectures."

Fortunately, Walther Gerlach was also teaching physics in the institute. At the time Gerlach was an energetic, enthusiastic thirty-three-year-old *Extraordinarius* (associate professor) who was up to date on all the recent experimental activities in atomic physics. He also kept abreast of the attempts, or more precisely the failures, to explain these experiments on the basis of the "old" quantum theory.

During his second semester at the university Hans took Gerlach's Advanced Experimental Physics course.[4] Though labeled experimental, it was not a laboratory course. In it Gerlach presented what was known at the time in atomic physics and thus gave Hans a glimpse of the exciting new developments in physics and their challenges. In the course he read Max Born's *Der Aufbau der Materie,* the expanded version of the lecture Born had given in November 1919 on the occasion of the fiftieth anniversary of the German Chemical Society. Born and Gerlach had been colleagues in Frankfurt until

Born's departure for Göttingen, and Gerlach had been impressed by Born's mastery of theoretical physics. Born's monograph is interesting—and important—because in it Born succinctly presented what was then understood about the constitution of atoms, molecules, and solids, including the properties of nuclei. Its importance lies in the fact that Born's analysis of the binding energies of atoms and molecules, and of the mechanical and elastic properties of solids—even though based on the old quantum theory—had led him to the conclusion that the forces responsible for the cohesion of atoms, molecules, and solids were all electromagnetic in origin. Bethe's interest in calculating the binding energy of polar crystals and of metals in 1928 and 1929 on the basis of the new quantum mechanics had its origin in his reading of Born's lectures.

Bethe at the time also read Leo Graetz's *Die Atomtheorie.* It consists of six lectures on recent developments in atomic theory. This little monograph is a perceptive, readily accessible exposition of what was known in atomic and nuclear physics in 1920. In it Graetz gives a detailed description of the apparatus used to produce beams of electrons and X-rays, as well as an original and insightful account of the Bohr-inspired models that had been used to try to account for the structure and the spectra of simple atoms and molecules.[5]

Gerlach's course also introduced Bethe to Sommerfeld's *Atombau und Spektrallinien,* then in its fourth edition and already the bible of atomic physics. Bethe remembered paying close attention to the experimental facts and data presented in the book and believed that he invested very little time in its theoretical exposition, except for the Bohr-Sommerfeld theory of the hydrogen atom. Gerlach's lectures and Born's and Sommerfeld's books made it clear to Hans that physics could satisfy his curiosity and his interests.[6] Unfortunately Gerlach left Frankfurt at the end of the 1924–25 academic year to become the professor of experimental physics in Tübingen.

In his second year at Frankfurt University Hans attended the lectures of Erwin Madelung, the forty-four-year-old theoretical physicist who had replaced Max Born.[7] He thought that Madelung "gave good courses that covered classical physics and the old quantum theory up to the description of the hydrogen atom using the Bohr–Sommerfeld quantization rules." He characterized Madelung's lectures as "not as inspired as Gerlach, but interesting." Madelung knew theoretical physics very well and his lectures convinced Hans that this was the field for him.

Tanzstunde

In the fall of 1925, at the beginning of Hans's second year of university studies, Werner Sachs, who had moved to Heidelberg when his father had accepted a professorship there, came back to Frankfurt to study at the university.

Bethe did not indicate the reason for Werner's return. Very probably the political ambiance at Heidelberg University was a factor, as the story of Walter Elsasser, a physicist of their generation, indicates. The Elsasser family's financial situation had deteriorated after World War I, and Elsasser decided to go to Heidelberg University in the fall of 1922 and live at home in order to minimize expenditures. He knew that he wanted to become an experimental physicist and therefore enrolled in Philipp Lenard's lecture course. Lenard had won the Nobel Prize in 1906 for his work on the photoelectric effect. Heidelberg University thereafter offered him a chair and built him a magnificent institute. By 1922 Lenard was also well known as an active participant in right-wing political movements. Politics was certainly not on the mind of the eighteen-year-old Elsasser as he sat in the large lecture room waiting for his first class in physics to begin. What happened thereafter became etched in his memory:

> Every seat in the hall was taken. In walked Professor Lenard wearing an impeccably tailored suit: to his left breast there was fastened a silver swastika of gigantic proportions, perhaps ten centimeters square. This was most unusual . . . in spite of war and revolution Germany . . . still remained a place of law and order. A distinguished senior professor was most certainly not expected to brandish symbols of political extremism in class. But the students thought otherwise. They applauded intensely. They clapped, and then they shouted: they kept on clapping and shouting, on and on and on. How long this continued I cannot say precisely but it was certainly the most dedicated and loudest ovation I ever witnessed in my life, before or after.[8]

Elsasser was devastated. Since the hall remained full at later lectures, he became convinced that the initial response had not been a staged event. He felt that the German academic youth had voted, and had clearly voted for the swastika. During the course of the year he was urged by concerned friends to leave Heidelberg, since the following year he would come in direct contact with Lenard in laboratory courses and Lenard was known to have attacked "non-aryan" students physically. Elsasser went to Munich in the fall of 1923.

Werner Sachs must surely have been aware of the political climate at Heidelberg University. Being a science major he probably took Lenard's course. He must have witnessed scenes similar to the ones Elsasser had experienced. His father, who was the professor of immunology there, most certainly was aware of the political situation at the university. He may have talked to Werner about it and encouraged him to leave Heidelberg.

When Werner came back to Frankfurt he and Hans took most of their courses together. It is not clear whether what happened in Heidelberg was discussed between them. What is clear is that in 1925 these political issues had not as yet surfaced overtly in Frankfurt. The Frankfurt faculty and student body were different from those at Heidelberg, reflecting the very different character of Frankfurt and its university. Also, the bond between Hans and Werner had been cemented earlier, and differences in ethnic background and religious roots were not relevant factors in their friendship.

During his two years at Frankfurt University Hans's social world expanded. Both he and Werner courted Fips (Emma) Boehm, a friend of Werner's sister, Ilse. Hans had met Fips during one of his visits to the Sachs family in Heidelberg in 1923. Because Fips and her friends wanted dancing lessons, their parents arranged a weekly *Tanzstunde* that made the rounds of the houses of the participants. The first of these dance meetings, which took place between 1924 and 1925, was not a particularly happy experience for Hans, who was still shy with girls. Furthermore, the other young people in the *Tanzschule* knew one another, and Hans felt like an outsider.

The second *Tanzstunde,* a year later, was very different: "It was wonderful."[9] A few of the parents had organized this second *Tanzstunde* in the fall of 1925. The gathering was to be a way for their daughters to meet suitable young men. Among the girls who attended the second series of lessons in the fall of 1925 were Fips Boehm, Käte Feiss, and Hilde Levi—all of them either sixteen or seventeen years old. "I was attracted to Hilde," recalled Bethe, "much to my mother's disgust. She did not like Hilde, but did like Fips."[10] Hilde Levi in her *Autobiography* described the group, which consisted of "6 or 8 girls and as many boys," as "a circle of close friends [who] took dancing lessons, went skiing together during winter vacation and walked in the woods during the summer." And she noted that Hans became one of her two "closest companions."[11]

Hilde's parents had been among the organizers of the second *Tanzstunde.* The somewhat older boys in the group were Hans, Fritz von Bergmann, Werner Sachs, Erwin Strauss, and Heinrich zür Strassen's younger brother, Konrad. The weekly dancing lesson was the formal way of bringing the twenty young people together. And once they got to know one another and discovered that they liked each other, they had fun together—proper, chaste fun. They had dancing parties; saw and read plays throughout the year; went on a skiing trip that winter and on hikes the following summer. Thus Hans and Erwin Strauss became good friends by virtue of spending their vacations together in the Swiss and Austrian Alps as members of the circle.[12] The *Tanzstunde* was also responsible for Hans's choice of roommate—Fritz von Bergmann—for Hans's second year at Frankfurt University.

The second *Tanzstunde* was a very successful social endeavor, with Hans partaking in its summer and winter activities even after he left for Munich.[13] In fact, the *Tanzstunde* circle of friends was enlarged by the addition of some young people from Munich. Hans and Werner Sachs "ran the show" for the group. They issued a journal, *Die Menckenke,* after a slang word for a mixed soft drink that was popular at the time, which recorded their activities. Hans wrote most of it with some assistance from Werner Sachs.

Hans was in charge of skiing. Every year from 1926 until 1933 members of the group went skiing over Christmas and New Year's. Hans always organized the trips and made all the decisions about where to go, which trains to take, where to stay, and for how long. Usually the group went to the Engadin in Switzerland.

Fips Boehm, Werner Sachs, Hilde Levi, and Hans were regular members of the skiing party.[14] At times, this foursome was enlarged by other members of the *Tanzstunde,* or by other friends. One year Hilde's brother came along and shared a room with Hans, later remarking how unkempt Hans was at the time. Hilde Levi commented that what went on was "quite proper": "the boys slept together and the girls slept together."[15] Hilde remembered one amusing incident from one of these trips in the late 1920s. They had missed the train that was to take them home, and Hilde, aware that her parents would be worried when she did not show up at the expected time, wanted to call home. Hans offered to do so. Hilde's mother answered the phone. "Here is Hans," he said in his deliberate, methodical manner. Wanting to sound responsible and mature, he used his gravest intonation. "You needn't worry," he told Hilde's mother, at which point she nearly fainted, having visions of her daughter lying in the hospital with several broken bones. "What happened?" she asked with trepidation. "We have only missed the train," Hans answered—which made Hilde's mother very happy.

The *Tanzstunde,* particularly the second one, was an important transition in Bethe's life. His emotional maturation up to that point can be characterized by three stages. The first was an emotionally constraining one. Growing up in Strassburg and Kiel, he had very little interaction with other children of his own age, and his mother was the dominant person in his life. The second stage began with his coming to Frankfurt, attending *Gymnasium* there, going to Kreuznach and Odenwaldschule, and then coming back to Frankfurt and becoming close friends with Werner Sachs, Paul Berlitzheimer, and Heinrich zür Strassen. During this period Hans achieved a measure of emotional independence. The third stage began with Hans, now eighteen, attending the *Tanzstunde* and dancing and socializing with girls. The second round of *Tanzstunde* made him realize that—despite his shyness and his social insecurities—he could be friends with young women. It was a revelatory experience that allowed him to enter into young manhood with a measure of social confidence, even as he dealt with the turmoil at home stemming from the breakup of his parents' marriage.

Anna's breakdown and Hans having to attend to her emotional needs seemed to prevent some of the closeness that had existed between them as mother and young boy from being transformed into a more mature relationship. This intimacy, so evident in the letters Hans wrote his mother during the crisis, had adverse consequences. It prevented Hans from opening up to his friends—male or female—and may have given him a distorted view of what to expect in his relationships with women friends. And since he truly considered his mother "his best friend," he wanted her approval of his relationships with women. Thus writing his mother on October 26, 1926, about his attempts to get closer to Käte, one of the *Tanzstunde* "Mädchen," he indicated,

> She [Käte] has per se the same huge need for communication as I, and she has the same inhibitions. In spite of our many serious conversations, she has even now, in October, never told me anything about herself, and similarly I little about myself. The conversations remain general and although per se I like that very much, something is missing, the complete honesty

> maybe—no, that is poorly put, the uninhibitedness of what is said . . . I have always the feeling with her, we both want to be more intimate, and cannot quite. That is different between you and me. We can get together and be very close.
>
> With Werner it is in many ways similar . . . Recently we have again talked about the deepest problems, but it remains so completely impersonal, with the difference from Käte that I do not feel any need with him to change that.[16]

On the one hand Hans was becoming more self-contained, yet on the other he yearned for the possibility of opening up. Shortly after sending his mother this letter Hans wrote to her again:

> Yesterday evening I was with Werner and we had a long discussion about girls. He is altogether against "philosophical discourse," and I think basically against Fips and Hilde and the dance lessons altogether . . . In a sense I am glad that we are going separate ways and am of the opinion that he does not have me so much under his thumb anymore. On the other hand it is accompanied by pain, or more exactly envy, in that he is much more satisfied with his way than I with mine, but there is no way out of that.

If by the beginning of 1927 the *Tanzstunde* circle had lost some of its attraction for Hans, nonetheless he recognized that his friendships with its members, whatever the limitation of the interactions, had made it possible for him "to have become so close" to both his parents individually during their crisis.[17] But this closeness had its limitations. His desire to open up and be close to someone gnawed at him. It seemed that he could do that only with his mother.

Bildung

Hans never thought of himself as Jewish. His mother repressed her Jewish roots and told Hans of the circumstances under which she had converted only when he was in his late teens. Most of Hans's close friends after he had moved to Frankfurt—Werner Sachs, Erwin Strauss, Paul Berlitzheimer, Hilde Levi—came from secular, assimilated Jewish families. This affinity continued when Hans went to Munich to Sommerfeld's seminar, where he met Rudolf Peierls, Edward Teller, and Margarete Willstätter. It was repeated in Rome and later in

Copenhagen, where deep friendships with George Placzek and Victor Weisskopf were initiated.

One factor which helps explain this affinity is that Hans was much taken by the warmth, openness, and liberality of the families of these friends. In addition, these assimilated Jewish families—the Berlitzheimers in Frankfurt, the Sachses in Frankfurt and Heidelberg, the Wallachs in Munich—welcomed him.[18] And in particular, he admired the matriarchs who headed these households and probably became enamored of some of them. One such was Ella Philippson Ewald, the wife of Paul Ewald, whose *Assistent* he became in 1929 and whose daughter, Rose, he married in 1939. Another was his cousin Barbara Fitting Levi, the elder daughter of Anna's sister, Bertha, who had married Carl Fitting, an artillery officer stationed in Strassburg. The Bethes' connection to the Levis was a complicated one. Bertha died of scarlet fever in 1900. Her two daughters—Barbara, who was born in 1894, and her sister Caroline, who was born in 1897—were subsequently raised by Carl and his mother. Carl Fitting then married Fanny Levi. Fanny had a brother, Friedrich Wilhelm Daniel Levi. Friedrich Levi, whose family lived in Mulhouse, Alsace, became a mathematician well known for his work in abstract algebra, geometry, topology, set theory, and analysis. He had received his doctorate in 1911 with Heinrich Martin Weber at Strassburg University. He served his mandatory military service in the artillery branch of the German army in 1906–1907, where he probably met Carl Fitting. Later called up during World War I, serving under Carl Fitting, he was awarded the Iron Cross and was discharged as a lieutenant. Friedrich was evidently the link between the Levis and the Fittings. In 1917, Friedrich married Barbara Fitting, with whom he had three children: Paul, Charlotte, and Suzanne. In 1920 Friedrich accepted a professorship in Leipzig. Hans became close to the Levis and visited them fairly often while in Munich.

There exists a remarkable wedding picture of Barbara and Friedrich. On the one hand it highlights the assimilationist movement on the part of a certain segment of German Jewry, and on the other testifies to the attractiveness of women of Jewish descent to a certain segment of German society.[19] As noted, each of the two women Hans's father married was of Jewish descent, and Hans chose likewise. Besides the family climate of warmth, openness, and liberality, there are other social and cultural factors that help explain why so many of Hans's friends were of Jewish descent. First of all they shared a common trait that he described as follows: "They all were much more outgoing, much more connected with the world, and that was what I was missing." Furthermore, all his German Jewish friends came from highly assimilated families or from families that had recently converted, for whom being German and being taken as German was important. For all of them intermarriage was totally acceptable. They thought of themselves as German first and foremost.

Walter Elsasser's memoirs again shed light on German secular Jewish attitudes in the first part of the twentieth century. Elsasser was born in 1904 in Mannheim, then a large, thriving commercial town on the Rhine some sixty kilometers south of Frankfurt. His parents were well educated. Walter's father had earned a Doctor of Laws degree from Heidelberg University, and the family was well off before World War I. His parents were of Jewish descent. They had converted to Protestantism when Walter was quite small. He was confirmed in the Evangelical Church when fifteen years old. In his autobiography Elsasser indicated that until the end of World War I he had "no notion whatsoever of Jews as a separate group, either as a religion or as a nationality . . . It was quite easy for a boy to grow up without ever having heard of Jews as a distinguishable element." He continued,

> But looking back, I can clearly see that my parents must have enjoined their relatives, all Jewish, with whom they were, on the whole, on quite good terms not to mention Jewish matters to me or my sister, and it is also clear that these relatives complied . . . My parents always had some Jewish friends, but I became aware of their Jewishness only much later in life; earlier they were only individuals or family friends. Judaism, or any religion for that matter, was just not discussed in my house . . . They were children of the nineteenth century, an age that had not precisely invented Reason, but had told them that Progress inevitably would soon do away with such ancient minorities as the Jews, and they believed it uncritically and unreservedly.[20]

The same must have been true of Hans's upbringing. He was close to his grandmother Amalie and to the Kuhn side of the family, and never thought of them as Jewish or having Jewish roots.

It is the notion of *Bildung* that can help explain Hans's affinity to and commonality with his Jewish friends, and conversely the close friendships of essentially totally assimilated young men and women like Hannah Arendt,

Hilde Levi, Rudolf Peierls, Victor Weisskopf, Edward Teller, and George Placzek with liberal Christians.

All of Hans's friends came from *Bildungsbürgertum* families, the educated and cultural bourgeoisie. The concept of *Bildung* became deeply imprinted into German culture at the end of the eighteenth century. It was a response to the quest for harmony in a Germany "touched by the industrialization of Europe, [and] frightened by the French Revolution."[21] It was also an attempt to meld the *Volkisch* element in German culture with the "rational" legacy of the Enlightenment. As elaborated by Wilhelm von Humboldt, an important contributor to the development of the concept of *Bildung* and its associated notions of self-cultivation and self-formation, aesthetics was the foundation for the harmony that *Bildung* imparted to an individual.[22] The *gebildete* person was compared to a harmonious work of art.

> The aesthetic was linked to the intellectual faculty, and both activated the moral imperative that resided in man.
>
> The beautiful as the essence of aesthetic education was not romanticized but understood through reason . . . The beautiful, in accordance with the Greek ideal, was conceived as harmonious and well proportioned, without any excess or false note which might upset its quiet greatness. Beauty was supposed to aid in controlling the passions, not in unleashing them, emphasizing that self-control which the bourgeois prized so highly. The ideal of Greek beauty transcended the daily, the momentary; for Humboldt, as for Goethe and Schiller, it symbolized the ideal of a shared humanity towards which *Bildung* must strive. This beauty was a moral beauty through its strictness and harmony of form; for Schiller it was supposed to keep humanity from going astray in cruelty, slackness, and perversity. *Bildung* was not chaotic or experimental but disciplined and self-controlled.[23]

With the attainment of the harmonious *gebildete* personality would come an enlarged perspective and balanced judgment that would allow a person to apprehend the whole truth, indeed to obtain "a living picture of the world properly unified."[24]

Bildung became emblematic of the German educated middle class. To be a *Bildungsträger* came to imply membership in both class and nation. The writings of Theodor Lessing, Johann Herder, Goethe, and Schiller gave *Bildungsträger* their sense of cultural identity. *Kulturträger* saw themselves as expressing intellectually the unity of the German-speaking states—prefiguring their eventual political unification. The self-development aspect of *Bildung* allowed some Jews to free themselves from the constraints of the highly regulated Orthodox communal life and adopt a way of life that would be common to them and other Germans as individuals.

The concept of *Bildung* acquired a more nationalistic and political tone following the Napoleonic French military victories at the beginning of the nineteenth century. Universalism became associated with the French occupation armies and internationalism seen as a way of keeping Germany weak, under the hegemony of the Great Powers.[25] Yet at that same time, as a result of the industrialization and urbanization taking place, the educated middle classes clamored for reforms that would give them a greater political voice and demanded that the state support *Bildung* in the *Gymnasia* and universities. Prussia was the first to acquiesce, recognizing that educational reforms would not only secure middle-class allegiance, but also provide the state with loyal civil servants. Indeed, for the German Christians of the new *bürgerlich* class that was vying for power, the concept of *Bildung* fulfilled certain vital functions: it helped legitimize them by differentiating them from the upper and lower classes, and it facilitated the creation of an elite that provided better civil servants for the Prussian state.[26]

As head of the section for Religion and Education in the Prussian ministry of education from February 1809 to June 1810, Wilhelm von Humboldt formulated the guidelines that would make *Bildung* the objective not only of the newly founded university of Berlin but of the entire Prussian educational system.[27] In Humboldt's program, the state would assume financial responsibility for education throughout the nation. He envisioned that in elementary school students would learn basic skills; in high school they would be taught to be intellectually independent. A student was to be considered mature when "he had learned enough from others to be able to learn by himself." The main function of the university was to join students and professors in an *elite* academic community based solely upon mutual self-cultivation through learning (*Wissenschaft*). University professors would be free to teach what they desired (*Lehrfreiheit*) and students to pursue the course of study they chose (*Lernfreiheit*) without the state interfering in any way.[28] By congregating students in a community steeped in *Wissenschaft* and giving them freedom to interact with their peers, Humboldt hoped they would acquire models to emulate and cultivate. By midcentury *Bildung* was institutionalized in the *Gymnasium* and the universities, and overseen by a new mandarinate: the university professors and, much lower in the pecking order, the *Oberlehrer,* the *Gymnasium* teachers.[29]

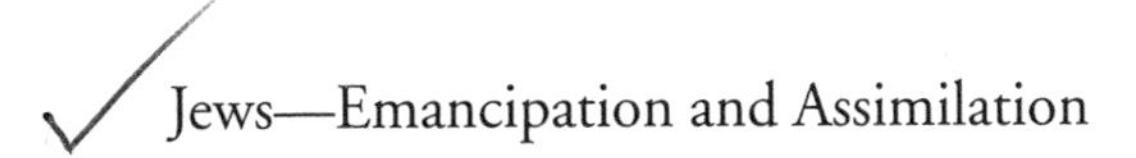

Jews—Emancipation and Assimilation

The French Revolution, and more specifically, the resolution of September 28, 1791, of the French National Assembly that explicitly recognized Jews as full citizens of France, was a watershed in the history of the Jews of Europe.[30] The German states conquered by Napoleon adopted constitutions that reflected the French model. Although Frederick William III signed a decree of emancipation on March 11, 1812, that granted the Jews of Prussia full civil rights and similar ones were promulgated in other German states,[31] nonetheless German Jews had to struggle for their emancipation until 1871, when their legal equality became incorporated in the constitution of the newly formed German Reich. It is often asserted that whereas French Jews received their equality with the Revolution and subsequently "had to prove that they were worthy of it," German Jews were "promised equality as the ultimate reward for their successful integration and acculturation into German society."[32]

Wilhelm von Humboldt had set the pattern. He was influential in determining the course of Jewish emancipation in Prussia.[33] His contacts with the cosmopolitan, emancipated, wealthy German Jews of Berlin and his friendship with Henriette Herz and the other Jewish members of her circle had made him favorably disposed to the idea of granting complete rights of citizenship to the Jews of Germany.[34] In 1809 Humboldt had prepared a paper recommending equal political rights for Jews. The subsequent Prussian Edict of Emancipation of 1812 did in fact grant Jews almost civil equality and made it possible for them to enter Prussian educational institutions. Although Jews could not become civil servants unless they became Christian, and hence could not become professors at Prussian universities without converting, they could now attend *Gymnasia* and universities and become doctors and lawyers.

In his 1809 memorandum Humboldt emphasized the need for uniform action throughout all the German states. His motivation was a combination of idealism and realism. As a modern nation, Prussia should be committed to the emancipation of the Jews. It needed to consolidate its power and have sovereignty over all its subjects without the mediation of social or religious institutions, and therefore could not tolerate the autonomous corporate status of the Jewish community. Uniformity of practice in all the German states was required so that Jews in other states would not immigrate to Prussia to enjoy the greater privileges there.[35] Humboldt regarded Jews not as a people but as a multitude of individuals, and equality was to lead to the disappearance of the Jews as an ethnic group.[36] This became the official government position. In return for Jewish emancipation, German Jews would have to disavow Jewish peoplehood—a price many secularized Jews were willing to pay.[37] But the price of emancipation—becoming autonomous individuals in a bourgeois society—raised the problem of what it meant to be Jewish. The answer was provided by *Bildung* and by the Reform movement in Judaism.

The Enlightenment concept of *Bildung* largely determined the postemancipatory Jewish identity. Emancipated Jews interpreted the message of the Enlightenment as attributing to all men equal potentiality. The criterion of equality was not to be their religious or national heritage but their *Bildung*. The concept of *Bildung* seemed ready made for their needs: everybody could attain *Bildung* through education and self-development. *Bildung* was to be the means by which German Jews—men *and* women—were to integrate themselves into German society.[38] In fact, by the end of the nineteenth century women in assimilated German Jewish homes often were able to aspire to the same goals and given the same opportunities as men.[39]

For secularized German Jews the ideas of *Bildung* and Enlightenment melded.[40] *Bildung,* which nurtured the individual personality, was the mechanism by which this was accomplished. For these secularized Jews, *Liberal Judaism*—as Reform Judaism was called in Germany—became the institutional framework during the nineteenth century for the expression of communal and ritualistic religious sentiments.[41] Liberal Judaism had begun as a movement of German Jews of high social status who wished to make synagogue services more decorous. The early Reformers also wished to make the traditional educational system, which consisted of studying the Bible, the Talmud, and their commentators, more morally edifying. Liberal Judaism evolved to become the means to express skepticism of traditional, orthodox Jewish religious beliefs, which included rigid regulations presumed to be of divine origin that governed every aspect of life: what prayers should be said when and where, what could be eaten, what clothes could be worn, marriage, sexual intercourse, divorce, inheritance, and so on. Liberal Judaism became the channel for confronting how to be a German citizen and a moral and upright Jew in the face of the higher criticism of the Bible, evolution, industrialization, urbanization, and more generally, modernism. The *Bildung* ideal attracted German Jews of the Liberal faith because they saw it as a means to surmount the gulf between their own history and German tradition, enabling them "to stress what united rather than divided."[42]

Liberal Judaism espoused a vision of mankind in which national and religious differences would be transcended. Cosmopolitanism and universality, as evoked by the writings of the Prophets, became central components of the tenets of Liberal Judaism, and in particular its tenets as formulated by Abraham Geiger, considered the leading light of Liberal Judaism, and by the somewhat more conservative Ludwig Philippson, an influential Magdeburg Liberal rabbi who opposed assimilation. Philippson founded and was the editor of the widely read and influential *Allgemeine Zeitung des Judentums.* Philippson authored a bilingual illustrated edition of the Bible with commentaries that explored the larger ethical, moral, and humanistic implications of the biblical text. In 1848, deeply involved in the revolutionary struggles taking place in Germany, Philippson wrote that he no longer considered the existing extensive discrimination against Jews to be a special case of prejudice, but part and parcel of the cause of the revolution. To his fellow Germans he declared, "We are and only wish to be Germans! We have and only wish to have a German fatherland. We are no longer Israelites in anything but our beliefs—in every other aspect we very much belong to the state in which we live."[43]

The revolutions failed, but the participation of German Jews in both the barricades and the National Assembly gave them the sense that they were finally becoming Germans. In particular, German Jews affiliated with the Liberal congregations joined together with political liberals of all Christian faiths. It was the beginning of German Jews' association with the liberal political parties.[44]

In the aftermath of the 1848 revolution, Prussia reaffirmed its status as a "Christian state" in its new constitution, and Jews there continued to be excluded from high positions in government, in the judiciary, and in universities. Other states likewise re-established the discriminatory practices that had been suspended during the revolutionary takeovers.

The Franco-Prussian War of 1870–1871 was a turning point in German and in French cultural history. After the defeat of France in August 1870 what had been a Franco-Prussian war became a Franco-German war by virtue of the unification of the German states, and the perception of the hostilities was nationalized. In a perceptive essay Silvia Cresti notes that after the founding of the Third Republic in 1870 French Jews defined their national identities in political terms, based on the concept of citizenship, whereas German Jews defined theirs in cultural terms, based on their German culture.[45] Thus French Jews could express their commitment to France through political involvement in institutions and make Judaism an exclusively private matter separated from the public sphere. In Imperial Germany on the other hand, to be German meant to participate in *Deutschtum,* an ideology that homogenized the identity of the entire *Volk. Deutschtum* came to define German identity, an identity in which culture, ethnicity, and Christian thought and practices were entangled and intertwined. *Deutschtum* posed grave problems for German Jews. Their Humboldtian notion of *Bildung* and their commitment to Liberal Judaism's universalism and historicized rationality made their accommodation to the nationalized version of *Bildung* and the exclusiveness of German identity difficult—if not impossible.

However, after the unification of the German states and the adoption of the Constitution of the Reich in 1872, legally sanctioned discrimination against Jews was eliminated.[46] Thereafter, baptism was no longer required for appointment to a professorial position, but conversion was a more effective and successful course of action than remaining Jewish. The rate of conversions grew, aided by the fact that already at the first conference of Liberal rabbis held in Brunswick in 1844, a motion had been adopted that stated—contrary to the tenets of Orthodox Judaism—that marriage between Jews and Christians, in fact marriages between adherents of monotheistic religions generally, were not prohibited "provided that the laws of the state permit parents to raise the children of such unions also in the Jewish faith."[47] Intermarriage became a passage to total assimilation and conversion. The Christian churches' ban on mixed marriages had been a barrier. But after civil marriage had been established in the *Kaiserreich* in 1875, the situation changed dramatically. Mixed marriages were now possible without religious conversion. Between 1876 and 1880 4.4 percent of Prussian Jews took non-Jews as marital partners. By 1890 the rate was 9.4 percent. It reached 13 percent just before World War I and climbed to 20.8 percent in 1920. Similar figures hold for Imperial Germany as a whole.

Germany's population increased dramatically from 41 million in 1871 to

65 million in 1910, but the Jewish population remained slightly over half a million. Its low fertility rate reflected its urban, well-educated, middle-class character. Thus, the percentage of Jews in the general population decreased from 1.25 percent in 1871 to 0.93 percent in 1910.[48] At the turn of the century, of those who registered as Jewish, over 80 percent belonged to Liberal congregations.[49]

The Reform movement in Germany fostered the enabling conditions for the amazing creativity of Jews in all branches of the cultural and intellectual life of Germany and the German-speaking part of the Austro-Hungarian Empire.[50] Liberal Judaism in Germany during the nineteenth century played a role similar to Unitarianism in Great Britain during the eighteenth century. Erasmus Darwin, Charles Darwin's paternal grandfather, had characterized Josiah Wedgwood's Unitarian beliefs, in which the Trinity and the divinity of Jesus had been discarded, as "a featherbed to catch a falling Christian."[51] These beliefs had made many Unitarians, and the members of the Lunar Society in particular, skeptical of traditional religious views, and made possible their belief in progress, Reason, and in the value and efficacy of scientific knowledge. In Germany Liberal Judaism became a way station for assimilation and for conversion, creating job opportunities in academia and government. But even after they had converted, baptized Jews were still socially and culturally excluded from German mainstream society and could not become guardians of traditional, conservative values. They thus came to populate the center and left of the cultural and political spectrum.

Thorstein Veblen's 1919 thesis of skepticism and marginality, which accounts for the exceptional post-Emancipation Jewish cultural creativity in Germany, becomes more convincing in light of the double alienation of secular German Jews: their alienation from Orthodox Judaism, with its strictly enforced system of laws and regulations, and their exclusion from the ranks of German conservatism.[52]

For emancipated German Jews *Bildung* had originally represented the capacity to become "free, autonomous person[s], living in harmony with like-minded men, in a spirit of tolerance, solidarity and friendship." Later in the nineteenth century it became detached from the individual but remained "a kind of religion" even for those assimilated Jews who had severed their ties with Liberal Judaism. Interestingly, Shulamit Volkov, one of the leading social historians of German Jewry, notes that the moral code of these highly assimilated families—for example that of Hilde Levi and Werner Sachs—was typified by "strictly controlled emotions and conventional sexual mores" that encompassed a defense of the values of the family.[53]

By mid-century the belief in individualism and the potential of human reason—central tenets of *Bildung*—had become infected by recurrent waves of neo-Romanticism and eroded by nationalism with its myths and symbols.[54] Nationalism, in fact, became the dominant German ideology, not only in the popular mind but also among the mandarins. But emancipated German Jews resisted the narrowing of the vision. The more the older concept of *Bildung* came under attack, the more it became identified as the central doctrine of secularized German Jewry. The notion of *Bildung* as the faith of secular, assimilated Jews, the defining feature of their Jewish identity, was reinforced after Bismarck's rise to power and the militarism that followed. In fact, the original concept of *Bildung* then served as a rallying point for conservative opposition to university reform.[55] But for many Jews, assimilation meant a continuing commitment to the Humboldtian notion of *Bildung*, to a liberal outlook on society and to a liberal position in politics. Needless to say, if there was this idealistic component underpinning assimilation, there was also a pragmatic component to conversion: liberal, assimilated Jews desired the social standing of being a high civil servant or a university professor, status impossible to obtain as a Jew. Additionally, it was hoped that these civil service positions would bring with them social acceptance and complete assimilation.[56]

But these hopes were thwarted. After the unification of Germany, Bismarck, as chancellor of the new state and prime minister of Prussia, embarked on a far-reaching policy of creating and consolidating the empire. He nurtured the rapid industrialization it demanded and encouraged the consequent rapid urbanization with its attendant squalor. However, the sharp economic downturn that dogged the latter parts of the 1870s, sparked by the crash of the Austrian stock market in 1873, affected all classes and produced a general political shift to the right. It also generated a new nationalism that promoted a view of the German nation that was more explicitly racist—now a racism based on science, namely on biology.[57] And accompanying this veer to the right was a palpable antiliberal mood in the public sphere and a marked increase in anti-Semitism.[58]

In 1878 Bismarck severed his ties with the liberal faction of the electorate, the National Liberals. And as the state grew more powerful and more conservative, its policies transformed universities from institutions committed to the Humboldtian ideal of *Wissenschaft*—in which students were to be motivated by intellectual curiosity and the acquisition of knowledge for knowledge's sake—into ones in which students were taught and trained for professions in ever more specialized fields.

Concomitantly, the mandarinate became much more conservative. Nonetheless, there remained a subset that adhered to the original Humboldtian vision. Its members were principally mathematicians, physical scientists, and medical scientists; and among the latter were many Jews or descendants of converted Jews such as Eduard Hitzig, Julius Sachs, and J. Richard Ewald.[59] The reason for this was that between 1853 and 1873—in part as a result of the Helmholtz, Brücke, and Du Bois–Reymond 1847 manifesto and the subsequent political activities of Rudolf Virchow—professorships of physiology were established at almost all the German universities, and these provided new niches for academically inclined converted Jews. And even though they now were civil servants, many of them retained their Enlightenment-inspired, liberal outlook.[60]

Being a physiologist, Hans's father Albrecht came in contact with this circle of liberal, assimilated, or baptized Jews. His own liberal political, moral, and social views resonated with those of these secularized German Jews, the *Kulturträger* of enlightened, cosmopolitan *Bildung*. Undoubtedly this affinity was a reason for his accepting a position in the medical school at Frankfurt, a university founded with generous financial support from the local Jewish population. And as noted, Albrecht twice married women of Jewish descent.

Hans similarly found attractive the warmth, liberality, and cosmopolitanism of Paul Ewald's household. Ella, Paul's wife, was the matriarch of the household, and was responsible for much of its welcoming atmosphere. Ella came from a long line of liberal, emancipated Jews: the Philippsons.[61] Ludwig Philippson was a distant relative of hers.

Ludwig Philippson was one of the respondents to the essay Heinrich von Treitschke published in the November 1879 issue of the *Preussische Jahrbücher*, the widely read periodical he edited. In it Treitschke had reviewed and assessed the current political situation in Germany and addressed the troublesome "Jewish question" that had become a widespread and wide-ranging national issue. Treitschke's article stimulated many responses from the German Jewish community. The review and the responses to it became known as the Berlin Anti-Semitism Dispute. It was "a dispute on relevance, meaning, and origins of the antisemitic movement that was . . . emerging at the time" and was a watershed. It made clear to the German Jewish community what a good many Christian German intellectuals thought being a "German Jew" should mean.[62]

Treitschke was a highly respected historian and political theorist who had strongly supported the Prussian-initiated unification of Germany. He was a member of the *Reichstag,* a professor at the University of Berlin, and had been considered liberal. In his article he attributed the deeply disturbing antiliberal mood that had swept the country in the recent past as an adverse reaction to "humanitarianism and Enlightenment" that became translated into "religious earnestness." The new anti-Jewish movement was one indication of this turn to the right. And this was the reason the "Jewish question" had arisen repeatedly in the recent past. What made it a *German* Jewish question, Treitschke asserted, was the mass immigration of Polish Jews.

Treitschke's review was interpreted as supporting the cause of those groups and individuals who had recently organized an aggressive anti-Jewish movement throughout Germany, and particularly in Berlin.[63] The rejoinders to Treitschke's deeply troublesome article attempted to understand why the liberal Treitschke had seemingly given support to anti-Semites and to find realistic, persuasive responses that took into account the consolidation of Germany into a national state. It became an occasion on the part of some of the responders to present "an extended and sophisticated argument about the concept of the nation, its relation to culture, the relevance of religion for culture, and likewise that of 'race' for culture and nation."[64]

In his article Treitschke had averred that part of the solution to the "German Jewish problem" consisted in German Jews' concession to the following demands: "that they become Germans, feel themselves simply and justly as Germans, regardless of their faith and their old sacred memories, which all of us hold in reverence; for we do not want thousands of years of Germanic civilization to be followed by an era of German-Jewish mixed culture."[65]

Many of the assimilated, Bildung-committed German Jews acceded to the demand. Many of them did become "simply" Germans—and did so by converting.[66] But as Anthony Kauders has emphasized, the increasing number of conversions in the period from 1890 to 1933 is a manifestation of the insecurity German Jews felt. And neither those who converted nor those who remained Jews recognized or understood the depth and scope of the new anti-Semitism of the 1890s. Nor had they anticipated the scientific racism that emerged with advances in genetics. Unlike the Judeophobia of the past, which had focused on the "proper" place of the Jew in German society and had perpetuated Christian stereotypes of Judaism and the Jews, the aim of racialist thinking was to ascertain essential differences between "German" and "Jew," differences that could not be eradicated through assimilation or conversion.[67]

During the early 1880s murderous pogroms took place in southwestern Russia. Though aware of the intensity and magnitude of the Russian pogroms, German Jews were convinced that such deadly outbursts could never happen in Germany, given its culture and economic well being. The Russian Jewish intelligentsia viewed the matter very differently. They saw the new wave of anti-Semitism in Germany as proof that emancipation did not work, and that assimilation and conversion were not the solutions to the Jewish question, neither in Germany nor anywhere else in Europe.[68] For them "autoemancipation" with the birth of a Jewish state was the answer. Thus was Zionism born.[69]

The period between 1890 and 1930 has been characterized as a second transition period in German Jewish intellectual life during which numerous traditional theological and religious terms and practices lost their meaning. It witnessed an associated rejection of Reform Judaism. And besides Zionism, the influx of eastern Jews into Germany introduced mysticism and Hasidism into German Jewish life. And all these elements threatened the "rational" German Jews.[70]

Anti-Semitism and the Natural Sciences

As the natural sciences and mathematics accrued ever-greater importance during the second half of the nineteenth century, many young assimilated Jews were attracted to studying them. Their families recognized the opportunities these fields presented and nurtured the talents of any male offspring showing ability in them. Furthermore, the universalism of the sciences and mathematics resonated with the internationalism and cosmopolitanism of their *Bildung* tenets, as did the norm of personal merit in these disciplines with their commitment to individualism. Critical to attracting assimilated Jews to these fields was the understanding that the knowledge produced was value-free and that there were objective, discipline-based criteria for excellence, originality, and achievement. Being outstanding, creative, original, and productive in mathematics or in the sciences—and recognized and accepted as such by the leading authorities and practitioners in these fields—was seen as trumping all supposed deficiencies, including Jewish roots.[71] There was no denying the outstanding contributions to *German* mathematics or *German* physics of Carl Gustav Jacobi or Heinrich Hertz. But it *was* claimed in certain quarters that Heinrich Heine's writings could not be considered *German* literature nor could Felix Mendelssohn Bartholdy's compositions be deemed truly *German* music. In the sciences and mathematics it was believed possible to hierarchically order the talents and research of the practitioners, and to be able to assert objectively that a given work was more important and generative than some other. The professors at the elite institutions—Berlin, Munich, and Göttingen—had considerable influence with both government officials and their colleagues at the other universities when recommending candidates to positions. Thus they could override existing discriminatory practices. This allowed Hermann Minkowski and Walter Kaufmann to be appointed at Göttingen, Leo Graetz at Munich—but not Paul Ehrlich at Berlin—not to mention all the assimilated and baptized Jews who were able to obtain appointments in the basic medical sciences during the last third of the nineteenth century.

The situation in theoretical physics was somewhat different. As a discipline it became institutionalized in Germany only at the end of the nineteenth century. The recommendation that Albert Einstein be appointed an *Extraordinarius* (associate) professor of theoretical physics at the University of Zurich in 1908 was not accepted because of anti-Semitism. He did get an appointment at the Eidgenössische Technische Hochschule (ETH). The subsequent recognition of his research and the enormous impression he made on his colleagues at the first Solvay Conference in 1911 led Max Planck and Walther Nernst in 1913 to recommend to the Prussian Academy of Sciences that Einstein be elected a full member and offered a research professorship.[72] His general theory of relativity and the corroboration of its prediction of the bending of light in its passage near the sun by Arthur Eddington and Frank Watson Dyson in 1919 catapulted him to fame. Even at that time some considered him as great a physicist as Newton. Lev Landau, in his famous ranking of physicists, put Einstein in a class by himself, ahead of Newton. Many of the outstanding young theoretical physicists who came into their own just after the invention of quantum mechanics—Bloch, M. P. Bronstein, Klein, Landau, Oppenheimer, Peierls, Jacques Solomon, Teller, Weisskopf, and Eugene Wigner—had Jewish roots. Very probably, Einstein was for many of them a role model and one of the reasons they were attracted to theoretical physics. This generation of physicists matured after World War I, and their political, cultural, social and intellectual contexts were very different from those of the earlier generations. The opportunities available to them in Weimar Germany and elsewhere were likewise different. Bethe's personal history makes that clear. When asked in what way he accounted for the extraordinary creativity in theoretical physics that came out of Germany at that particular time, he responded,

> Well, that's of course one of the big questions that many people have raised. The answer that I have heard most often and that I believe most is that Germany after the 1918 breakdown had to reorient completely. Before that people, very many bright people would go into the army or into business. Business wasn't much to speak of in those days, and if they went into academics, they would go into classics, but here was a vacuum. So I think it was fairly natural that many of the bright young people went into physical sciences, physics, mathematics, chemistry, and once there was a critical mass, it generated more. It started with people like Heisenberg and Pauli. But in my case, I think I just felt attracted to it, not knowing that other young people had done it.

Hans's choice of career, although influenced by the political and cultural climate, had more direct roots. The most important person in Hans's life during his second year at Frankfurt University turned out to be the thirty-year-old Karl Meissner, "a very good spectroscopist," who had replaced Gerlach.[73] Meissner had studied physics in Munich with Wilhelm Röntgen and Arnold Sommerfeld. During the winter semester of 1925–26, Hans took Meissner's course on experimental physics and was "discovered" by him. Early in the course, the students were asked to calculate the moment of inertia of various bodies. Given Hans's mastery of and facility with calculus, he could do this very easily and very quickly. Meissner was most impressed by Hans's abilities, and a few weeks into the course he told him, "You must not stay here in Frankfurt. You can't learn enough here with Madelung. You must go to a better university. You have to go to Munich and study with Sommerfeld." He wrote a letter to Sommerfeld recommending Hans, which made Sommerfeld accept him in his seminar when Hans presented himself to Sommerfeld in April 1926. "Meissner essentially sent me away" is the way Bethe characterized it.

3

Becoming Bethe

> If you want to find out anything from theoretical physicists about the methods they use, I advise you to stick closely to one principle: don't listen to their words, fix your attention on their deeds.
>
> Albert Einstein, *Methods in Theoretical Physics*

Hans's father played a significant role in shaping the young Hans's character and his scientific outlook, but it was Arnold Sommerfeld who first molded him as a theoretical physicist. In telling the story of Hans's studies with Sommerfeld, I make use of the concept of images of scientific knowledge and body of scientific knowledge introduced by Yehuda Elkana.[1] Images and content of knowledge constitute a schematic, malleable distinction that focuses on two interconnected aspects of scientific knowledge. The body of scientific knowledge includes those components directly related to the subject matter of any given scientific discipline. In physics, this is made up of the experiments, instruments, methodologies, theories, models, and open problems. Images of scientific knowledge are concerned with aspects of the body of scientific knowledge, questions about which cannot in general be answered by the body of knowledge itself. In theoretical physics, the relative importance and status of various subdisciplines at a given time would be a reflection of the image of scientific knowledge then held by the members of that community. Similarly, the claim at a given time by a particular theorist, based on his professional expertise and standing, that a particular open problem is the most important problem to be addressed and that the solution should be undertaken in a certain way, would be indicative of the image of physics knowledge to which he was committed. The perceived relationship between mathematics and physics that Bethe inherited from his association with Sommerfeld is the particular image of science that I will concentrate on.

Munich

In June 1958, Werner Heisenberg delivered the festival oration at the celebration of the 800th anniversary of Munich, the city of his birth and of his youth. He told his audience that for him Munich brought to mind "the Ludwigstrasse, bathed in sunlight from the Siegestor to the Feldherrnstrasse; the view from the Monopteros across the flowery meadows of the English Garden; *The Marriage of Figaro* at the Residenztheater; the Dürers at the Pinakothek; the train crammed with skiers on its way to the Schliersee and the Bayrischzell; and finally, the beer tent at the Oktoberfest arena, crowned with the Bavarian lion."[2] Indeed, at the beginning of the twentieth century when Heisenberg was growing up, Munich was a bustling, thriving metropolis with broad boulevards and beautiful parks boasting huge water fountains. Its distinctive architecture was the legacy of having been the hub of a successful counter-reformation during the seventeenth and eighteenth centuries that the kings of Bavaria had nurtured aesthetically. The Jesuits had come to Bavaria during the sixteenth century and established schools throughout the region to educate a new generation of priests to attend to the needs of the masses. They also brought with them the Baroque style of architecture. The opulence and lavishness of the decoration of the churches and the cathedrals that they built stand in stark contrast to the simplicity of Protestant churches. The extravagance of the ornamentation of the Theatinerkirche, erected in Munich at the end of the seventeenth century, illustrated the conviction of Ignatius Loyola, the founder of the Jesuit order, "that the most effective way of making converts to Catholicism, or of winning back those who had left it, was to appeal not to their reason but to their senses, by overpowering them with impressions that would dazzle and move them, sweep them up into the heart of the drama of redemption, and give them a glimpse of the beauty and magnificence of God's throne and the joy that awaited the faithful."[3]

Located in the foothills of the Alps, which at the beginning of the seventeenth century were still covered with dense forests, Munich became the site of an active lumber industry. The crafts also played an important role in shaping the city: gold smithing, brass founding, book printing, and the making of stained glass were important components of the city's economy. Being located relatively far from industrial raw materials, Munich did not experience the industrial revolution the way cities in the Ruhr did during the second half of the nineteenth century. Instead of large factories and steel plants, a tool and optical industry prospered there.[4]

Munich's and Bavaria's dependence on agriculture and on tool- and instrument making was one of the reasons the city took an interest in the development of the natural sciences at its university. Chemistry took pride of place during the nineteenth century.[5] By the beginning of the twentieth century all the professorial chairs in the sciences were occupied by distinguished researchers whose reputation extended well beyond the confines of Munich and Bavaria.[6] In 1901, shortly after his discovery of X-rays and the award of the first Nobel Prize in Physics, Wilhelm Röntgen accepted the professorship of experimental physics at the university. In 1906 Arnold Sommerfeld became professor of theoretical physics there.[7]

Arnold Sommerfeld

On January 28, 1949, Arnold Sommerfeld received the Oersted Medal of the American Association of Physics Teachers for notable contributions to the teaching of physics. In comments preceding the award, Paul Kirkpatrick, the chair of the award committee, noted the textbooks that Sommerfeld had written and the outstanding students that he had trained. He characterized the various editions of Sommerfeld's *Atombau und Spektrallinien* [Atomic Structure and Spectral Lines] that had appeared during the 1920s as the "guidebooks" to the unfamiliar paths that had to be charted in order to gain an understanding of atomic structure. Kirkpatrick also reminded his audience that Sommerfeld's famous *Wellenmechanische Ergänzungsband* [Wave Mechanics, Supplementary Volume], published in 1929, in which he presented the recent developments in the new quantum mechanics, "was perhaps the most influential of the early scholarly interpretation of wave mechanics. In many American universities teachers were soon mediating it to their classes far ahead of the appearance of an English translation."[8] Furthermore, Sommerfeld's *Lectures on Theoretical Physics,* the content of his six-semester presentation of all of theoretical physics, "have contributed notably to the teaching of physics abroad."[9]

Kirkpatrick went on to note that even more impressive than the books Sommerfeld had written were the exceptional students who had come to work with him and their subsequent achievements. From among the scores of physicists who had been trained in whole or in part under Sommerfeld, Kirkpatrick could list many for whom "the surname is quite sufficient identification": Debye, Ewald, Laue, Bragg, Pauli, Heisenberg, Hertzfeld, Brillouin, Landé, Wentzel, London, Heitler, Houston, Laporte, Bethe, Eckart, Pauling, Condon, Stueckelberg. And he added that the list could easily have been extended.

It might seem unusual for the American Association of Physics Teachers to select a foreigner to be the recipient of its most prestigious award. Moreover, Sommerfeld was not only a foreigner, he was also a German, and less than four years had elapsed since the termination of World War II. But Sommerfeld was indeed one of the most influential physics teachers between the two world wars. Also, he never condoned the activities of the National Socialists during the Hitler era. In addition, Sommerfeld had special ties to the American physics community. He had made several visits to the United States during the 1920s and had helped strengthen physics there.[10] And as the above list of surnames indicates, many American students had come to study with him in Munich. To this list could be added the names of quite a few others who in 1949 held important positions at American universities: Lloyd P. Smith, the chair of the physics department at Cornell; Julius Stratton, then the provost at MIT; also Isidor Rabi, the chair of Columbia's physics department and an influential governmental adviser.

Arnold Sommerfeld was a remarkable physicist. He was nominated for the Nobel Prize over seventy times but never received the award, although he very probably deserved it for his extension of the Bohr quantization rules and his investigations of the fine structure of hydrogen. In Munich he built the most outstanding school of theoretical physics of the first third of the twentieth century.[11] He was born on December 5, 1868, in Königsberg in East Prussia and attended the town's Altstädtische Gymnasium, where his classmates included Hermann Minkowski and the cousins Max Wien and Wilhelm (Willy) Wien. Sommerfeld remained in close contact with all three of them, and Willy Wien became his colleague in Munich in 1920 upon the retirement of Leo Graetz. The young Arnold was an outstanding student, displaying particular aptitude for Latin and Greek. Yet, after passing his *Abitur* in 1886 he enrolled at the University of Königsberg to study mathematics. His teachers were Ferdinand Lindemann (who later gave a proof of the transcendental nature of π), Adolf Hurwitz, and David Hilbert, and the brilliance of their courses kept him in Königsberg.

Königsberg was one of the first German universities to establish an Institute of Theoretical Physics and to create a full professorship in the subject.[12] Franz Neumann was the first occupant of that chair, and he was succeeded by Paul Volkmann when he retired in 1876. Sommerfeld entered Königsberg University just when Heinrich Hertz had carried out his famous experiment on the propagation of electromagnetic waves. This led to the abandonment of the action-at-a-distance formulation of electromagnetism that had been advocated by Neumann and taught at the German universities. Emil Wiechert, who was a *Privatdozent* in the institute, introduced the young Sommerfeld to Maxwell's description of electromagnetic phenomena using Ludwig Boltzmann's text. Wiechert was a brilliant, unusually versatile scientist, and Sommerfeld came under his spell.[13]

In 1891 Sommerfeld submitted his doctoral dissertation in mathematics: "The arbitrary *(willkürlichen)* functions of mathematical physics," a work that he completed in a few weeks' time after he had conceived the idea for it. The subject matter of his thesis—the representation of arbitrary functions by complete, orthogonal sets of functions—remained a topic of great interest to him throughout his life.[14]

In the fall of 1893, after a year's military service, Sommerfeld went to Göttingen. Although he accepted a position as Theodor Liebisch's assistant in the Mineralogical Institute, his primary interest remained mathematics and what Felix Klein called "physical mathematics," a field not to be identified with mathematical physics but rather with mathematics relevant and applicable to physics and engineering problems. Physics and engineering thus became sources of inspiration for mathematics.[15] Felix Klein was then the dominant and dominating figure in Göttingen, and the person there responsible for the blurring of the line between pure and applied mathematics.[16] Sommerfeld was overwhelmed by Klein's lectures and by his "grand personality." Sommerfeld considered Klein his real teacher in mathematics and mathematical physics and the person who shaped his views and approach to these fields.[17] In his *Autobiographical Sketch* Sommerfeld acknowledged that "the impression I received of F. Klein's imposing personality through his lectures and in conferences with him was overpowering . . . I have always regarded Klein as my teacher, not only in mathematical, but also in mathematical-physical matters, and in the conceptual interpretation of mechanics."[18] It might be added that Sommerfeld also adopted Klein's demeanor as a *Geheimrat.* He too became a *Bonze;* in German a *Bonze* denotes an influential, authoritarian personage.[19] Max Born in his obituary for Sommerfeld commented:

> Well can I understand this feeling of devotion and awe which Klein inspired, as I came under his influence about eleven years later, at a time when Klein's magical powers as a teacher were perhaps at their summit. In fact, I found him too olympian, his lectures too perfect, his seminar discussions too encyclopaedic, and I preferred the more human personalities and formally less accomplished but livelier teachings of Hilbert and Minkowski. But they were not yet in Göttingen in 1893 when Sommerfeld appeared there, hence it is no wonder that he fell completely under Klein's spell. Many a time later we have discussed the merits of these three giants of mathematics, but Sommerfeld never wavered in his preference for "the great Felix."[20]

Sommerfeld became Klein's assistant in 1894. His responsibilities included the supervision of the *Lesezimmer* (the Mathematical Reading Room) and its library, and writing up Klein's lectures for the use of students. Sommerfeld thus gained an intimate knowledge of Klein's style and method of lecturing, and these became the model for his own lectures. He also took part in the seminar that Klein ran. In the seminar Klein would "distribute among his students certain portions of the broader field in which he himself was engaged, to be investigated thoroughly [by them] under his personal guidance and to be presented in final shape at one of the weekly meetings."[21] In the process he transformed the seminar into a highly competitive community but also into a communicative community in which the spoken word often carried more weight than did information conveyed in written texts.[22] This was the German version of what was happening at the University of Cambridge, where the Tripos examinations in maths required students to solve technically demanding mathematical problems and to produce knowledge on paper.[23] Klein's approach, like that at Cambridge, proved to be a watershed for the discipline of mathematical physics. In Munich Sommerfeld would later provide his students with a *Lesezimmer* and run his seminar the way Klein had done, except that by the 1920s he would assign the supervision of the preparation of the students for their presentations to one of his *Assistenten.* And like Klein, Sommerfeld also became a great lecturer. Rudolf Peierls, who joined

Sommerfeld's seminar in Munich during the academic year 1926–27, commented that "His introductory lectures were a model of clarity."[24]

After coming to Göttingen in 1886 Klein had embarked on a program to strengthen German technology through a greater reliance on foundational science, particularly physics, and a greater use of mathematical techniques.[25] During the academic year 1895–96 Klein gave a series of lectures to high school teachers on the theory of the gyroscope, out of which grew the four-volume treatise *Die Theorie des Kreisels,* on which Klein and Sommerfeld collaborated over the next decade. The first two volumes, written while Sommerfeld was still in Göttingen, emphasize the mathematical aspects of the subject. The third and fourth volumes, which were completed in 1910 after Sommerfeld had gone to Munich, deal with applications to geophysics, astronomy, and engineering, and rely much more on physical arguments. The joint work reflected the change in the commitments of both authors: that of Klein to strengthen and enrich applied science and engineering by the use of advanced mathematics, and that of Sommerfeld to be a leading contributor to the enterprise.[26] Incidentally, Klein and Sommerfeld's work on the gyroscope found a novel application during World War I: the gyroscope was used as a navigational instrument for the German submarines that wreaked terrible havoc on Allied shipping.[27]

In 1895 Sommerfeld became a *Privatdozent* in mathematics with a *Habilitation* on diffraction theory that Poincaré praised as "a very important piece of work." Two years later, following in the footsteps of Klein, one of whose first academic positions had been an appointment at the Munich *Technische Hochschule,* Sommerfeld accepted the position of professor of mathematics at the Mining Academy of Clausthal in the Harz Mountains. Clausthal is not very far from Göttingen, so Sommerfeld remained in close contact with Klein and Hilbert. In 1900 he moved to the Aachen *Technische Hochschule* when he was offered the chair in technical mechanics. At both of these institutions he proved himself a most useful member of the staff, both in his teaching and in his research.[28]

Though very successful in his applied mathematical work, Sommerfeld confessed that he really was not a "professor of technology, but a physicist."[29] What had changed him from regarding himself as a mathematician to becoming a physicist was Klein's appointing him in 1899 editor of the highly respected and widely read physics volumes of the *Encyclopädie der Mathematischen Wissenschaften,* a charge he executed with great distinction from 1898 until 1928. His editorial responsibilities placed him at the center of an extended network of eminent physicists and mathematicians. He thus came to know such leading theoretical physicists as Ludwig Boltzmann, Hendrik Lorentz, Max Planck, Lord Kelvin, and Willy Wien, to cite only the most well known. His interactions with them inspired Sommerfeld to become a theoretical physicist.

In 1906 Sommerfeld accepted the chair in theoretical physics at the University of Munich, to which he had been nominated by Röntgen and recommended by Lorentz, Boltzmann and Willy Wien. The position had been vacant for four years following the departure of Ludwig Boltzmann to Vienna. In his *Autobiographical Sketch* Sommerfeld remarked that from the very beginning he made every effort through his seminars and colloquia to found in Munich "a nursery of theoretical physics."[30] Theoretical physics at that time was a new specialty. It had emerged as a separate academic discipline in the German-speaking universities only in the last third of the nineteenth century. The increasing dependence of the physics community on theorists to teach the more advanced and mathematical portions of the field was one of the factors that helped establish theoretical physics as a subdiscipline. Also, as physics became perceived as the foundation for all the other sciences and the driving force of technological development, the study of physics was encouraged and the number of students enrolling in physics seminars and institutes increased.[31] By the end of the nineteenth century almost every German university had an institute of physics housed in a separate building, and most of them had an associate professorship in theoretical physics to satisfy the teaching needs of the institution. Often these positions were filled by promising young experimenters who took them while waiting for a full professorship to open up elsewhere.[32]

Until the end of the nineteenth century the directors of the various physics institutes in the German universities were all experimenters. The director of an institute had essentially complete control over its activities. Separate theoretical physics institutes only came into existence at the turn of the century. Sommerfeld's institute in Munich was one of the first. Moreover, it was not unusual to house some laboratories within the institute for theoretical physics since theoretical physicists still did experiments at that time. One of the conditions for Sommerfeld's acceptance of the professorship in Munich had been the creation of an Institute for Theoretical Physics that he would direct, and a guarantee of an adequate budget to hire several assistants. Sommerfeld also had insisted that the institute include facilities for supporting experimental work connected with his theoretical investigations of X-rays—this in part to prove that he was a physicist and not an applied mathematician. He argued—and convinced both the faculty and the ministry—that it was important for a theorist to keep in close touch with real phenomena. The institute that was established for him did include experimental facilities and a machine shop. During the first decade of his tenure at Munich Sommerfeld directed an extensive program of experimental research in both hydrodynamics and X-ray diffraction to check his own theories, and even supervised doctoral dissertations in experimental physics.[33]

At first, Sommerfeld's institute was located in the "Augustinestock" on Neuhauserstrasse. It was the building in which the Bavarian Academy of Arts and Sciences had met, and which it had vacated when it moved into new quarters. Then when the university completed building an extension along Amalienstrasse, the institute moved there in 1910, occupying part of the ground floor and all the basement. It was now close to Wilhelm Röntgen's Institute of Experimental Physics.[34] Sommerfeld's institute consisted of four offices, a lecture hall that could accommodate about sixty people, an equipment room for the lecture demonstrations, a workshop, a darkroom, and four experimental and storage rooms that were located in the big basement. The space allocated to theoreticians consisted of two rooms: one was Sommerfeld's office, and the other a large room in which his *Assistenten* and his students worked and in which the seminars were held. The main experimental activities carried out in the institute were concerned with two areas of great interest to Sommerfeld: the onset of turbulence in fluid motion in an open channel and the properties of X-rays. By 1911 he had two *Assistenten:* Peter Debye, who had come with him from Aachen, and Walter Friedrich, who worked on X-rays. The seminar consisted of about ten students working on doctoral dissertations under Sommerfeld, and another ten or so "predoctorals" who worked closely with Debye. Sommerfeld had already adopted Felix Klein's model of the *Doktorvater.* If a student in the seminar wished to do his thesis with him, Sommerfeld would first assign him a small problem whose answer he knew. If the student successfully completed the assignment in a reasonable amount of time and gave proof of independence, inventiveness, and imagination, Sommerfeld would recommend a subject and a more extended problem as a thesis.[35] Paul Ewald, who was a student in the institute from 1908 until 1912, recalled later that physics discussions often also took place in the Café Lutz in the Hofgarten, when the weather permitted under the shade of the chestnut trees, and otherwise indoors. This was the general rallying point of physicists after lunch for a cup of coffee and the tempting cakes. Once these were consumed, the conversation which might until then have dealt with some problem in general terms, could at once be followed up with diagrams and calculations performed with pencil on the white smooth marble tops of the Café's tables—much to the dislike of the waitresses who had to scrub the table clean afterwards. Sommerfeld and his friend R. Ehmde (Professor at the Technical University, and well known for his pioneer work on stellar atmospheres) and also others like the mathematicians Herglotz, Caratheodory, Schoenflies came to this unofficial centre of exchange of physical ideas and news. For the younger members of the group it was most exciting to watch research in the making, and to take sides in the first tentative formulation of experiments and theory."[36]

Sommerfeld retained close ties with Göttingen. While at Clausthal and Aachen he frequently visited Hilbert and Minkowski's seminar. Later he was invited to all the workshops and conferences initiated by Klein and Hilbert. The esteem Sommerfeld was held in by Hilbert is indicated by the fact that after Minkowski's death, when Hilbert ran by himself the physics seminar he and Minkowski had overseen together, almost all of Hilbert's physics assistants—Paul Ewald (1912–1916), Alfred Landé (1916–1919), and Lothar Nordheim (1922–1928)—came from Sommerfeld.

Sommerfeld was one of the first defenders of Einstein's theory of special relativity. Following the publication of Einstein's paper in 1905 Sommerfeld included an exposition of the theory in his lectures. After hearing Minkowksi's lecture on the four-dimensional representation of the special theory at the *Naturforscher und Ärzte Versammlung* in Köln in 1908 Sommerfeld wrote two lengthy papers on four-dimensional vector and tensor analysis.[37] The current formulation and terminology of special relativity are those that Sommerfeld introduced in these papers. He became deeply interested in Einstein's development of general relativity and likewise kept abreast of Hilbert's attempts to formulate a covariant theory of gravitation. Sommerfeld was also one of the first supporters of Einstein's photon concept and of Einstein's views of the quantum of action. His distinction as a theoretical physicist was recognized early. He was one of the youngest physicists invited to the 1911 Solvay Conference and asked to make a presentation, titled "The Quantum of Action and Non-Periodic Molecular Phenomena." His pragmatic proclivities were already evident there. In a comment following Planck's exposition of "Radiation Theory and the Quanta," Sommerfeld expressed his belief "that Planck's new hypothesis of emission quanta, as well as the original hypothesis of the quantum of energy should be considered as a form of explanation [a model] rather than as physical reality."[38]

At first Sommerfeld was somewhat uncertain and undecided about Bohr's 1913 theory of the hydrogen atom but quickly became an enthusiastic advocate. He extended the Bohr quantization conditions, and his theory of the fine structure of the hydrogen atom, a seminal extension of Bohr's theory, was a "crucial" calculation. The first edition of his important and widely acclaimed *Atombau und Spektrallinien* appeared in 1919. It went through several editions and became the bible of the old quantum theory.

In Göttingen Sommerfeld had come to recognize the value of being a member of an active, stimulating, intellectual community and of the importance of research *and* teaching in training students.[39] Sommerfeld's seminar in Munich reflected this awareness. It soon became recognized as an excellent training center for becoming a theoretical physicist. Einstein in 1908, while still working at the Bern patent office, wrote him: "I assure you that, if I were in Munich, and time would permit it, I would attend your lectures in order to perfect my mathematical-physical knowledge."[40] Shortly after Einstein met Sommerfeld at the September 1909 meeting of the *Gesellschaft deutscher Naturforscher und Ärzte* in Salzburg he wrote to him to tell him that he understood why "students like you so much. Such a beautiful relationship between teacher and students is surely unique. I will strive to emulate your example."[41] And Einstein told his colleague Jakob Laub that he was "completely entranced" by Sommerfeld and thought him to be "a splendid man."[42] There developed between the two men a deep respect and warm friendship.

Sommerfeld was indeed an outstanding teacher. He invested a great deal of time and effort in his teaching duties and instructional activities. He became the *Doktorvater* of some thirty doctoral students, and among them were many of the outstanding theorists of their generation: Peter Debye, Paul Ewald, Paul Epstein, Gregor Wentzel, Wolfgang Pauli, Werner Heisenberg, Otto Laporte, Albrecht Unsöld, Herbert Fröhlich, and Hans Bethe. *Doktorvater* had a special meaning. In the traditional German family the patriarchal father was called *Hausvater*. He exercised legal, economic, and disciplinary powers over all the members of his household. Before Bismarck's unification of Germany, the ruler of a state was the *Landesvater* to the *Landeskinder*, the citizens of his land; and similarly, in the German universities the *Doktor-*

vater was responsible for his doctoral students, and they accountable to him. Though very much the *Herr Geheimrat,* Sommerfeld was—in his early days in Munich—also close socially to his students. Later, as his administrative duties and professional responsibilities became more time-consuming, he became somewhat more distant socially, but always approachable regarding physics. And if as *Doktorvater* Sommerfeld felt responsible for the successful completion of the dissertations of his doctoral students, his sense of responsibility extended far beyond this for the people who had done their *Habilitation* with him. Thus in 1931 Sommerfeld not only helped Bethe find a suitable position but supported him with his own private money until he did find a job.[43]

Lord Rayleigh in his biography of J. J. Thomson listed three ingredients that are necessary to form a scientific school:

> The first is a stimulating leader, one who is not only abounding in energy and ideas, but also one who can without too great an effort throw himself into the difficulties of others. This requires a peculiar kind of versatility not always easily combined with great powers of concentration on any one line of thought. Then again it is necessary to have a productive line of investigation opening up; lastly it is necessary to have the right kind of pupils. These must be men of the not very common kind of ability which makes a scientific investigator: they must not be too young: and they must be provided with means of subsistence while the work goes on.[44]

Sommerfeld's school fulfilled all these requirements. Sommerfeld's reputation as an outstanding teacher and researcher was established early. So many gifted students wanted to study with him that he could choose the best among them. His success at attracting outstanding students and subsequently being able to obtain university positions for them—as limited as such positions were—magnified his appeal and influence. His seminar became the outstanding school for the production of theoretical physicists in Germany until the advent of quantum mechanics, and a new generation of theoreticians who had been his students—Debye, Ewald, Heisenberg, Pauli, and Wentzel—became professors. Shortly thereafter, however, Hitler changed the character of German universities, and Germany lost its dominance in theoretical physics.

In his interview with Charles Weiner in 1963 Bethe noted that one of the reasons for Sommerfeld being such a successful *Doktorvater* was that

> he always knew thesis topics to suggest. He was so much in the midst of the development of atomic physics that there were always a dozen things he wanted to have done. This is perhaps a trait similar to my own—he was in-

> terested in the application of the theory, and I think it's in the application of the theory that the largest number of problems exist that can be done by graduate students . . . Sommerfeld could very accurately gauge the ability of the students, and he gave difficult problems to the abler students and simpler problems to the less able students, and he had a list of problems of various difficulty from which he just picked. And I think this was very different from most of the theoretical physicists in Germany at least. There's a story about Planck—that somebody came to him and asked him for a thesis topic, and Planck is supposed to have answered: "Well, young man, I would gladly give you one, but if I knew a good topic to work on, I would do it myself."[45]

Sommerfeld knew his limitations. He recognized early that he was not able to formulate foundational theories, but that he was a master craftsman in applying the instruments of his formidable mathematical toolkit. In a letter to Einstein in January 1922 he told him, "I can only advance the craft of the quantum, you have to make its philosophy." He saw himself as a problem solver, and derived huge satisfaction from the solutions of the many physical problems he tackled using the considerable mathematical techniques at his disposal, in particular the use of complex variable analysis. Suman Seth characterizes his physics as a "physics of problems."[46]

Seth suggests that between the publication of the first edition of *Atombau und Spektrallinien* in 1919 and the fourth in 1924, Sommerfeld moved from understandings of the atom in terms of models to understandings in terms of phenomenological, empirical rules, and that Pauli drew heavily on this approach in formulating his phenomenological exclusion principle.[47] However, after the advent of Schrödinger's wave mechanics Sommerfeld clearly reverted to a faith in *Modelmässigkeiten,* given the extraordinary success in applying wave mechanics to the structure of atoms.

When Bethe came to Munich in 1926 Sommerfeld was mostly concerned with problems in atomic physics, and many of his *Doktoranden* were doing problems whose solutions would find applications in other fields, such as crystallography and astrophysics. The formation of Sommerfeld's school and its success reflected the strengths and powers of the young Sommerfeld. Michael Eckert insightfully notes and stresses the fact that in his formative years, 1893–1906, Sommerfeld was at the crossroads of the fields that nurtured the emergence of theoretical physics intellectually and institutionally: mathematics, mechanics, and physics. Each of these fields had complementary aspects that helped mold theoretical physics. The distinction between theoretical physics and mathematics derives on the one hand from the kind of mathematics used in theoretical physics and the way it is used—pragmatically, not necessarily rigorously, and made to yield numbers that can be compared with empirical findings—and on the other hand, from the fact that at the beginning of the twentieth century mathematics was becoming an autonomous discipline and distancing itself from physics. It was the wealth of practical problems Sommerfeld addressed during his Aachen years—diffraction problems, gyroscopes, and electric motors—and his previous applied mathematical orientation as Klein's *Assistent* and as his collaborator in their joint work on gyroscopes that shaped his broader research approach and that of his School.[48] It was in physics—with its interdependence of theory and experiment—that Sommerfeld's approach to the choice and solutions of problems came to fruition. Carl Eckart characterizes Sommerfeld's style as follows: "an eagerness to display the mathematical tractability of the problems across the entire spectrum of physical specialties, from mechanics to atomic theory. This breadth enabled Sommerfeld to have so many doctoral students, and it was mirrored in their research themes during the first years of his 'nursery' for theoretical physics."[49]

Bethe would follow in Sommerfeld's footsteps. Bethe's physics likewise became a physics of model-based problems. His mathematical toolkit became as extensive as Sommerfeld's, his ability to explore the complex plane as far reaching, and the breadth of his interests similarly stretching across the entire spectrum of physical specialties. And like Sommerfeld, Bethe became an "unphilosophically disposed" master of broad domains in theoretical physics.[50]

Mathematics was at the center of all of Sommerfeld's research, and this in all fields. His stay in Göttingen, and thereafter his close contacts with Klein, Hilbert, and Minkowski, had made him, like them, a believer in a pre-established harmony between mathematics and physics. For Hilbert pre-established harmony came to mean a pre-established unity of mathematics and the possibility of stipulating a set of axioms from which all of mathematics might be derivable. For Klein and Sommerfeld, it came to mean the existence of (invariance) principles that constrain all physical theories, a viewpoint Sommerfeld communicated to all his students. Later, for Einstein it came to mean the existence of a fundamental equation from which all of physics could be derived.

It is to this topic, pre-established harmony, that I turn next.[51]

Pre-established Harmony

Einstein first stated publicly what the "pre-established harmony between mathematics and physics" meant to him in his famous encomium to Max Planck in 1918, in which he asserted that "the supreme task of the physicist is to arrive at those universal laws from which the cosmos can be built up by pure deduction." He went on to say that at any given time out of all conceivable possible systems of theoretical physics a single one "has always proved itself superior to all the rest."[52] Moreover that system is uniquely determined by the world of phenomena "in spite of the fact that there is no logical bridge between phenomena and their theoretical principles; this is what Leibniz described so happily as a 'pre-established harmony.'"[53] For Einstein this longing "to behold this pre-established harmony [was] the source of the inexhaustible patience and perseverance" with which he "devoted himself to the most general problems of physics, refusing to let himself be diverted to more grateful and more easily attained ends."[54]

Felix Klein, David Hilbert, and Hermann Minkowski, the mathematicians who molded Göttingen mathematics in the two decades around the turn of the twentieth century, often spoke of the pre-established harmony between mathematics and physics. Lewis Pyenson has indicated some of the roots in German culture of this (extended) Leibnizian notion of pre-established harmony.[55] Gottfried Leibniz was the figurehead for the German *Gelehrtenwelt,* was venerated as the founder of the Berlin *Akademie,* and became and remained pre-eminent as a philosopher.[56] Leibniz's philosophy, with its emphasis on continuity, consistency, and harmony, resonated with Klein's ambitious plans for reforming the *Gymnasium* curriculum and placing mathematics at its core.

Leibniz's philosophy enjoyed a revival in the last third of the nineteenth century.[57] He believed that all truths are analytic and that truths that appear to be synthetic have been assigned that status because of our limited understanding, but that these truths would eventually be properly recognized as analytic with the growth of our knowledge. This belief later resonated with those who saw the impressive progress in human understanding during the nineteenth century as a step toward perfect understanding. The hope that at least for mathematics all truths could be proved to be analytic is what underlay Gottlob Frege's, Georg Cantor's, and Bertrand Russell and Alfred North Whitehead's efforts to ground all mathematics on logic.

Another reason for Leibniz's appeal to German scientists was that by the end of the nineteenth century many of the leading German scientists were deists.[58] Leibniz in his *Theodicy* had maintained that the world is governed by immutable laws, imposed on it by God in its initial creation. This was in sharp contrast to Newton, who had God intervening in the operation of Nature and providentially rectifying the course of the planets to guarantee their stability.[59] The possibility of understanding the world in terms of immutable, timeless laws of nature—to be able to formulate cosmological, physical, chemical, biological theories in which theistic considerations played no role—was an attractive approach to the sciences.

Yet another reason for Leibniz's appeal to theoretical physicists and mathematicians was that the foundations of his philosophy—the principle of continuity, the principle of contradiction, and the principle of sufficient reason (that nothing is without a reason)—all had roots in mathematics.[60] Similarly, the principle of least action—the guiding concept in formulating dynamical laws in physics—was very close in spirit to the guiding concept in Leibniz's theodicy.[61] The principle of continuity was originally a generalization of the property of numbers, "that they can be continued without end, and divided without limit," and thus connected with the infinitely large, "in which everything is contained," and to the infinitely small, "of which everything is made up." Since everything "contained" was created by God and now exists in the mind of God, the mutual harmony of everything real is guaranteed by its original constitution. This harmony was "once for all established for all times, it is pre-established."[62]

Hilbert spoke of the "pre-established harmony between nature and mathematics" in his famous Paris lecture at the Second International Congress of Mathematics in 1900 in which he formulated the twenty-three outstanding mathematical problems to be solved in the twentieth century.[63] In that lecture, Hilbert asserted that this pre-established harmony was based in the unity of mathematics, which he considered to be "an indivisible whole, an organism whose vitality is conditioned upon the connection of its parts."[64] Eventually, for Hilbert pre-established harmony came to mean the possibility of the human mind to base all of mathematics on a finite set of consistent axioms.

Minkowski, in his famous 1908 Köln lecture in which he expounded his four-dimensional formulation of special relativity, stressed that the validity *without exception* of what he had called the "world-postulate"—the equivalence of all inertial frames for the description of all physical phenomena, mechanical *and* electromagnetic—was "the true core (*der wahre Kern*) of the new image of the world, which, discovered by Lorentz and further revealed by Einstein, now lies open in full light of day." A four-dimensional formulation that stipulated the invariance of the line element

$$c^2\,dt^2 - dx^2 - dy^2 - dz^2$$

was its natural mathematical description. And he ended his lecture with the statement, "In the development of its mathematical consequences there will be ample suggestions for experimental verifications of the postulate, which will suffice to conciliate even those to whom the abandonment of old-established views is unsympathetic or painful, by the idea of a pre-established harmony between pure mathematics and physics."[65]

Already in a talk in 1907 Minkowski had declared that the principle of relativity—that the laws of physics are the same for all inertial observers, and that this fact is expressed mathematically as their invariance under Lorentz transformations—is to be viewed as a truly new kind of physical law, in that it is a requirement imposed on equations that are sought to describe observable phenomena.[66]

In his *Autobiographical Notes* Einstein stressed the fact that it was Minkowski who had put forth the requirement of what has become known as "symmetry dictating interaction." Before him the symmetry exhibited by the Maxwell-Lorentz equations, that is, their covariance under Lorentz transformations, was a derived, secondary property. Minkowski inverted the process: Lorentz invariance—the symmetry—was the starting point and the primary attribute: the field equations describing the phenomena were required to satisfy that symmetry, in other words, to be covariant with respect to the symmetry. To express this succinctly, before Minkowski theory building proceeded from experiments to appropriate (particle and field) equations describing the phenomena, and thereafter certain symmetry properties of the equations were noted. After Minkowski it is the symmetry, or invariance, that dictates the form of the equations and the mathematical expression of the interactions.

Others, in particular Poincaré, had expounded similar views.[67] Poincaré was committed to a "physics of principles" which demanded that particular (model) theories, what Einstein would later call "constructive theories," be compatible with the general principles. This was of central importance in his construction of particular representations of physical phenomena. The principle of relativity, according to which "the laws of physical phenomena must be the same for a stationary observer as for an observer carried along in a uniform motion of translation; so that we have not and cannot have any means of discerning whether or not we are carried along in such a motion," which Poincaré stated explicitly in 1904, was one such principle.[68] In addition, like Minkowski, Poincaré had explicitly demanded that Lorentz covariance constrain all interactions: gravitation and electromagnetism, as well as the forces responsible for the so-called Poincaré stress in an electron that prevented it from being blown apart by electrostatic repulsion.

I have focused on Göttingen—Klein, Hilbert, and Minkowski—because that was Sommerfeld's community.[69] It was this mathematical tradition—through Hilbert and his assistants, and Sommerfeld and his School—that made symmetries and invariance a *Denkstile* at the center of this theoretical physics *Denkkollektive*.[70]

Minkowski died in 1909. Max Born, who had become his assistant, was charged with editing and publishing Minkowski's papers on electrodynamics. The original typescript of Minkowski's 1907 lecture was edited for publication by Sommerfeld.[71] Lewis Pyenson notes that Sommerfeld altered the manuscript at several places. The most important one concerned Einstein's role in the development of the principle of special relativity: "Sommerfeld was unable to resist rewriting Minkowski's judgment of Einstein's formulation of the principle of relativity. He introduced a clause inappropriately praising Einstein for having used the Michelson experiment to demonstrate that the concept of absolute space did not express a property of phenomena. Sommerfeld also suppressed Minkowski's conclusion, where Einstein was portrayed as a clarifier, but by no means as the principal expositor, of the principle of relativity."[72]

In his papers that analyzed Hilbert's motivation and intent in axiomatizing mathematics and physics Leo Corry convincingly indicated that both Hilbert and Minkowski believed that when formulating the mathematical scaffolding of physical theories in general, certain universal principles must be postulated. An example is the principle of relativity: the requirement of covariance of the equations of motion under homogeneous Lorentz transformations.[73] Invariance under space and time translations are likewise such principles: the requirement that the "absolute" position of the origin of the coordinate system cannot be a relevant variable in the description of phenomena, and similarly the "absolute" time that marks the beginning of an experiment cannot be a relevant variable; only time intervals can enter the description. Time translation invariance will entail the conservation of energy, space translations the conservation of momentum, if it is postulated that the equations of motion for the closed system are to be obtained from a principle of least action, and that the Lagrangian which describes the system is a scalar, i.e., has the proper covariance properties. The importance and centrality of the Lagrangian formalism stems from the fact that it is a mathematical theorem, known as Noether's theorem, that the invariance of the Lagrangian under a global, position independent, symmetry transformation implies a conservation law.

For Hilbert and Minkowski "general principles"—as for example, Lorentz covariance—were such that should new empirical data force changes in the particular (constructive) theory that had been advanced to represent the phenomena, the general principles would remain unchanged and would again impose stringent constraints on any new representation.[74] Similarly Einstein came to see Lorentz covariance for special relativity and general covariance for general relativity as such "general principles." In fact, Einstein explicitly differentiated between what he called principle theories and constructive theories: "Constructive theories in physics attempt to build up a picture of the more complex phenomena out of the materials of a relatively simple formal scheme from which they start out."[75]

Physical theories should be constructive. They are "bottom-up" theories. Such theories describe phenomena in terms of underlying mechanisms, as in the case of statistical mechanics. Such theories are both intuitive and complete and when formulated they explain the phenomena. However, at times the theorist is limited to advancing nonconstructive principles—such as the second law of thermodynamics and the invariance principles of special and general relativity—that only impose general restrictions on possible physical processes. "Principle-theories employ the analytic, not the synthetic method. The elements which form their basis and starting point are not hypothetically constructed but empirically discovered ones, general characteristics of natural processes, principles that give rise to mathematically formulated criteria which the separate processes or the theoretical representations of them have to satisfy."[76] Einstein had been deeply impressed by the physical results that derived from the requirement of Lorentz invariance and strove to enlarge Lorentz invariance. This, together with the equivalence principle, led him to the general theory of relativity.[77] It was his demand that the field equations that were to describe gravitation be invariant under arbitrary, local, *position dependent* transformations, and that in the limit of weak fields they yield the Newtonian description of gravitation that eventually led him to his field equations. These equations were generally covariant, meaning that their form remained unchanged under arbitrary transformation of the space and time coordinates, and they incorporated Einstein's principle of equivalence, which asserted the equivalence of inertia and gravitation. The principle of equivalence's starting point had been the observation that it is impossible to distinguish physical phenomena observed in the presence of a homogeneous gravitational field from those observed in a uniformly accelerated coordinate system. Euclidean geometry was not valid in the latter frame, hence Einstein's adoption of Riemannian geometry, and his translation and generalization of the principle of equivalence into a requirement on the metric tensor that, on the one hand, describes the inertio-gravitational field and, on the other, the geometry of space-time.

The symmetry exhibited by general relativity came to be called a gauge symmetry.[78] General relativity was the first instantiation of a theory in which such a (gauge) symmetry is used to derive a theory of interactions. A gauge symmetry dictates interactions, and all the presently known interactions, electroweak and hadronic, are presently described by gauge theories.[79] This is why Einstein is given credit by the physics community for the formulation of this demand. It also gives credit to Einstein for imposing symmetry principles as constraining all constructive theories. But this is a Matthew effect at work.[80] Credit should go to Hilbert and Minkowski.

For Hilbert, Minkowski, Einstein, and also for Sommerfeld, the belief in the pre-established harmony of mathematics and physics meant belief in the existence of such universal principles. Einstein believed that the laws of thermodynamics were universal principles. But when he expressed his conviction that thermodynamics would "never be overthrown," he added the qualifying phrase "within the framework of applicability of its basic concepts," since he himself had had the major share in establishing the limits of that framework.[81] The caveat when asserting that thermodynamics and Lorentz covariance are universal principles is to recognize their domain of applicability. For thermodynamics this means that one is describing equilibrium phenomena involving systems with a huge number of microscopic particles; systems that have come to a state in which they, at the thermodynamical, macroscopic level of description, do not change with time; systems whose initial conditions are irrelevant; systems which have no history.[82] Similarly, for special relativity: one cannot describe phenomena under the terms of special relativity unless the matter present has negligible effect on the gravitational field (so that the space-time framework can be considered as rigid).

The special role that mathematics played in Sommerfeld's craft and practice was part and parcel of his training in Königsberg and Göttingen. His close friendship with Einstein added a new dimension to the role of mathematics in formulating physical theories. After inventing general relativity, Einstein had rejected the possibility of arriving at laws of nature by induction from experience, and since he did not believe in *a priori* knowledge,[83] he had to give an answer to this question: "How does the physicist know that he is on the right track when trying to arrive at representation of nature?" This question was of immediate importance and relevance, for starting in 1919 Einstein had embarked on a program to unify gravitation and electromagnetism that was to isolate him from the physics community. The answer he gave was that "Nature is the realization of the simplest conceivable mathematical ideas." Furthermore, he came to believe that the "creative principle" in this enterprise resided in mathematics.[84]

Like Einstein, Sommerfeld believed in the pre-established harmony of mathematics and physics. But unlike Einstein, Sommerfeld in striving for theoretical descriptions of physical phenomena never severed his ties to experimental findings and always thought they would yield the clues that would allow a mathematical description that deciphered them. Thus in 1919 he wrote Einstein, "You ponder the fundamental questions concerning the light quantum. Because I do not feel the strength in me for that, I content myself with looking into the particulars of the quantum magic in spectra."[85] In Sommerfeld's exposition of special relativity to *physicists,* pre-established harmony of mathematics and physics, like for Minkowski, meant that symmetry (in this case Lorentz invariance) dictated the mathematical form of the equations representing physical phenomena. In Sommerfeld's public lectures "pre-established harmony of mathematics and physics" became an assertion of a commitment to a Pythagorean vision. Thus in 1933 in his James Scott Lecture, delivered to the Edinburgh Royal Society, Sommerfeld made explicit his views regarding the shortcomings of the logical positivists in their banishing of metaphysics. "There is surely some metaphysics, when we speak of the laws of nature as independent of us and immutable, 'dauernder als Erz'" (more durable and lasting than bronze). And to describe more accurately the timeless, "eternal" character of physical laws, Sommerfeld called on mathematics.

It seemed to him that Plato's assertion that God is a geometer was "truer today than ever before." Furthermore, echoing somewhat Einstein's views that that the differential equations he obtained to describe gravitation were found by looking for the simplest equations that Riemannian geometry would yield, Sommerfeld asserted, "We see evermore clearly that the most general mathematical formulation is simultaneously the most fructiferous."[86] He indicated that Nature is a much better mathematician than we are, and recalled that in his *Atombau und Spektrallinien* he characterized the quantum numbers encountered in the spectral analysis of atoms as a revival of the mystical Pythagorean belief in the harmonies of the spheres.[87] Given the extended discussions of the particle-wave duality of matter after the advent of quantum mechanics it is interesting to note that Sommerfeld went on to state that he was satisfied *(befriedigt)* with this seemingly unresolved duality. He thought that this physical duality is a useful contribution to one of the deepest philosophical questions: namely, the relation between matter and mind, body and soul. Echoing Leibniz, he thought that there existed a parallelism regarding a "psycho-physicalism pre-established harmony" that would account for the mind-body duality and the "mathematics-physics pre-established harmony" regarding the particle-wave duality.[88]

In his encomium on the occasion of Sommerfeld's eightieth birthday in 1945 Pauli averred that Sommerfeld had given his students a vision of what Kepler had meant with his assertion that "geometry is the archetype of the beauty of the world." He too pointed to the preface of the first and subsequent editions of *Atombau und Spektrallinien* where Sommerfeld had written "What we are nowadays hearing of the language of spectra is a true music of the spheres within the atom, chords of integral relationships, an order and harmony that becomes ever more perfect in spite of its manifold variety . . . All integral laws of spectral lines and of atomic theory spring originally from the quantum theory. It is the mysterious *organon* on which Nature plays her music of the spectra, and according to the rhythm of which she regulates the structure of the atoms and nuclei."[89] Pauli commented that "It is as though there was here an echo of Kepler's search for the harmonies in the cosmos, guided by the musical feeling for the beauty of just proportions in the sense of Pythagorean philosophy."[90]

Recalling Pythagoras and Kepler added an acceptable metaphysical, somewhat mystical, but very traditional and respectful gloss to the Sommerfeldian practice of theoretical physics. For Sommerfeld theory became identified with mathematics; and mathematics and the possibility of mathematical representation were the scaffolding for his rationalism. For him mathematics had epistemological *and* ontological content, and advances in mathematics and theoretical physics added to that content. All of Sommerfeld's students obtained from him a rigorous training in the use of mathematics in physics, and most of them also took over a belief in the efficacy (thus epistemological) *and* ontological content of mathematics.

Sommerfeld's students became theoretical physicists—though at times they were also practicing applied mathematicians. Partial Differential Equations in Physics was the topic that regularly concluded Sommerfeld's six-term series of lectures. Sommerfeld later published a slightly updated version of the course. In his introduction Sommerfeld stressed that he was dealing with *physical mathematics* and not with mathematical physics; in other words, "not with the mathematical formulation of physical facts, but with the physical motivation of mathematical methods." And he added, "The oft mentioned 'pre-established harmony' between what is mathematically interesting and what is physically important is met at each step and lends an esthetic—I should like to say metaphysical—attraction to our subject."[91]

Most of Sommerfeld's students believed in some form of the Leibnizian pre-established harmony. Heisenberg and Pauli explicitly mentioned their belief in it in some of their writings.[92] Pauli made the metaphysical connection explicit. Although his initial philosophical inclinations were toward positivism, a mental breakdown and his subsequent analysis of the episode with the help of Carl Jung made him turn toward mysticism and made him recognize and accept commitments to possibly nonrational elements in one's metaphysics. These elements manifested themselves in his exposition of the interpretation of the formalism of quantum mechanics. For him quantum mechanics recognizes that the outcome of an experiment on an individual microscopic system may be completely undetermined. However, the results of repeated identical experiments on identically prepared, identical microscopic systems are nevertheless determined by statistical laws. For example, in the famous double-slit experiment, even though the separate electrons do not have any interactions with each other, nonetheless they manifest a lawful statistical behavior. Pauli invoked an analogy with the pre-established harmony of Leibniz's windowless monads to try to elucidate this irreducible fact about the behavior of microscopic and sub-microscopic systems. Thus in a letter (dated September 26, 1949) to his colleague and former student Markus Fierz, Pauli emphasized the connection to Leibniz's monads: "The statistical behavior of identical individual systems, that have no contact with one another (windowless monads), and which are seemingly not causally constrained, is nonetheless understood as an ultimate, irreducible fact in quantum mechanics."[93]

Bethe likewise became committed to a belief in the pre-established harmony of mathematics and physics. He accepted the Kantian legacy that it is impossible to describe the world in its totality, and that all we can achieve are partial, delimited descriptions and representations; and that the part of the physical world which is accessible to us is made legible by mathematics. Mathematics was the language in terms of which the "intellectual mastery of nature" was to be achieved. He was very much aware of Hilbert's legacy regarding pre-established harmony. Thus when reviewing the English language version of the well-known "Courant-Hilbert" *Methoden der Mathematischen Physik*, written in the early 1920s, "a period when mathematicians and physicists were clearly diverging," Bethe commented that when it was published in 1924 "it succeeded admirably in its purpose of making physicists aware of the essential unity of the mathematical methods which they were employing in widely differing fields."[94]

To believe in a pre-established harmony of mathematics and physics is to commit oneself to a version of mathematical Platonism. The latter is the metaphysical view that abstract mathematical objects exist independent of any context, that their existence is independent of us, of our language, thoughts, and practices. The attribution of abstractness to mathematical objects means that they are to be considered nonspatiotemporal and not causally efficacious entities. Platonists believe that just like chemical elements, stars and planets exist independent of us; and Platonists do not address the issue of the context dependence of these entities. Numbers, sets, functions, topologies, and so forth also exist independent of us. And just as statements about chemical elements, stars, and planets are made true or false by their (seemingly) objective properties, so are statements about numbers, sets, functions, and other mathematical elements. Mathematical truths are therefore discovered, not invented. For Platonists, reality is not exhausted by the physical world. Therefore, if mathematical Platonism were true, it would imply that we have knowledge of abstract objects since we undoubtedly possess mathematical knowledge. This, in turn, would pose a great challenge to naturalistic theories of knowledge.

The mature Bethe's position in these matters can be summarized as follows: He rejected the independence thesis, or better put, he didn't understand what it meant to assert that mathematical objects exist *independent* of us. He recognized the indispensability of mathematical objects when formulating and articulating our best physical theories—for example, the theories of relativity and quantum field theory—and their "fructiferous" qualities.[95] At the emotional level Bethe's belief in pre-established harmony replaced what in other people would be deism. It gave him the "reassurance he needed for stability of spirit."[96] At the pragmatic level pre-established harmony for him translated into indispensability and generativity. By virtue of their being indispensable and generative he believed that mathematical entities had a special status, that they could be considered *as if* abstract. But Bethe was agnostic as to whether one ought to have an ontological commitment to mathematical objects. And having been introduced to the complexity and diversity of the living world by his father, he did not believe that pre-established harmony translated into the existence of fundamental laws from which everything could be derived, as Einstein believed, nor that it was "the supreme task of the physicist . . . to arrive at those universal laws from which the cosmos can be built up by pure deduction." He did not believe that such laws could be formulated. Bethe would have agreed with Sommerfeld's assessment: "I was always content if I could explain mathematically a certain complex of phenomena, without worrying too much that there are other things that do not fit in."[97] Bethe's belief reflected Sommerfeld's influence but also Bethe's recognition of where his strengths lay.

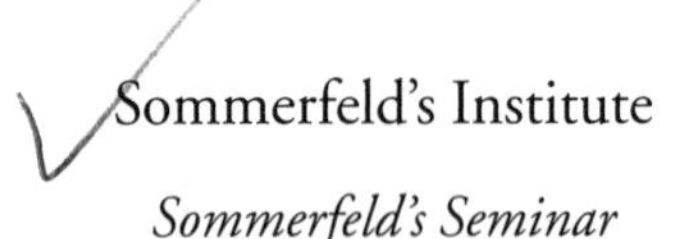

Sommerfeld's Institute

Sommerfeld's Seminar

Hans arrived in Munich at the end of April 1926, the beginning of the summer semester of the academic year 1925–26. The first thing he did was to look for a place to live and found "a rather nice one near the English Garden."[98] The university is on one side of the English Garden; Bethe's room was on the other, on Lerchenfeldstrasse. He could bicycle from there to the university in ten minutes.

Bethe recalled that before seeing Sommerfeld he was mostly concerned with whether Sommerfeld would allow him to register as a student in his seminar, that is, accept him as a doctoral candidate. He couldn't quite remember his first impression of Sommerfeld when they met except that Sommerfeld "was of very short stature" and had a Prussian demeanor. Sommerfeld's large, balding head with its lofty forehead towered over his smallish body; and as he always stood very erect he looked like "a Hussar officer," which was the way Pauli described him.[99] Hans presented him the letter of recommendation that Meissner had written, as well as a letter from his father, who had met Sommerfeld "casually at some place" in the rear of the Western Front during World War I. "I don't know what Sommerfeld did there, and he didn't pay any attention to my father's note, but he did to Meissner's, and said okay, you may come."

When Hans first met Sommerfeld he was a real *Bonze:* a forceful and charismatic personality who knew his standing in the profession and his status in the institutional framework. But though he was very much the *Herr Geheimrat,* the atmosphere in the seminar was characterized by the intellectual give-and-take between him and his students. During Bethe's stay in Munich Sommerfeld was always addressed as Herr Professor Sommerfeld. Bethe recalled the possibly "apocryphal" story about the encounter between James Chadwick and Sommerfeld in the early 1920s. Chadwick had been invited to Röntgen's institute for a two-month stay and "knew a little bit of German, but not very much, and on first meeting Sommerfeld addressed him as Herr Professor. Later Chadwick came to see Sommerfeld again and addressed him as "Herr Geheimrat Herr Professor Sommerfeld" and the story is that Sommerfeld then said, "Your German has improved tremendously."[100] To Hans's generation Sommerfeld gave the impression of being "rather unapproachable" because he was so very busy.[101] But Bethe quickly added, "he was somewhat unapproachable, and at the same time was very approachable. His seminar was wonderful, and he really took care of his students."

It is a measure of the man that Sommerfeld never felt threatened by the brilliant students he trained. He learned from them and from his *Assistenten* and collaborated with them. After they left the seminar they kept him informed of the latest developments. His students fashioned new tools that he, the master craftsman, learned to use. He thus continued to grow and to adapt to the new topography of theoretical physics.

An insight into Herr Geheimrat Professor Sommerfeld's dealing with students at the time Hans came to his seminar can be gleaned from Sommerfeld's interaction with the eighteen-year-old Subrahmanyan Chandrasekhar ("Chandra"), whom Sommerfeld met during his trip around the world in 1928.  Chandra, the third child in a well-educated Brahman family, was born in

1910. His uncle was the Nobel laureate Chandrasekhara Venkata Raman. He was first educated at home and already at a very young age showed unusual talents in mathematics. While growing up he avidly studied mathematics and physics. In 1927 he read the third edition of Sommerfeld's *Atombau und Spektrallinien,* which had been published in 1922 and whose translation into English had appeared in 1923. In the fall of 1928 Sommerfeld visited India and gave a lecture at Presidency College in Chennai (Madras), where Chandra was a student. Chandra went to the hotel in which Sommerfeld was staying and introduced himself. He told Sommerfeld that he was interested in physics and that he had read his book. Sommerfeld was clearly impressed by Chandra's mastery of the content of the third edition and by his passion for physics. He informed him that "the whole of physics had been transformed after the book had been written" and referred him to the discovery of wave mechanics by Schrödinger and other new developments resulting from the work of Heisenberg, Paul Dirac, Pauli, and others. Sommerfeld then gave him the galley proofs of the papers he had just written on electron theory of metals.[102] Sommerfeld's papers were an exposition of the application of Fermi-Dirac statistics to the conduction electrons in metals. Three months later Chandra sent Ralph H. Fowler a manuscript titled "Compton scattering and the new statistics." Fowler and Neville Mott made some suggestions to improve the paper, which Chandra readily accepted. It was published in the *Proceedings of the Royal Society of London* in August of 1929. Fowler was so impressed by what Chandra had done that he invited him to come to Cambridge for his doctorate once he finished his undergraduate degree. In 1930 Chandra sailed to England. His encounter with Sommerfeld was the beginning of a brilliant career that culminated with a Nobel Prize in 1983. Late in life, Chandra told Kameshwar Wali, his biographer: "From a purely scientific view, the most crucial incident [in my life] was my meeting with Sommerfeld when he visited Madras in 1928."[103]

In many ways Sommerfeld had done the same for the young Pauli and the young Heisenberg when they came to him. And Bethe received a similar welcome. Upon being admitted to the seminar he went to Sommerfeld's weekly meeting with his doctoral students and assistants. Schrödinger's papers on wave mechanics were just then being analyzed and discussed. Heisenberg's matrix formulation of quantum mechanics and its extension and generalization by Born, Heisenberg, and Jordan had appeared in the previous fall, but its abstract character and its unfamiliar mathematics had limited its acceptance. Schrödinger's approach used familiar mathematical methods that resonated with Sommerfeld. And as had been the case with Felix Klein, Sommerfeld had his *Doktoranden* and *Assistenten* make presentations of Schrödinger's papers as they were received in Munich to be published in the *Annalen der Physik.* Appendix C briefly describes the development of quantum mechanics.

"I am eternally grateful," Bethe commented, "that I came in just at that time. I never needed to learn all the contortions of the old quantum theory. We discussed Schrödinger's papers, and every seminarian had to give a talk, and it fell on me to talk about Schrödinger's second paper on perturbation theory, and I learned that very well, and I have used it ever since."[104] In his interview with Thomas Kuhn in 1964 Bethe emphasized that "in the beginning" Sommerfeld and Wentzel "were just interested in that you could do all this [with Schrödinger's wave mechanics], and one of the fascinating things was the Stark effect [the perturbation of the spectrum of an atom by the presence of an electric field]—that one could do the Stark effect and do the intensities . . . I think this first seminar was simply fascinated by the fact that you could do all this." And the reason for Sommerfeld's preference of Schrödinger over Heisenberg was because "we could calculate with Schrödinger."[105]

Bethe commented further that "it was the job of Sommerfeld's *Assistent,* Karl Bechert, who worked on the theory of spectra, and was very good at that, to coach each of the seminar speakers before he gave his seminar. But, I think I thought that I could do without. In my first semester in the seminar we simply went through Schrödinger's papers, one by one, in great detail, I think it took all semester." The fact that Hans felt he didn't need Bechert's help to make his presentation indicates that he had taken the measure of the other students in the seminar and had come to the conclusion that he could trust his assessment of his own abilities. It also indicates that contrary to some of the statements he made in his oral interviews, the training he had obtained in Frankfurt University, particularly in mathematics, but also in the courses he took with Madelung and Gerlach, had equipped him with a powerful set of mathematical tools. His training had given him an excellent introduction to classical physics and to the experimental side of atomic physics. His self-confidence grew in the seminar: "I felt very competent after reading Schrödinger's papers. I did follow the literature, though I don't recall precisely how carefully. I'm sure I read papers by Heisenberg and I'm sure I later read his papers on the helium atom, and the uncertainty principle. I know I followed the literature very carefully after my doctor's degree. I read most of the papers in the *Zeitschrift für Physik* on quantum mechanics or its applications. I read many papers in the *Proceedings of the Royal Society,* many papers in the *Physical Review.*"

In characterizing further the atmosphere of the seminar Bethe commented that "Sommerfeld would of course know exactly what was going on, but he would always ask stupid questions, and that made everybody feel at home, and so everybody asked questions. It was a very instructive seminar." Another characteristic of Sommerfeld's seminar when Hans came there as a graduate student ought to be noted: it was self-contained; there were very few outside speakers who came to lecture. And except for the American postdoctoral students who came there to learn—Pauling, Condon, and Rabi, for example—all the students were German.

Hans's first semester in Munich was taken up by the seminar mastering Schrödinger's papers, two courses given by Wentzel, and Sommerfeld's lectures on the partial differential equations of physics.[106] The latter was the first of Sommerfeld's lecture courses that Hans took, and he would later characterize it as "the best" of Sommerfeld's six courses—"best" in terms of content. All of them were carefully prepared and crafted, and delivered with characteristic aplomb. Walter Elsasser was a student in Munich for three semesters starting in the fall of 1923 and took three of Sommerfeld's lecture courses. He characterized the one on mechanics as "the best university classes [he] ever attended."

> Sommerfeld lectured without a textbook or without notes given out to the students . . . He never used an eraser in his classes. He would stand in front of a very large fourfold blackboard that covered one of the walls in the classroom . . . In a meticulous hand he would write formulas as he developed them; sections were separated from each other and some formulas emphasized by boxing them in with straight lines. At the end of the hour, the boards were almost filled with formulas, and no matter how difficult the mathematical process, it seemed a major esthetic experience.[107]

The subjects of the courses that Hans took with Wentzel were molecular spectra and dispersion theory. Dispersion theory is concerned with the problems of the scattering of light by atoms, the dependence of the scattering on the frequency of the radiation and the energy levels of the atom, and more generally, with the theories and models that help explain the passage of light and X-rays in matter: their reflection, diffraction, refraction, and so on. In his interview with Charles Weiner Bethe recalled,

> The most impressive staff member [in Sommerfeld's institute] at that time was Wentzel. I remember that Wentzel gave a beautiful course on dispersion theory, starting with the old Lorentz-Drude dispersion theory and going through the Kramers-Heisenberg paper [that had led to Heisenberg's formulation of quantum mechanics in 1924] and the dispersion theory as deduced from the Schrödinger equation. So we discussed Schrödinger's papers. In fact, I read Heisenberg's papers [on quantum mechanics] only about a year later. My first introduction to Heisenberg's theory was in Schrödinger's paper about the connection between his theory and Heisenberg's. I only read the papers leading up to Heisenberg's theory, like Bohr, Kramers and Slater many years later.

Even though he only took two courses with him, Wentzel influenced Bethe deeply.[108] After the advent of quantum mechanics in 1925, Wentzel's research was concerned with ascertaining how far the familiar methods of the "old" quantum mechanics could be recaptured or rederived within the new matrix and wave mechanics. In 1926, after Born in his formulation of the wave-mechanical description of collision processes had advanced his interpretation of Schrödinger's wave function as a probability amplitude, Wentzel worked on the formulation of the wave-mechanical description of radiation-induced transitions. In an important paper published in the fall of 1926, Wentzel gave the first wave-mechanical treatment of the photoelectric effect in atoms. He had also been the first to indicate that to the lowest approximation, the scattering of charged particles by a Coulomb field was again given by the formula that Rutherford had derived on the basis of classical mechanics. It was in Wentzel's course that Bethe had first been exposed to Born's crucial work on scattering theory and learned the approximation methods to deal with its mathematical formulation. Sommerfeld's lectures on wave mechanics gave Bethe the technical tools to master Schrödinger's papers and to improve on some of their results. Wentzel's course introduced him to all the problems concerned with the interaction of electromagnetic radiation and matter that the new quantum mechanics had to address—this at a time when the electromagnetic field was still described classically.

1926–27

The long summer vacation of 1926 was spent climbing mountains in the Italian Tyrol with Hans's uncle Erich and uncle Martin, and with his cousin Inge,

Martin's daughter. When Hans came back to Munich for the fall semester he and Fritz von Bergmann, one of the young men he had gotten to know in the *Tanzstunde* who had come to Munich to study medicine, rented two nicely furnished rooms from an elderly couple in a house located midway between the university and the medical school. They got along well, and Fritz introduced Hans to his circle of friends.

Most of Hans's time that second year in Munich was taken up attending and participating in Sommerfeld's seminar; taking his lecture courses; fulfilling his duties as teaching assistant in Sommerfeld's lecture course on mechanics, which in part entailed correcting the students' papers; and taking various courses in experimental physics and mathematics to fulfill the requirements for his degree in theoretical physics. He took a course by Constantin Carathéodory on the mathematical foundations of thermodynamics, and thereafter studied Planck's book on the subject "because Carathéodory's presentation was very formal." He also took an advanced course on complex variables given by Oskar Perron. He enrolled in courses on physical chemistry with Kasimir Fajans, which he found very interesting although he didn't think "they were quite that good." He also took two of the required lecture courses on experimental physics and the required advanced experimental course that Willy Wien gave in the Institute for Experimental Physics.[109] He found Wien's course "quite interesting" and subsequently of great help to him. He told Kuhn that, having taken them, "I think I know a little bit how well you can do an experiment, what troubles you get into, what accuracy you can expect and so on. I am very fond of designing experiments, interpreting experiments, and I think I know a little more than most theoretical physicists what can be done and what cannot be done."[110]

During that year Hans attended Sommerfeld's lectures on electrodynamics and on optics, read "from cover to cover" Sommerfeld's books on these subjects, and did all the exercises in them and those assigned in the courses. Sommerfeld's *Optics* includes an exposition of his solution of the problem of the diffraction of electromagnetic waves at the edge of a perfectly conducting wedge bounded by two intersecting semi-infinite planes. L. D. Landau and E. M. Lifschitz in their *Electrodynamics of Continuous Media,* a very sophisticated, advanced exposition of electrodynamics, characterized Sommerfeld's presentation in his *Optics* of his "very complex mathematical theory" of diffraction "as beyond the scope of [their] book," a book that takes the student through some very intricate and difficult contour integrations. During that second year Hans also attended Robert Emden's lectures on the astrophysics of the sun. And occasionally he would go to the colloquia in Wien's Institute for Experimental Physics because there were no joint experimental-theoretical colloquia.

Relations between Wien and Sommerfeld were cordial on the surface. Sommerfeld acted as co-editor of the *Annalen der Physik.* He was responsible for theoretical physics submissions, even though Max Planck edited the *Annalen* with Willy Wien. But there were clearly tensions between Sommerfeld and Wien. Walter Heitler, who was a student in Munich from 1924 to 1926, working on a doctorate with Karl Hertzberg, noted that Sommerfeld and Wien would always sit in the first row of any colloquium they would jointly attend, but at opposite ends of the row.[111] One reason for the tension between them was probably due to their political differences. Though very conservative, Sommerfeld never had any sympathy for the Nazis. According to Heitler, already in 1924 Wien had made explicit his support of Hitler. Walter Elsasser, who had left Heidelberg for Munich because of Philipp Lenard's affinities for the Nazis, corroborates this. The Nobel laureate Wien had built a quite substantial laboratory for those days, and had five *Assistenten* working for him. Elsasser had gone to Wien's institute in the fall of 1924 with the intention of getting a doctorate in experimental physics. Early in his third semester in Munich, Elsasser was approached "somewhat officiously" by one of Wien's *Assistenten,* who pointed out to him certain things that evidently were not apparent on the surface: "Every single member of the faculty with the exception of the director and himself was a card carrying member of the Nazi party." He advised him that "it might be decidedly better" if he were to study at another university, "Göttingen for example," which was not only very good but was also "full of Jews."[112] Bethe was to confront such overt anti-Semitism in Sommerfeld's institute only in 1933.

The full list of the courses that Hans took in Munich is given in Appendix D. It is interesting to note that during Bethe's second term in Munich, even though he was taking what surely was a very heavy load of physics courses, he found time to take a course in psychology and one in politics. The course in psychology was probably to help him understand better the unfolding events at home—the separation and divorce of his parents—and to gain some further self-understanding; and the one in politics to make him a more educated citizen.

Attending Sommerfeld's lectures on his current research in the fall of

1926 proved to be a very stimulating and rewarding experience for the young Bethe. Sommerfeld at the time was writing up his *Wellenmechanischer Ergänzungsband*,[113] the first thorough exposition of wave mechanics and of some of its applications, and he scheduled a set of special lectures in order to present a good deal of the advanced materials to the seminar. Hans assiduously attended them. To Kuhn Bethe remarked, "This was the greatest characteristic of the Sommerfeld group: we were not made aware of great puzzles. We were given the impression that there was a wonderful new tool. Now you could do things, now you could solve all these interesting problems like all the complicated atoms and so on and chemistry."[114]

Peierls characterized Sommerfeld's teaching methods in a similar way: "It was characteristic of Sommerfeld's positive attitude [that he believed] that one learned more from the successful solutions of difficulties than about the mysteries that remained. He was completely fair in listing the contradictions. It was just a matter of emphasis."[115]

Reflecting on the seminar and on Sommerfeld's lectures, Bethe added,

> [Sommerfeld] never let you forget that physics was an empirical subject, and that, although in doing theoretical work you were concerned with laws that could be expressed in mathematical terms, you always had to be clear about their empirical basis. He was also a master at mathematical techniques, and managed to make very sophisticated methods transparent. But his greatest strength lay in the way he could guide research students. He always managed to find problems that were interesting and difficult enough to be worth a serious effort, yet not too difficult for the student . . . Sommerfeld was very pragmatic. I had that tendency even before coming to the seminar. There was practically no discussion of the general problems of quantum mechanics. I suppose we must have discussed the uncertainty principle when it came out. I can't believe that we didn't, but I don't remember it . . . From Sommerfeld, I got the general attitude that problems are there to be solved, and quantum mechanics is ready to solve any problem.

Bethe thus corroborated Suman Seth's characterization of Sommerfeld's physics as "physics of problems." The fact that there was not much discussion of the uncertainty principle and other philosophical issues concerning the new quantum mechanics in Sommerfeld's seminar is a reflection of Sommerfeld's pragmatic inclination and of his mathematical approach to problems. But an additional comment is in order.

Sommerfeld had attended the September 1927 Como meeting at which Bohr had presented his complementarity principle.[116] Heisenberg's uncertainty principle, Bohr's complementarity principle, and more generally the conceptual and interpretational framework of quantum mechanics were the central concerns of the fifth Solvay Congress held in Brussels just one month later, in October 1927.[117] Lorentz, the organizer of the congress, had chosen the theme of the conference: "Electrons and Photons." For the first time since World War I the twenty-nine invited participants included German scientists, and in particular all the major theorists who had contributed to the developments of quantum physics and the formulation of the recent wave and matrix mechanics: Planck, Bohr, Debye, Einstein, Paul Ehrenfest, Born, Louis de Broglie, Hendrik ("Hans") Kramers, Heisenberg, Pauli, Dirac, Schrödinger—*but not Sommerfeld.*

The reason for the exclusion of Sommerfeld from the conference had to do with his political views and his political activities during World War I. Sommerfeld was very conservative and had been very nationalistic. During the war he had advocated the annexation of Belgium by Germany and in 1915 had signed the "Proclamation" that Willy Wien had drawn up.[118] Wien, who believed that England was the Reich's worst enemy, accused British physicists of having appropriated discoveries made in Germany and demanded in his "Proclamation" that physicists have nothing to do with British journals.[119] The wounds of that conflict among the Entente countries, particularly France and Belgium, had not healed by 1927.[120]

Planck, who had signed the "Manifesto of the 93 Intellectuals" repudiating the accusation that the German army had committed atrocities against Belgian civilians, had been invited to the 1927 Solvay Congress. But the invitation was likely extended because in an open letter to Lorentz in the spring of 1915, a letter published in several newspapers, he courageously repudiated the "Manifesto" itself after he had learned from Lorentz about the brutality of the invasion and what had happened after the German Imperial Armies invaded Belgium in August 1914. One of the first casualties of the German's Belgian campaign had been the University of Louvain, one of the oldest European universities, established by papal charter in 1425. The shelling of the city by the German army had burned the university's library. Over 1,000 incunabula and some 300,000 books were destroyed. The bombardment of an open city and the sacking of the library came to symbolize the irrationality of war and kindled a sense of outrage among the Allies. But the offense and the injury were aggravated by the October 1914 manifesto. Planck further redeemed himself by trying to ease the plight of Belgian academics during the war.[121] Sommerfeld did not sign the "Manifesto of the 93 Intellectuals" as probably he was not considered distinguished enough at the time to be asked to be one of the signatories. Even so, his stand regarding the annexation of Belgium was evidently not yet forgiven in 1927 by the French and Belgian physicists who oversaw the Solvay activities.[122] Their objections were responsible for Sommerfeld not being invited.

Thus Sommerfeld was not present at the exchanges between Bohr and Einstein at the 1927 Solvay Congress, nor did he participate in the expanded exposition of their views in the proceedings of the conference.[123] Though Sommerfeld was pragmatic, it is nonetheless difficult to believe that he would not have reported on the exchanges between Bohr and Einstein and their meaning for the interpretation of quantum mechanics had he been in attendance at the meeting.

In 1949 the eighty-year-old "unphilosophically inclined" Sommerfeld wrote up his lecture notes on optics and published them. In the book he had occasion to comment on the dualistic nature of light. His position can be stated as follows: As long as there is no logical inconsistency, nor any contradiction in the quantum theoretical attribution of wave and corpuscular properties to matter and light—and Heisenberg's uncertainty principle guaranteed this—we must accept the new understanding of the world pragmatically:

> Light has a *dual nature;* it presents us with either its corpuscular or wave aspect depending on the particular question we are posing. It is wrong to ask which of these is the *true* one. As far as we know today, *they are both on an equal footing* and only both aspects taken together are capable of representing the nature of light completely. One speaks therefore, not of a *duality* of light but, more appropriately of its *complementarity* . . . It is clear that this complementarity overthrows the scholastic ontology. What is truth? We pose Pilate's question not in a skeptical, antiscientific sense, but rather in the confidence that further work in this new situation will lead to a deeper understanding of the physical and mental world.[124]

Reminiscing about Sommerfeld Bethe commented that Sommerfeld was in frequent contact with Heisenberg and with Pauli and had very good relations with them.

> Sommerfeld was the only person for whom Pauli had respect. He was very deferential to Sommerfeld, and always addressed him as Professor Sommer-

feld. Heisenberg was not so deferential, but he liked to have Sommerfeld's approval. Sommerfeld had good, but not close relations with Bohr. They were on friendly, but rather distant terms. He had very good relations with Born, and he went to Göttingen at least once a year. But he never suggested to me to go to Bohr. He never suggested that I should go to Göttingen. He never suggested to any of us to go to either place—Copenhagen or Göttingen—for any length of time.

Sommerfeld's Free-Electron Theory of Metals

In the fall of 1926 Pauli had made a crucial calculation in which he tentatively applied Fermi-Dirac statistics to describe the collective properties of the conduction electrons in a metal. The model of a metal he used was the one that Drude and Lorentz had formulated at the beginning of the century.[125] In the Drude-Lorentz model the mobile negative carriers were electrons that were assumed not to interact with one another, and their velocity distribution when at temperature T was described by a Maxwell-Boltzmann velocity distribution that stipulated that the number of electrons per unit volume with velocity in the range between $\boldsymbol{v}$ and $\boldsymbol{v} + d\boldsymbol{v}$ was proportional to

$$f_{MB}(\boldsymbol{v}) = n\left(\frac{m}{2\pi k_B T}\right)^{3/2} e^{-mv^2/2k_B T},$$

where n is the number density of the electron gas, $n = N/V$, N being the number of electrons in the volume V. The positively charged ions were assumed to remain fixed at their equilibrium positions. A quantum mechanical description of the "free" conduction electrons constrains the states that are available to them. Each electron can occupy the one particle state that corresponds to being free in the volume V of the metal. The Pauli exclusion principle limits the occupancy of these states (which are characterized by their momentum and a two-valued spin quantum number) to one electron in each state. The combined effect of wave mechanics and of the exclusion principle is to replace the classical Maxwell-Boltzmann distribution by the Fermi-Dirac distribution

$$f_{FD}(\boldsymbol{v}) = \frac{(m/\hbar)^3}{4\pi^3}\frac{1}{e^{(\frac{1}{2}mv^2-\mu(T))/k_B T}+1},$$

with $\mu(T)$ (the chemical potential) determined by the requirement that

$$n = \int f_{FD}(\boldsymbol{v})d\boldsymbol{v},$$

with n the number density. Thus at absolute zero there is one electron in each of the possible states up to some maximum momentum. Since for most of the electrons the nearby states are occupied, only electrons near the top of the distribution are perturbed by any external electric or magnetic field, or by an increase in the temperature. This is also true at finite temperatures T, when $k_B T \ll \mu(T)$. This was the basis of Pauli's explanation of the weak paramagnetism of metals: at low temperatures the temperature dependence of the Fermi-Dirac distribution is such that only electrons near the top of the distribution are excited, and since these are but a very small fraction of the total number of electrons, they contribute but little to the magnetic properties of the metal at that temperature.[126]

In the spring of 1927 Sommerfeld attended a meeting organized by the physics faculty of the University of Hamburg at which he delivered a lecture, the "Present State of Atomic Physics." At the railway station before he was departing to go back to Munich Pauli showed him the proofs of his paper on Fermi-Dirac statistics that he had submitted to the *Zeitschrift für Physik.* In that article he had adduced the weak paramagnetism of metals as a test case of whether electrons obey Fermi-Dirac statistics. He did not state explicitly that electrons do in fact obey Fermi-Dirac statistics; and although he pointed out that at low temperatures the electron gas is highly degenerate—meaning that almost all the levels up to the Fermi energy are filled except near the very top of the distribution—he did not indicate how this explained why the conduction electrons gave no appreciable contribution to the specific heat of a solid. Sommerfeld immediately appreciated the implications of Pauli's ideas and recognized that one could make further applications to other metallic properties using Pauli's insights, thereby removing many of the outstanding difficulties of the older theory.[127] He saw Pauli's paper as a fundamental advance in solid state theory.[128]

Upon his return to Munich Sommerfeld set to work on the application of Fermi-Dirac statistics to the theory of conduction electrons in metals and "gave some advanced lectures on that topic. In these lectures, he reported pretty much what he was doing, day by day."[129] Hans attended these lectures faithfully and kept detailed notes on them. They were his introduction to modern solid state physics. As in the Drude-Lorentz model Sommerfeld assumed that the conduction electrons in a metal did not interact with one another or with the ions. They were thus described by the quantum states of a "free" particle in a box of volume V (the assumed size of the metal), but obeyed the Pauli principle and therefore Fermi-Dirac statistics. In his lectures Sommerfeld indicated how Fermi-Dirac statistics explained the smallness of the contribution of the electrons to the specific heat of metals and what he considered its particular triumph: the explanation of the Wiedemann-Franz law.[130]

That the electrons could be described as "free" was clear from the success of the model and its good agreement with experimental data. The problem was to explain why the Sommerfeld model described some of the properties of metals so well, since electrons do interact with one another and with the ions that constitute the metal. Bethe recalled that he was dubious of the approach and only became convinced of its validity after he had derived the fact that electrons in a periodic potential had "quasi-free" wave functions, and after he had become acquainted with Felix Bloch's work on the conduction of electrons in metals.

The Davisson-Germer Experiment

In April 1927 Clinton Davisson and Lester Germer published in *Nature* their experimental findings on electron diffraction by single crystals of nickel. The data indicated, as Louis de Broglie had suggested in his doctoral dissertation, that electrons did behave like waves.

De Broglie had submitted his thesis, in which he attributed wave properties to corpuscular matter, to the Faculty of Science of the University of Paris in late November 1924. During the examination Jean Perrin had asked de Broglie whether these wave properties of matter could be experimentally verified. In his reply de Broglie indicated that "this should be possible by diffraction experiments of electrons by crystals."[131] Martin Klein, who investigated the response of the theoretical physicists to de Broglie's ideas, came to the conclusion that "de Broglie's arguments were not compelling for the majority of theoretical physicists, who [as a result of Compton's experiments] already had more than they could handle in the wave-particle duality for radiation."[132] However, this was not the case for Einstein, who had been sent a copy of de Broglie's thesis by Paul Langevin. Einstein, who at the time was involved in extending Bose's ideas on the statistics describing black body radiation to gases composed of atoms, immediately recognized the originality of de Broglie's ideas. He included a discussion of de Broglie's hypothesis in his second paper on the quantum mechanical description of gases.[133] Einstein also wrote Max

Born about de Broglie's ideas, and Born in turn obtained a copy of his thesis for the Göttingen library.

One of the seminars that Walter Elsasser attended when he came to Göttingen in the fall of 1924 was Hilbert's longstanding one, "The Structure of Matter." But because of Hilbert's illness, Max Born was in charge of it.[134] One of the first presentations in that seminar was by one of Born's students, Friedrich Hund, who reported on a paper by Davisson and Charles Henry Kunsman that had recently appeared in the *Physical Review.*[135] Davisson and Kunsman had directed a beam of low-energy—monoenergetic—electrons at a polished plate of platinum in a high vacuum and had measured the intensity of the electrons that had been back scattered from the plate as a function of angle. They found that the angular distribution exhibited maxima and minima—"a surprising and quite mysterious result." Hund reported that Born had suggested to him that the observed maxima and minima might be the result of the electrons interacting with the various electronic shells of the platinum atom and had asked him to investigate the matter further. Hund as yet had not carried out any calculations. The ensuing discussion captivated Elsasser—then a twenty-year-old graduate student in James Franck's experimental physics institute. Elsasser left the lecture with the distinct impression "that something rather significant for atomic structure might be hidden behind these intriguing experiments."[136] Not long thereafter Elsasser found in the library Einstein's two papers on Bose statistics applied to gases. In them Einstein had indicated that under certain conditions gases behaved like assemblies of waves and had referred to de Broglie's thesis that had just been published. To his surprise Elsasser discovered that the library had a copy of this thesis.

Elsasser proceeded to carefully study de Broglie's thesis. At Franck's suggestion he submitted a paper to *Die Naturwissenschaften* indicating that de Broglie's idea of attributing to electrons moving with velocity v a wavelength $\lambda = h/mv$ might explain the findings of Carl Ramsauer in his scattering experiments of slow electrons by atoms in 1922 and 1923 and those of Davisson and Kunsman in 1923 on the reflection of electrons from metallic surfaces. Since the de Broglie wavelength of slow electrons was of the order of 10^{-8} cm, the size of atoms and of interatomic distances, these experiments could be interpreted as diffraction effects. Elsasser's paper was sent to Einstein to referee, and he urged its publication. Elsasser went on to attempt carrying out an electron diffraction experiment but could not overcome the technical difficulties associated with obtaining pure crystals and a sufficiently high vacuum.[137]

In the summer of 1926, Clinton Davisson took a vacation in England and made a point to attend the Oxford August 1926 meeting of the British Association for the Advancement of Science to meet colleagues and friends. He there heard Max Born's lecture, "Physical Aspects of Quantum Mechanics," in which Born referred to Elsasser's paper that had suggested that Davisson and Kunsman's 1923 paper was evidence for the wave properties of electrons.[138] Not being aware of de Broglie's thesis or of the quantum mechanical papers of Heisenberg, Born, and Jordan, nor of those of Schrödinger, Davisson was startled by what he heard. Actually, since carrying out the Davisson-Kunsman experiment, Davisson, with the assistance of Lester Germer, had carried out experiments on the scattering of slow electrons (< 100 eV) on single crystals of nickel. In the experiment a collimated monoenergetic beam of electrons was directed at the (111) face of a nickel single crystal, which could be rotated about an axis parallel to the beam. The elastically scattered electrons from the surface were detected with a movable Faraday cup. The experimental procedure consisted in measuring the intensity of the scattered electrons for one setting of the crystal and collector as their energy was varied. They observed peaks in their angular distribution.

Davisson showed some of the preliminary results of these experiments to Max Born, James Franck, and Douglas Hartree, and they impressed upon him the importance of his results as confirmatory evidence of de Broglie's ideas. On the voyage back to the United States in the beginning of October 1926 Davisson assiduously studied Schrödinger's papers, determined to carry out single crystal experiments with high accuracy. The first of three papers describing the results of such experiments carried out by Davisson and Germer after Davisson's return from England was published in the April 19, 1927, issue of *Nature.*[139]

Their model of a crystal was the standard one adopted to explain X-ray diffraction from crystals. The crystal was considered built up by identical layers of atoms, with the plane of each layer parallel to the surface of the crystal, and each successive layer displaced relative to the previous one by a distance $\boldsymbol{a}_3$. Crystal structure meant that the atoms in each plane display a translational symmetry, the position of each being specifiable by a vector $l_1\boldsymbol{a}_1 + l_2\boldsymbol{a}_2$, with l_1 and l_2 integers.[140] X-ray scattering experiments had in fact shown that pure solid nickel is a face-centered cubic crystal. “Face centered” means that the crystal can be imagined to be constituted of cubic unit cells each having a Ni atom at each corner and one in the center of each face, with none inside the cube. As in the explanation of X-ray diffraction Davisson and Germer thought of their Ni crystal as being equivalent to a series of plane gratings piled on top of one another, separated by a distance that depended on the particular orientation of the surface face. For electrons of a given energy impinging normally on a face of the crystal, the angle θ at which diffraction peaks were to be observed—if the electrons had wave properties characterized by the de Broglie wavelength $\lambda = h/mv$, with the velocity determined by the voltage they have been accelerated with—could then be calculated by analogy with the diffraction of X-rays by using the diffraction grating formula $n\lambda = d\sin\theta$, with d the distance between lines of the grating and n an integer. Several orders of diffracted beams should be observed (for $n = 1,2,3 \ldots$) for that particular d. Since the atoms in the next plane parallel to the surface are displaced with respect to the one above, the d' for that grating is different and maxima ought to be observed at θ' such that $n\lambda = d'\sin\theta'$. The grating distances, d, d', were in fact taken from X-ray data. In their *Nature* article, Davisson and Germer expressed the results of their experiments in a tabular form in which they compared the angles at which maxima occurred with their low-energy electrons and with X-rays of the same wavelength. The angles agreed for electrons falling perpendicularly on the nickel surface. In the case of electrons falling obliquely on the nickel crystal surface, the observed peaks at a given angle and given energy were considered the result of constructive interference from electron waves reflected from different atoms at different points in successive layer planes *inside* the crystal (the von Laue viewpoint). For the oblique case an ad hoc “contraction factor” of 0.7 had to be applied to the wavelength of electrons inside the crystal in order to obtain agreement with the Bragg reflections.

Sommerfeld read this paper by Davisson and Germer. Because the diffraction of waves by crystals was of very special interest to Sommerfeld ever since the 1912–1914 von Laue–Bragg interpretation of X-ray diffraction by crystalline structures,[141] Sommerfeld gave a series of lectures on the Davisson-Germer experiments and their findings in his seminar. After the final one, Sommerfeld asked Hans to try to explain why in their experiments with obliquely incident electron beams Davisson and Germer could not attribute to the electrons the de Broglie wavelength $\lambda = h/mv$ corresponding to the velocity, v, with which they impinged upon the crystal.[142] Undoubtedly, Sommerfeld also wanted to see whether he would accept Hans as a *Doktorand.*

By the time Bethe was ready to write up his interpretation of the Davisson-Germer data three papers had appeared in the United States addressing the issue: Eckart 1927, Patterson 1927, and Zwicky 1927. Bethe was aware of Arthur Lindo Patterson's paper, for he cited it in his paper on the subject, which he submitted to *Die Naturforschungen* in late July 1927. It was published in November of that year. Carl Eckart, in his paper, had proposed that ascribing a refractive index to the electron-crystal interaction was the most straightforward way to account for the discrepancy in the oblique case. But Bethe had not seen Eckart's paper, and like Eckart, he too proposed describing the motion of electrons when in the crystal by introducing an index of refraction. The Schrödinger equation assigns a wavelength

$$\lambda = \frac{h}{\sqrt{2m(E-V)}}$$

to an electron with energy E when in a *constant* potential V. In analogy with the index of refraction for the electromagnetic waves, Bethe defined the index of refraction for electrons as

$$\mu = \frac{\lambda_{\text{vacuum}}}{\lambda_{\text{crystal}}} = \frac{v_{\text{crystal}}}{v_{\text{vacuum}}} = \frac{(E-V)^{1/2}}{E^{1/2}} = \left(1-\frac{V}{E}\right)^{1/2}$$

where V is the (constant) potential the electron experiences inside the crystal. For X-rays $\mu < 1$, and Bethe uncritically accepted that this was also true for electron waves. This meant attributing a longer wavelength to the electron when inside the metal. Bethe thus concluded that V is positive and approximately equal to 20 volts—which was difficult to reconcile with other descriptions of electrons in metal. In fact, a positive V would expel electrons from the inside of the metal. Nonetheless, Bethe wrote up his findings and published them.[143]

In his interviews with Charles Weiner in 1967, and with Lillian Hoddeson in 1981, Bethe commented that "[the X-ray crystallographer] A. L. Patterson and others in the United States found the same solution [as I had], so nobody paid attention to my paper"[144] and added that he was "rather dismayed" when he found out that he had been scooped. This was "a great disappointment for a physicist who had just published his first paper."[145] Bethe was very ambitious. Being "scooped" had clearly been a blow to him. That three physicists were able to publish before him was bad enough. In addition, it turned out that in his paper Bethe had made the dubious physical assumption that V inside the crystal was positive.

It is quite remarkable that Bethe was able in less than three months to write a paper on a subject which required him to have a thorough understanding of all that was known at the time about the structure of crystalline solids and to become highly skilled in the interpretation of data from X-ray diffraction experiments to determine the structure of crystals. What had made this possible? Undoubtedly, necessary conditions for this were Bethe's off-scale ability to absorb a huge amount of technical data and synthesize it into a coherent whole, his remarkable memory, and his great powers of concentration. But there were other enabling conditions. The courses he had taken with Wentzel had taught him how the quantitative description of the scattering of electromagnetic radiation off single atoms and molecules could be used to explain the dispersion of light in gases. And of great importance was the fact that during the summer 1927, while working on the problem, he was taking Heinrich Ott's course on X-ray crystallography. Ott had been a Sommerfeld *Doktorand,* and his thesis in 1924 dealt with the theory of crystal structure as determined by X-ray scattering. Ott had stayed on in Munich as one of Sommerfeld's *Assistenten,* and had also done his *Habilitation* with Sommerfeld. He was a *Privatdozent* when Bethe came to Munich. The course Bethe took with Ott was based on the article Ott was writing for the *Handbuch der Experimentalphysik* on the determination of the structure of crystals by X-ray diffraction.[146] Ott's completed article, published in 1928, was 322 pages long and covered every aspect of X-ray crystallography: the description of crystal systems in terms of lattices with basis and the associated lattice planes; the symmetry of space lattices; a full exposition of the geometry and mathematical properties of the corresponding reciprocal lattices; the scattering and structure factors from single atoms; the scattering from two- and three-dimensional gratings; the kinematical and dynamical theory of X-ray diffraction due to von Laue, Bragg, and Ewald; a detailed exposition of the instruments and apparatus used in X-ray diffraction experiments; the determination of crystalline and molecular structure from X-ray diffraction data; and much else.[147] It is clear that Ott's course taught Bethe a great deal about crystal structure, and more important, how to interpret the angular distributions and intensities of the scattered beams as functions of energy and angle of incidence in the X-ray case—knowledge that was immediately applicable in the electron diffraction case. Bethe's articles and his thesis reflected this knowledge, but he did not acknowledge Ott for his contributions. At the end of the first *Naturwissenschaften* paper Bethe stated: "This problem was posed to me in connection of a course of lectures (*Kolleg*) by Herr Prof. Sommerfeld, to whom I want to express my thanks for his helpful interest during the progress of the work." This acknowledgment reflects the necessity and importance of thanking the institute director in the German institutional context. It is not clear whether Sommerfeld did more than occasionally express interest. In his interview with Thomas Kuhn Bethe indicated that he received very little help from Sommerfeld while doing his thesis. He remarked, "I didn't see much of Sommerfeld aside from the seminar. I saw him once when he assigned the thesis and twice while I was doing the thesis and not much outside that . . . There were occasional outings when we went to the café in the Hofgarten but I don't think I was involved in that [until later] when I came back [from Stuttgart in 1929], and that happened maybe two or three times a semester." Even though the assumption $\mu < 1$ was physically implausible, Sommerfeld liked Bethe's first paper and asked him to "do a general theory of electron diffraction in crystals as a thesis, and recommended Ewald's theory of the diffraction of X-rays by a crystal as a model." Sommerfeld knew well Ewald's dynamical theory of X-ray diffraction—the papers Ewald published in 1916 and 1917 in the *Annalen der Physik* on the topic had been Ewald's *Habilitationsschrift* with Sommerfeld. Ewald thereafter was a *Privatdozent* in Sommerfeld's institute until 1921, where he gave several *Spezialvorlesungen* on the dynamics of crystal lattices and the diffraction of electromagnetic waves by crystalline solids.[148]

It was only after Bethe had almost completed his doctoral dissertation, which consisted in the working out of a detailed quantum mechanical dynamical theory of electron diffraction—in which the periodic potential that an electron experiences when inside the metal was taken into account—that Bethe submitted his second *Naturwissenschaften* article. In this paper, submitted on May 11, 1928, a consistent kinematical model was advanced. Since the article was designed for nonspecialists, Bethe still described the Ni crystal by the Drude-Lorentz model. But he now assumed that the incident electron when inside the Ni crystal experiences a *negative* constant potential, the same potential that prevents the conduction electrons in the metal from escaping. As in the photoelectric effect, this potential is related to the "work function" of the metal, the amount of energy necessary to extract an electron from the metal. Since the impinging electrons experience a negative potential within the crystal, their kinetic energy is greater there, and therefore their de Broglie wavelength is shorter than outside the crystal. This explained the discrepancy in the angular positions of the observed maxima, and good agreement was found with the data of Davisson and Germer's experiments with electrons up to some 360 eV assuming the same inside constant (negative) potential V of about 15 eV.[149]

Electron and X-ray Diffraction

Having been assigned by Sommerfeld the subject of his doctoral dissertation and the approach to be taken in addressing it, Bethe studied "quite diligently" Ewald's papers on the foundation of crystal optics and his dynamical theory of X-ray diffraction. Paul Ewald, in three papers written while he was a soldier in the German army serving as an X-ray technician on the eastern front, had formulated a dynamical, microscopic theory of X-ray scattering by a crystal. Von Laue in 1931 declared that these papers "belong to the masterpieces of mathematical physics."[150]

When recalling his thesis Bethe commented that for him the Davisson-Germer experiments had proven the wave nature of electrons. They had indicated that electrons were "free" inside the Ni metal and behaved very much like X-rays. These conclusions stemmed from the fact that the experimental data could be phenomenologically explained by ascribing to the scattered electrons a *constant* de Broglie wavelength outside and inside the crystal. When inside the crystal the phenomenological model assumed that the electrons did not interact with the "free" valence electrons but only with the ions. A dynamical description therefore had to explain how the ascription of the electron seeing a constant negative potential inside the crystal could be reconciled with the more accurate picture of the electron experiencing a *periodic* potential there.

It is also interesting to note that when writing his thesis Bethe believed that the theory of electron diffraction by metallic crystals was "simpler" than the theory of X-ray diffraction since electrons are described by the Schrödinger equation. One is then dealing with a scalar field and not, as in the case of X-rays, with electric and magnetic fields that are vector fields: "So it was easy to translate Ewald's dynamic theory into electron diffraction." What is to be noted is that Bethe—and Sommerfeld—at the time considered the Schrödinger equation as describing a wave field, and that the wave-mechanical description of the "diffraction" of electrons by crystals was amenable to a mathematical description identical with the one given by Ewald in his classical description of X-ray diffraction by crystals, in which the X-rays are represented as an electromagnetic wave of a given wavelength interacting with the dipoles making up the crystal. Born's scattering theory based on the Schrödinger equation was interpreted as stating that every element of volume d^3r in which the potential V (due to the crystal) acts gives rise to secondary ψ waves whose amplitude at unit distance from d^3r is proportional to $V\psi d^3r$ where ψ is the amplitude of the Schrödinger wave at d^3r; and if the potential V is weak, ψ may be taken as the amplitude at d^3r of the unperturbed incident wave. This is Born's first approximation, the initial approximation Bethe used. But in fact the interaction of an electron with the ions of a crystal and with the "gas" of valence electrons is much stronger than that of electromagnetic waves and makes the justification of the underlying assumptions of the electrons moving freely inside the crystal much more difficult to justify.

Like Ewald, Bethe adopted a model for the description of Davisson-Germer electron-crystal scattering experiments that assumed the atomic ions in the crystal to be located on a Bravais lattice. The crystal was assumed semi-infinite, the plane $z = 0$ being the boundary of the crystal. The model thus attempted to describe a real experiment with an incident beam on the $z = 0$ plane.[151] In the model the combined effect of the ions and the conduction electrons was assumed to generate, for $z > 0$, a periodic potential,

$$V(\boldsymbol{r}) = \sum_n v_n(\boldsymbol{r} - \boldsymbol{R}_n),$$

where $v_n(r - R_n)$ is the potential generated by an ion at the position R_n on the lattice. Outside the crystal, i.e., for $z < 0$, $V(r) = 0$. As in Ewald's model, the $v_n(r - R_n)$ represented not the potential of isolated ions but that of ions immersed in a sea of conduction electrons. And since Bethe was considering the specific case of the scattering of electrons off a monoatomic metal, all the $v_n(r)$ were the same $v(r)$.[152] This potential was static, assumed not to be affected by the passage of the incident electron in the neighborhood of the ion. The form of $v(r)$ was stipulated and given as a component of the model. Part of Bethe's dissertation consisted in a derivation of an appropriate potential $v(r)$ for Ni ions that would be used in the model. The analytical representation of $v(r)$ was to be simple enough so that it would be possible to calculate structure factors and the intensities of the scattered beams.

The Hamiltonian that defined the model is given by

$$H = -\frac{\hbar^2}{2m}\nabla^2 + V(\boldsymbol{r})$$

with

$$V(\boldsymbol{r}) = \sum_n v(\boldsymbol{r} - \boldsymbol{R}_n)$$

for $z > 0$. The model thus entailed describing quantum mechanically the motion of an electron in a periodic potential. Bethe established that the wave function inside the crystal by virtue of the symmetry of the potential

$$V(\boldsymbol{r}) = V(\boldsymbol{r} + \boldsymbol{l}) \quad ; \quad \boldsymbol{l} = l_1\boldsymbol{a}_1 + l_2\boldsymbol{a}_2 + l_3\boldsymbol{a}_3$$

must be of the form $\exp(i\boldsymbol{k}\cdot r)\ u_k(r)$, with $u_k(r)$ having the periodicity of the potential, a result independently arrived at by Felix Bloch when he addressed the problem of the electrical conductivity of metals. Bloch showed that such a wave function implied that electrons behaved like free particles with an altered energy-momentum relation.

Given the nature of the solutions for $z > 0$ and $z < 0$, their symmetry properties, and the boundary conditions they had to satisfy in order to describe scattering, Bethe, like Ewald, established that for certain directions and energy intervals of the incident electron there would be peaks in the intensity of the beams of reflected electrons in certain directions; this by virtue of the translational periodicity of the crystal in the plane $z = 0$. Bethe also established that for certain incident directions and energy intervals there did not exist any propagating wave functions for the electron inside the crystal. However, Bethe did not recognize the connection between the forbidden intervals and the gaps between the energy bands of electrons in metals. This was realized only later, after Bloch's work had received wide acceptance.

A brief, somewhat technical exposition of Bethe's thesis that highlights what he had learned from Ewald's dynamical theory is given in Appendix E. It is based on the formulation of Bethe's work given by von Laue in his 1931 paper in which he criticized Bethe's assumption that a crystal could be considered semi-infinite and the potential inside the crystal exactly periodic through out. In fact, Bethe was aware of some of the problems associated with assuming the particular form $V(\boldsymbol{r}) = \sum_n v(\boldsymbol{r} - \boldsymbol{R}_n)$, and in his thesis discussed some of the limitations of this model. Patterson and Fritz Zwicky in their articles had pointed out that the incident electron induces a charge on the surface of incidence that is not taken into account by assuming the potential $V(\boldsymbol{r}) = \sum_n v(\boldsymbol{r} - \boldsymbol{R}_n)$ for $z > 0$. This is an important perturbation for low-energy electrons. Similarly, it was not clear how many layers of the crystal electrons with energies between 50 eV and 300 eV could penetrate, and whether the potential in these surface layers is accurately modeled by $V(\boldsymbol{r}) = \sum_n v(\boldsymbol{r} - \boldsymbol{R}_n)$. To validate some of the idealizations regarding the mathematical formulation of his model and some of the approximations he made in his calculations—this to derive numbers to compare with empirical data—Bethe limited his considerations to electrons whose energies were such that the ratio of their de Broglie wavelength, λ, to the characteristic distances, a, related to the structure of the crystal—a could be a length characteristic of the potential $v(r)$ of the ion, or it could be an interionic distance—was such that λ/a was very small. This allowed him to represent, as a first approximation, the wave function of the electron inside the crystal as a superposition of plane waves that incorporated the symmetry of the potential.

Bethe's results concerning the angular location of the maxima as a function of angle and of energy were in agreement with the experimental results of Davisson and Germer. But this was not the case for the intensity data. The difficulties that Bethe experienced in his approach stemmed from the fact that, unlike X-rays, low-energy electrons interact strongly with the ions of the solid. First-order perturbation cannot describe accurately these interactions, and multiple scattering and inelastic processes must be considered to obtain agreement between theory and experiment.[153]

Being very ambitious Bethe wanted to consider approximations higher than Born's first to see whether this would produce better results, but he "went too far."

> I tried to get a second approximation, which is of no use, except that I was told that recently it has been used by somebody in X-ray diffraction. There the second approximation is more sensible because the interaction is much smaller, and apparently you get quite good results from the second approximation for the case when there are three beams interfering. The usual case that people deal with is the case of two beams, an incident one and a reflected one. More than that, three beams, is a complicated case, which made trouble for Ewald all his life. One of his former collaborators showed in the last five years or so that if you use second order approximation you can do very well on the three beam case, but that was many years later. I got immersed in this second approximation to no good effect, and I should have

> realized that the interaction of electrons with the lattice was much too strong to use second order effects. And that I realized only many years later, and I then formed the opinion that if you can't get the result by first order perturbation theory, it isn't any use to find second order, except if the first order is zero . . . So, I solved the assigned problem. But I wasn't terribly happy about the thesis because I had gotten into this quagmire. Pauli afterwards made that very evident to me. When Pauli first met me in the spring of 1929, he said: "Ah, so you are Mr. Bethe. After Sommerfeld's tales about you, I had expected much better from you than your thesis."

But Bethe added: "I guess from Pauli that was a compliment."

Bethe's assessment may have been too unforgiving. It was clearly a very important piece of work, and it taught him a great deal about solid state physics, laying the foundations for a good deal of his subsequent work in that field. More specifically, it laid the foundations for his 1933 review article with Sommerfeld of solid state physics in the *Handbuch der Physik*. Equally important was the fact that the thesis made him realize that, although his abilities were such that he could tackle most problems of interest to the physics community, before he undertook trying to unravel a given problem he had to feel that he could solve it to his own satisfaction. He had to feel confident that he had the necessary tools to do so. This implied that in the future he would choose the problem he would address. There was another lesson to be learned from the thesis, one that became apparent to him only somewhat later, which was the importance of interacting with the experimenters conducting the experiments he was trying to model and explain. Bethe did most of the research connected with his thesis by himself. Sommerfeld and perhaps Ott were the only persons he interacted with, and he did so infrequently. Pauli's criticisms would have been less stinging had he known that what he did was of great value to experimenters—or could have been of much greater value had he chosen to benefit from their input.

Very little low-energy electron diffraction (LEED) experimental work was done following Davisson and Germer's studies until the 1960s due to the technical difficulties connected with surface cleanliness and the maintenance of a good vacuum.[154] The value of Bethe's approach to low-energy elastic electron scattering was rediscovered in the early 1960s when methods of producing very clean surfaces and maintaining ultra high vacua (less than 10^{-10} Torr) became standard.[155] Low-energy—50 eV to 1 keV—electron diffraction became the best way to study the properties of surfaces. Surface physics matured into a recognized subdiscipline and became an important field of physics application. Knowledge gleaned from this subdiscipline found wide application in the manufacture of many electronic devices. Work in this field also eventually led to the creation of field emission and field ion microscopes and to the scanning electron microscope.

The thesis took half a year longer to complete than Sommerfeld had expected because Bethe wanted "to polish it a little more." He also set aside some time to study for his oral examinations. The examiners for his orals were Wien, Fajans, Perron, and Sommerfeld. Bethe saw each one separately for about an hour in the course of single day. Wien's examination was "hard." But Bethe later recalled that "apparently Wien was well impressed by me. Here was at last a theoretical physicist who knew something about experimental physics, even if he was not able to do experiments. I was pretty well informed on experiments, especially modern experiments. And so I could answer his questions satisfactorily." The examinations of Sommerfeld and Perron were "easy," and that of Fajans "extremely pleasant and easy."[156]

Bethe obtained his doctorate—summa cum laude—in July 1928. The thesis consisted of his work on electron diffraction in crystals and was subsequently published essentially complete in the *Annalen der Physik*. Sommerfeld, when submitting the dissertation, commented: "When in the beginning of last year [1926] the epochal work of Davisson and Germer appeared, I lectured about it in my special topics course and entrusted Mr. Bethe with its theoretical evaluation . . . His results were so admirable that he published two preliminary communications in *Die Naturwissenschaften*. Mr. Davisson himself agreed with Bethe's conclusions . . . The work represents fundamental progress in the theory of electrons in metals."[157]

Social Environment

Bethe didn't have much of a social life during his first semester in Munich, and most of his social interactions were with Werner Sachs, who had come to Munich to continue his studies in chemistry there.

> I had more social life when I returned to Frankfurt occasionally. I saw mostly my friends from Frankfurt.
>
> I lived in a room, and went out for lunch and for dinner, and I stuck pretty much to myself. I had a room in different places at different times. There were a few older people, with whom I had social contact, among

> them the Wallachs. The Wallachs were a relative of Hilde Levi, who had been one of the girls in our dancing circle, and they invited both Sachs and me several times. Also once a year, Sommerfeld invited most of his people to his home for the evening; and that was very nice, quite a big party. From time to time, maybe once a month or so, Sommerfeld and his people went to a cafe closer to the center of town, and there were very nice discussions in that context . . . I went to the theater from time to time, mostly with Sachs, and then occasionally to a movie, not very often.
>
> Later on, I worked very much in the evening after supper, in fact mostly. Certainly in the later years when I came back to Munich from Frankfurt and Ewald, I didn't eat dinner out. I ate lunch out, which is the main meal, and then I bought myself a pound of bread and a pound of tomatoes, and a quarter pound of ham, and went home and ate that. I didn't quite eat the pound of bread, but a large part of it.

Regarding the quantity of food consumed, Bethe's younger sister, Doris Overbeck, noted the following: "Bethe visited us in Frankfurt in the summer of 1938. [I was six years old at the time.] I remember he got up very late every morning, he drank lots of water before the meal was beginning, and he could eat in large quantities. I have been told he vied [competed] sometimes—when he was a student—with his friend Fritz von Bergmann or other fellow students, who could eat the most number of rolls or potato dumplings, and not seldom [i.e., often] he won."[158]

There may not have been much "social life" in Munich, but nearly every weekend, Bethe and Werner Sachs, often together with Erwin Strauss or Fritz von Bergmann or both, would go to the Alps near Munich. This was easy to do as there were fast trains to get there. They would get up very early, take the train, hike in the mountains all day—"nothing adventurous, but very nice scenery"—and return to Munich late in the evening. "There was a special feeling associated with these walks and it was very important to do that, to get ready for the next week's work."

But these outings were not without danger. In December 1928 on one of their sorties Werner, Rudolf (Rudi) Peierls, and Bethe were caught in a blizzard high up in the mountains, and by the time they got down Bethe had a bad case of frostbite in his legs. He got himself to Baden-Baden where his mother and her doctor nursed him back to health. It took over two months until he was able to walk again, but the time was well spent: he did an extended and difficult calculation for his thesis, studied Bieberbach's *Theorie der Differentialgleichungen,* and read a great deal, in particular Lion Feuchtwanger's *Jud Süss* and Sigrid Undset's *Kristin Lavransdatter*—novels that moved him deeply.[159]

Bethe's social world had become enriched in the fall of 1926 with the arrival of Rudi Peierls to Sommerfeld's seminar. His encounter with Peierls was the beginning of a deeply meaningful, lifelong friendship.

Rudolf Peierls

Peierls was born on June 5, 1907, in Oberschöneweide, a western suburb of Berlin along the Spree River, whose main industry at the time was the cable factory of the Allgemeine Elektrizitäts-Gesellschaft (AEG). Rudi's father, Heinrich Peierls, a true *pater familias,* was the managing director of the factory. Emil Rathenau, the father of Walther Rathenau, was the founder of AEG. He had recognized the importance of Edison's electrical inventions, had bought some of his patents, and with them had laid the foundations of the German electrical industry. Rudi's mother, née Elisabeth Weigert, was a cousin of his father, and Rudi remembered her as "a gentle person and very attractive."[160] The Peierls lived in a large apartment in a house owned by AEG next to the factory. They maintained an active social life, with music-making an important component of it. The parents were Jewish, but converted to Lutheranism in 1905, supposedly to allow the children to make their own choices. Rudi was baptized a Protestant when he was born.

Rudi was the youngest of the three children in the family; his brother was eight years older and his sister six. Rudi did not remember his mother very well, even though when he was held back from entering elementary school for a year because he was wearing glasses, she taught him at home every day. During World War I she was very busy with volunteer work, and Rudi did not see much of her. After the war she suffered from depression, became ill with Hodgkin's disease, and died in 1921. Shortly after she died a lady came to live in the house to look after the household, "and before long [Rudi's] father married her."[161]

Although he never became very close to his stepmother, he did become attached to the family of her sister, who was married to Ludwig Fulda, then a well-known poet and playwright.[162] Fulda's son was the same age as Rudi, and they saw a good deal of each other. Rudi spent several summers vacationing with the Fuldas in their house in the Italian Alps, with Rudi and Fulda's son doing a lot of hiking and climbing together.

"Without exerting himself" Rudi was an outstanding student in both the local elementary school and local *Gymnasium*.[163] Mathematics was his favorite subject, and he "easily" passed his *Abitur.* He wanted to become an engineer because of his fascination with, and serious study of, machinery, railroads, motorcars, airplanes, and radios, but his father was convinced that he would not be a good engineer because he had to wear glasses—"an engineer has to have good eyesight"—and because he was clumsy with his hands—"an engineer must be able to make things with his hands."[164] Being an obedient son he accepted his father's views and decided to become a physicist, the profession he thought was nearest to being an engineer. Fritz Haber, who was a friend of his father, suggested that he become an experimentalist, but Rudi decided to keep an open mind.[165] He spent the six months after graduating from *Gymnasium* until the start of his university studies as an apprentice in the research department of the telephone company that was a subsidiary of AEG. He there learned the use of power drills, lathes, and milling machines, and spent time designing automatic circuits that included relays. He found all these activities very satisfying. The person he worked for told Rudi's father that he would hire Rudi anytime as a physicist. Rudi thereafter had his father's blessing.

Because his parents felt that at eighteen he was not yet ready to leave home, Rudi enrolled in the University of Berlin. And because first-year students were not allowed to take the experimental laboratory course, most of the courses he took were lectures on mathematics and physics. He rated the lectures by Max Planck "the worst" he ever attended. However, the introductory physics course by Walther Nernst was stimulating, and Walther Bothe's lecture on X-rays introduced him to the old quantum theory. The mathematics courses turned out to be most useful, in particular, Issai Schur's lectures on modern algebra.

Already as a first-year university student Peierls gave evidence of a characteristic feature of his personality: he needed social contacts and throve through intellectual interactions. Sabine Lee in her introduction to Peierls's private and scientific correspondence points to this aspect of his personality. Lee notes that Peierls's wife characterized him as an "intellectual tennis player, not a golfer." He became a member of the Berlin University mathematics and physics *Fachschaft,* the club that looked after the mathematics library and organized discussion sessions as well as tutorials by older students to help younger ones.

After two semesters at the University of Berlin Peierls knew that he wanted to be a theoretical physicist. Moreover, he felt he was old enough to leave home, and since his parents no longer objected, in the fall of 1927 he went to Munich to study with Arnold Sommerfeld. He arrived in Munich just as Sommerfeld in his seminar was expounding the new methods of wave mechanics, which eventually appeared in print as *Atombau und Spektrallinien: Wellenmechanischer Ergänzungsband.*

The twenty-year-old Bethe had been a student in Sommerfeld's seminar for just one semester when Rudi, aged nineteen, came to Munich. But by that time Sommerfeld had already recognized Bethe's unusual abilities and had asked him to grade the problems assigned in the recitation section attached to the lectures on mechanics he was giving during the fall semester. Rudi was the outstanding student in that class. So at first Bethe was Peierls's teacher rather than a fellow student. However, as Peierls noted in his autobiography, *Bird of Passage,* the difference in age and in seniority became inconsequential, and they grew to be very good friends—skiing, hiking, and dancing together. Thus in October 1927, on Rudi's invitation, Bethe went to Berlin for a week. Some sixty-five years later he still remembered that visit vividly: exploring Berlin and its museums and parks, noting its vices "which did not tempt him," and going almost every day to the theater.[166]

Judging from the letters between them it is only in 1929, when Rudi was writing his doctoral dissertation with Heisenberg, that Rudi and Bethe become each other's sounding board and critic in physics, and that their friendship became all encompassing. As they both were in Munich during the 1926–27 academic year, except when Bethe was in Baden-Baden recovering from his frostbite, their correspondence during that year dealt primarily with social matters. Bethe was concentrating on research connected with his thesis and evidently did not discuss the content of his dissertation with anyone.

When asked why the chemistry between Rudi and him worked so well, Bethe answered: "Maybe because he was very outgoing, he was very eager to discuss physics, and I was more restrained. I didn't talk easily. And here was a man of my age, *talking* about the things that interested me, much on the same level as I. I think the main thing was that he was so, felt so comfortable in talking." Bethe, for all his self-confidence in physics, was still socially reticent.

Rudi became Bethe's "dearest" friend throughout their lives. But Peierls was not Bethe's sole comrade. Edward Teller became Bethe's "closest" friend from 1935 until 1943—from the time both had emigrated to the United States until both came to Los Alamos to work on the atomic bomb project. They had first met when Edward joined Sommerfeld's seminar in the spring of 1927.

Edward Teller

Edward was born on January 15, 1908, in Budapest, and some of his earliest memories are of the Danube and the beautiful bridges over it. His family lived on the eastern side of the river, in Pest, next to the Supreme Court building and parliament. He characterized his parents as "kind" and loving. Although his sister, Emmi, two years older, was his father's favorite, his mother doted on him, and he could do no wrong in her eyes. His father was a successful lawyer and a "reserved, pleasant" but stern *pater familias.* Edward could never talk easily with him, but Edward's love of mountains was apparently inspired by his father's enjoyment of them and their joint excursions there. Edward's mother, "a small woman, barely five feet tall," was overprotective of her children and a "worrier." She was also a very gifted pianist. It is possible that she wished to become a professional pianist before she met Edward's father, but she lacked the self-assurance required for performing for large audiences and would only play for members of her family. Teller's love of music grew from listening to his mother play Beethoven piano sonatas. He himself became an accomplished pianist.[167]

The family was Jewish, to some extent observant, and assimilated. Hungary before World War I was somewhat different from other European countries in that it had been more hospitable to Jewish immigration than the other Slavic countries and more liberal in its treatment of its Jewish population. The attempt by Hungary in 1848–49 to free itself from Habsburg rule had been crushed by Franz Joseph with the help of tsarist Russia. But the subsequent Austrian weakness stemming from its defeat by Germany in 1866 led to the so-called *Ausgleich* (Compromise) of 1867, whereby the dual monarchy of Austria and Hungary was created. Thereafter, the ruling Hungarian gentry made a tacit agreement with the assimilated, educated component of the Jewish population in order to modernize Hungary. Skilled and educated non-ethnic Hungarians would provide the technical expertise to raise Hungary to the standards of Western Europe, and the landowning gentry, who controlled parliament and the administrative bureaucracy, would provide the legislative experience and support for the expansion of the economy. István Deák, a historian of eastern and central European history and politics, comments that the resulting success of Jews and other immigrants was quite impressive: "Although they made up less than 5 percent of the population before 1914, Jews created, owned, and managed the majority of Hungarian heavy industry and mining, and nearly every one of the great banks. They were equally successful in commerce, small entrepreneurship, the liberal professions, and all aspects of culture and the arts. By the beginning of the twentieth century they had also made significant inroads into the state bureaucracy, the judiciary, the officer corps, and large landownership."[168]

There was of course anti-Semitism originating from that part of the population who begrudged the Jews their wealth and blamed them for their own comparative poverty. However, the government and some of the press did criticize overt acts of anti-Semitism.[169] Outwardly, the Tellers seemed a conspicuous success story. But evidently a wealthy and esteemed family did not translate into a calm and serene childhood for Edward.

Looking back when writing his *Memoirs* late in life, Teller indicated that, as a child, he had "an almost chronic bad conscience . . . [and] worried almost all the time that some absurdity or another [he had committed] was an offence" and that at times he was "overwhelmed" by guilt. As a child he had "such a terrible fear of the dark" that as a grownup he was often amazed "that he no longer had it." He learned to speak late, but was precocious otherwise. By the age of four he had an "unrelenting desire to understand" and "to find patterns and regularities in the world." By the age of four and a half he was "consistently spending time thinking about numbers." "Finding the consistency of numbers" was his first memory of "feeling secure."[170]

Edward was six years old when World War I broke out. The events of the last year of the war, when it had become apparent that Germany and Austria-Hungary would lose the war, and those of the period after the Armistice when the consequences of that defeat were palpable, became deeply etched in his memory. He vividly remembered watching civilians wearing tiny chrysanthemums and soldiers with flowers in the barrels of their guns marching toward parliament in support of Mihály Károlyi, the leader of the republican government that had been formed just before the war ended. Károlyi's ascension to power had marked the beginning of an independent, democratic Hungary. But that government was short lived. The stipulations of the peace treaty were made known shortly after the Armistice, and the new democratic government was unable to survive its terms. Hungary was to cede two-thirds of its territories to Romania, Czechoslovakia, and Yugoslavia and, in addition, pay heavy reparations. From a multiethnic nation of 21 million people Hungary was to be reduced to one with fewer than 8 million, and with almost half of those who were ethnic Hungarians forced to live under foreign rule. In mid-spring of 1919 the Communist Party under the leadership of Béla Kun took over. Every aspect of society and the economy was impacted by Kun's assumption of power. Edward's father could no longer practice law because lawyers were considered extraneous and out of place in a "good" society. Two soldiers moved into the Teller's apartment, which was considered to have "extra space." Though deeply apprehensive of the Kun government the Tellers managed to survive the brief period that the Communist Soviet was in power.

In August 1919 Admiral Miklós Horthy, a "reactionary conservative," led an army assembled in Romania into Budapest, overthrew the Béla Kun government, and reinstated the constitution of the old kingdom of Hungary—but prevented the king from returning. Horthy thereafter pursued a terrifying policy of retribution against those who had been associated with the Communist government. This included many Social Democrats, as the Social Democratic party had initially supported the Béla Kun government. Over 5,000 people, most of them Jews, were executed, and many more emigrated to other countries.

Edward and his family had a further dreadful experience stemming from the consequences of the Allies redrawing the boundaries of Hungary after its defeat. In the summer of 1919 Edward's father had taken the family to Lugos, where Edward's mother had been born and where some of her family still lived. After the Armistice Lugos was ceded to Romania. Edward's mother was now considered a Romanian citizen and so were Edward and his sister, being the children of a Romanian. They were not allowed to leave the country without governmental permission. The three of them thus were stranded in Lugos for some eight months, before Edward's father, who had returned to Budapest after bringing the family to Lugos, could extricate them.

The Horthy dictatorship was overtly anti-Semitic. Edward had become conscious of anti-Semitism during the last year of the war. His parents were Jewish, and the family celebrated Yom Kippur and fasted on that day. His father also said "prayers for his parents on Saturdays and on all the Jewish holidays."[171] But otherwise religion was not discussed in the household. When Edward started *Gymnasium* at age ten, over half the students in the class were of "Jewish descent," with Edward being one of the half-dozen students who had not converted. Although he was the butt of anti-Semitic comments and name calling by Christian fellow students, he felt that it was no different from the ethnic slurs made against boys with other backgrounds. But toward the end of the war "the Jews" became singled out as being responsible for the imminent defeat, and the young Edward began to recognize the racism and deep prejudice inherent in many aspects of Hungarian anti-Semitism. He wondered "whether being a Jew really was synonymous with being an undesirably different kind of person."[172] Before the fall of Béla Kun in the summer of 1919 his father, who had foreseen Kun's demise, told him that anti-Semitism would follow: "Too many of the communist leaders are Jews, and *all* the Jews will be blamed for their excesses."[173] This was indeed the case. Some of the members of Edward's family who had been Social Democrats and government officials were killed in the Horthy purges.

The intense discrimination against Jews by the Horthy regime deeply affected Edward's education. He had entered *Gymnasium,* the famous Minta, when he was ten, and his first few years there were "miserable." He displayed outstanding talents in mathematics and other subjects but was discriminated against by some of his teachers and by his fellow students for being Jewish. He contracted scarlet fever and had to stay out of school for a semester. However, during his last years at the Minta, his abilities were recognized by both his teachers and classmates, and life there became more bearable. He helped students who had difficulties in mathematics and earned their affection and admiration. He became good friends with some of his Christian classmates. He was particularly fond of Mici Schütz-Harkány, the sister of his classmate Ede, and eventually he married her.[174] Also, during the last two years at the Minta he became acquainted with "three young men from the Jewish community in Budapest who were studying and working in Germany as scientists": Eugene Wigner, John von Neumann, and Leó Szilárd.[175] Whenever they were in Budapest he would meet them and discuss physics and mathematics with them—

and it was they who gave him a vision of the good life: being a professor who taught and did research in either mathematics or science. Doing very well on the Matura, the examination to be passed in order to graduate from *Gymnasium,* and winning the physics and mathematics prize in the prestigious national Eötvös competition gave Edward the confidence needed to undertake university studies.

Anti-Semitism in Hungary had convinced Edward's father that Edward must study in Germany. Although Edward wanted to study mathematics, his father was adamant that he take "a more practical course" that would offer more job opportunities, and they compromised on Edward studying chemistry.[176] Teller was only seventeen when he graduated from the Minta, and his mother insisted he stay in Budapest for another year. He therefore enrolled at the Budapest Institute of Technology. The following year he went to Karlsruhe to study chemistry at the Technical Institute. But like Bethe he quickly discovered that his laboratory skills were wanting, and so he took a fair number of courses in physics and mathematics. Hans Mark introduced him to quantum mechanics, and Paul Ewald to electrodynamics and crystallography.[177] Ewald also invited him to his home, where he met Ewald's family, including the ten-year-old Rose Ewald, who would later become Bethe's wife.

After his father had consulted a distant relative, Felix Ehrenhaft, the "famous-infamous" professor of physics in Vienna who had claimed that he had discovered particles that carried a fraction of the charge of an electron, he allowed Edward to undertake a career in physics. In early May 1928 Edward enrolled at the University of Munich to study with Sommerfeld. Edward at the time was seeing much more of Mici, who by then had finished two years of studies of mathematics at the University of Budapest and was teaching mathematics at the Odenwaldschule, the school that Bethe had attended ten years earlier.

On July 14, a week after he had returned to Munich from visiting Mici at the Odenwaldschule, Edward was on the way to meet friends to go on a hike. He "absentmindedly" rode the trolley past the stop for the railway station. With his rucksack on, "full of cheer and invulnerability," he jumped off the front platform of the trolley, lost his balance, and fell in such a way that the wheels of the trolley severed the lower part of his left foot.[178]

Edward underwent a lengthy hospital convalescence after an operation that fused some of the bones of his amputated leg. In his *Memoirs* Teller recounted that one visit stood out:

> A student two years my senior, and thus a more professional person, was kind enough to come to the hospital. Hans Bethe was then studying under Sommerfeld and preparing for his PhD examination. He had the finest memory and the most comprehensive knowledge of theoretical physics that I had ever encountered. I was touched and honored that he would take the time to visit me; but once in my room Hans didn't know what to say or do. He was stiff and uncomfortable in the presence of my injury, and he made his escape as soon as he felt it polite to do so.[179]

Bethe didn't remember the incident, and denied visiting Teller in the hospital. Whichever story is correct, what can be said with certainty is that for Bethe its occurrence—or not—was very significant. If in fact what occurred was as Teller remembered, Bethe blamed himself for his awkwardness. If his own recollection was correct, he blamed himself for his insensitivity: for not having visited Teller and for not being more supportive during a very difficult time.

4

Beyond the Doctorate: 1928–1933

Among Bethe's many strong points, three have been decisive in his growth as an outstanding physicist. The first was his ability to recognize his strengths and his limitations. From early on he was very much aware that he was not creative in the way Heisenberg, Dirac, or Schrödinger were. He was not able to formulate radically new ideas, but, on the basis of foundational approaches of others, he could critically analyze and extend theories and formulate generative idealizations and models that would test their validity. The criteria that determined the validity of a theory always had an empirical component. Certainly one of the characteristic features of Bethe's research in physics is its close relationship to experiments and to experimental practice. His second insight into his abilities was that he came to know *when* to undertake a problem by evaluating the empirical data and theories at hand. The third was being able to assess realistically whether he possessed the knowledge and the mathematical tools to tackle and solve the problem, and whether he could at the outset say to himself: "I can do that."

After completing his thesis on the topic that Sommerfeld had assigned to him—electron diffraction by crystalline solids—until late in life Bethe himself would determine the problems he worked on. For him, addressing problems in physics required that fairly stable and accurate experimental data be available, that the empirical findings display sufficient coherence to be modeled, and that the foundational theories to address the problem were at hand. "Foundational" meant that the theory that was believed to apply to that particular domain of nature had been formulated. Nonrelativistic quantum mechanics in the case of atoms, molecules, and solids; hole theory to calculate pair production of electrons and positrons; Fermi's β-decay theory to explain nuclear radioactivity: Bethe never believed that these theories were "fundamental" but instead were approximate limited theoretical accounts. He always was aware that empirical "facts" were theory dependent. Data were obtained by instruments. The interpretation of the data depended on the foundational theories that had gone into the building of the instruments and explained their operation. And he knew intuitively what Poincaré had stated explicitly, namely that facts do not speak: "Science is built up of facts, as a house is built of stones; but an accumulation of facts is no more a science than a heap of stones is a house."[1]

The above-mentioned traits were in evidence when Bethe became concerned with problems in cosmic ray physics in 1933 and 1934. His entry into that field and his attempts at interpreting data regarding cosmic ray showers came only after the discovery of the positron and its connection to Dirac's hole theory had been established by compelling cloud chamber evidence.[2] The same was true in 1934 when he began working in nuclear physics: only after the discovery of the neutron and of the deuteron, and the publication of Heisenberg's and Ettore Majorana's papers applying quantum mechanics to the problem of nuclear structure, did he enter that field. Similarly, at the Washington Conference in 1938, listening to what was known about the constitution of stars and of nucleosynthesis he came to the conclusion then and there that he "could do that," in other words, solve the problem of energy generation in stars given his knowledge of nuclear physics and stellar structure.

After completing his doctorate Bethe spent a semester in Frankfurt as Madelung's *Assistent,* and then a semester in Stuttgart as Paul Ewald's *Assistent.* In Ewald's institute he became aware of the wealth of atomic and solid state problems he could tackle and solve, given his mathematical toolkit and his knowledge of quantum mechanics.

In Munich Bethe had learned what was going on in physics by reading the principal physics journals: the *Zeitschrift für Physik,* the *Annalen der Physik,* the *Physical Review,* and the *Proceedings of the Royal Society.*[3] Coming to Stuttgart in the spring of 1929 as Ewald's *Assistent* gave Bethe a novel perspective of doing physics: that of doing it in an environment where theoretical and experimental physicists constantly interacted with one another. Contacts between experimenters and theorists working on problems in crystal physics were the norm in Stuttgart. And by virtue of Ewald's international standing as a crystallographer there was a stream of foreign visitors coming to Stuttgart reporting on their latest work and on that of their colleagues. Additionally, Ewald's institute was much more informal than Sommerfeld's, and Ewald was also much more social than Sommerfeld. Bethe throve in this new environment.

After being Ewald's *Assistent,* Bethe went back to Munich to do his *Habilitation* with Sommerfeld. For it he formulated the quantum theory of the energy loss of fast charged particles in their passage through matter (by their excitation and ionization of the electrons of the atoms that made up the matter). His theory became the point of departure for all the subsequent investigations of the subject, Bethe himself making key contributions as new mechanisms for energy loss—Compton scattering (the scattering of photons by electrons), *Bremsstrahlung,* pair production, and others—became relevant mechanisms with increasing energies. Following this, a Rockefeller fellowship took him to England and then to Rome. The British cultural and social context at the Cavendish and especially his contacts with Patrick Blackett and with Nevill Mott were transformative experiences. The same was true for Rome and his interactions with Fermi. In his interviews with Charles Weiner, with Lillian Hoddeson, and with me, Bethe would aver that he molded himself as a physicist after Sommerfeld and Fermi. But equally important, I believe, were the social settings and atmosphere he discovered in Stuttgart, Cambridge, and Rome, and the friendships—professional and social—he made there. He thereafter wanted very much to do physics in an environment like the one he found in Cambridge and Rome.

This chapter tells the story of Bethe's peregrinations from Munich to Frankfurt, then to Stuttgart and back to Munich, and thereafter to Cambridge and Rome.

Assistent to Madelung and Ewald

At the end of the 1928 summer semester Sommerfeld, who was going to turn sixty in December of that year and did not want the event to be celebrated, took a sabbatical leave to go on a lecturing trip around the world during the academic year 1928–29. Sommerfeld did not make any arrangements for Bethe to have a position in Munich. Moreover, since Wentzel had accepted an associate professorship in Leipzig the previous year, Munich was without a senior theorist, and therefore not that attractive a place for Bethe to stay. The news that Bethe was available did travel, for he received an offer of an assistantship from Madelung in Frankfurt, and somewhat later an offer from Friedrich Hund in Rostock. He decided to take the position in Frankfurt.[4] One of the reasons for his acceptance of the Madelung assistantship was that he knew many young people in Frankfurt. That city offered a much more satisfactory social life. Thus, in his letter to Rudi Peierls in December 1928 he commented that he had "found out that it is nicer to go the theater with a girl-friend than on your own" and that he had become "such a snob" that "he was going to every single premiere in Frankfurt: e.g. [Kurt Weill's] *Driegroschenoper,* [Ferdinand Bruchner's] *Krankheit der Jugend,* [Georg Kaiser's] *Lederköpfe.*" He also became a devotee of masked balls that year.[5]

When Bethe came to Frankfurt in September 1928 his father invited him to live with him. He accepted and did so for personal and not financial reasons. He commented that after he obtained his doctorate he was always self-sufficient as far as money matters were concerned. "I didn't get any subsidy from my parents."

During the summer of 1928 Hans's father had met Vera Congehl, the woman who became his second wife. Vera, like his first wife, also had Jewish roots.[6] Vera was some twenty-six years younger than Albrecht. Bethe recalled that when he was living with his father in the fall and winter of 1928 his father had an intense correspondence with Vera because he was very concerned about the age difference between them. "I don't believe we ever talked about it. He did tell me that he had met a woman who pleased him, but that's all I knew." After his father married Vera in 1929, Hans initially was "quite unhappy about this remarriage" and was somewhat "hostile" to her. He thought that the same was true for Vera. But, as he told Charles Weiner in 1966, "I learned to like my stepmother quite well, and now I'm on the best of terms with my stepmother, who is only ten years older than I."[7] Albrecht had two children with Vera. The older one, Doris, was born in 1933, and the younger one, Klaus, in 1934. Bethe kept in touch with both of them, and used to visit them later in life whenever he went to Germany.

Bethe's official duties as an *Assistent* in Frankfurt consisted of being one of the instructors in the elementary physics laboratory; and evidently his "experimental facilities were just sufficient for that."[8] Madelung himself expected very little from him, basically only "to chat with him two hours a day every morning" about quantum mechanics and about some of Madelung's "philosophical-physical specialties."[9] Walter Elsasser, who became Madelung's *Assistent* in the summer of 1931, insightfully described him in his autobiography:

> [Madelung] had written an excellent thesis [with Max Born on the binding energy of polar crystals] . . . but [he] did not have much drive, and he did not publish another original piece of research in his life . . . He felt uneasy about his place in the profession but was very honest about it: he told me this whole story in the first few days of my presence. Otherwise, Madelung was a very sensible man, who also [showed] character in dealing with the nationalist extremists when he had to fulfill administrative functions for the university.[10]

Madelung tried to interest Bethe in his approach to understand the relativistic equation for the electron that Dirac had published earlier that year, but Bethe found Madelung's interpretation unconvincing. He had taken "Dirac directly,"[11] and had studied carefully Darwin's analysis of Dirac's equation and his application of the equation to the case of an electron in a hydrogen atom. What is to be noted is how Bethe kept himself up to date regarding everything of importance happening in theoretical physics. In a letter to Peierls (dated February 12, 1928), after he had been with Madelung for four months, Bethe revealed that "For most of the day I am on my own in the institute, which coming from Munich feels a bit strange to me as there it was the other way around . . . Being on my own is not good for me, by the way. I work less than I would if I shared the office with others. I walk around the room, and sometimes don't even think about physics or the Fourier integrals that I'm working on at the moment."[12] After Bethe had finished his thesis he became very interested in the application of group theoretical methods to quantum mechanical problems. He had been somewhat depressed because he was unhappy about his attempts to extend his calculations on electron diffraction to higher orders and commented: "When I collected myself after my Ph.D., my first task was to study and understand group theory . . . that was rather natural, because after all, my thesis was about crystals and their effect on electron waves." He had learned the group theoretical classification of crystals in his course with Ott, and though he had made limited use of group theory in his thesis when dealing with the consequences of the translational symmetry of crystals for electronic wave functions, he was aware that the applications of group theory to the other symmetries of a crystal could yield important insights and results. But this required the deepening of his knowledge of the representations of the crystal groups in the linear vector spaces built from the quantum mechanical descriptions of the states of physical systems. Madelung told him to study the 1923 textbook by Andreas Speiser, which he did.

Bethe had obtained a yearlong appointment in Frankfurt, but found Madelung not very stimulating.[13] So when Paul Ewald invited him to come to Stuttgart he accepted. He went to Stuttgart in late April 1929 after obtaining Madelung's consent. Hans had met Ewald in the spring of 1928, before he had finished his doctoral dissertation. Ewald had invited him to come to Stuttgart and give a talk on electron diffraction by crystals at a sectional meeting of the German Physical Society. As Hans's father was an acquaintance of the Ewalds, he was invited to stay in their house for a couple of nights.[14] Bethe remembered that visit with great pleasure.

Some of what went on behind Ewald's invitation to come to Stuttgart is made explicit in the letter Ewald wrote to Sommerfeld on December 20, 1928:

> Dear Sommerfeld,
>
> I must already trouble you with domestic matters today, just after you have arrived in California, and even request an answer as quickly as possible.
>
> As I already wrote you, Madelung, who was able to offer Bethe an *Assistent*ship, has kidnapped him from me. As Fues's[15] position with me will be unfilled as of April 1, I turned to Bethe to inquire whether he would not like to come now. Since Madelung's *Assistent,* Lanczos, is returning [to Frankfurt], Bethe is not disinclined to accept, especially since a very nice collaboration would certainly be available here for both of us.
>
> However I recall that you yourself had designs on Bethe and of course I do not want to take Bethe away from you if you need him right away. I will then have to look around for another *Assistent.* Since doubtlessly, Bethe and I would quickly come to an agreement and it would not take very long for him to familiarize himself with the work, I would be agreeable to taking him only for one semester, the summer 1929. The question for which I am requesting an answer as soon as possible is, whether you think you already need Bethe in the summer or whether I can hire him for the summer.

Sommerfeld answered that he would need Bethe only upon his return to Munich in the fall of 1929.

Regarding his stay in Stuttgart as Paul Ewald's *Assistent* Bethe commented:

> Stuttgart was wonderful. Stuttgart was wonderful mainly because of the Ewalds.
>
> Paul took me into the family. His was for the first time a house which was really hospitable. There were many guests in the house.
>
> Paul and his wife were extremely kind to me and had me at their house many times, and we went on walks together, and occasionally I was given the job of taking the two older children—aged 15 and 12 respectively—for a walk. The younger one, Rose, afterwards became my wife, the other, Lux, was a boy. In a personal way this was about the happiest time I had in Germany.

Stuttgart was also "very nice and very good" scientifically:

> Ewald knew everybody and invited interesting people to come to Stuttgart. One of them was Douglas Hartree. As I was very interested in the quantitative side of atomic spectra, and then of crystals, Hartree's method of treating atoms immediately appealed to me. I guess I had read about his method before he came to talk. But, then, his talk was very illuminating, and I liked that very much. I think he must have stayed for two or three days. So there was Hartree, and then there was a visit by Sam[uel] Goudsmit, who also impressed me very much. He explained the vector model very simply . . . much more simply than Wigner and von Neumann. And Goudsmit was very cheerful, and really a pleasure to have around. So there was much more life in Stuttgart than even in Munich.

Hartree's Self-Consistent Field Method

By 1928 it was known that variational methods could give fairly accurate results for the ground state energy and wave function of two-electron atoms, and could be extended to yield good results for some excited states.[16] However, it was recognized that for complex atoms this approach became inaccurate or prohibitively unwieldy. Douglas Hartree developed an intuitive and manageable method to calculate the properties of complex atoms.[17]

Recall that the states of the hydrogen atom are specified by four quantum numbers as a consequence of the fact that the Coulomb potential the electron experiences is spherically symmetric. The principal quantum number, n, takes on integer values 1,2,3, . . . and determines the "radius" of the orbital (as the quantum state is sometimes called); the orbital quantum number l, which can take on integer values from 0 to $n - 1$, determines the magnitude of the orbital angular momentum; the quantum number m, which takes on integer values from $-l$ to $+l$, determines the orientation of the orbital; and the fourth quantum number, s, determines the orientation of the spin and takes on the value $+1/2$ or $-1/2$. In general, in the case of a spherically symmetric potential each energy level $E(n,l)$ is $2(2l + 1)$ fold degenerate. For the Coulomb potential, because of a latent higher symmetry, the energy of the orbital depends only on the principal quantum number, n, with each n level being $2n^2$ fold degenerate.

In the case of a multielectron atom with nuclear charge $+Ze$ the effects of the repulsive interactions between the electrons must be taken into account. If one approximates the effect of all the other electrons on a given electron by a spherically symmetric potential, the level structure is again determined as in the hydrogenic case by four quantum numbers but with the energy of the orbital depending on the l value. The levels must then be filled with Z-electrons but in accordance with the Pauli exclusion principle that stipulates that no orbital specified by a quantum number (n,l,m,s) can be occupied by more than one electron. The degeneracies of the quantum states specified by the n,l,m,s quantum number then account for the shell structure of atoms, as exhibited by the Mendeleev table of the elements. Totally filled shells, when a set n,l is completely filled, as in the case $Z = 2,10,18,36,54$ and 86, correspond to the noble gases. When the shells are partially filled there are various "rules" that determine the level structure. These rules reflect the further assumptions concerning the forces an electron in an unfilled shell experiences. Goudsmit talked about the phenomenological derivations of such rules.

Hartree's lectures concerned the calculations of the structure of atoms starting from the Schrödinger equation for a Z-electron atom. Hartree had just published an article in which the wave function of an n-electron atom was approximated as a product of n one particle wave functions, and the latter determined variationally. Hartree had imposed the Pauli principle by requiring that no more than two electrons (with opposite spins) could be in the same one particle state.[18] In this approach each electron moves in the attractive Coulomb field of the nucleus and the repulsive Coulomb field of all the other electrons as described by their charge density. More precisely stated, the repulsion between the electrons is calculated as if their charge density were spread out as determined by their calculated wave functions. This process is repeated until the calculated wave functions reproduce the potential that had given rise to them. One thus spoke of the Hartree method as being "self-consistent."

To illustrate the method consider the helium atom: the atom with a nucleus of charge $+2e$ and two electrons.[19] In the full quantum mechanical approach the energy levels of the atoms are determined by the Schrödinger equation

$$H\psi(r_1,r_2) = E\psi(r_1,r_2) \tag{1}$$

with the Hamiltonian

$$H = -\frac{\hbar^2}{2m}\left(\nabla_1^2 + \nabla_2^2\right) - \frac{2e^2}{r_1} - \frac{2e^2}{r_2} + \frac{e^2}{|\boldsymbol{r}_1 - \boldsymbol{r}_2|} \tag{2}$$

Hartree assumed that $\psi(r_1,r_2)$could be approximated by a product wave function $\psi_1(r_1)\psi_2(r_2)$, with $\psi_1(r_1)$ corresponding to electron 1 moving in the field of the nucleus, i.e., in the potential $-2e/r_1$, and in the *average* field of electron 2. The potential created by electron 2 was calculated by assuming that its charge density at the point $\boldsymbol{r}$ is given by $-e|\psi_2(r)|^2$. The first electron thus moves in a field in which its potential energy is

$$V_1(r) = -\frac{2e^2}{r} + e^2\int d^3r' \frac{|\psi_2(\boldsymbol{r}')|^2}{|\boldsymbol{r}-\boldsymbol{r}'|}. \tag{3}$$

Its wave function is therefore determined by the Schrödinger equation

$$\left(-\frac{\hbar^2}{2m}\nabla_1{}^2 + V_1(\boldsymbol{r}_1)\right)\psi_1(\boldsymbol{r}_1) = E_1\psi_1(\boldsymbol{r}_1). \tag{4}$$

Similarly, the second electron moves in a field in which its potential energy is given by

$$V_2(r) = -\frac{2e^2}{r} + e^2\int d^3r' \frac{|\psi_1(\boldsymbol{r}')|^2}{|\boldsymbol{r}-\boldsymbol{r}'|} \tag{5}$$

and its wave function is determined by the Schrödinger equation

$$\left(-\frac{\hbar^2}{2m}\nabla_2{}^2 + V_2(\boldsymbol{r}_2)\right)\psi_2(\boldsymbol{r}_2) = E_2\psi_2(\boldsymbol{r}_2). \tag{6}$$

In the application of the method one "guesstimates" V_1, then solves for ψ_1; thereafter V_2 may be calculated. Given this calculated V_2, ψ_2 may ,be determined from the Schrödinger equation it satisfies. This in turn allows the computation of a new V_1. The procedure is repeated until the wave functions are such that the V's reproduce themselves—and "self-consistency" is achieved.[20]

The situation is somewhat simpler for the ground state of the helium atom, since then ψ_1 and ψ_2 are the same function and the potential $V(r)$ is the same for both electrons.

Hartree's lectures were the stimulus for a set of papers Bethe wrote while in Stuttgart. Partly in order to learn how to obtain reliable atomic form factors, quantities that could be calculated from accurate atomic wave functions and were essential to Ewald and his research team when interpreting the data from X-ray scattering, Bethe read Hartree's two papers and gave a talk on them. In the first of his publications written while in Stuttgart dealing with analytic approximations to electronic wave functions, Bethe compared the electronic charge distribution in helium when computed according to various methods: taking hydrogenic wave functions for each electron; taking hydrogenic wave functions describing the electrons as moving in the Coulomb field of an "effective" nuclear charge Z' (because of the screening of the nuclear charge $Z = 2$ by the other electron) and determining Z' by a variational method; calculating the charge distribution with the electron wave functions determined by Hartree's method.[21]

The variational approach also led to a paper that Bethe was very proud of. In it he calculated the energy levels of a two-electron atom with the lowest nuclear charge—the hydrogen negative ion, H^-.[22] This was the first calculation of the electron affinity of hydrogen and the first time anyone had proved that the hydrogen negative ion is stable, that is, that its energy is lower than the energy of the hydrogen atom. Bethe also established that H^- had only one bound state. Later on, Egil Hylleraas calculated the energy levels of H^- much more accurately.

The electron affinity of hydrogen is a very important number in astrophysics, because in the relatively cool surface region of stars, H^- ions are abundant, and the absorption and emission of radiation is due to the formation and destruction of these negative hydrogen ions.[23]

This group of papers displayed a characteristic feature of many of Bethe's articles. They usually consist of a statement of the particular problem that is to be addressed; a presentation of the model that is assumed to represent the problem that is addressed; an account and explanation of the way the problems encountered are solved; and a description of the solution. Often, readers of a paper by Bethe could delude themselves and say, "I could have done that."

Many times this is true. But they could not have done it in the time it took Bethe to do it, nor with as few errors in the calculations as he made. In addition, they might not have been able to overcome the obstacles he faced and overcame. It was only after they had read Bethe's paper and had been shown the way to do it that the material became accessible to them. Victor Weisskopf, who is regarded as one of the great twentieth-century physicists, put the matter succinctly: "Bethe's strength was that of finding simple and general ways to deal with complicated situations . . . Every paper of Bethe's (there are some 330 titles) bears the characteristic marks of his insight and special way of attacking problems. He differs from most of us in his universality and his encyclopedic knowledge."[24] The sheer number of papers that Bethe wrote, their simplicity, their attention to detail, their seminal character, and the range of problems addressed simply overwhelms.

The papers dealing with hydrogenic and helium wave functions were only a small part of Bethe's research while in Stuttgart. Most of his efforts while there went into deepening his knowledge of group theory and exploring the possibility of explaining the structure and symmetries of crystalline solids. His interactions with Ewald were crucial factors in this. The ninety-year-old Bethe recalled that while he was in Stuttgart "Ewald and I got along very well. He wanted to learn quantum mechanics. He made me give lectures once a week about quantum mechanics."

A comment is in order regarding this recollection of Ewald by Bethe. A month before Bethe came to Stuttgart, on March 15, 1929, Ewald had delivered an invited lecture to the Faraday Society, the professional society of physical chemists in Great Britain, entitled "Some Modern Developments of Wave Mechanics and Their Bearing on the Understanding of Crystal Structure."[25] In his presentation Ewald had summarized "the Schrödinger picture of the atom" and had reviewed the wave mechanical formulation of "Pauli's exclusion principle." In the section of his lecture entitled "Resonance Principle: Exchange Energy" Ewald outlined the implications of the exclusion principle, in particular the effect of the antisymmetry of electronic wave functions on the energy of particular electronic configurations in atoms, molecules, and solids. He went on to discuss "Unsöld's and Pauling's Work on Polar Crystals"; "London-Heitler's Application of Exchange Energy to H_2"; and in particular "Exchange Phenomena in Metals." In this last part of his lecture he referred to Bethe's thesis on electron diffraction.

It is clear that Ewald already knew quantum mechanics, though perhaps not at the technical level that Bethe did. The intent of Ewald's Faraday Society talk was made clear in the section where he referred to Unsöld's and Pauling's work on polar crystals. In his classic work on the stability of polar crystals Born, using classical mechanics, had assumed that the forces acting between neighboring ions in a polar crystal such as NaCl could be approximated by a function of the form $(a/r^2)-(b/r^n)$, where the first term represents the electrostatic attraction between the ions when separated by a distance r, and the second term their repulsion at short distances—the latter to account for the stability of the crystal. Ewald wanted Born's force law justified by quantum mechanics, and in the lecture indicated that the following questions needed to be answered: To what extent was the force between the Na^+ and Cl^- ions like the Heitler-London homopolar bond responsible for the binding of the hydrogen molecule?[26] How important are exchange effects, since X-ray diffraction had indicated that the ions in the NaCl crystal behaved as if they had charge $+e$ for the Na^+ ion, and $-e$ for the Cl^- ion? Was the repulsive force properly characterized by an r^{-n} dependence, or would an exponential dependence (such as $be^{-\kappa r}$) be a more accurate representation? Ewald had referred to what Unsöld, Pauling, and Hermann Brück had done to justify the validity of assuming spherical symmetry for the force law. Ewald pointed out that since the electronic structure of the ions was affected by the crystal symmetry of the solid, the ions thus acquired a dipole moment and the contribution of this dipole–dipole interaction ought to be included in the binding energy. He also raised the question of whether the particular crystalline structure that a solid assumed could be calculated; that is, whether the cohesive energy could be calculated quantum mechanically to sufficient accuracy to determine for which crystal structure the cohesive energy would be a minimum.[27]

Madelung and Ewald had both made important contributions to the calculation of the binding energy of polar crystals. Madelung (1918) had calculated the lattice energy resulting from the Coulomb interaction of the ions for simple cubic lattices. In other words, he had computed the contribution of

$$\sum_{j;j\neq i} \pm\frac{e^2}{r_{ij}} = -\frac{e^2}{R}\sum_{j;j\neq i} \mp\frac{R}{r_{ij}} = -\alpha\frac{e^2}{R}$$

where R is the nearest neighbor distance. The constant α,

$$\alpha = \sum_{j;j\neq i} \mp\frac{R}{r_{ij}},$$

has in fact become known as Madelung's constant. It depends only on the crystal structure. Ewald in 1921 formulated a very elegant technique and a rapidly converging approximation for calculating binding energies of polar crystals. It made use of the symmetry property of the lattice, and of the fact that the charge density, $\rho(r)$, of the ions gave rise to an electrostatic potential, $\phi(r)$, which is related to the charge density by Poisson's equation

$$\nabla^2\phi(\boldsymbol{r}) = 4\pi\rho(\boldsymbol{r}).$$

The Fourier transform of $\phi(r)$ and $\rho(r)$ are thus related by

$$\tilde{\phi}(\boldsymbol{k}) = -\frac{4\pi}{k^2}\tilde{\rho}(\boldsymbol{k}).$$

It turns out that binding energy calculations are greatly facilitated by relating these Fourier transforms to ϕ and ρ expanded in terms of the reciprocal lattice vectors.

For a point charge located at the origin, $\rho(r) = e\delta(r)$, hence $\tilde{\rho}(\boldsymbol{k}) = 1/(2\pi)^3$ and therefore $\tilde{\phi}(\boldsymbol{k}) = -\frac{4\pi e}{k^2}$ and $\phi(r) = e/r$. Bethe made extensive use of this identity in his *Habilitation* paper on the passage of fast charged particles through matter. It allowed him to obtain closed form expressions for cross sections and energy loss per unit length.

Undoubtedly Ewald talked to Bethe about the content of his 1921 paper and of his Faraday Society lecture, and very probably gave him a copy of his Faraday lecture, for while in Stuttgart Bethe tackled all the problems Ewald had mentioned in it. However, Bethe did not thank nor give credit to Ewald for suggesting them or give a reference to Ewald's 1921 paper in his *Habilitationsschrift*.

Several factors—sociological and personal—were responsible for why Bethe didn't acknowledge Ewald's help. James Conant, in his autobiography, provided insights into the reasons for this behavior. In 1925, as a young, ambitious, and very successful professor of organic chemistry at Harvard, Conant took a semester off to visit Germany to find an answer to the question, "What had made German science, particularly organic chemistry, so fruitful?" His answer was that three factors were responsible for the preeminence of German science: First, there was only one full professor in each subject at each university. Second, contrary to the practice at American universities, a "young man" (and there were only men!) "was never promoted within the university where he was teaching." And third, there was an unofficial yet permanent hierarchy of institutions, with Berlin, Munich, and Göttingen at the top. Promotion consisted in being called to a better university. The system was intensely competitive, with the rate of progress through the various steps—from *Privatdozent* to full professor and director of an institute—dependent on the judgment of the senior professors at the top universities. One way the most prominent *Privatdozenten* were evaluated was by inviting them to present papers at the annual meeting of the regional professional societies. The professors listened and made note of the originality of the research presented and of the quality of the presentation. Devastating questions from the professors would be the norm and "would be met by a vigorous show of approval by the audience." The intense rivalry and competition among universities and among the individuals—the professors, the *Privatdozenten*—had resulted in a "ruthless intolerance of mediocrity and showmanship . . . Not even within a single university was there any room for give-and-take among individuals of equal rank." The coldhearted, insensitive, merciless demand for excellence became the justification and maintenance of the no-holds-barred, inhuman academic world that had been created.[28]

Bethe was *very* ambitious and wanted to indicate to the physics community at large his great powers, and to point out that what he had done he did all by himself. Bethe wanted to obtain an academic position. However, as the number of university positions open to a young theorist was extremely small, he had to prove himself exceptional to obtain any one of them.[29] Given the competitive environment he found himself in, it might have been the case that at that stage of his life Bethe did not have sufficient empathy with his colleagues to have been aware that acknowledging Ewald's helpfulness in the matter would be deeply meaningful to Ewald.

Thanking colleagues for discussions—if there were any—was then not part of the German tradition when writing physics papers in German periodicals. Thus there are no acknowledgments in Heisenberg's *Zeitschrift für Physik* papers on quantum mechanics in 1925 and 1926, nor in Bethe's *Annalen* paper on electron diffraction. But in his 1927 *Zeitschrift für Physik* paper on the uncertainty principle, written after a year in Copenhagen, Heisenberg acknowledged "being indebted for heartfelt thanks to Prof. Bohr" for informative discussions.[30] Bohr had a different attitude toward acknowledging the help of colleagues in scientific papers. Its origin is his interaction with Ernest Rutherford in Manchester in 1913. The interdependence of the experimenters working with Rutherford in Manchester and at the Cavendish made for a different community than physicists working in a German institute. At the Cavendish, expressing indebtedness to co-workers was the norm, and explicitly acknowledged in publications.[31] Bohr's stance regarding acknowledgments very probably also reflects the Danish cultural context. Bohr and his younger brother Harald were deeply impressed by the friendly, inquisitive, cooperative spirit displayed by everyone at the meetings his father, a physiologist, hosted at their house. The philosopher Harald Høffding and other colleagues of his from the university attended these meetings, where all subjects were discussed—literature, music, psychology, and philosophy—and both brothers were allowed to attend when very young.[32]

The German theoretical physics community condoned—in fact nurtured—assertive, rude, conceited behavior. Bethe himself recalled an incident corroborating this: "There was a physics colloquium in Frankfurt during the time I was an *Assistent* to Madelung, in which the speaker, who was a very dignified person, said a lot of nonsense, and so I piped up afterwards and said, 'That's all wrong, and clearly it goes this way,' which was not at all friendly, but my mother was very proud of that." Similarly, Bethe told of attending a meeting of the *Deutsche Naturforscher Gesellschaft* (in Danzig or Königsberg) in 1930 at which Emil Rupp spoke and claimed to have observed the effect of electron polarization at quite low electron energy and to have found maxima that required half integral orders of reflection. This seemed "absurd" to Bethe, and during the discussion of the paper he "spoke up," dismissing its claims in no uncertain terms. "I was vigorously beaten down by another member of the Rupp's ARG (*Allgemeine Elektricitäts-Gesellschaft*) research team, Dr. Ramsauer (an excellent physicist), who said, in essence, 'How does such a young pipsqueak dare to attack the work of a member of our laboratory?'"[33]

Bethe continued learning group theory and its applications in quantum mechanics in Stuttgart. After finishing his thesis he had studied the papers of Wigner and von Neumann on the subject, and while he was an *Assistent* with Madelung he had mastered the content of Speiser's book. Bethe's ambition while in Stuttgart was to calculate quantum mechanically the binding energy of a simple polar crystal such as sodium chloride or cesium chloride, as well as the binding energy of a simple metal such as sodium, and to determine their lattice structure—the problems that Ewald had posed. The aim was to calculate the total energy of the solid as a function of atomic volume (at the absolute zero of temperature). The minimum of the total energy would then determine the lattice constants of the solid, its compressibility (the change in volume produced by a unit change of pressure), and other properties.

When Bethe undertook these problems, the binding of polar crystals was explained classically as being due primarily to the electrostatic attraction between charged spherical ions. In the classical approach, the short-distance repulsion between the ions was called *van der Waals repulsion* in analogy with the repulsion created by impenetrable hard spheres in van der Waals's model. This short-distance repulsion was, as noted earlier, usually approximated by a short-range potential of the form $+b/r^n$. Thus the binding energy could be calculated from the ionic charges, the van der Waals repulsive potential, and the ionic distances.[34] In 1927 Heitler and London formulated a quantum theoretical model of the homopolar binding of molecules, such as the hydrogen molecule H_2. In their perturbative approach to the H_2 molecule, they considered it as being formed from two hydrogen atoms. When the two hydrogen atoms were far apart, separated by a large distance R, the atomic electrons were described by hydrogenic wave functions. In the stable H_2 molecule the two electrons have opposite spin, in which case the Pauli principle allows the electronic charge density to be concentrated between the two protons.[35] This configuration is energetically favorable, since the electronic charge density then screens the charges of the protons and reduces their mutual repulsion, making binding possible. The model also allowed the calculation of the short-distance repulsion.

Bethe intended to combine the Hartree approximation to electronic structure in atoms with the Heitler-London approximation to homopolar binding in molecules. When the inter-ionic distance was large compared to the size of an atom Bethe assumed the zeroth order approximation of the electronic wave functions of an atom inside the crystal to be the wave functions of the free atom. Bethe then argued as follows:

> There are two reasons to expect a perturbation of the free atom on its inclusion in a crystal: on the one hand, the atom will enter into exchange of electrons with the other atoms of the crystal, i.e. its permutation group is changed. This exchange effect will have to be treated in a way quite analogous to the case of a molecule, there being at most a purely quantitative difference in regard to the number of neighbors with which the exchange can take place. On the other hand, there acts on the atom in the crystal an electric field arising from the other atoms which has a definite symmetry which removes the directional degeneracy of the free atom.[36]

An electric field of definite symmetry will cause a splitting of the energy levels of the unperturbed atom that is analogous to the Stark effect splitting and will be characteristic of the electric field the atom finds itself in. Thus in a crystal, the levels, which for an isolated, free atom exhibit the degeneracies permissible under the full rotation group (a symmetry of empty space), exhibit the reduced symmetry of the point group appropriate to the atom's crystalline environment.

Bethe undertook to analyze the specific perturbations of the free atom that are due to the symmetry of the crystal and to apply group theory to determine the splitting of the energy levels of atoms in crystals by virtue of the periodic potential they experienced.[37] He did so for the case of weak, of intermediate, and of strong crystalline electric fields. This resulted in a long paper published in the *Annalen der Physik* in 1930, which David Mermin and Neil Ashcroft, two leading solid state physicists, called "monumental" in their article "Hans Bethe's Contribution to Solid-State Physics."[38]

It was the first application of group theory to solid state physics—and one of the most cited papers of Bethe. In an evening lecture in 1968 Bethe commented that he did this work "on the splitting of atomic levels when the atom is inserted into a crystal, into a site of given symmetry" because he had studied Speiser's book on group theory, and "you can't really understand something unless you apply it and work with it yourself. So since Wigner had done all the important things with group theory, I thought the only thing that remains to be done is to take an atom in a crystal of various symmetries and see how the energy levels will look there. I am told that people have used that paper, but I have never seen what has come out of it."[39] This paper on term splitting in crystals was meant as a preparation for determining better electronic wave functions to calculate the binding energy of a three-dimensional crystalline solid. Investigating descriptions of the valence electrons was the first order of business. Bethe's Heitler-London approach assumed that the behavior of a valence electron in the neighborhood of any one ion is determined by the potential $U(r)$ of that ion and that the effects of the other ions could be neglected as a first approximation—an approximation valid if the ions are far apart. This "free" atom electronic wave function is then determined by the Schrödinger equation

$$\left(-\frac{\hbar^2}{2m}\nabla^2 + U(r)\right)\phi(\boldsymbol{r}) = E\phi(\boldsymbol{r}).$$

If r_n denotes the position of the nth ion, an electron's wave function in the neighborhood of ion n will be given by $\phi(r - r_n)$. In his thesis Bethe had deduced that the translational symmetry in a periodic potential implied that electronic wave functions were of the form $e^{ik\cdot r}u_k(r)$, with u_k having the periodicity of the lattice. Bethe now assumed that the wave function of an electron inside the crystal could be approximated as

$$\psi_k(\boldsymbol{r}) = \sum_n e^{i\boldsymbol{k}\cdot\boldsymbol{r}_n}\phi(\boldsymbol{r}-\boldsymbol{r}_n).$$

This approximation became known as the "tight-binding" approximation, and the approach had been used by Felix Bloch for the treatment of electrical conductivity. Bloch furthermore had established that for every state of an electron in the free ion there exists a band of energy in the crystal. The method was elaborated by John C. Slater and others. Bethe never published his results, but they were incorporated into the Sommerfeld and Bethe *Handbuch* article on the electron theory of metals. Mott and Jones adapted Bethe's presentation in the *Handbuch* article and disseminated his presentation in their textbook.[40]

The calculations that could determine the lattice structure of a particular crystalline solid require approximations as well as numerical computations to an accuracy impossible to achieve with a slide rule, Bethe's usual calculational tool, or with the hand-cranked calculating machines then available. The project had to be abandoned. But Bethe did apply his approach in a one-dimensional model for which, using what became known as the *Bethe ansatz*, he was able to obtain both the eigenfunctions and eigenvalues for that N body problem. I shall devote a later subsection of this chapter to the Bethe ansatz.

In the summer of 1929 Bethe gave a talk on the splitting of the energy levels of atoms in crystals at a meeting of the southwest section of the German Physical Society in Freiburg—the first meeting of the full society that he attended. It was there that he first met Pauli. He was also introduced to Robert Oppenheimer at that meeting. Oppenheimer had just finished writing his paper with Born on an approximation that describes how to separate the electronic motion from the nuclear vibrations and rotations in molecules, a paper that became quite famous. Bethe remembered his encounter with Oppenheimer as being "quite unpleasant."

After Ewald heard Bethe's lecture on the subject he pointed out to him that the crystals formed from rare earth salts had very sharp spectral lines, the reason being that a rare earth atom is usually ionized down to the 4f-shell electrons and that this shell has a very small radius. These inner-shell electrons lie deep in the atom, and so they are not much perturbed by the electric field of the neighboring atoms, and the electric field they produce is spherically symmetric. The small radius of the ion also implies that it can be approximated as a point charge and that the group theoretical methods Bethe had developed would be applicable to such an idealized model. Ewald suggested that Bethe look at what happens when a magnetic field is present—i.e., at the Zeeman effect of these sharp lines of rare earth salts—which Bethe proceeded to do. He wrote a paper on the subject, completed in Rome during his stay there in Fermi's institute in the spring of 1930.[41] He later commented that he had been "very rude," for he did not thank Ewald in the paper for the suggestion.

In Stuttgart Bethe made the acquaintance of Herman Mark, a friend of Paul Ewald, who was directing the I. G. Farben laboratory in Mannheim, which was investigating the diffraction of electrons by gases such as CCl_4.[42] Mark invited Bethe to come to Mannheim and give a talk on the quantum theory of electron scattering off molecules and the applications of electron scattering to the determination of atomic and molecular structure. This Bethe did. As his knowledge of X-ray scattering off molecules could readily be transcribed to the electronic case, he could point out that the interference of the scattering by the various atoms in the molecule would give rise to scattering with varying intensities in different directions. Moreover, the angular distribution depended on the orientation of the molecule. To compare theoretical predictions with experimental intensity one therefore had to average over all molecular orientation. Nonetheless, as in the X-ray case, Bethe suggested that it should be possible to deduce interatomic distances from the observed broad interference maxima.

The visit was Bethe's introduction to an industrial laboratory and to applied physics. Bethe clearly gave a fine talk from which Mark learned a great deal, for Mark and his associates went on to obtain experimental electron diffraction results that gave the structure of molecules in the gaseous state.

In the summer of 1930, Debye organized in Leipzig a conference, "Electron Interferences," to which he invited Bethe to talk about electron diffraction by crystals.[43] Bethe for some reason declined—he might still have been in England—and Herman Mark took his place and gave a talk on the scattering of electrons and X-rays off atoms, molecules, and crystals.[44] The second part of his presentation was essentially a report on Bethe's *Annalen der Physik* paper on electron diffraction.

Habilitation

In the summer of 1929, upon his return to Munich from his world trip, Sommerfeld, in true *Bonze* fashion, wrote a postcard to Ewald: "Bethe is my student. I need Bethe. I want him back immediately."[45] As noted earlier, Sommerfeld and Ewald had agreed that Bethe was to return to Munich in the fall but that Ewald was permitted to "keep" him during the summer semester. Bethe recalled that he was very reluctant to return to Sommerfeld, for he had been "happier than [he] had been for years" in Stuttgart. But Sommerfeld offered him several inducements for returning to Munich. Bethe outlined them in a letter to Peierls:

> Stuttgart-Heidelberg railroad 27 VII 29
>
> . . . Beginning October 1 I have sold myself to Sommerfeld—yes!—and in return I had a few sweets put into my mouth, namely: 1) a stipend for travel for one year, of which I need to go only for 1/4 year to New York [to stay with Clinton Davisson at Bell Labs]—Sommerfeld could not ignore my aversion to long visits to experimental environments—and otherwise to Cambridge and probably Rome. 2) *Habilitation* for the next semester. After I had explained that I was attracted to Stuttgart, among other things, by the prospect of *habilitieren* as soon as possible, i.e. by the end of "S.S." [summer semester] 1930, Sommerfeld had no alternative but to promise it a half year sooner. 3) The next available position in Munich. However this last-mentioned promise is not worth anything because probably no position will be open soon and, besides, these kinds of promises can be forgotten. Nevertheless I was so satisfied with what I achieved in the bargaining over the 130 Mark per month, that I did not look any further at my work on the exchange energy in ionic crystals . . .[46]

And so, Bethe dutifully went back to Munich in the fall of 1929 and undertook his *Habilitation* with Sommerfeld. In fact, from 1929 until 1932, when he accepted a position in Tübingen, Munich was his headquarters. His home until he went to Cambridge in the fall of 1930 was the apartment of his great aunt, Jeanette Seligmann, a sister-in-law of Anna's mother, who lived in Munich.[47]

To support Bethe financially, Sommerfeld got him a fellowship from the *Notgemeinschaft der Deutsche Wissenschaft* that amounted to roughly 200 to 250 Marks a month, enough to live on.[48] Sommerfeld also made available to him an office in the institute: a half-room at the top of the staircase from the basement.

For his *Habilitation* Bethe returned to Born's important papers on collision theory. He had read Born's first two papers on collision theory shortly after they were published in June and July 1926 and had subsequently mastered the content of all three.[49] He found the general theory expounded therein "very useful, clearly right, good, and interesting." And since "you can't really understand something unless you apply it and work with it yourself," Bethe therefore sought an interesting problem, one in which the solution was relevant to experimentalists and in which he had to apply Born's theory to compute cross sections for the processes involved. "There existed a first application of Born's work to the elastic and inelastic collisions of electrons in hydrogen that had been made by Elsasser in a rather straightforward manner." And more important, Elsasser, in a paper submitted to the *Zeitschrift für Physik* in late August 1927, had made the first quantum mechanical calculation of the energy loss per unit path length of a charged particle in its passage through a gas of hydrogen atoms as a function of the charged particle's energy, thereby introducing Bethe to Bohr's classic work and the subsequent contributions by Ralph Fowler and others to the subject. All these earlier papers dealt with the problem using classical physics for the description of the motion of the projectile charged particle and described each of the bound electrons in the target atoms by harmonic oscillators whose natural frequencies corresponded to the frequencies of the light emitted when the electron made a transition from one of its stationary states to another.

Thus once again Bethe's and Elsasser's paths crossed. Elsasser had gone to Göttingen in the fall of 1925 and had been working in James Franck's laboratory, but not making much progress and becoming quite frustrated because of his inability to tool the handmade devices and instruments to carry out his experiments. In the summer of 1926, Born, who had recognized Elsasser's theoretical inclinations and was impressed by his talents in that direction, approached him, probably at Franck's suggestion, to ask whether he would consider becoming a theoretical physicist and doing a thesis with him. Moreover, Born felt that Elsasser had been "sitting around long enough" and told him that he would give him a thesis that was not too difficult technically, allowing him to get his doctorate in a relatively short time. After consulting Franck, Elsasser accepted Born's offer. Born gave him the simplest application of his recently developed collision theory: to calculate the elastic and inelastic collision cross sections for electron-hydrogen scattering, this in lowest Born approximation in which the wave functions of the incoming and of the scattered electron are approximated by free particle wave functions; in other words, by plane waves.[50]

According to Elsasser, it was Oppenheimer, who was then one of Born's *Doktoranden,* rather than Born—who was unapproachable and seemed uninterested—who helped him overcome whatever technical difficulties he encountered.[51]

The Passage of Charged Particles through Matter

In 1929, when Bethe addressed the problem of the passage of charged particles through matter, the subject had a long history. The absorption of β-rays by matter had been investigated by J. J. Thomson soon after he had identified the "rays" and could make beams of them.[52] Rutherford extended this work after he came to the Cavendish to work with Thomson in the mid-1890s and also investigated the absorption of α-rays by matter. When it became clear that α-rays consisted of positively charged particles, William Henry Bragg indicated that α-particles, when traveling through a gas, should have a definite range, depending on their initial velocity. This led to the carrying out of an extended series of experiments to determine the ranges of α-particles in different gases and metals as a function of their initial velocity.[53]

The analysis of the scattering and stopping of charged particles in their passage through matter played a seminal role in the development of atomic and quantum physics. Bohr became acquainted with the subject during his stay with Rutherford in Manchester in 1912. He wrote two papers on the absorption of α- and β-rays, one in 1913 and the second in 1915, that assumed Rutherford's atomic model. These papers became the starting point of all subsequent investigations. Bohr maintained a lifelong interest in the subject and used it as a test of the methods of atomic mechanics.[54] He did so because the explanation of the experimental findings requires knowledge of the atomic structure of the entities that make up the atoms of the gas or solid being traversed, as well as assumptions about the dynamics of the interactions between the projectile and the target atoms.

In a paper written in 1948 reviewing the historical development of the theoretical understanding of penetration of atomic particles through matter, Bohr stressed the importance of these phenomena and commented that their treatment "has been an important test of the methods of atomic mechanics and has not least offered instructive lessons as regards the extent to which the application of classical mechanical concepts is adequate and to what points proper quantum-mechanical analysis is required."[55]

Once the particulate nature of electrons, protons, α-particles, and nuclei was established, and the Rutherford model of the atom accepted, the basic mechanisms for energy loss by a proton or an α-particle in its passage through matter were assumed to be either

a. elastic collisions of the incident particle with the nuclei of the atoms composing the matter. Due to the electrostatic Coulomb repulsion between the nucleus of the atom and the projectile, the projectile will be deflected. The angle through which it is deflected depends on its velocity and its impact parameter and is described by Rutherford's formula. For high velocity charged particles elastic collisions are a small effect as far as energy loss is concerned.
b. inelastic collisions between the incident charged particle and the atomic electrons (by virtue of their Coulomb interaction) whereby the atom makes a transition to an excited state or is ionized. The energy necessary for these processes comes from the kinetic energy of the projectile, which is thereby slowed down.

The problem is thus to calculate the excitation and ionization of the atoms. This is done by calculating the perturbation of an atomic electron by the moving particle.

Bohr in his 1913 paper had derived an expression for the energy loss per unit length of a charged particle of mass M, charge ze, and velocity V, traversing matter whose density of electrons is n. Bohr's theory formed the point of departure for Bethe's investigations.

Bethe's Calculation

After reading Elsasser's paper Bethe believed that he could improve on Elsasser's results, including Elsasser's calculation of the energy loss of electrons in their passage through a gas of hydrogen atoms, and in fact generalize it to more complex atoms and to solids. Born's collision theory allows the calculation of the probability amplitude for the projectile having its momentum changed from $\boldsymbol{k}$ to $\boldsymbol{k}'$ and the atomic electron making a transition from an initial state m to a final state n by receiving an amount of energy $(k^2 - k^{2'})/2m = E_m - E_n$. Having an expression for the probability of the transition, and more specifically, for the cross section, $\sigma_n(E_n - E_0)$, for an inelastic collision leading to the final atomic state n with energy E_n, with $E_n - E_0$ the energy loss by the projectile in this collision, Bethe could calculate the average energy lost by the projectile in traveling a distance dx in a region of space with a specified number of atoms per unit volume, atoms whose electronic structure is known. The energy loss per unit length is given by

$$-\frac{dW}{dx} = \sum_n N\sigma_n(E_n - E_0)$$

with N the number of atoms per unit volume.

In his talk in Trieste in 1989 Bethe indicated that "Out of [Elsasser's] calculation there came rather long and unwieldy formulae, getting worse the higher the quantum number of the excited state, and there was no way to foresee how terrible these formulae would be when you came to the excitation of states in the continuous spectrum. I thought . . . one should be able to do it more simply." What Bethe did

> was to find a simple transformation which made it all very much simpler. In fact, it was similar to one that I had already used in my thesis, but I understood it much better in this paper. Essentially, it was the following: In Born's theory you find that the scattering is given by the Fourier transform of the potential, and I showed that you could transform this into the Fourier transform of the charge of the electrons, which then connects much more directly with the wave function. And using that trick, I found it very easy to calculate both elastic and inelastic cross-sections, and I found a nice sum rule for the inelastic cross-sections, which then permitted me to calculate the total energy loss in inelastic collisions. This, then, gives you the stopping power of a material for a fast charged particle, for which the Born approximation is applicable. So this paper, which I consider one of the four best papers I have written, is about the stopping power of materials for fast-charged particles.[56]

In the introduction to his *Annalen* paper Bethe stated the aims and the limitations of his calculations clearly: they were intended to obtain the energy loss per unit length of a *fast* charged particle in traversing matter due to *incoherent* inelastic processes. Fast particles meant particles of positive or negative charge whose velocity was *large* compared with the velocity of atomic electrons. Bethe stressed that the approximation method used, the first order Born approximation, was only valid for such high-energy particles—but under those conditions the results were "dependable."

In first order Born approximation,[57] in which incident and scattered particle are treated as free, unperturbed by the potential $V(\boldsymbol{r})$ they find themselves in, the amplitude for the elastic scattering of a particle of momentum $\hbar\boldsymbol{k}$ to a state of momentum $\hbar\boldsymbol{k}'$ by the potential $V(\boldsymbol{r})$ is given by

$$\begin{aligned} G(\boldsymbol{k}',\boldsymbol{k}) &= \int d^3r \exp(-i\boldsymbol{k}'\cdot\boldsymbol{r})V(\boldsymbol{r})\exp(i\boldsymbol{k}\cdot\boldsymbol{r}) \\ &= \int d^3r V(\boldsymbol{r})\exp(i\boldsymbol{q}\cdot\boldsymbol{r}) \end{aligned} \tag{1}$$

where $\boldsymbol{q} = \boldsymbol{k} - \boldsymbol{k}'$ is the change in the momentum of the scattered particle. For the scattering of a particle with charge ze by a hydrogen atom, $V(r)$ is given by

$$V(\boldsymbol{r}) = \frac{ze^2}{|\boldsymbol{r}-\boldsymbol{r}_1|} - \frac{ze^2}{r}$$

where $\boldsymbol{r}$ is the coordinate of the incident particle and $\boldsymbol{r}_1$ that of the bound atomic electron. For the case of inelastic scattering of charged particles from a hydrogen atom, the amplitude for the transition of the atom from the state m to n, while the incident particle of momentum $\hbar\boldsymbol{k}$ is scattered to a state of momentum $\hbar\boldsymbol{k}'$, is given by

$$G(\boldsymbol{k}',\boldsymbol{k})_{nm} = \int\int d^3r d^3r_1 \exp(-i\boldsymbol{k}'\cdot\boldsymbol{r})\overline{\psi_n(r_1)}\left(\frac{ze^2}{|\boldsymbol{r}-\boldsymbol{r}_1|} - \frac{ze^2}{r}\right)\exp(i\boldsymbol{k}\cdot\boldsymbol{r})\psi_m(r_1).$$

Instead of integrating over $\boldsymbol{r}_1$ first in the term stemming from the $e^2/|\boldsymbol{r}-\boldsymbol{r}_1|$ interaction, as Elsasser had done, Bethe integrated over $\boldsymbol{r}$ first using the fact that

$$\int d^3r \exp(-i\boldsymbol{q}\cdot\boldsymbol{r})\frac{1}{|\boldsymbol{r}-\boldsymbol{r}_1|} = \frac{4\pi}{\boldsymbol{q}^2}\exp(-i\boldsymbol{q}\cdot\boldsymbol{r}_1),$$

so that the contribution of that term is given by

$$G'(\boldsymbol{q})_{nm} = \frac{4\pi}{\boldsymbol{q}^2}\int d^3r_1 \overline{\psi_n(r_1)}\psi_m(r_1)\exp(i\boldsymbol{q}\cdot\boldsymbol{r}_1),$$

with $\boldsymbol{q} = \boldsymbol{k} - \boldsymbol{k}'$. For $m = 0$, that is, for the case that the initial atomic state is the ground state of hydrogen, this amplitude is relatively easy to evaluate.

For elastic scattering off the state n, $\overline{\psi_n(r_1)}\psi_n(r_1)$ is just the charge density $\rho_n(r_1)$, so that the contribution of this term to the scattering amplitude becomes

$$G'(\boldsymbol{q})_{nn} = \frac{4\pi}{\boldsymbol{q}^2}\int d^3r_1\, \rho_n(r_1)\exp(i\boldsymbol{q}\cdot\boldsymbol{r}_1) \tag{2}$$

Upon relating (by Poisson's equation) the charge density $\rho_n(r_2)$ to the potential $V_n(r)$ it creates, Eq.(2) reduces to the form given by Eq.(1).

$G(\boldsymbol{q})_{nm}$ is the form factor corresponding to the transition m to n. After inserting appropriate normalizing and flux factors $|G(\boldsymbol{q})_{nm}|^2$ yields the differential cross section for the transition $m,\boldsymbol{k} \rightarrow n,\boldsymbol{k}'$. It clear that the above readily generalizes to the scattering of a charged particle by an atom with nuclear charge Z and Z-electrons, in which case the scattering amplitude becomes

$$\begin{aligned} G(\boldsymbol{k}',\boldsymbol{k})_{nm} &= \int\cdots\int d^3r d^3r_1 d^3r_2 \cdots d^3r_Z \\ &\quad \exp(-i\boldsymbol{k}'\cdot\boldsymbol{r})\overline{\psi_n(r_1,r_2,\cdots r_Z)}\left(\sum_{j=1}^{Z}\frac{ze^2}{|\boldsymbol{r}-\boldsymbol{r}_j|} - \frac{Zze^2}{r}\right) \\ &\quad \exp(i\boldsymbol{k}\cdot\boldsymbol{r})\psi_m(r_1,r_2,\cdots r_Z) \\ &= \frac{4\pi ze^2}{q^2}\int\cdots\int d^3r_1 d^3r_2 \cdots d^3r_Z\, \overline{\psi_n(r_1,r_2,\cdots r_Z)}\left(\sum_{j=1}^{Z} e^{i\boldsymbol{q}\cdot\boldsymbol{r}_j} - Z\right) \\ &\quad \psi_m(r_1,r_2,\cdots r_Z) \end{aligned}$$

with the final integrations being over the coordinates of the atomic electrons only. The quantity

$$F_n = \int\cdots\int d^3r_1 d^3r_2 \cdots d^3r_Z\, \overline{\psi_n(r_1,r_2,\cdots r_Z)}\left(\sum_{j=1}^{Z} e^{i\boldsymbol{q}\cdot\boldsymbol{r}_j}\right)\psi_n(r_1,r_2,\cdots r_Z)$$

is the form factor of the state ψ_n encountered in the theory of elastic X-ray scattering. (For inelastic scattering, the scattering by the nucleus coming from the term

$$\int\cdots\int d^3r_1 d^3r_2 \cdots d^3r_Z\, \overline{\psi_n(r_1,r_2,\cdots r_Z)}\, Z\psi_m(r_1,r_2,\cdots r_Z)$$

vanishes owing to the orthogonality of the eigenfunctions ψ_m and $\psi_{n'}$.) Bethe could thus relate electron scattering to incoherent X-ray scattering. For elastic scattering the scattering amplitude reduces to

$$G(\boldsymbol{k}',\boldsymbol{k})_{mm} = \frac{4\pi z e^2}{q^2} \int \cdots \int d^3 r_1 d^3 r_2 \cdots d^3 r_Z \, \overline{\psi_m(r_1, r_2, \cdots r_Z)} \left(\sum_{j=1}^{Z} e^{i q \cdot r_j} - Z \right) \psi_m(r_1, r_2, \cdots r_Z)$$

$$= \frac{4\pi z e^2}{q^2} \left(F_{mm}(q) - Z \right)$$

where $F_{mm}(q)$ is form factor for scattering off the state m, well known from X-ray scattering.

Bethe observed that he could make use of sum rules when calculating total cross section for excitation to all states since $\sum_n \left| G(\boldsymbol{q})_{nm} \right|^2$, by the completeness of the wave functions involved, yields a δ function in the $r_1, \ldots, r_Z$ variables, and the result is a relatively simple expression. Bethe could manipulate the operators in the expression for $G(\boldsymbol{q})_{nm}$ by using their quantum commutation relations, which were deducible from the fundamental canonical commutators of the position and momentum operators, and the sum evaluated by using the completeness of states.[58] The importance of the method lay in the fact that all this could be done without determining individual wave functions, which would have involved the impossible task of solving the Schrödinger equation with all interactions.[59] Bethe's insight and approach proved to be very generative, finding many applications in numerous fields, such as in current algebra investigations of collision processes and quantum chromodynamical calculations of jet events.[60]

The next practical problem became how to approximate the charge densities that entered in the calculations in order to obtain numbers that could be compared with empirical data. Here Bethe made extensive use of the Thomas-Fermi method for finding the density of electrons in an atom. This method is a good approximation when the number of electrons in the atom is large and the atom is in its lowest quantum state. Bethe then obtained the following expression for the energy loss per unit path length of a fast heavy particle:[61]

$$-\frac{dW}{dx} = \frac{4\pi n Z^2 e^4}{m V^2} \log \frac{\bar{q}_{\max}}{\bar{q}_{\min}}$$

with $\bar{q}_{\max}$ and $\bar{q}_{\min}$, average quantities related to the $b_{\max}$ and $b_{\min}$ in Bohr's formula.[62] Bethe thus had derived an expression for the loss of energy by a fast charged particle in its passage through matter per unit length traveled using a quantum mechanical description of the atomic processes involved. Like

Bohr's formula, it did not depend on the mass of the particle, and its dependence on the properties of the atoms making up the matter could readily be calculated. Its applicability, however, was restricted primarily to heavy fast particles with $M \gg m$, because electrons of large energy lose most of their energy by radiation.

This paper for the *Habilitation* was written up in the spring of 1930 and submitted to the *Annalen der Physik* on April 3, 1930.[63] The title of the article is "The Theory of the Passage of Fast Particle Waves through Matter." The abstract of the paper gives an indication of what he had accomplished:

> The theory is developed in detail for collisions of charged particles (of mass M and charge Z) with hydrogen atoms, and as far as possible with complex atoms. The following are computed:
>
> a. the angular distribution of the elastically and inelastically scattered particles
> b. the excitation cross sections for the excitation of the optical and X-ray levels by electron collisions, the sum of all inelastic and elastic collisions, and also the number of the primarily and secondarily formed ions, the velocity distribution of the secondary electrons, and finally
> c. the slowing down (*Bremsung*) of the colliding particles by gas atoms.
>
> The agreement of the theory with experiment is satisfactory to good.

Throughout the paper, whenever possible, Bethe had compared theory with experimental results. At the end of the paper he thanked "Dr. Kulenkampff for his valuable advice regarding the experimental literature."

The formalities connected with the *Habilitation* and becoming a *Privatdozent*—and with it the right to teach at a university—require, besides writing an acceptable paper, presenting a number of "theses." These claims are then published. Subsequently, every member of the science faculty can attend the candidate's defense, attacking these claims and attempting to prove them wrong. According to Bethe, the ceremony is "only a formality and usually the faculty is very gentle."[64] Bethe's defense took place on May 3, 1930. The list of "Thesen" can be found in Appendix F.

Bethe's *Habilitationsschrift* is a stunning piece of work in what it encompassed, synthesized, and created. Moreover, the research and calculations were carried out in less than four months! The originality and importance of the article is attested by the fact that from 1960 until 1975 it was cited over 500 times, and over a hundred times during the next two years—this forty-five years after the article had been published!

His impressive accomplishment was achieved in an extremely competitive environment. As noted earlier that setting did not nurture qualities of collegiality in young *Privatdozenten*. After Born had read Bethe's paper he sent him "an appreciative letter." Bethe evidently replied to it as follows: "It is rather a pity that you did not see the connection between the collision problem of material particles and the well-known scattering of X-rays." In his autobiography written in 1978, Born commented that the letter was one "an angry teacher would write to a feeble pupil, not that of a young scholar to a much older one" and that it took some time to get over his anger. And he added, "Experiences like this taught me to consider myself not to be a first-class physicist like Einstein, Bohr, Heisenberg, and Dirac." That Born should have come to that conclusion because of his interactions with young researchers is a terrible indictment of the German academic milieu.[65]

Bethe's *Habilitationsschrift* became the basis of a good deal of the material to be found in Bethe's *Handbuch der Physik* article on the one- and two-electron problem in quantum mechanics.[66] Bethe extended his treatment of energy loss to include relativistic corrections and to encompass the energy loss by highly energetic electrons (for which *Bremsstrahlung* becomes a crucial mechanism for energy loss) and by high energy photons (for which Compton scattering and pair production become important mechanisms for energy loss). The comparison of the expressions obtained for the energy loss by high energy electrons with the experimental data obtained using cosmic ray electrons of a few hundred MeV traversing lead plates became a way of testing the limits of quantum electrodynamics.

A Rockefeller Fellow in Cambridge

In the fall of 1929 Sommerfeld, as promised, had recommended Bethe for a Rockefeller Foundation Travelling Fellowship. A yearlong fellowship was awarded to Bethe soon thereafter. Bethe felt enormously rich, for the fellowship's stipend was $150 a month, twice as much as his monthly salary as an *Assistent*. As Davisson was not enthusiastic about Bethe visiting Bell Laboratories, Sommerfeld suggested that he go for five months to Cambridge and for five months to Rome.

Bethe spent the fall semester of the academic year 1930–31 in Cambridge. He stayed in a pension on Castle Street that George Gamow had recommended. The room he rented was very beautiful, but he found the food "abominable."

Nominally, Bethe was under the aegis of Ralph Fowler, a distinguished theoretical physicist who had made major contributions to statistical mechanics and to astrophysics. Fowler was also Rutherford's son-in-law and had an office in the Cavendish, and thus had a great deal of contact with the Cambridge experimentalists. This was in contrast to the other Cambridge theorists such as Dirac and Nevill Mott, who were members of the mathematics department and usually had offices in the various colleges. Fowler got Bethe an office "somewhere," politely listened to him expounding on his work on the splitting of energy levels in crystals "but was not especially interested."

Bethe's reactions and impressions of the places he visited in England were recorded in the letters he wrote to Sommerfeld.[67] He had come to England in early October, but his first letter to Sommerfeld was dated November 2:

> Please excuse that I have not written and thanked you sooner—for after all you are responsible for the fact of my stay here, and I experience it as extremely pleasant. All the people here at the Institute are particularly kind, and especially Fowler concerns himself with my *Seelenheil* ["well-being"][68] . . .
>
> The large number of official and unofficial colloquia here are especially agreeable, all of them on an extraordinary high level. Nuclear physics is of course the main topic at the official meetings, and I was surprised how far one has already come in this field, especially with respect to the determination of nuclear levels.
>
> Besides that, the "Kapitza Club" meets once a week in a private room in one of the colleges—with about 20 people discussing more general physics.[69] [Peter] Kapitza participates very actively and wants to know "why" every two minutes. A particularly interesting report was by Fowler about the overthrow by Milne of Eddington's theory of stellar interiors: Eddington had taken a singular solution of the equilibrium equation for the star's interior, which is generally known to assume much too high a value for the absorption-coefficient. All other solutions seem to give much higher densities (nearly densely packed nuclei and electrons) and temperatures (10^{11}–10^{12} degrees) for the center of the star. Fowler thought "That's just the nice hell one desires to have for the formation of nuclei etc." . . .
>
> You can see that much is going on here, and perhaps the most beautiful

> thing here is that one can so easily attend these discussion groups, or alternatively, Fowler introduces me everywhere and is "awfully sorry" when I miss some such event as recently a lecture by Eddington, in which he computed the mass ratio of proton and electron as $136^2/10$ ($10 = 1/2 \times 4 \times 5$, $4 =$ the number of dimensions of the world).[70]

The Cambridge experience proved to be both stimulating and liberating. Bethe found the openness and cordiality of two of his British hosts, Ralph Fowler and Patrick Blackett, most engaging and attractive. The thirty-three-year-old Blackett, at the time a fellow of Kings College and one of the most active and versatile experimentalists at the Cavendish,[71] described himself as a physicist as follows:

> The experimental physicist is a Jack-of-all trades, a versatile but amateur craftsman. He must blow glass and turn metal, . . . carpenter, photograph, wire electric circuits and be a master of gadgets of all kinds; he may find invaluable a training as an engineer and can profit always by utilizing his gifts as a mathematician. In such activities will he be engaged for three quarters of his working day. During the rest he must be a physicist, that is he must cultivate an intimacy with the physical world. But in none of these activities, taken alone, need he be pre-eminent, certainly not as a craftsman . . . ; and not even in his knowledge of his own special field of physics need he, or indeed perhaps can he, surpass the knowledge of some theoretician . . . The experimental physicist must be enough of a theorist to know what experiments are worth doing, and enough of a craftsman to be able to do them. He is only pre-eminent in being able to do both.[72]

Bethe and Blackett had met earlier that year when Blackett spent the summer in Berlin. They liked and admired one another, and so Blackett made a point of often inviting Bethe to his house. Bethe found Blackett's wife, Caroline, "who was originally from Italy, a very charming woman." Blackett was quite liberal in his political outlook, and Bethe remembered a particularly "irreverent" discussion about the king at his house. The younger physicists at the Cavendish also often asked Bethe to come dine with them at "high table" at their college, so that Bethe felt very much part of the community. He was deeply impressed by the conversations of the Fellows when they gathered after dinner in the Common Room of the College. They talked about literature, plays, music, past experiences, flying, and politics, and "it all seemed very rational, and very easy, and it was a revelation, after the stiffness and conservatism of the German universities. Sommerfeld was a very nice man, but he was very conservative."

Bethe went to listen to Dirac's lectures on quantum mechanics and quickly realized that the lecture format followed Dirac's book almost word for word, "although interspersed with some additional pretty examples."[73] Writing to Sommerfeld Bethe indicated, "I am quite taken with [Dirac's] method of first giving a rigorous [*exakte*] foundation for quantum theory, and only then calculating. I think, however, that someone who has heard nothing of quantum theory will visualize only extremely little when hearing of abstract bracket, | >, symbols and equally abstract "observables" instead of electrons and atoms which he presumably expects to hear something about."[74] Bethe decided that he didn't need to listen to Dirac as he studied his book. A friend of Bethe, W. Riezler, a young experimentalist from Munich who didn't speak English well, attended Dirac's lectures because he would then have the book, could study it, "and at the same time, listen how Dirac pronounced its content" and thus learn English.

Bethe had studied English in *Gymnasium,* and spoke it "modestly well."[75] When he first came to Cambridge people could understand him, but he could hardly understand anyone else, except for the physicists. He could always understand Fowler, who spoke with a loud voice, and also Nevill Mott because Mott made an effort to talk slowly and distinctly. Gradually he learned to understand what most people were saying. After approximately a month in England, Bethe went to a restaurant in London, in which there happened to be people who were speaking German sitting at the next table. Bethe found that he could not understand them at that distance. So he "decided" that he "now could understand English equally well."[76] The incident is interesting because it reveals something about Bethe's relation to German, his mother tongue. His decision, after spending a month with British physicists, that he "could now understand English equally well" as German suggests that thereafter he no longer had any special emotional attachment to German, even though until the late 1930s German remained the language in which Bethe could express himself most easily. Thus, until the outbreak of World War II in September 1939, his extensive correspondence with Rudi Peierls is in German.

When Bethe was visiting Cambridge, the Cavendish was *the* outstanding nuclear physics laboratory in the world.[77] But Bethe didn't interact very much with Rutherford, Chadwick, or Cockcroft, the nuclear physicists there.

> I listened to a few talks, seminars or colloquia, it [nuclear physics] seemed to me a very messy subject, and I would at that time have agreed with Max Born's secretary, who was supposed to type a manuscript about nuclear physics, and interchanged the "n" and the "u." What then existed was entirely empirical, and made no sense to me at all. I knew [Gamow's] theory of alpha-decay, and believed that. Beta decay was a mess, . . . a great mess, and the attempts to explain the energy levels of the heavy radioactive nuclei seemed to me totally ad hoc, which they were.

In his interview with Charles Weiner and Jagdish Mehra in 1967 Bethe expressed himself more bluntly:

> In 1930 I was repelled rather than attracted by nuclear physics. It seemed to me really groping in the dark for energy levels and not having enough evidence to give a quantum mechanical description . . . So in 1930 I did not work on nuclear physics while in Cambridge at all, and I had no idea that it ever would interest me. In fact, I began to be interested really only after Chadwick's big paper in 1933, plus Heisenberg's paper, plus emigrating to England in the fall of 1933 and being very closely associated with Peierls.[78]

Bethe did hear a talk by Rutherford in which he made fun of Eddington's theory to derive the value of the fine structure constant and the ratio of the electron and proton mass.[79] And this in Eddington's presence! Bethe also met Cockcroft, who "was very nice." Cockcroft showed him the accelerator he and Walton were building, and told him what he intended doing with the machine once it was built.

While in Cambridge Bethe attended the meetings of the $\nabla^2 V$ Club and of the Kapitza Club and partook in the discussions.[80] His British hosts, recognizing Bethe's abilities and accomplishments, had asked him to speak both at the Kapitza Club and at the London Physics Club.[81] The meetings of both these clubs were devoted to discussions of advances in physics. The gatherings were informal and conducted so that the expertise and the insights of the experimentalists were to benefit the theorists, and conversely. W. Lawrence Bragg and Hartree invited him to Manchester to give a colloquium there; and Bethe did so in December 1930.

Kapitza invited Bethe to his laboratory, the Mond, where he had built magnets that could produce magnetic fields of the order of a hundred kilogauss lasting for a fraction of a second. Kapitza had found that under those conditions the resistance of metals increased greatly. Bethe had been deeply involved in Sommerfeld's program to explain the thermal and electrical conductivities of metals,[82] so Kapitza very much wanted Bethe to explain his experimental findings. Bethe did do a calculation, but the results he got did not agree with Kapitza's experimental results.[83] And so after telling Kapitza of his results, he gave up on the problem. Kapitza was also interested in the structure of the energy levels of atoms in crystals, and in the effect of magnetic fields on them—what Bethe was mostly working on at the time—and Bethe gave a lecture on that subject at the Mond.

"What Physics Was All About"

Of all Bethe's scientific interactions in England, his discussions with Blackett and with Nevill Mott were the most consequential. Blackett, who with Giuseppe Occhialini was then making the cloud chamber an extremely versatile and accurate instrument to study the interactions of high-energy radiation in its passage through matter, told Bethe, "Look here, you have a theory of the energy loss of charged particles going through matter. We have wanted for many years to get a reliable quantitative theory of the stopping of alpha particles. Your qualitative results are no good to me. I want to know this energy loss quantitatively, and I want to know it so accurately that I can measure the range of a particle [in a cloud chamber picture] and from that range deduce the energy of the particle. Can you please extend this to get a range-energy relation." Blackett knew of W. E. Duncanson's paper (which Bethe had mentioned in his *Annalen* paper) in which the range had been calculated on the basis of the theory Bohr had formulated in 1913 and 1915 using classical physics. Blackett asked Bethe to redo these calculations using quantum theory and to obtain range-energy relations for charged particles in different gases, liquids, and solids. And so Bethe was led to generalize his theory to these cases and obtain formulas for directly measurable quantities, such as the range of a charged particle of a given mass, charge, and energy in a given substance. The formulas developed in Cambridge were subsequently made more accurate by incorporating relativistic corrections,[84] and the refinement of the theory to include relativistic effects, pair production, *Bremsstrahlung,* straggling, multiple scattering, and so on kept Bethe's interests and those of some of his students busy for many years thereafter.[85]

Nevill Mott

Bethe was very taken by the free, easy, and intense interactions of the experimentalists at the Cavendish. Nevill Mott, a young theorist who shared Bethe's interests in physics, offered proof that this could also be the case with theorists.[86] Bethe spent part of almost every day discussing physics with Mott. They became good friends and frequently went on excursions together. "We walked together a great deal."

The tall and lanky Nevill Mott was a year older than Bethe.[87] His father had been a student of J. J. Thomson at the beginning of the century and became a science educator and administrator in public and private schools. His mother had likewise obtained scientific training at Cambridge. In fact, his parents had met in the Cavendish. After attending a preparatory school where his mathematical skills became evident, Nevill won a scholarship to Trinity College in Cambridge in 1924 to study physics. Under the influence of Dirac and Fowler he learned quantum mechanics by reading the original papers of Heisenberg, Schrödinger, and Dirac, and by 1927 was publishing papers in the *Proceedings* of the Cambridge and London Royal Societies. In characteristic Cambridge fashion, Mott never got a doctorate, feeling that his Trinity fellowship was a better degree.

When Bethe met him, Mott was recognized as an outstanding young theorist by Bohr, Fowler, and Rutherford. Cambridge University Press had just published Mott's book on wave mechanics, the content of which was a polished version of the lectures he had given in the winter of 1929–30 in the physics department of the University of Manchester.[88] He had earlier written several papers on the application of Born's collision theory to the scattering of electrons by atoms and their nuclei. Mott was the first to prove that the cross section that Rutherford had derived classically for the scattering of a (nonrelativistic) charged particle by a Coulomb field was also valid quantum mechanically by giving an exact solution to the problem.[89] The quantum mechanical description of scattering, of energy loss by charged particles in their passage through matter, and the electric and magnetic properties of an electron as described by the Dirac equation were some of Mott's interests.

While in Copenhagen in 1928, Mott had derived the cross section for the scattering of electrons by free electrons and of α-particles by He nuclei that revealed the exchange effect due to the indistinguishability of the particles during their interaction—a purely quantum-mechanical phenomenon. Mott had *predicted* that the cross section for electron–electron scattering at 45° would be half as large as the Rutherford value by virtue of their fermion character, and that of α-particle–α-particle scattering at 45° would be twice as large as the Rutherford value by virtue of their boson character.[90] Chadwick verified the prediction in 1930.

Mott's paper "On wave mechanics of α-ray tracks," which had been stimulated by Charles Galton Darwin, became a classic. In it he posed the question: "How it is that a cloud chamber produces a well defined trajectory for an α-particle emitted from a radioactive nucleus, given that the α-particle emerges from the nucleus in an unspecified direction and is described wave-mechanically by a spherical wave?" By considering the interaction of the α-particle with two separate atoms (described by using a many-body wave function), Mott showed that both atoms could be ionized only if the original α-particle and the two atoms lay very close to a straight line containing all of them. In other words, the first collision may occur at any point, but thereafter the track is determined.[91]

In 1928 a grant from the Department of Scientific and Industrial Research had allowed Mott to go for six months to Copenhagen and for six months to Göttingen. As attested by his letters to his mother, Bohr and Copenhagen were transformative experiences.[92] In his autobiography Mott put it thus: in Copenhagen

> I learned what physics was all about, how it was a social activity and how a teacher should be with students and how beautiful physics could be. For us theorists the whole atmosphere was as different as could be from that in Cambridge. We came to the institute at ten and left at about six. We were in and out of each others' room all day and so was Bohr. No one would ever keep an idea to himself. In fact I might say [Bohr] gave to us theoreticians a life such as Rutherford created for his "boys," the experimentalists at the Cavendish.[93]

Mott conveyed to Bethe that way of doing physics. Even though Ewald had treated Bethe as a peer, the German culture made it impossible to have that kind of rapport. Even Bethe's interactions with Peierls had been constrained by that culture. The letters between Bethe and Peierls until 1929—when Rudi began working on his doctorate with Heisenberg in Leipzig—dealt primarily with social and cultural matters. Heisenberg, for whom Bohr became a father figure and from whose deep influence he eventually had to free himself, brought to Leipzig the way of doing physics that Mott had described. As Bloch indicated in his article "Heisenberg and the early days of quantum mechanics," he, Giovanni Gentile, Peierls, George Placzek, Fritz Sauter, Weisskopf, and Gian Carlo Wick, the postdoctoral students who were with Heisenberg in 1931, were indeed "in and out of each others' room all day" and so was Heisenberg; and "no one ever kept an idea to himself." Bohr's interactive way of doing theoretical physics, which he in turn had observed in Rutherford's laboratory while in Manchester in 1912 and 1915, became Heisenberg's way when he went to Leipzig. Peierls was heavily influenced by that environment. Thereafter Bethe and Peierls became closer "social" friends and also very close "professional" friends. Especially after Peierls married Eugenia Kannegiesser, Bethe could open up with Rudi regarding his feelings in a way that he could not and did not with most of his other professional friends. This distinction was a real one. Nevill Mott became a professional friend, but Bethe would not open up to him. Peierls, and later Teller and Weisskopf, became both.

The Hoax

The British context evidently allowed the shedding of some of the stiffness that a German education bestowed upon scholars, for the first issue of *Naturwissenschaften* in 1931 opened with a startling *Kurze Originalmitteilung* entitled "Concerning the Quantum Theory of the Absolute Zero of Temperature." The note was signed by G. Beck, H. Bethe, and W. Riezler, all three postdoctoral fellows at the Cavendish Laboratory.[94] It came on the heels of Eddington's attempt to explain the numerical value of the fine structure constant, and was probably inspired by the lecture they had heard Rutherford give in which he had made fun of Eddington's theory. The value of the fine structure constant $\alpha = e^2/\hbar c$ at the time was close to 1/137. Eddington believed that $\hbar c/e^2$ should be exactly equal to 137 and formulated a theory in which this was so.

Noting that $273 = 2 \times 137 - 1$, the three of them set out, tongue in cheek, to make fun of Eddington by concocting a theory that claimed that the temperature of absolute zero, -273 K, was related to the inverse of the fine structure constant $1/\alpha = \hbar c/e^2 = 137$, and then had it published in *Naturwissenschaften.* To give the note credence, it referred to the properties of crystal lattices at low temperature, a field in which Bethe was being recognized as having done important work. Their note read

> Let us consider a hexagonal lattice. The absolute zero of the lattice is characterized by the fact that all degrees of freedom of the system are frozen out, i.e. all inner movements of the lattice have ceased, with the exception, of course, of the motion of an electron in its Bohr orbit. According to Eddington every electron has $1/\alpha$ degrees of freedom where α is the fine structure constant of Sommerfeld. Besides electrons our crystal contains only protons and for these the number of degrees of freedom is obviously the same since, according to Dirac, a proton is considered to be a hole in a gas of electrons. Therefore to get to the absolute zero we have to remove from the substance per neutron (= 1 electron plus 1 proton; our crystal is to carry no net charge) $2/\alpha - 1$ degrees of freedom since one degree of freedom has to remain for the orbital motion. We thus obtain for the zero point temperature $T_0 = -(2/\alpha - 1)$ degrees. Putting $T_0 = -273$ K, we obtain for $1/\alpha$ the value 137 in perfect agreement within the limits of accuracy with the value obtained by totally independent methods. It can be seen very easily that our result is independent of the particular crystal lattice chosen.

Since in those days papers in respected scientific journals were read with an absolute trust in the honorable intentions of the authors and the integrity of the editors, it took a while for the community to realize that *Naturwissenschaften* had been "had" and that the paper was a prank. Arnold Berliner, the editor of *Naturwissenschaften,* was not amused. Nor was Sommerfeld.

Riezler had gone back to Munich over the Christmas vacation and had attended a physics colloquium in the institute. At the end of the colloquium, Sommerfeld came over to him and asked him, "Well, what about that paper you wrote with Bethe and Beck." With a straight face Riezler explained what they did. Sommerfeld pressed him: "Well, that's really quite strange. You talked about degrees of freedom and degrees of temperature. They are not the same thing." So then Riezler began laughing, and admitted that it was a hoax, and Sommerfeld became infuriated. He wrote Bethe an angry letter asking him how he could do this, especially since Berliner had been instrumental in getting Bethe his stipend from the *Notgemeinschaft der Deutsche Wissenschaft,* the German national science foundation of that time.

Berliner demanded an apology, and on March 6 there appeared the following "Correction" in *Naturwissenschaften*: "The note by G. Beck, H. Bethe and W. Riezler, published in the January 9 issue of this journal, was not meant to be taken seriously. It was intended to characterize a certain class of papers in theoretical physics of recent years which are purely speculative and based on spurious numerical agreements. In a letter received by the editors from these gentlemen they express regret that the formulation they gave to the idea was suited to produce misunderstanding."

If Beck, Bethe, and Riezler were somewhat contrite, George Gamow, who was in Copenhagen at the time, felt that pranks were his province and that any opening for one should not be allowed to pass. Max Delbrück, who roomed together with Gamow at the Pension Have, recalled that on the day after Berliner's "Correction" appeared, Gamow formulated his plan. He would wait until another seemingly outrageous paper appeared in *Naturwissenschaften,* and he would then write to Berliner to persuade him that he had again been victimized by a prankster and should publish another retraction. Gamow did not have to wait long. On April 4 there appeared a note by A. V. Das, "The Origins of Cosmic Penetrating Radiation," which seemed to fill Gamow's bill. He got Léon Rosenfeld and Pauli to agree that each one of them would write to Berliner to deplore the morals of the young generation, communicate their outrage concerning the Beck-Bethe-Riezler scandal, express satisfaction at seeing the Berliner correction, and profess their dismay at seeing Berliner victimized again. Gamow and Rosenfeld did write to Berliner. Pauli, who evidently had agreed to write his letter during a dinner at which a fair number of bottles of wine had been consumed, got cold feet when in a more sober mood.

Clearly Bethe found England a very congenial place. He had been quite unhappy in Germany, for most of the people around him there were "arch conservatives" who were dissatisfied with life in general and constantly ranted about the terrible disgrace of the Versailles treaty. He found people in Cambridge "relaxed, talking about much more general and more interesting things." He also discovered in Cambridge that he could make friends much more easily than in Munich. He had formed close friendships during his *Gymnasium* days, but had made no other friends in Munich—except for Rudolf Peierls. "So Cambridge was a revelation that people could be human and I had very good time there."[95]

A Rockefeller Fellow in Rome

In February 1931 Bethe went to Rome. He arrived by train and took a room in the Hotel Massimo d'Azeglio on Via Cavour, not far from the physics institute. After a few weeks in Rome Bethe wrote to Sommerfeld on April 9, "The best thing in Rome is unquestionably Fermi. It is absolutely fabulous how he immediately sees the solution to every problem that is put to him, and his ability to present such complicated things as quantum electrodynamics simply . . . I am now actually sorry that I cannot stay here longer, or as the case may be, that I did not come here for all of the Rockefeller-time." Writing to Peierls a few weeks later, Bethe indicated that he was "excited about Rome." "You can go to the city over and over again, and always find something beautiful. I am more and more enthusiastic about antiquity and less about Renaissance and Baroque churches, which do not correspond to my idea of what a church should look like. Great (and very important) is the food. You can make a discovery tour just through the menu, and the 53 different kinds of macaroni, calamari, and artichokes taste much better than the mutton and cabbage at Cambridge."[96]

As he had done in his letter to Sommerfeld, he sang Fermi's praises to Peierls. He thought Fermi had not been appreciated enough by the physics community. Fermi's "ability to summarize any problem is amazing; he can immediately tell whether or not a paper makes sense, . . . and his judgment of the theoretical and experimental (!) literature is virtually infallible. Besides that, he is the first boss who is really interested in me and my work."[97] In Rome Bethe was exposed to a much freer and more informal mode of interaction between Fermi and his students than had been the case between professors and students in Munich. In his interview with Charles Weiner and Jagdish Mehra in October 1966 Bethe characterized Fermi's institute "as a very nice small place where all of the people working there, half a dozen or so, talked to each other constantly, so there were constant discussions. Fermi was very free in telling everybody . . . what interested him and what he was proposing to do . . . We mostly talked German. Fermi spoke German very well."[98] However, when invited to Fermi's home, English was spoken because that was the only language which was common to Fermi, his wife Laura, and Bethe.

Bethe noted that when discussing physics everyone in the institute spoke freely; but everyone was very guarded regarding other matters, particularly cultural and political issues. When asked in his interview with Judith Goldstein in 1982 whether he felt the presence of fascism on that first visit to Rome, Bethe answered, "You couldn't help it; it was everywhere. The police were everywhere. It was very visible. I was very much aware of it. I was very much aware that I should be cautious in whatever I said . . . Italy was very much in the fascist hands, and that I found very repulsive. I didn't want to participate in the life of Italy, except for talking to these physicists."[99] Bethe learned very little Italian, just enough to get by on the street and in a restaurant, but he formed several friendships while there. Franco Rasetti and Emilio Segrè, two of Fermi's associates, invited him to accompany them on their weekly Sunday excursions exploring Rome's antiquities.

Bruno Rossi, a very fine experimentalist who held an appointment in Florence, invited him to come there and give a lecture. Rossi had a command of fast electronic circuits. He had designed coincidence experiments that had essentially proved that the primary cosmic radiation must be composed of corpuscular entities and was not electromagnetic radiation. Blackett had gotten Bethe very interested in cosmic rays, and Rossi cemented that interest.

A month after Bethe's arrival in Rome, one of Fermi's young men found him a room in a pleasant boarding house, the Pensione Haeslin, near the Piazzale di Porta Pia. In May Bethe invited Käte Feiss, one of the *Tanzstunde* young women with whom he was "quite in love," to join him in exploring the antiquities in Rome. She accompanied him on his three-day visit to Florence, where he gave his lecture. Thereafter they traveled to Siena and then to the south of Italy: Naples, Pompeii, and Capri. Käte's visit lasted for a "beautiful 3 weeks." He proposed marriage to her but was refused. Bethe in his recollections commented that during the entire visit they stayed "in separate rooms."[100]

Though only five years older than Hans, Fermi became another great formative influence, in addition to Hans's father and Sommerfeld. Fermi helped Bethe free himself from the rigorous and exhaustive approach that was Sommerfeld's hallmark. From Fermi Bethe learned to reason qualitatively, to obtain insights from back-of-the-envelope calculations, and to think of physics as easy and fun, as challenging problems to be solved. And Fermi's proclivity in dealing with concrete problems resonated with Bethe's own tendencies in that direction. As he recalled later,

> You would come to Fermi's office and say "Well I have this problem . . ." and explain in one sentence what you were interested in. And so Fermi would say "Now let's see. Let's sit down." And for fifteen minutes or so he would think about the problem in a general way, and at the end of the fifteen minutes you had the rough outline of the answer and then you could go back either to your lab or paper and put in the mathematics and figure it out. And so Fermi gave me the wonderful method of doing things quickly and easily . . . Sommerfeld never did that. Sommerfeld said "Well here is the title of your problem, now you do it" and then you had to put in differential equations and if possible Bessel functions. For Fermi that didn't matter. You just did the mathematics the best way that came to your mind, and the physics was clear by the time you started.[101]

Bethe's insights into Fermi's way of doing physics, which he communicated to Segrè in 1970, merit further quotation. They indicate the depth of Fermi's influence on him.

> My greatest impression of Fermi's method in theoretical physics was of its simplicity. He was able to analyze into its essentials every problem, however complicated it seemed to be. He stripped it of mathematical complications and of unnecessary formalism. In this way, often in half an hour or less, he could solve the essential physical problem involved. Of course there was not yet a mathematically complete solution, but when you left Fermi after one of these discussions, it was clear how the mathematical solution should proceed.
>
> This method was particularly impressive to me because I had come from the school of Sommerfeld in Munich who proceeded in all his work by complete mathematical solution. Having grown up in Sommerfeld's school, I thought that the method to follow was to set up the differential equation for the problem (usually the Schrödinger equation), to use your mathematical skill in finding a solution as accurate and elegant as possible, and then to discuss this solution. In the discussion finally, you would find out the qualitative features of the solution, and hence understand the physics of the problem. Sommerfeld's way was good one for many problems where the fundamental physics was already understood, but it was extremely laborious. It would usually take several months before you knew the answer to the question.[102]

Bethe noted that Fermi's mathematical skills were top-notch, but that he did not care to waste effort on inessential calculations. He approached problems pragmatically, using estimation and always keeping the essentials in view.

> By working in this manner he clarified the problems very much, especially for younger people who did not have his great knowledge. For instance, his formulation of quantum electrodynamics is so much simpler than the original Heisenberg and Pauli that it could be very easily understood. I was very much intimidated by the Heisenberg-Pauli article, and could not see the forest because of the trees. Fermi's formulation showed the forest.[103]

Bethe noted another of Fermi's characteristics as a physicist: "When a paper came which had an abstract, Fermi would read the abstract, and then he would say, 'Now, let's see how one could do that.' And that was lovely. And that's the way I used to work. From the time of my arrival to Cornell to the second World War, I did that quite often, almost regularly with the papers that I was interested in." Bethe's craftsmanship as a physicist became an amalgam of what he learned from these two great physicists and teachers, combining the best of both: the thoroughness and rigor of Sommerfeld with the clarity and simplicity of Fermi. He had learned from them how to balance rigorous analysis with approximate methods. Yet the result became more than amalgamation, in that Bethe could do things that neither Fermi nor Sommerfeld could. And whatever he subsequently did always manifested those elements that constitute discrimination, judgment, and taste in the creation of new scientific knowledge.

When Bethe came to Rome the twenty-nine-year-old Fermi was recognized as an outstanding theoretical physicist. He had made major contributions to quantum mechanics, and in particular, had formulated the statistics obeyed by particles obeying Pauli's exclusion principle—in other words, particles that have odd half-integer spin and that are now referred to as *fermions* (in contrast to *bosons,* particles with integer spin or no spin that obey the Bose-Einstein statistics). He had also given a formulation of the quantum mechanical description of the electromagnetic field that was transparent and readily usable. Recollecting his experiences in Rome in 1931–32 the ninety-six-year-old Bethe had this to say:

> Meeting the great man was a shock. This was no "highbrow" in an ivory tower. Coming into his office, you found a man who looked like other Italians on the street—but was busily doing some algebra on quantum electro-

> dynamics. The door would be open, and one or more of his colleagues would drop in whenever they felt like it. The informality was a great change from the structure of the German universities to which I was accustomed . . .
>
> Each one of us was working on his own experiment or theoretical problem. But we talked physics all the time. If we disagreed, we would refer the problem to the "pope," Fermi. He was after all, "infallible in matters of the faith."[104]

Benito Mussolini had made Fermi a member of the Accademia d'Italia, which conferred upon him the title of Excellency. This title was evidently occasionally very useful. Fermi's Institute of Physics was at the time located on top of a hill on a side street off Via Cavour, at 89A Via Panisperna, a beautiful narrow cobblestone paved road joining the Mercati Traianei and Santa Maria Maggiore. The institute shared the hill with the Ministry of the Interior. Every once in a while, this government department would hold a meeting to which many high-level officials would be invited. Two *carabinieri* would then guard the entry of the joint road to the institute and the ministry. On one such occasion Fermi drove up in his somewhat dilapidated car dressed in his usual informal attire. One of the *carabinieri* stopped him and asked him who he was. He answered that he was "the driver for his Excellency Fermi." "We cannot let you through," the guard told him. To which Fermi retorted, "His Excellency would be very annoyed if he cannot find me there." Thereupon the *carabinieri* let him through. When, with a gleam in his eyes, Fermi told that story to Bethe, he stressed that everything he had stated was true.

Enrico Fermi

Fermi was born in Rome on September 29, 1901, and grew up there. His father had been a high official in the Ministry of Communications. His mathematical aptitudes were recognized early and were nurtured by his family. At age ten he struggled to understand why the equation $x^2 + y^2 = R^2$ represented a circle and succeeded after great efforts. A little later he studied a 900-page exposition of physics written in Latin in 1840 by a Jesuit priest. That he read it from cover to cover is attested to by the numerous marginal notes in his handwriting. At age thirteen he taught himself projective geometry from a book an associate of his father had lent him and solved all the problems in it without encountering any difficulty. He was endowed with an exceptional memory and remembered the contents of all the books in physics and mathematics he had studied. At the age of eighteen he stated categorically, "I studied mathematics with passion because I considered it necessary for the study of physics, to which I want to dedicate myself exclusively."

In 1918 he won a fellowship to attend the Scuola Normale Superiore of Pisa, then the outstanding Italian institution of higher education in the sciences. The fellowship also allowed him to be a student at the University of Pisa, where he obtained his doctorate in physics in 1922. By that time he had studied on his own and mastered Hermann Weyl's exposition of general relativity in *Raum, Zeit, Materie*; Appell's *Mécanique Rationelle*; Poincaré's *Théorie des Tourbillons*; and Nernst's *Theoretische Chemie.* In addition, he had acquired an extensive and deep knowledge of mathematics. After he obtained his doctorate he spent a year in Rome and became acquainted with Senator Orso Mario Corbino, the professor of experimental physics and the director of the Physics Institute at the University of Rome. Corbino immediately recognized Fermi's exceptional talents and used his influence to advance and promote Fermi's career. In 1923 Fermi received a scholarship from the Italian government and spent some months in Göttingen with Max Born, who had just been appointed director of its Institute of Theoretical Physics. He was not particularly happy there and did not become a member of the inner circle around Born and Pauli. In 1924 he was awarded a Rockefeller Foundation Fellowship and went to Leyden to work with Paul Ehrenfest. The following year he became a "Libero Docento," which allowed him to teach at Italian universities. He accepted a lectureship in mathematical physics and mechanics at the University of Florence, a post he held for two years. In the fall of 1926 Fermi was appointed Professor of Theoretical Physics at the University of Rome, a chair that Corbino had established with Fermi in mind. Fermi there created an outstanding school of physics. His first students were Emilio Segrè, Edoardo Amaldi, Ettore Majorana, Bruno Rossi, Gian Carlo Wick, and Giulio Racah—who all became outstanding physicists.

Physics Research while in Rome

Before going to Rome Bethe had accepted writing an article for the *Handbuch der Physik* concerning one- and two-electron atoms. In April Sommerfeld invited him to join him in writing "The Theory of Metals" for the *Handbuch.* On April 25 Bethe replied

> I thank you for the offer that we together write the *Handbuch* article about electrons in metals. In itself that would naturally attract me very much, but I fear just as you yourself do, that I will load myself up with too much writing. Here in Rome I will naturally not begin to write so as to properly utilize the time here, and two *Handbuch* articles (the first one I presumably owe to you as well!) from July to December would mean a nearly continuous occupation with writing, and that is a bit much. It would be a different matter if the article were not due until April 32. Then the matter would attract me sufficiently to say "yes," as I would certainly learn a lot, and the honorarium is very decent. Under the conditions given I would prefer to decline the offer, especially as I know that you are of the same opinion.

Adolph Smekal, the editor of the *Handbuch,* accepted having the article submitted in March 1932, so Bethe committed himself to writing it after all.

Of Wave Functions

When Bethe came to Rome in 1931 he was doing some of the range calculations that Blackett had asked him to do. As Bethe wanted to compute the range quantitatively, he needed the wave functions of electrons bound not only in hydrogen and helium, but also in complex atoms. A Hartree-Fock wave function was the obvious choice for the approximate wave function but, as Bethe noted in his reminiscences of his stays in Rome,

> A Hartree-Fock calculation was beyond my capacity. When I consulted Fermi, he suggested that I start from the electron distribution in his statistical atom.[105] I took his suggestion, and calculated the electrostatic energy made by the [Fermi-Thomas] statistical-atom distribution and the nucleus. Then I calculated the electron wave functions in this electrostatic potential. Fermi decided to include my result in his handwritten encyclopedia, which he called *ψarum* ("Of wave functions"). I was very proud.[106]

Bethe Ansatz

While in Frankfurt Bethe had worked on a quantum mechanical "toy" model consisting of a linear chain of *N* atoms that had but two levels (described by a spin 1/2 variable), in which each atom interacted only with its nearest neighbors. Bethe's interest in statistical mechanics, the branch of physics whose aim is to account for the thermal macroscopic properties of systems in terms of the dynamics of its components, had been stimulated by the success of Sommerfeld's theory of metals. In that theory the conduction electrons were assumed to be "free," obeying Fermi-Dirac statistics. The challenge was to justify the assumption of freely moving conduction electrons beyond what Bloch had shown. Bethe's model was designed to extend Heitler and London's theory of the hydrogen molecule to a whole chain of atoms and be a model of a one-dimensional metal composed of atoms with a single valence electron outside closed shells. The title of the paper, "Zur Theorie der Metalle. I," reflects this.[107] The particular mathematical form of the model was probably inspired by an article of H. Ludloff in the *Zeitschrift für Physik,* as indicated by Bethe's letter to Sommerfeld on June 30.[108]

> I have begun to calculate ferromagnetism. The work of Ludloff about entropy at absolute zero in one of the latest issues of ZS is the starting point. One can easily find the general eigenfunctions for the Ludloff problem—chain of atoms in the s-state without spin—, and from these one can immediately read off Ludloff's eigenvalues. (L. had only guessed, albeit correctly, the eigenvalues for the general case in which more than two atoms of the chain have an electron angular momentum directed toward the right.) I now have a method for treating the problem of ferromagnetism "exactly" (i.e., in the first approximation of the London-Heitler method). The complication in this case—chain of atoms in the S-state with spin—lies in that one obtains different energies according as how many pairs of parallel spins are adjacent to each other. So far I have carried through the case where two spins point to the right and all others left.

Bethe was able to solve his model exactly.

The Hamiltonian for the model is

$$H = \frac{J}{2} \sum_{i=1}^{N} (\boldsymbol{\sigma}_i \cdot \boldsymbol{\sigma}_{i+1} + 1)$$

where $\boldsymbol{\sigma}_i = (\sigma_x, \sigma_y, \sigma_z)_i$ are the Pauli matrices representing the spin 1/2 atom at the site *i.* There are *N* atoms in the chain, and to avoid endpoint effects Bethe imposed periodic boundary conditions such that $\boldsymbol{\sigma}_{i+L} = \boldsymbol{\sigma}_i$. When the constant *J* is negative, parallel spins are energetically favored, and the model corresponds to the Heisenberg ferromagnet. If *J* is positive, antiparallel spins are favored, and the model corresponds to the Heisenberg antiferromagnet.

Consider the case when *J* is negative. Then the state of lowest energy is that state in which all the atoms have their spin up. Since $J_z = \frac{1}{2} \sum_{i=1}^{N} (\sigma_{iz})$ commutes with the Hamiltonian the eigenfunctions of *H* can simultaneously be made eigenfunctions of J_z. If the N atom state in which the n spins $x_1,x_2, \ldots, x_n$ are down is denoted by $|x_1,x_2, \ldots, x_n>$, the eigenfunctions corresponding to J_z having the value $1/2(N - n)$ can be represented by linear combinations of these basis states

$$|\psi> = \sum_{1 \le x_1 \le x_2 \ldots x_n \le L} a(x_1, x_2, \ldots, x_n) \,|\, x_1, x_2, \ldots, x_n > .$$

By analyzing the eigenfunctions for $n = 1,2,3, \ldots$[109] Bethe's inspired insight was to recognize that the amplitudes $a(x_1,x_2, \ldots, x_n)$ are of the form

$$a(x_1, x_2, \ldots, x_L) = \sum_{P \varepsilon S_n} A_P \exp\left(\sum_{j=1}^{n} i k_{P_j} x_j \right)$$

where the sum P is over all $n!$ permutations of the integers 1 to n, and the coefficients A_P are given by

$$A_P = \varepsilon_P \prod_{1 \le i < j \le n} \left(1 - 2\exp(k_i) + \exp(k_i + k_j)\right)$$

where ε_P is the signature of the permutation. For $|\psi>$ to satisfy the periodic boundary conditions and be a solution of $H|\psi> = E|\psi>$, the variables k_i must satisfy the system of equations

$$\exp(ik_i L) = (-1)^{n-1} \prod_{l \ne j} \frac{\left(1 - 2\exp(ik_l) + \exp(k_l + k_j)\right)}{\left(1 - 2\exp(ik_j) + \exp(k_l + k_j)\right)}$$

for $j = 1,2, \ldots, n$. These nonlinear equations for the wave numbers k_j have become known as the Bethe ansatz equations. The energy eigenvalues are expressed in terms of the k_j

$$E = JN + 2J \sum_{j=1}^{n} (\cos k_j - 1).$$

Bethe demonstrated that his ansatz generated 2^N energy eigenvalues, and therefore that it produced the complete energy spectrum of the system described by the Hamiltonian.

As should be apparent, obtaining the solution was a brilliant achievement.[110] Thierry Giamarchi in his 2004 monograph, *Quantum Physics in One Dimension,* introduced his discussion of the Bethe ansatz as follows: "Bethe-ansatz provides a nice analytical solution of various one-dimensional systems. It is very elegant but quite complicated method, so only a very basic description can be done in the context of this book."[111] Murray T. Batchelor, in a very informative article in *Physics Today* in January 2007, reviewed the Bethe ansatz after seventy-five years. The intent of his article was summarized in its abstract, namely, that Bethe's 1931 solution which "lay in obscurity for decades . . . now finds its way into everything from superconductors to string theory." Its conclusion was that the "Bethe ansatz is alive and well."[112] Gerald Brown, in the introduction to the volume *Hans Bethe and his Physics,* which he edited, stated that he thought that Bethe's ansatz paper "probably has had the most influence over a wider variety of fields of any of Hans' works."[113]

Starting in the mid-1960s, when the method was rediscovered, the ansatz made possible exact solutions of a host of one-dimensional quantum problems and of two-dimensional classical statistical mechanics problems. The ansatz could solve relevant quantum many-body problems for which perturbative approaches had failed. It opened new fields of "pure mathematical" research. Thus its generativity and fruitfulness became firmly established. The recent experimental advances in creating one- and two-dimensional optical lattices and in the manipulation of atoms in such lattices, and also in the creation of two-dimensional materials such as graphene, suggests that the Bethe ansatz will continue to be exploited in the solution of relevant problems for a long time to come.

The Handbuch Articles

Bethe's Rockefeller fellowship terminated at the end of June 1931. The Rockefeller board offered Bethe another year, which he accepted but for only part of the year, and he postponed assuming it for six months. Bethe would return to Rome in spring 1932, but in the meantime was obligated to write two lengthy reviews for the supplement to volume 24 of the *Handbuch der Physik* that Adolf Smekal was editing.[114] One of these reviews was to present the state of knowledge concerning one- and two-electron atoms; the other, on electrons in metals, was to be written with Sommerfeld.

During that second trip he renewed his friendship with Rasetti and Segrè, and got to know Edoardo Amaldi, who had just come to Rome to study with Fermi. But his closest professional colleague was George Placzek.

Placzek was born in 1905 in Brno, Moravia, in what is now the Czech Republic but which in 1905 was part of Austria-Hungary. He was the oldest of the three children of Alfred and Marianne Placzek. His parents were rich, assimilated Jews, fully integrated into the mixed German-Czech culture of Brno. Placzek attended the Deutsches Staatsgymnasium in Brno, then enrolled in the University of Vienna, but also spent some time in Prague taking courses at Charles University. He obtained his doctorate in physics from the University of Vienna in 1928. Thereafter he spent three years in Utrecht working with Hendrik Kramers, and in 1930–31 went to Leipzig to work with Heisenberg. He there became good friends with Peierls, Weisskopf, and Wick. Wick sang the praises of Fermi, and in 1931–32 Placzek was in Rome, and Bethe became his close friend.[115] They shared a room in the institute. "It was a big room, . . . with very big windows, and we had an extremely long table. I sat at one end, and he sat at the other end," Bethe noted. They would eat lunch and dinner together.

> I would usually come shortly before noon. I always was a late riser, and became an even later one there. I used to work often until four in the morning in the hotel room, and then go to sleep and get up around noon and go to the lab, skip breakfast, and go to lunch with Placzek.
>
> We would take the bus to the center of town, and find a place where to eat. He would say "I want to have something very light," and then in the end he always settled on something like pork chops, roast goose, or something which was obviously not light at all.

Bethe further commented,

> Placzek had the highest standards of anybody I have ever known, both in ethics and in science, and his high standards in science were very detrimental to his productivity. He never was satisfied with anything he himself did.
>
> I was writing my one and two electron problems, and he was writing an article for another encyclopedia, Marx's *Handbuch der Radiologie,* in which he finally produced an article which has been used quite a lot.[116] But it was certainly a very difficult birth. He would have a piece of paper in front of him, and then in the middle of the piece of paper somewhere, he would write something in very large letters, and that was all. Then he would disappear into the library and investigate the history of the scattering of light, going back to the early 1800, and the library in Rome was especially good at that. I'm surprised it didn't contain any Roman books. Placzek would delve for the whole day then into some paper of 1830 or so, in which somebody had done something about the scattering of light, all as part of his article about scattering of light: Rayleigh scattering and the Raman effect. And he wanted everything proved very precisely in his own work. After

having filled a few sheets of paper, he would then talk to me at dinner, and say, "I have again discovered a scientific paradox," and he would explain that over dinner. We would then try to resolve it, sometimes successfully, and then he would continue the next day. And so, his article progressed very slowly, and he constantly said, "Well, someday Mr. Marx will come here"—Marx was the editor of that Encyclopedia—"and somehow tell me to deliver the manuscript immediately." I think the manuscript was overdue already by a year or two, and in fact, at the end when Placzek had gone to Copenhagen, after Rome, Mr. Marx, or his emissary, did come, and took whatever Placzek had written, which was very incomplete, and said "well, never mind, that it is incomplete, I'll take what you have," which made Placzek very unhappy. Then somehow he managed, in proof, to add the second half. This is very characteristic of Placzek.

When Bethe was asked by Karl Scheel, the overall editor of the *Handbuch der Physik,* to write an article on the quantum mechanical description of hydrogen and helium—undoubtedly because of high recommendations from Sommerfeld—he was very happy to do so. It very much interested him, and "it was very good money. Springer paid sixteen marks per printed page, which was a lot of money in those days, and salaries were very meager, so it was quite useful to have this additional money."[117]

In his interview with Charles Weiner in 1963 Bethe explained what he had tried to do in the one- and two-electron article in the *Handbuch*:

I wrote down everything I knew. The beginning was fairly obvious, to derive the wave functions of hydrogen by the standard methods and then discuss the wave functions. I added a little bit, calculating various averages, which might be useful. On helium . . . I put in everything I knew, starting with the early attempts of Heisenberg's using perturbation theory and then leading up to Hylleraas' work [using variational methods], which gives much better values for the energy of the various levels. Then I discussed the Stark effect and the Zeeman effect, again using mostly the literature. I supplemented the literature on the Zeeman effect in writing more about the Zeeman effect of the fine structure. And I think also in the case of the Stark effect I calculated how you go from the normal unperturbed Dirac state of the hydrogen atom to the Stark effect, which means a change of the quantum number. So I filled in wherever I thought there was a gap. Then I filled in a lot when it came to the optical transition probabilities. Many of them had been calculated. I supplemented the tables. I presented the tables in a different fashion, which made them a little more useful to the experimenters, giving tables of oscillator strength and transition probabilities, per unit time. Then I wrote down a number of conclusions from this: just how does the hydrogen atom behave when it is in an excited state? And I discussed the complicated questions about the intensity of spectral line, and in particular, how to find out just how many atoms are in the excited state, and that's particularly difficult in the case of hydrogen because the excited states are degenerate, and so you have to know how many are in each of the degenerate states and each state of different *l*-value. I found a lot of things in the theory where the basis existed but where the material had really not been put in a form useful to the experimenter, and I tried to do that. One of the original points I put in with regard to transition probability had to do with the Stark effect in hydrogen. The $n = 2$ state of hydrogen is a quite complicated state (it has a 2s state and a 2p state, and there is a fine structure between 2p 1/2 and 2p 3/2). Now, when you have an electric field, then the s and the p state get mixed but they start out at different energies. So I discussed just how the mixture takes place in the presence of the fine structure splitting and also in the presence of the strong radiative probability, which exists in the 2p state and not in the 2s state. The 2p state constantly decays in making the Lyman line, and the 2s state is metastable. Now, this work later on was very useful to Lamb when he discovered the Lamb shift because it told you about the behavior of these states under the influence of all these forces. Then there were the problems of continuum radiation on which Sommerfeld had worked very much, and I took Sommerfeld's papers and other papers, Wentzel's and Stauber's and so on, and again tried to exploit them to get things useful for the experimenter. There's the photoelectric effect, which is very important for X-ray absorption. How do you actually calculate X-ray absorption? What these people had done was to take the simple atom with just one electron around it in the lowest shell, in the k shell, and calculate the transition probabilities to the continuous spectrum where the k electron is removed. But that's not what a real atom is like. So I put in the corrections for the so-called screening; and discussed different kinds of screening, outer electrons and inner electrons and what effect does that have on photo-electric absorption. So this is the kind of thing I did with the existing theory. Then there came the collision theory and . . . the theory of stopping power . . . Part of that is in the *Handbuch* article . . . and part of it, in the paper "Stopping of Electrons of Relativistic Velocities," which I did in Rome.

In their chapter on Bethe's career in the *Festschrift* dedicated to him on the occasion of his sixtieth birthday by his students and friends, Robert Bacher and Victor Weisskopf commented that the one- and two-electron article in the *Handbuch* was unusual in several respects.[118] First of all, it was most unusual in that the 1959 revised edition, written in collaboration with Edwin Salpeter, was almost identical with the original version.[119] In fact, not much had to be added in order to bring it up to date after twenty-five years, apart from new experimental results and the presentation of the refinements stemming from the renormalization program in quantum electrodynamics. These refinements allowed the circumvention of the divergence difficulties the theory had been plagued with during the 1930s and made possible the calculation of the energy difference in the 2s-2p levels of hydrogen, the Lamb shift, and the calculation of the corrections to the magnetic moment of the electron predicted by the Dirac equation. Bacher and Weisskopf also commented that the article appeared "unusual in its breadth" in that "It contains under its modest title all essential parts of atomic and molecular physics: not only is there an exhaustive study of the one-and-two electron spectra, transition probabilities, etc., but also relativistic effects, fine and hyperfine structure, collision processes, Stark and Zeeman effects, radiation phenomena, electromagnetic field couplings, etc. In short, it is a textbook of nonrelativistic and relativistic quantum mechanics."[120] Indeed, Bethe's one- and two-electron *Handbuch* article was *the* presentation of quantum mechanics from which an entire generation of physicists learned how to address the problems of immediate relevance to experimentalists during the 1930s.

Bethe told the following story regarding how he was asked by Sommerfeld to help him write the article on electrons in metals. When Sommerfeld was giving his seminars on the application of Fermi-Dirac statistics to electrons in metals he stressed that the Fermi-Dirac statistics implied that the electrons near the top of the distribution have considerable kinetic energy, on the order of 10eV for a dense metal. When Bethe worked on his thesis dealing with electron diffraction, he made the suggestion to Sommerfeld that the observed work function of a metal—the minimum energy to photo-eject an electron from a metal—was the difference between the negative potential energy he had introduced in his *Wissenschaften* paper and the positive Fermi energy. "This is not accurate enough to calculate the work function, but it explains the phenomenon qualitatively. After this suggestion, Sommerfeld considered me a great expert in solid state physics, which I was not. But his favorable estimate led to our joint article of the electron theory of metals, . . . a report that gave me a great deal of pleasure."[121]

Bacher and Weisskopf's assessment of the review of electrons in metals in the *Handbuch* was the following: "It contains the basis of what today is solid state physics. It is astounding how clear, simple and complete Bethe's presentation was in the year 1933 when this branch of physics was in its infancy. We find an exhaustive discussion of Brillouin-zones, of radiative processes in metals, of dia-, para-, and ferromagnetism, of all kinds of thermoelectric and galvanomagnetic effects, and of course, a thorough discussion of the theory of conductivity."[122] In fact, the electrons in metals article is probably a more impressive achievement than the one- and two-electron article. Bethe took the most important recent development—the approach that Wigner and Seitz had formulated to deal with the problem of the structure of metals, their cohesion, band structure, and electronic and thermal conductivity—and made it the basis for a coherent and comprehensive treatment of the subject. In the process he extended their method, presented much new material regarding the Brillouin zones of metals whose crystal symmetries were known, and extended his own calculations regarding electron diffraction, cohesion of metals, spin waves, transport properties, and so on.

The amount of experimental and theoretical materials that Bethe had to critically analyze, synthesize, extend, and present in an understandable and generative manner is staggering. The influence of the article can be gauged by how much the text Mott and Jones wrote in 1936 incorporated materials found in Bethe's *Handbuch* article and how often they reference him. Mott himself indicated that "We [Mott and Jones] based our book very much on Bethe's *Handbuch* article, and tried to extend its influence by sorting out the differences between real materials, and making approximations and using intuition whenever we liked."[123] The importance of Bethe's contributions to the development of the quantum theory of the solid state can be assessed by the number of times he is referred to in Lillian Hoddeson's *Out of the Crystal Maze* and others' histories of the subject: only Bloch has more citations.

It took a little over a year and a half for Bethe to complete the two *Handbuch* articles, and this constituted the major part of his published work during this period. Bethe indicated to Weiner, "I did a little work on the side. But I would say two-thirds of my work during each of these years went into writing these articles." What some of the other third of his work consisted of was revealed in the letter Bethe wrote to Sommerfeld in April 1932 during his second visit to Rome, to be discussed in the next section. A further insight into how Bethe wrote his *Handbuch* article with Sommerfeld is obtained from Bethe's February 2003 interview with David Mermin, a distinguished theoretical physicist and a colleague of Bethe at Cornell, who together with Neil

Ashcroft wrote a widely acclaimed exposition of solid state physics that became the leading textbook on the subject.

MERMIN: In thinking about this conversation, . . . I looked at your long article with Sommerfeld on the electron theory of metals. And I was struck by how big it is. It makes a 300-page, small-print book. And the other thing that struck me was how much duplication there is between your 1933 book and my 1976 book with Neil Ashcroft on solid-state physics—how much was already known in 1933.

BETHE: Shows the stability of the theory.

MERMIN: Yes. The other thing that struck me was that it took me and Neil eight years to write our book, and I believe it took you . . .

BETHE: It took me one year.

MERMIN: Yes. And the question is how? Why did it only take you one year to write such a massive. . . ?

BETHE: Well, there was a lot of it known at the time, and it was very close to our daily work.

MERMIN: But that book is almost 300 pages of very small print with many figures, and I presume you were doing other things at the time.

BETHE: Yes, I was. By that time my real interest was nuclear physics, which started with Chadwick's discovery of the neutron in '32.

MERMIN: Are you the kind of writer who can just write things once and it's perfect?

BETHE: I was at that time. A friend of mine, George Placzek, claimed that I just had two stacks of papers—one was white and the other I had written on—and I wrote continuously and turned it over to the other pile which had the finished product. Fermi worked the same way.

MERMIN: And in addition, I assume you didn't sleep very much?

BETHE: Oh, I slept a great deal.

MERMIN: Really? I'm very impressed.

BETHE: Eight hours a day.

MERMIN: I noticed that there's a footnote to chapter 1 by Sommerfeld saying that he wrote the first chapter on free electrons, which is 35 pages or so, and that you wrote the rest of the book. Nevertheless, the book is Sommerfeld and Bethe, and not Bethe and Sommerfeld.

BETHE: Well, Sommerfeld was a famous physicist . . . Sommerfeld was known all over the world, so in order to show that this was a serious book, Sommerfeld had to come first.[124]

One of the reasons Bethe was able to write his two *Handbuch* articles in less than two years was that his method of gathering information was reading: "Listening might get me interested, but couldn't get me full knowledge of it, as reading gave me." And when reading scientific papers, "In most cases, I would actually follow the calculation, and do the algebra. I read the papers with great intensity." After he had read a paper, worked it out, corrected any mistakes in it, and extended it if necessary, "it was stored." And once stored, the information stayed with him. If he had to use that material, he didn't have to start at the beginning of the paper: in fact, he would be able to come in at the place where needed, and very often use the material "without looking at the paper." "It was not a photographic memory. I couldn't have told you, this formula is in the middle of the page on the left-hand side, some people can, but it was a logical memory." This was the case for physics and mathematics, "not nearly as much for poetry or literature." Thus when writing the one- and two-electron article for the *Handbuch* he had at his fingertips "the energy levels of hydrogen and helium, including fine and hyperfine structure, their wave functions in lowest approximation," and much more.

Clearly, one of the traits that made possible Bethe's remarkable accomplishments was his amazing memory. His wife, Rose, characterized this aspect of his brain as a "filing cabinet and a computer." A filing cabinet from which essentially nothing got lost and from which accurate information could always be readily retrieved, and simultaneously, a thinking machine that could draw the logical and quantitative conclusions from the explicit assumptions and data of the input with unerring accuracy. He himself admitted that this was the case—at least until he was in his sixties. He became aware that not every person had this facility "quite late." Only in Sommerfeld's seminar did he notice that "some other members didn't have quite the same facility."

When the one- and two-electron *Handbuch* article is taken together with the solid state one, the two articles are paradigmatic of a central concern in Bethe's research: understanding the properties of systems possessing a large number of degrees of freedom in terms of the properties of the constitutive entities and their interactions. The atomic, molecular, and solid state systems considered in the *Handbuch* articles were the first cases where the interactions between the constituents are thought to be well approximated to first order by static two-body potentials—the Coulomb potential, the spin-orbit magnetic interaction, the spin-spin interactions—and the dynamics governed by non-relativistic quantum mechanics. In lowest approximation, the model for such atomic and molecular systems usually adopted by Bethe was based on Hartree's intuitive argument that an individual particle would respond to the average field generated by interactions with surrounding particles, generalized to automatically take into account the Pauli principle. These same assumptions were later invoked by Bethe when first trying to understand the structure and properties of nuclei.

By writing his two *Handbuch* articles Bethe came to see how the constituents of the atomic world—atoms and molecules—with their unfamiliar and at times strange rules of behavior and unfamiliar logic could be seen to give rise to the macroscopic world, whose structures follow classical rules with their familiar logic and causality. Or perhaps better stated, Bethe came to see how the classical world could be seen to "emerge" from the quantum one; "emerge" because the classical world had properties not attributable to its microscopic constituents, and derivable in special situations only by virtue of approximations that were mathematically difficult—if not impossible—to justify.

Through these *Handbuch* articles, calculating and understanding "mathematically" became for Bethe "knowing" and having an "intuition" on how the microscopic world operated. The atomic world acquired a reality for him, but a reality different and distinct from that of the macroscopic world.

Nuclear Physics

Bethe's interest in nuclear physics was sparked while in Rome in 1932. During Bethe's first visit to Rome in 1931, Fermi "was completely a theoretical physicist without any question. But in 1932 while he was still doing theoretical work, he was determined to go into experimental nuclear physics."

Fermi had become extremely puzzled by the experiments by Walther Bothe in Germany and by Frédéric Joliot in France which indicated that a "radiation" is emitted when beryllium is bombarded by alpha particles, and that that radiation had none of the properties of any known radiation. The radiation did not consist of charged particles because it went through matter very easily. Nor did it consist of gamma rays because its absorption in materials like paraffin was as strong or stronger than its absorption in the same thickness of lead. Had it been gamma rays, the difference would have been huge, with lead absorbing the radiation much more strongly. In addition, this radiation seemed to be able to eject protons from nuclei, which would also be very puzzling were the radiation gamma rays: you would expect electrons to be ejected when gamma rays interact with atoms and not protons. Thus it was a most peculiar radiation. Bethe averred that "The existence of this radiation persuaded Fermi to go into nuclear physics, and in 1932, in the spring while I was there, Fermi determined that he would work experimentally on nuclear physics." "But nuclear physics at that time was still too unclear to appeal to me. Later in 1932 Chadwick found the right solution—I don't know exactly which month—and the right solution was the neutron. He very soon did a series of really quite ingenious experiments showing beyond any question that this was a neutral particle of a mass very close to that of the proton."[125]

Only thereafter did Bethe start working in nuclear physics. A month after having read Chadwick's article Bethe wrote to Sommerfeld from Rome on April 20, 1932:

> I have, following the momentary fashion, occupied myself with Chadwick's neutron and have tried to calculate its collision-probability. For the collision of neutrons with protons or heavier nuclei the results were reasonable, whereas the probability of collision with electrons should be nearly zero which appears to contradict experience. I believe one must proceed in this way if one really wants to know something about the neutron: one makes diverse plausible "Ansätze" for the interaction energy with other particles and then calculates the scattering and compares with experiments. In any case the neutron is a highly interesting object and finally there is again something new in physics in principle . . .
>
> Fermi is doing experiments with neutrons, or as the case may be, making preparations for them. For that reason, I am sorry to say, I see him somewhat less than I did last year. It is also partly the fault of the *Handbuch,* because I have less to tell. Besides myself, Placzek and Teller are here, and until recently also [Fritz] London.[126] We have very many interesting discussions with each other and go on outings together as well.

Quantum Electrodynamics

During his stay in Rome in the spring of 1932, Bethe and Fermi collaborated on a paper involving quantum electrodynamics (QED). It was from Fermi's articles that Bethe learned quantum electrodynamics, the field to which Bethe made some of his most seminal contributions. Before turning to that subject let me briefly recall the history of the subject.

Most of the empirical data that channeled the development of quantum mechanics stemmed from atomic spectroscopy. The quantum mechanical rules for the description of systems of microscopic particles as formulated in the years 1925 and 1926 consisted in promoting the dynamical variables describing the system classically into operators satisfying certain commutation rules. By this operation, the initial classical particle language that described the system had to be complemented with a wave picture language. One of the theoretical challenges was the description of the emission and absorption of light by atoms. The problem was addressed not by considering the realistic case of a weak radiation field—for whose description the photon concept is central—but by considering the case of a strong field, in which the energy density is so high that a huge number of photons are present and the discrete nature of the electromagnetic field is averaged out. Such a field can be considered as a given classical external field, and its scalar and vector potentials appear in the Schrödinger equation in the same way as the Coulomb potential—as given functions of space and time. And so one can calculate transition probabilities as a function of the intensity of the light wave and also obtain the probability for the inverse transition: stimulated emission by the light wave. Einstein's relation between absorption, stimulated and spontaneous emission then allowed one to obtain the transition probability for the latter.

The initial step extending the formalism to encompass the interaction of charged particles with the electromagnetic field when both are treated dynamically and quantum mechanically was taken by Dirac in 1927. In that first paper the electromagnetic field was described as an infinite assembly of massless, spin 1 bosons. For Dirac, "particles" (whether they have a rest mass or, like photons, were massless) are the "fundamental" entities. During the initial formulations of quantum mechanics "particles" such as electrons were conceived very much like they had been in classical physics: immutable objects with essentially fixed properties such as their mass, spin, and electric charge but which under certain circumstances exhibited wave properties as characterized by their de Broglie wavelength. By virtue of their indistinguishability "particles" were described collectively by wave functions with special symmetry properties: the wave function describing a system of identical particles with zero or integral spin remains unchanged under the interchange of any two particles, whereas that describing odd half-integral spin particles changes sign under such a transposition.

Interestingly, in his first QED paper Dirac, still believing in "particle" conservation, introduced a vacuum that consisted of an infinite number of zero momentum, zero energy photons and considered the emission of a photon by an electron in an atom as a photon having made a transition from the vacuum to a state of momentum and energy appropriate for total energy and momentum conservation in the transition.[127] It is only in his second paper on QED that photon creation and annihilation are irreducible events: the photon did not "exist" before the atomic electron made its transition.[128]

In contrast to Dirac, Pascual Jordan championed the point of view that fields constituted the "fundamental" substance: that they were the basic ontology. Jordan advanced a unitary view of nature in which both matter and radiation were described by wave fields. The quantization of these wave fields then exhibited the particle nature of the quanta and thus elucidated the mystery of the particle-wave duality. An immediate consequence of his approach was an answer to the question: "Why are all particles of a given species (as characterized by their mass and spin) indistinguishable from one another?" The answer was that the field quantization rules made them automatically thus, since the "particles" were excitations of the same underlying field.[129]

By the early 1930s it had become clear that particle creation and annihilation—as first encountered in the quantum mechanical description of the emission and absorption of photons by charged particles—was the genuinely novel feature of quantum field theory (QFT). Up to that time the description of microscopic phenomena had been predicated on a metaphysics that assumed conservation of the number of "particles": that material particles were neither created nor destroyed. Dirac's quantum electrodynamics was the first step in the elimination of that preconception.[130]

Similarly, Dirac's "hole theory" was the first instance of a relativistic quantum theory in which the creation and annihilation of matter was an intrinsic feature.[131] Dirac had formulated hole theory after recognizing that the equation he had advanced in 1928 to describe a *relativistic* spin 1/2 particle, besides possessing solutions of positive energy, also had solutions with negative energy. In the quantum theory—in contradiction to the classical situation—the states with negative energy could not be ignored, for transitions to them cannot be ruled out. In 1930, in a paper entitled "Theory of Electrons and Protons," Dirac noted that in quantum electrodynamics "transitions can take place in which the energy of the electron changes from a positive to a negative value even in the absence of any external field, the surplus energy, at least

mc^2 in amount, being spontaneously emitted in the form of radiation." He also pointed out that the negative energy difficulty is a common feature of all relativistic theories, and observed that "an electron with negative energy moves in an external field as though it carries a positive charge," which result had suggested to some people "a connection between the negative energy electron and the proton."

Dirac then advanced the hypothesis: "that there are so many electrons in the world that all the states of negative energy are occupied except perhaps a few of small velocity . . . Only the small departure from exact uniformity, brought about by some of the negative-energy states being unoccupied, can we hope to observe. We are therefore led to the assumption that the holes in the distribution of negative electrons are the protons." To answer the objection that the proton is much heavier than the electron, Dirac speculated that the interactions among the negative energy particles might change their mass.[132] Thus conceived, the vacuum was a region of space that is in its lowest possible energy state. As one fills up the negative energy states, a lower and lower total energy is obtained. When all the negative energy states are filled, the minimum energy is obtained. One thus spoke of a sea of negative energy electrons, with one electron in each of the negative energy states of a given momentum and spin.

When Dirac advanced his hole picture he was aware that the theory was symmetrical between positive and negative energies, so that a "hole" should have the same mass as an electron. In 1978 Dirac recalled that at that time the only positively charged particle that was known was the proton. Furthermore, it was believed that all of matter was to be explained just in terms of electrons and protons: "I just didn't dare to postulate a new particle at that stage, because the whole climate of opinion at that time was against new particles. So I thought this hole would have to be a proton . . . [and] I published my paper on this subject as a theory of electrons and protons."[133]

Objections by Oppenheimer, at the time a young professor of theoretical physics at the University of California in Berkeley, by Igor Tamm in Moscow, and by Hermann Weyl in Göttingen forced Dirac to alter his views. Oppenheimer and Tamm both computed the rate of annihilation of protons and electrons into γ-rays in Dirac's hole theory when protons corresponded to holes, and found that the instability of the hydrogen atom predicted on this picture would readily be observable (and in sharp disagreement with the observed stability of the atom). Weyl, on his part, pointed out that the inversion symmetry of the Dirac equation demanded that a hole have the same mass as the electron. In a talk before the British Association for the Advancement of Science (BAAS) in Bristol in 1930, Dirac indicated that Oppenheimer had suggested that "all and not merely nearly all" of the negative energy states for an electron are occupied, so that a positive energy electron can never make a transition to a negative energy state.[134]

Dirac adopted this proposal in 1931 in his paper, "Quantised singularities in the electromagnetic field," and noted that "A hole . . . would be a new kind of particle, unknown to experimental physics, having the same mass and opposite charge to an electron. We should not expect to find any of them in nature, on account of their rapid rate of recombination with electrons, but if they could be produced experimentally in high vacuum they would be quite stable and amenable to observation."[135] Blackett, who was working at the Cavendish Laboratory in Cambridge at the time, told Dirac that he and Giuseppe Occhialini had evidence for this new kind of particle. Dirac's recollections of the interactions between him and Blackett at the time were quite "intimate" and intense.[136] But Blackett did not want to publish his results without further proof, and "while he was obtaining the corroboration, Anderson quite independently published his evidence to show that the positron really existed."[137]

The history of the conceptual developments associated with the discovery of the positron provides a vivid illustration of the struggles involved in overcoming ingrained ways of seeing the world. What is striking is the conservatism of the leading theorists, such as Dirac, Pauli, and Bohr, on the issue of postulating the existence of new particles, while at the same time advocating revolutionary stands in theory formation. It reflects the fact that physics was to be an empirical science, and postulating the existence of an unobserved particle was too speculative.[138]

Bethe kept abreast of all these developments, but only started doing work in these fields after his second visit to Rome, after mastering Fermi's approach to quantum electrodynamics.

Fermi and Quantum Electrodynamics

In 1926, shortly after the publication of Schrödinger's wave mechanics papers, which Enrico Fermi had carefully studied, he wrote a short article entitled "Arguments pro and con the hypothesis of light quanta." In it he indicated that "at the present time the state of science is such that one can say that we lack a theory that gives a satisfactory account of optical phenomena." He listed the experiments that were convincingly explained by light being assumed to be constituted of (particlelike) photons such as the photoelectric effect and the Compton effect, and the principal ones that were readily understood in terms of the wave theory, namely, interference and diffraction. The challenge, Fermi asserted, was to elucidate the processes involved in the interaction of light with matter at the atomic level and to give intuitive explanations of optical phenomena at this microscopic level.[139] Fermi met the challenge.

In a series of papers from 1929 to 1932 he formulated a relativistically invariant description of the interaction between charged particles and the electromagnetic (e.m.) field, which treated both the particles and the e.m. field quantum mechanically. Note, however, that the charged particles were not treated field theoretically.[140] Fermi first devised a simple, readily interpretable, Hamiltonian description of charged particles interacting with the e.m. field, then indicated how to quantize this formulation, and thereafter showed how to exploit perturbation theory to describe quantum electrodynamic phenomena.

There were others who tackled these same problems, but none of their formulations—in particular, Heisenberg and Pauli's papers[141]—had the simplicity, *Anschaulichkeit* (clarity), yet thoroughness of Fermi's approach. Fermi's three accomplishments in his papers were: First, to provide a simple, visualizable way to describe the interactions between photons and charged particles. It was a formulation from which an entire generation learned how to think about quantum electrodynamic effects in atomic phenomena.[142] Second, he indicated under what circumstances the intuitive picture of a photon as a massless, spin 1 particlelike entity that moves with velocity c is appropriate. And, third, in a paper with Bethe published in 1932, he helped secure the perturbation theoretic picture that depicts the interaction between charged particles as stemming from the exchange of photons.

Early in 1929 Fermi had received a copy of the work of Weisskopf and Wigner on the theory of line width. The uncertainty principle indicates that the energy spread of an excited state of an atom that has a lifetime Δt was of the order $\Delta E \approx \hbar/\Delta t$. Weisskopf and Wigner had given a quantum electrodynamic account of the lifetimes of excited states of atoms and of the frequency spread of the spectral lines in their transitions to lower states. This work was of great interest to Fermi because very shortly after the advent of wave mechanics he had unsuccessfully addressed the problem of the lifetime of an excited state of an atom and that of the natural line breadth of the emitted radiation.[143] Fermi mastered all the details of the paper. In fact, Weisskopf and Wigner's article became the key for Fermi to explain interference and other undulatory light phenomena from a quantum electrodynamical, i.e., from a microscopic, viewpoint.

Amaldi, in his biographical introduction to Fermi's works, wrote that while Fermi was carrying out his research

> he taught his results to several of his pupils and friends including Amaldi, Majorana, Racah, Rasetti and Segrè. Every day when work was over he gathered the various people . . . around his table and started to elaborate before them, first the basic formulation of quantum electrodynamics and then, one after the other, a long series of applications of the general principles to particular physical problems. A striking feature of Fermi's method of working on a theoretical problem in public (so to speak) and of teaching at the same time, was the way in which he could say out loud what he was thinking, proceeding at a steady unhesitating pace; never going extremely fast, but never failing to make progress.[144]

Fermi lectured on his approach to QED in the course that he gave in Paris in April 1929, and his lectures are summarized in the *Annales de l'Institut Henri Poincaré.* In them he elaborated on the physical content of his formulation. He pointed out that the description of the electromagnetic field as coupled, forced, linear harmonic oscillators allowed an intuitive (*anschaulich*) view of electromagnetic processes. The principal thrust of the Paris lectures was to establish QED as a readily *usable* theory. To do so Fermi indicated how perturbation theory could be recast as a technique that allowed transition amplitudes to be analyzed and calculated in an almost algorithmic manner. Recall that until the spring of 1929 no one had given a fully quantum mechanical formulation of electrodynamic processes. The computation of the cross sections for the photoelectric effect, for Compton scattering, and so on—all depended on either semiclassical or correspondence principle approaches. What Fermi did was to indicate how all these processes could be given a fully quantum mechanical treatment. He demonstrated how perturbation theory should be handled to derive the cross sections for the processes.

A fuller version of these results was the content of the lectures that Fermi gave at the University of Michigan Summer School in Ann Arbor during the summer of 1930. All the topics addressed in these lectures had been worked out in his papers or those written by his students. They were included in his widely read and influential *Reviews of Modern Physics* article.[145] Causality was among the topics addressed. The fact that the emission of a light quantum by an atom at a time t could induce a transition in an another atom located at a distance r from the first *only* after a time $t + (r/c)$ had elapsed was a statement of causality. Bethe had already indicated this in 1930, making use of the photoelectric effect: if a photon is emitted at time $t = 0$, then an electron on an atom a distance r from the photon's source is not ionized before a time r/c has elapsed.[146]

Bethe made the following comments regarding the Fermi QED lectures at the Enrico Fermi memorial symposium at the Washington meeting of the American Physical Society on April 29, 1955, shortly after Fermi's death: "Many of you probably, like myself, have learned their first field theory from Fermi's wonderful article in the *Reviews of Modern Physics* of 1932. It is an example of simplicity in a difficult field which I think is unsurpassed. It came after a number of quite complicated papers and before another set of quite complicated papers on the subject, and without Fermi's enlightening simplicity I think many of us would never have been able to follow into the depths of field theory. I think I am one of them."[147]

Fermi's β-Decay Theory

Fermi's QED studies were essential preparations for his being able to formulate in 1933 his theory of β-decay. It had been recognized since 1915 that the nucleus was the site of all radioactive processes, including β-radioactivity. It was therefore natural to believe that electrons existed in the nucleus. Already in 1914 Ernest Rutherford had assumed that the hydrogen nucleus is the positive electron—he called it the H-particle—and he conjectured that nuclei were made of H-particles and electrons. During the 1920s the generally accepted model of a nucleus was that it consisted of the two elementary particles then known: protons and electrons. Rutherford in his Bakerian Lecture of 1920 had suggested that a proton and an electron could bind and create a neutral particle, which he believed was necessary for the building up of the heavy elements. However, if nuclei were assumed to be composed of protons and electrons, the Pauli principle made it difficult to understand the spin of certain nuclei, such as N^{14}. Similarly, should there be electrons in the nucleus, their magnetic moment—as determined by the hyperfine structure of atoms—ought to be much larger than the values determined experimentally, which are three orders of magnitude smaller than atomic moments. Confusion reigned, compounded by the difficulty in understanding β-decay.[148]

The process of β-decay, wherein a radioactive nucleus emits an electron (β-ray) and increases its electric charge from Z to $Z + 1$, had been studied extensively during the first decade of the century. If the process is assumed to be a two-body decay, i.e., if the decay consists in a nucleus undergoing the process $A^Z \rightarrow A^{Z+1} + e^-$, then energy and momentum conservation requires the electron to have a definite energy. However, in 1914 James Chadwick had proven that the energy of the electron that is emitted from a radioactive nucleus undergoing β-decay is not the same in each decay. The possible energies of the electrons form a continuous spectrum. As no particles other than the initial nucleus, the final nucleus, and the emitted electron could be observed in the process, β-decay seemed to violate energy conservation. Only at the maximum electron energy was energy conservation found to hold to the accuracy of the measurements in the experiment.

By the end of the 1920s no explanation of the continuous β-spectrum had proven satisfactory, and some physicists, in particular Bohr, were ready to give up energy conservation in β-decay processes.[149] Pauli was not. In December 1930 Pauli, in a letter addressed to the participants of a conference on radioactivity, proposed "a desperate remedy" to save energy conservation by proposing that "there could exist in the nuclei electrically neutral particles that I wish to call neutrons [later renamed neutrinos by Fermi], which have spin 1/2 and obey the exclusion principle . . . The continuous β-spectrum would then become understandable by the assumption that in β-decay a [neutrino] is emitted together with the electron, in such a way that the sum of the energies of the [neutrino] and electron is constant."[150] Fermi took Pauli's hypothesis seriously. After attending the 1933 Solvay Conference at which Pauli's neutrino had been discussed Fermi decided to study the formulation of quantum field theories in terms of creation and annihilation operators. Once he had mastered the formalism he told his collaborators in Rome: "I think I have understood them. Now I am going to make an exercise to check whether I really understand them, [to see] whether I can do something with them." To make sure, he developed his theory of β-decay, which, according to Segrè, Fermi thought to have been probably the most important work he did in theory. He believed it "would be the discovery for which he would be remembered."[151]

An initial note was submitted to *Nature* but rejected. The full statement of the theory was published in both the *Zeitschrift für Physik* and in *Il Nuovo Cimento* in 1934. Fermi's theory of β-decay marked a change in the conceptualization of "elementary" processes. In the introduction to his paper Fermi indicated that the simplest model of a theory of β-decay assumes that electrons *do not* exist as such in nuclei before β-emission occurs

> but that they, so to say, acquire their existence at the very moment when they are emitted; in the same manner as a quantum of light, emitted by an atom in a quantum jump, can in no way be considered as pre-existing in the atom prior to the emission process. In this theory, then, the total number of the electrons and of the neutrinos (like the total number of light quanta in the theory of radiation) will not necessarily be constant, since there might be processes of creation or destruction of these light particles.[152]

Fermi's paper on β-decay extends the recognition that particle creation and annihilation are the novel features of any relativistic quantum theory. But underneath it was still a particle conceptualization of processes. All the calculations concerning electron-positron pair creation and annihilation were done hole-theoretically. Bethe, when describing his understanding of Fermi's β-decay theory after he had read Fermi's paper, stated that whereas QED with electrons described by Dirac's equation allowed the possibility of the creation and annihilation of a *single* photon by virtue of the $\boldsymbol{\gamma}.(\boldsymbol{p} - e/c\, \boldsymbol{A})$ term, similarly Fermi's allowed the creation and annihilation of *two* new particles in the process $n \rightarrow p + e^- + \bar{\nu}$ as described by the interaction term $\int d^3x \psi_p^*(x) \Gamma \psi_n(x) \psi_e^*(x) \Gamma \psi_\upsilon(x)$.

Fermi's theory marks the birth of quantum field theory as applied to elementary particle physics.

Relativistic Collisions

The joint 1932 paper of Bethe and Fermi clarified the description of the interaction between two electrons when treated quantum electrodynamically. Both Helge Kragh and Roqué in 1991 have given thorough expositions of the background to this paper, namely the theory of Møller scattering—the scattering of electrons by electrons—and the controversies over the derivations by

Breit and others of the electron-electron interaction potential that included both magnetic and retardation effects.[153] Bethe himself described how the joint paper came about:

> By 1932, using Møller's theory to calculate the interaction of relativistic charged particles, I had calculated the stopping power of charged particles of relativistic velocity in matter. Fermi was somewhat interested in my result; he later improved it by including the dielectric constant of the matter. But his main interest was in the Møller interaction, which was just the first order in e^2 of the result of QED.
>
> So he proposed that we write a paper on the various expressions for the interactions of relativistic charged particles: the full result of QED, the Møller approximation, and Breit's interaction, which was valid to order v^2/c^2. We had soon done the algebra, which left only the writing of the paper.
>
> Fermi had no secretary, so he did the typing. He would speak every sentence in German—which he knew well from a year as a postdoc in Germany. I then had a chance to suggest corrections. I made very few, some in language, fewer in content; then he would type it. My job was to write the few formulas, which would then be inserted by hand into the typed manuscript. After we had each read over the whole paper, we sent it for publication.[154]

Some Metaphysics

While writing his two seminal *Handbuch* articles in Rome—where Fermi was remolding himself as a nuclear physicist and dealing with a new set of "elementary" entities—protons, α-particles, γ-rays, and nuclei—Bethe perceived that the revolutionary achievements of the new quantum mechanics in atomic, molecular, and solid state physics stemmed from the confluence of two factors. The first factor was a theoretical understanding: the representation of the dynamics of microscopic entities by quantum mechanics that had resulted from the synthesis of the work of Heisenberg, Born, Pascual Jordan, Dirac, and Schrödinger. The second was the apperception of an approximately stable ontology: electrons and nuclei.

"Approximately stable" meant that these particles—the electrons and the (nonradioactive) nuclei, building blocks of the entities (atoms, molecules, simple solids) that populated the domain being carved out—could be treated as *ahistoric* objects, whose physical characteristics were independent of their mode of production and whose lifetimes could be considered infinite. These entities could be assumed to be "elementary," (essentially) pointlike objects that were specified by their mass, their intrinsic spin, and by their electromagnetic properties such as their charge and magnetic moment. They could be treated as "elementary" because the energies involved in the environment where they were normally found and in the experiments analyzing their properties and structures never were sufficient to probe or alter the constitution of the nuclei. In addition, these "elementary" constituents were indistinguishable: all electrons are identical; all protons are identical; all (stable) nuclei in their ground state are identical; all sodium atoms in their ground state are identical. Their indistinguishability implied that an assembly of them obeyed characteristic statistics depending on whether their spin is an integer or half odd integer multiple of 1/2.[155] Furthermore, the simplest compound entities —atoms—have some plasticity that allows them to combine into more complex objects—molecules and solids, for example—many of whose properties could be accounted for quantum mechanically.

In contrast to classical physics, quantum mechanics reasserted that the physical world presented itself hierarchically and was to be understood in this manner.[156] The world was not carved up into terrestrial, planetary, and celestial spheres, but layered by virtue of certain constants of nature.[157] As Dirac emphasized in the first (1930) edition of *Principles of Quantum Mechanics,* Planck's constant allows the world to be parsed into microscopic and macroscopic realms.

Accepting this thesis implies a further emendation of the Kuhnian view of the growth of scientific knowledge insofar as it applies to physics. We have already seen that Bethe's adaptation of Ewald's prior theory of X-ray diffraction by crystalline solids to electron diffraction by crystals implied that there is much greater continuity in paradigm replacement than Kuhn predicated. Ontology, dynamical theories, and domain of applicability cannot be separated. So the identification of the neutron and the positron in 1932 were as "revolutionary" discoveries as the construction of quantum mechanics. One should not, as is usually done, characterize 1932 merely as an *annus mirabilis* by virtue of the discovery of the neutron and the positron but as a year in which a revolution occurred. The discoveries induced a gestalt switch, which freed the theoretical community from its fear of postulating new particles—particularly Pauli with regard to the neutrino and Dirac with regard to the positron. Neutrons, positrons, and neutrinos expanded the vision of ontology. They were psychologically generative: they made possible Fermi's theory of β-decay and later Yukawa's theory of nuclear forces that postulated the existence of an unobserved entity—what was later called a meson—whose exchange between nucleons was the origin of the short-range nuclear forces.

Political Upheaval

Finding suitable employment and making ends meet became a constant worry after Bethe had "habilitated" and had become a *Privatdozent.* His letters to Sommerfeld attest to this.

After Bethe's return to Munich in the fall of 1931 the fellowship from the *Notgemeinschaft der Deutschen Wissenschaft* together with the small fees he received as a *Privatdozent* were enough to meet his needs. As a *Privatdozent* he gave a course on collision theory. Eight or ten students attended, and the total honorarium was 100 marks, but the university subtracted successively "one thing and another thing and another thing," so that only something like fifty marks remained.

Sommerfeld put him in charge of looking after the American postdoctoral fellows that were coming to the institute: "Now you have been in England," Sommerfeld declared, "now you can take care of the English speaking visitors." It was then that Bethe first met Lloyd P. Smith, who was a young instructor at Cornell and had come to Munich as a National Research Council postdoctoral fellow during the academic year 1931–32. He worked with Bethe, who gave him the problem of using the Hartree-Fock approximation to determine the states of helium when one of the electrons is in an excited S state of high quantum number.[158] John Kirkwood was the other American postdoctoral fellow Bethe supervised. Kirkwood had obtained a Rockefeller International Fellowship and spent half the year in Leipzig working with Debye on ionic solutions and the other half in Munich working with Bethe on hydrogen in its solid phase.

In the fall of 1932, after his return from his second trip to Rome, Bethe obtained an appointment in Tübingen as an acting assistant professor *(Beauftragte Lehrkraft).* He arrived there in late September 1932 and found a nice room at a place that also offered breakfast. Bethe recalled that when he first came to the university Hans Geiger, the professor of experimental physics, showed him his experiments.[159] When introducing him to the members of his institute, Geiger made a point of stressing how accomplished a theoretical physicist Bethe was. That fall, in addition to a course on thermodynamics Bethe taught a course on quantum mechanics, which Geiger and some of his *Assistenten* attended. And shortly after coming to Tübingen he accepted two students to work toward their doctorate under his supervision.

> One of them, by the name of Henneberg, was a very good man, and his thesis is still a good piece of work which is mentioned in the literature from time to time and in textbooks. He worked at Siemens later on the electron microscope . . . and he was very highly regarded by his collaborators. He was also partly Jewish. I think he was one quarter, had one Jewish grandparent, and so mainly for this reason the Nazis thought they should put him into uniform and they sent him to the front and he was killed in the war.[160]

As his status was becoming recognized Bethe got invitations to give talks. Heisenberg invited him to the *Leipziger Tagung* (conference) he organized each year. The subject of the 1933 conference was magnetism, and it was to be held in mid-February. Bethe accepted and gave a talk on what became known as the Bethe ansatz. The conference was attended by many people from all over Germany, and Kapitza came from England.

But as far as Bethe was concerned, the best aspect of Tübingen was not its physics institute: it was the bus to Stuttgart.[161] In contrast to the train, which took an hour and a half, the bus got him to Stuttgart in an hour. He took that bus often to visit the Ewalds.

As he told Charles Weiner, "Tübingen was really a very unpleasant place to live in at that time because it was a hotbed of the Nazi party. I remember they had any number of victory celebrations after elections and after lots of events in the early days of the Nazi government. They came to power on the 30th of January in '33, and I kept myself very much out of the main stream of the city." Throughout the 1920s Bethe had kept abreast of the political developments in Germany.[162] He recalled that in the summer of 1925—after the end of hyperinflation and after France and Belgium had stopped their occupation of the Ruhr district—the political situation became "pretty satisfactory."

> The government of Germany was moderately conservative. Stresemann was a very important person, obviously quite a sensible man.[163] There were changes of government very often, but there didn't seem to be any instabil-

ity. However, in 1929, after the crash of the stock market in New York, that stopped.

While in Frankfurt and Munich I read the *Frankfurter Zeitung* every day. It was the best paper in Germany, and now the *Frankfurter Allgemeine Zeitung* is still one of the best papers but perhaps not the best.[164]

At the time I was on the left end of the non-socialist parties. My father was of the same opinion, and it was he who deeply influenced me in my political views. He had given electioneering talks in 1919 and the early twenties for the democratic candidates. He got along very well with the Social Democrats, who then were the leading power in the Frankfurt city council. He was once nominated for that council. The City Council was big, I think about a hundred members. He didn't get elected.

Once I was of voting age, which was in 1927, I voted in every election that I could. At first I voted for the liberal *Sozialdemokratische Partei,* and later in 1930 when things became bad and the *Demokratische Partei* merged with the *Staats Partei* I voted the for the *Staats Partei.*[165] As I still had my German citizenship I continued to vote even later, when I came back from England or the United States to visit my mother, if there happened to be an election.

It was a little bit dangerous for me to do that. There was one election in which you were asked: "Keep dem Führer. Nein Ja," and I voted Nein.

After 1929 the political situation in Germany became of ever greater concern to Bethe. In the fall elections of 1929 the Nazis got a hundred seats in Parliament, a gain from some twenty. This looked very ominous to Bethe. And the worrisome political situation was more apparent in Munich and in Bavaria than elsewhere. Thus when he came to Munich in the fall of 1931 after his first stay in Rome, when looking for a room Bethe felt "under the circumstances, that it was better for me to say that I was half-Jewish, and indeed, I think the first two places said no. And then I got a room at the house of a very conservative man, I don't think I ever saw him. I saw his daughter and his daughter said, "Oh, of course. That's fine." She was very nice to me.

By 1932 he discerned the handwriting on the wall. Writing from Tübingen, Bethe openly expressed his dismay and worries to Sommerfeld, "Otherwise there is not much new, other than that Mr. Stark is moving threateningly closer to the Ministry of Culture—in Prussia albeit still more than in Bavaria. Quite a catastrophe—nonetheless warm greetings from your grateful H. Bethe." The appointment of Hitler as chancellor on January 30 changed everything. Upon hearing of the enactment of the Civil Service Law on April 7, 1933, which forbade any "non-Aryan" or "politically unreliable" person from holding any state or federal governmental position, Bethe that same day wrote to Geiger to inquire about his status. Geiger, responding to his query, sent him a very cold note confirming that he had been dismissed. Immediately thereafter Bethe wrote Sommerfeld:

> 11 IV 33
>
> As so often, I must ask for your advice again today. You probably do not know that my mother is a Jewess: According to the new law concerning civil servants I am therefore "not of Aryan lineage" and consequently not worthy of being a civil servant of the German Reich.
>
> For the time being the law pertains only to the dismissal of current civil servants. It can be assumed that in the foreseeable future those people who have just been removed will not be made new civil servants, i.e. professors. It appears moreover that the "cleansing" will go very far, so that, for example, already in the next semester nothing will come of Tübingen. I have no idea whether my birth-error is known there and from where—in any case, upon inquiry as to what will happen, I received the enclosed letter from Geiger. I feel its brevity nearly insulting, and in view of its wording I no longer believe that I have many words to say in Tübingen. Actually I had thought that it would not be possible for the law to be "working" so rapidly and radically for the simple reason that there do not exist that many Aryan theorists.
>
> Ultimately, the question of the next or even the very next semester seems to me of secondary significance. The essential, it seems to me, is that with today's policies the chances for me to ever obtain a professorship in Germany have become very small. It is presumably not to be assumed that the anti-Semitism will become weakened in the foreseeable future, or that the definition of Aryan will change. For better or worse, I must draw the consequences and try to find a place somewhere in a foreign country. You will understand that I do not do this easily—I know for certain that I will not feel at home anywhere abroad and nowhere as happy as in Germany. But insofar as one can survey the situation I am left with only the choice of starving in Germany as a private scholar or leaving. And better to draw the consequences right away than to wait until the last reserves are used up!
>
> I now want to ask you, first, if you approve of this decision, and, second, what you advise me to undertake. Naturally I do not want to go head over heels and take any position that exists outside of Germany, but am prepared

to wait one or even two years for something suitable. However I would like the leading people to know that I want to.

Insofar as I know, a position for theoretical physics exists in *Manchester.* It has been occupied always only for brief periods and lastly by an exceptionally incompetent man—prior to that Mott was there. It has occurred to me to write to Bragg for that reason—I know Bragg personally and consider it likely that I could get that lectureship if the position happens to be unoccupied.—England is really very congenial, personally, and also because of the language. Naturally the British have many people themselves, but also quite a number of positions, stipends, etc. Since I know most of them personally, I also thought of writing to Fowler, if he can do anything for me (even though personal begging letters are not particularly pleasant). I only fear that while one obtains a temporary position relatively easily in England, it is very difficult to obtain anything better if one is a foreigner.

I also thought of Italy, i.e. Fermi. There are not many good people there and many positions still open, but also with little money. Besides I should have to learn the language better, which however is a "cura posterior."—I do not believe there is anything in other European countries. Bohr will first take care of the people of his close acquaintance (if he is at all able to do so), such as Bloch etc. I do not know to what extent there might be possibilities in Sweden, i.e., whether one can persuade Siegbahn that he needs a theoretician. France naturally needs theoretical physics badly, but is of course less congenial to me for political reasons. If an offer were to be made, I would of course not reject it.

This finally leaves America. It would doubtlessly be easy to find something suitable there, if the economic conditions were somewhat better than they are. All this is made more difficult under today's circumstances—but otherwise I think that great sympathy exists for the German Jews (and for those who have been additionally included according to German law, such as I). I have the greatest favor to ask of you in this connection: You have always had especially good and friendly relations with America, could you write to some of the appropriate people? In contrast to England and Italy, I do not know anyone in the U.S.A. well. It seems to me that in spite of the economic depression the U.S.A. still presents the greatest opportunity.

Secondarily the question presents itself as to what shall happen in the time until I have found something different. I am financially secure for the next two years on account of the *Handbuch.* Perhaps no one is permitted to employ me as "*Assistent,*" although I had the offer from Madelung for this summer. Naturally one cannot know either whether Heisenberg would be permitted to take me in the winter. Maybe he already has to dismiss Bloch now. (Incidentally I recommended Henneberg to Mr. Madelung for an "*Assistent*"—since he has the advantage that (as far as I know) he does not possess a Jewish mother or grandmother, and therefore his chances have improved. One can expect that the people whose ancestry is in order, can become somebody under today's policies even if they are not first class. Therefore I believe that one should no longer push Henneberg to become a middle-school teacher under all circumstances.)

What will probably happen is that in the next semester I contritely return to Munich. Financial assistance will hardly be available—certainly not from the "*Notgemeinschaft*"—and for the moment this is also not necessary. For the time being one must await what happens.

Presumably you will have gotten several more such letters. In any case many thanks in advance for all the effort which you need to make on my account. Best regards from your always grateful
Hans Bethe[166]

On April 17 while in Baden-Baden visiting his mother, Bethe got a postcard from one of his doctoral students telling him that he had read in the newspaper that Bethe had been dismissed from the university and asking him what he should do. Bethe requested him to send him the dismissal notice in the newspaper, which the student did. Near the end of April Bethe received an official letter from the minister of education of the state of Württemberg informing him that he had been dismissed but that his salary for the month of April would be paid.

Sommerfeld was able to help Bethe. He invited him to come to Munich and awarded him a fellowship until the end of the summer. The fellowship was privately funded and therefore not subject to the racial laws.

Bethe recalled that shortly after arriving in Munich after his dismissal from Tübingen he and an American visitor discussed the rumors concerning the Nazis having opened concentration camps, in particular, one in Dachau near Munich, and told one another anti-Nazi jokes. Sellmaier, the man in charge of the teaching laboratories in the institute, overheard their conversation and came over to warn them to be careful as there were some "strong" Nazis among Sommerfeld's doctoral students. The changed atmosphere was further brought home when Bethe received a notification from the Bavarian state authorities that it had cancelled his qualifications as a *Privatdozent* but that he could call himself "Dr. habit. Bethe." Although he could no longer teach regular courses during the summer semester, he offered to talk about Chadwick's neutron paper in the physics colloquium. A day or two before the talk he was advised by Sommerfeld and by Gerlach, who now was the professor of experimental physics in Munich, that he shouldn't give the talk because the Nazi students in Sommerfeld's seminar would demonstrate against him personally. "The racial, anti-Jewish laws had gone into effect on the 1st of April and I knew that there was something coming . . . I knew that I had to leave the country, but I thought that a talk was still in order." In May 1933 Heisenberg, who by virtue of the racial laws had lost Felix Bloch as his *Assistent,* offered the position to Bethe. Bethe wrote him that he would be delighted to accept, but that he was disqualified by the same anti-Jewish laws.

Sommerfeld knew many physicists the world over, and he found Bethe a position with Bragg in Manchester. The appointment was for one year in order to replace E. J. Williams who was going to Copenhagen on a sabbatical leave to work with Bohr.

During our interviews I asked Bethe: "Suppose the Nazis had not been anti-Semitic, but otherwise had behaved the same way: had incarcerated communists and other dissenters in concentration camps; had invaded Poland, and had gone to war against the Soviet Union." What would he have done? Bethe answered,

> I don't know. I would not have been a hero, like von Laue who took endangered friends over the border in his automobile. I would have tried to avoid military service, and might have done research for the war to escape service. I would not, unlike Heisenberg, have wished for German victory. I would have continued to read relativity and quantum theory, the forbidden parts of physics. Otherwise I would have tried to be inconspicuous, like many Germans. In other words, I would have tried to make as few compromises as possible, but to survive. That is what I believe. But I am grateful to my Jewish mother that her ancestry made it perfectly clear to me that I should emigrate.

In 1933 emigration was relatively easy. In fact the Nazi government encouraged the dismissed professoriate to leave Germany.[167] Later it became much more difficult. Bethe was allowed to take all his money, some 5,000 marks, out of Germany.[168] And in 1934, he was allowed to transfer the honorarium for his *Handbuch* article with Sommerfeld on solid state physics, some 3,000 marks, to a British bank.

Thus began a new life for Bethe. But before joining him in England let us consider the person and personality that is leaving Germany. I will do so by examining the imprint of the three individuals who thus far had molded him as a scientist: his father, Sommerfeld, and Fermi, and use Max Weber's essay "Science as Vocation," and Bourdieu's insights into the scientific field, to structure the assessment.

An Assessment of the Young Bethe

In September 1918, shortly before the end of the war, with the defeat of Germany imminent and internal revolutionary uprisings destabilizing the autocratic governance of Hindenburg and Luddendorf, Max Weber—by then a well-known political economist who had achieved national prominence through his efforts to write a constitution for a democratic German republic—delivered a lecture entitled "Science as a Vocation" to the students of Munich University.[169] By "science" Weber meant pure science—that is, science whose product is knowledge and the intention of its practitioners only to create and supply knowledge.

The main theme of his address was the question of the "meaning, significance and value" of scientific knowledge and how it differed from former times. To do so Weber first of all compared the "plutocratic" German institutional frameworks where scientific work was being done and scientific knowledge was being acquired and disseminated with the "bureaucratic" American institutions. He did so in order to assess the opportunities for the students to work as scientists in Germany in the future, and to decide whether science could still be a vocation rather than a profession.[170] As the contemporary responses to the lecture made clear, Weber also wanted to convey to his audience his belief that the antiscience and antischolarship temper that was prevalent in a very large segment of the defeated German population was symptomatic of "the cultural and political crisis facing modern Western civilization."[171]

For Weber, the contrast between the situation the students were facing and that of the past was due to the fact that, "In our time, the internal situation, in contrast to the organization of science as a vocation, is first of all conditioned by the facts that science has entered a phase of specialization previously unknown and that will forever remain the case. Not only externally, but inwardly, matters stand at a point where the individual can acquire the sure consciousness of achieving something truly perfect in the field of science only in case he is a strict specialist."[172]

What Weber was describing was true of Bethe's father: Albrecht Bethe had become a specialist in neurophysiology with wide interests in animal behavior. However, this was *not* yet so for Sommerfeld. When Bethe completed his doctoral work and *Habilitation* with him, Sommerfeld had a mastery of almost *all* of theoretical physics at the time. And the same was true of Fermi when Bethe came to Rome in 1930–31. Whatever Fermi or Sommerfeld did not know—whether in physics or in mathematics—they could acquire in short order. That situation was a reflection of certain epistemological and sociological factors: first, the scope of quantum mechanics and its success in explaining those parts of the microscopic world then accessible experimentally and the level of precision that could be achieved; second, the stability of classical physics when describing the macroscopic world at a certain level of precision; and finally, the newness of theoretical physics as a sub-discipline and the relatively small size of the theoretical physics community.

Having a mastery of almost *all* of theoretical physics would also be true for Bethe during the 1930s. The other outstanding young theorists of his generation—Bloch, Fröhlich, Heitler, Landau, Mott, Peierls, Tamm, Teller, van Vleck, Weisskopf, Wigner—had a mastery of all of physics. This enabled them to carry out "pure" research in all its subspecialties. The goal of theoretical physics was the intellectual mastery of *all* of physical nature.

Weber noted that what still was true, and what made science still "worthy" as in former times, was the fact that it could yet be pursued "with passionate devotion."[173] "Passionate devotion" to one's work was also true in the arts.[174] But, Weber went on to stress, science has a fate which profoundly distinguishes it from artistic work: Scientific work is harnessed to the course of progress. And this gave rise to a paradox. Science is "progressive" because it continuously asks new questions whose answers undermine the foundations upon which the questions were based. "Every (scientific) 'fulfillment' means new questions; it asks to be 'surpassed' and made obsolete." The situation thus demands "heroic resignation" on the part of scientists since they know that their work "will be out of date in ten, twenty, or fifty years." Weber therefore asked: What is the meaning of an enterprise that will never come to an end, since its "progress is in principle infinite"? Why should anyone commit himself or herself to such an endeavor? And what is the meaning of such a commitment to the members of the "intellectual aristocracy" who do pursue science as a vocation rather than as a specialized profession, and do so in a disenchanted, "intellectualized" world? By a "disenchanted" and "intellectualized" world Weber meant a world in which "there are in principle no *mysterious, incalculable powers at work,* but rather that one could in principle master everything through *calculation.*" Moreover, "The fate of our age, with its characteristic rationalization and intellectualization and above all the disenchantment of the world, is that the ultimate, most sublime values have withdrawn from public life, either into the transcendental realm of mystical life or into the brotherhood of immediate personal relationships between individuals." This then led Weber to ask whether science in a disenchanted world can have any value beyond its instrumental and practical ends and, more significantly, whether it could give answers to the fundamental questions Tolstoy had asked: What is the meaning of life? The meaning of the world?

Erich Kahler, in his essay on Weber's "Science as a Vocation," which he wrote shortly after its publication in 1920, phrased the questions succinctly: "Can and should pure science decide whether the things it tells us deserve, from our point of view, [to] be promoted (or combated) and thus respected and known, and to what extent, in what degree, and in what form?"[175] Weber had given a resoundingly negative answer to these questions in his essay: science cannot and should not do this. He had concluded, "Science is meaningless because it gives no answer to the only question important to us," namely: "What shall we do and how shall we live?" This because "Wherever . . . rational empirical knowledge has consistently carried out the disenchantment of the world and its transformation into causal mechanisms, there appears the ultimate challenge to the claims of the ethical postulate, that the world is a divinely ordered cosmos with some kind of ethically meaningful direction." But Weber noted that there were means at hand for scientists to live productively and come to terms with the disenchantment of the world and the seeming "meaninglessness" of the scientific enterprise, but they could do so only by accepting certain presuppositions which science *cannot* prove. Bethe's father had done that.

Albrecht Bethe was the embodiment of the scientist who recognized that the world had become disenchanted, accepted it and lived creatively in it, and molded it to make it meaningful and rewarding. As a young man Albrecht had given up the rituals of religion because they appeared to him artificial. Nor would he later on give support to any attempt to re-enchant the world by traditional institutional means. For Albrecht scientific work by virtue of being harnessed to progress resonated with his political outlook. For him science was still "worthy," because he could pursue it with passionate devotion. Although he knew that his work "will be out of date in ten, twenty, or fifty years" and was carried out in a "disenchanted," "intellectualized" world, nonetheless he believed that science had value beyond its instrumental and practical ends, even though it could not give answers to the fundamental Tolstoyan questions concerning the meaning of life and of the world. He did not believe that science *could* give answers to moral and transcendental questions, nor that one could go from "is" to "ought," in other words, to deduce a moral precept from a presumed fact about the natural world. For him the line between stating a descriptive fact and making an evaluative judgment about it was clear and not to be crossed.

Concerning the presuppositions of his science, Albrecht Bethe had rejected vitalism in its modern Drieschian garb. He did specialize in neurobiology but overcame this limitation by asking global evolutionary questions: How can animal behavior be understood evolutionarily in terms of neural development and makeup? The answers to the questions he posed had to explain how things—including mechanisms such as vision and intention—*came to be!*

Already in 1899 he had stated his *Weltanschauung* clearly and succinctly: "If one were not to find satisfaction in the search for knowledge, one would despairingly put one's hands in one's lap and say: It is too difficult for us humans." What is remarkable about this statement is that it captures and insists on the tragic element brought about by the disenchantment of the world—that same tragic component that permeates Weber's essay. Albrecht Bethe found satisfaction in his search for meaning by organizing his laboratory as a communal, collective enterprise—where men and women were treated equally, irrespective of status, religious affiliation, or social or ethnic background. And by his demeanor he endowed his institute with a charismatic quality.

Sommerfeld had imparted a similar message to Hans, perhaps in a more constrained fashion, for he was a true *Geheimrat,* a *Bonze,* and very conservative in his politics. He too had indicated to Hans that *how* science was *practiced* would be a significant factor in giving meaning to being a scientist. The communicative aspect of the seminar was one such indication.

Sommerfeld's belief that the laws of physics were universal—that the same laws applied on earth as in the stars, within the solar system as in distant galaxies—was an important presupposition. It was a presupposition that could be tested empirically. In fact it helped determine the intellectual agenda of the seminar. Sommerfeld kept an active interest in astrophysics, and one of his *Doktoranden,* Albrecht Unsöld, became an outstanding astrophysicist. His view of the universality of physics also incorporated the notion that physical laws were immune to cultural context: a Japanese or Indian physicist would derive the same laws. This became part of Bethe's outlook. Sommerfeld also gave Bethe a belief in a pre-established harmony between physics and mathematics. The successes and "fulfillment" that this belief had generated had given an almost transcendental *meaning* to theoretical physics and to being a theoretical physicist. This presupposition replaced former theistic or theological assumptions. It did so for Bethe. For him "pre-established harmony" did not imply that a *final* theory might exist—as Einstein believed; and despite impressive progress, the search might never come to an end. The successes of theoretical physics and the new questions these generated kept (and would keep) the enterprise going, gave it (and would give it) meaning, and *secured the presupposition further.*

Sommerfeld had also demonstrated to Bethe how to be a responsible teacher, *Doktorvater,* and mentor by his lectures and in his interactions with his students, *Doktoranden,* and *Assistenten.* He had also shown growth and courage in political matters. And he was there for Bethe when he needed his help after he was dismissed from Tübingen.

Fermi's imprint was no less pronounced in these matters. The simplicity of Fermi's approach to physics was mirrored in a simplicity of presentation of self, appearance, bearing, and demeanor—qualities Bethe would emulate. Fermi also demonstrated to Bethe how teachers and their students could interact generatively and cooperatively; how students would be imbued with deep dedication and devotion to the search for understanding physical phenomena by simplicity, straightforwardness, and the absence of pretension and showmanship.

To the three mentors who molded Bethe as a scientist—his father, Sommerfeld, and Fermi—should be added another: Paul Ewald. If Sommerfeld anchored Bethe's self-confidence by indicating that as a craftsman he had powers similar to Debye, Pauli, Heisenberg, and Wentzel, Paul Ewald reinforced this feeling by treating him as an equal. And Ewald also provided another model on how to run an institute that stressed cooperative interactions within the institute and cooperation with the physics community at large. Physics research could be carried out in cooperation with others rather than in competition with them.

In his lecture Weber had stated that "In the field of science only he who is devoted *solely* to the work at hand has personality."[176] "Personality" is perhaps best understood as "individuality," in the sense that Helmholtz had used that term: "Compare the work of two contemporary investigators even in closely-allied branches of science, and you will generally be able to convince yourself that the more distinguished the men are, the more clearly their individuality comes out, and the less qualified would either of them be to carry on the other's researches."[177] Albrecht Bethe, Sommerfeld, Ewald, and Fermi were all "personalities" in the sense of Weber's and Helmholtz's formulation: all four were ambitious and wanted to be a "personality." Bethe too—given his gifts, powers, and early accomplishments—molded himself "to make a light of himself" and made himself a "personality."

The theoretical physics community was—and still is—a hierarchically structured, (self)-organized community. In the 1910s and 1920s the number of full professorial positions in theoretical physics in Germany was very limited. Only the very best and the very brightest could expect to obtain such an appointment. The physicists at the top of the hierarchy determined who was the best and brightest. In Germany this made for intense competition and intensified the hierarchical institutional structure. The same was true in England, where a veritable class structure existed in the colleges of Oxford and Cambridge, in the Cavendish, and more generally in the British educational system. Similarly, Bohr's institute during the 1930s was a finishing school for those considered the best and the brightest young theorists of their generation. Its members looked with disdain on those who had been excluded from their circle because of what had been deemed ordinary accomplishments or poor recommendations. In writing its mythical history, the theoretical physics community extolled and enshrined some of the nastiness exhibited by its leading practitioners. The story of Pauli cutting a lecturer short with the statement "Not even wrong" never relates how the recipient of the barb felt. The tacit assumption is that if he (and everyone was a "he") could not take it, he clearly could not be, or become, an outstanding theorist.

Bethe very early recognized the competitive nature of the discipline and acquired all the tools—intellectual, cognitive, and psychological—to be successful. Both his father and Sommerfeld had given him the self-confidence needed to become an outstanding theorist. Undoubtedly, Bethe's own recognition that he could do everything Sommerfeld could, and in fact more, given his mastery of the new quantum mechanics, was another reason for his self-confidence. The joint authorship by Bethe and Sommerfeld of the solid state article in the *Handbuch der Physik* and Sommerfeld's recommendation that Bethe write the one- and two-electron article for the *Handbuch* were an expression of Sommerfeld's blessing. The articles were proof of Bethe's standing in the community: he was of the same stature as Hund, Wentzel, and Mott, the other "young" authors of Volume 24,1 of the *Handbuch.* They were all writing in the shadow of Pauli, who had contributed the lead article elucidating quantum mechanics.

When very young, Bethe acquired and cultivated a deep sense of integrity: he would not make any assertion, certainly not a scientific one, unless he had convinced himself of its validity, and had determined under what conditions the assertion was valid. He became disdainful of those who did not adhere to the same standards as he in this matter. Buttressed by a fully justified self-confidence in matters of physics, this disdain in the German context became at times expressed as conceit. As a young man Bethe thought of himself as very conceited. In 1988 he commented: "I was very conceited when I was young. I still am. But now I can hide it better."[178] We shall later observe some of the early manifestations of this "conceit" and the methods he used to hide it. The English, Italian, and later the Cornell milieu helped him do so. Probably Bethe's "conceit" was a mixture of ordinary conceit—a measure of self-importance and superiority—and disdain, for he was very much aware of what Bourdieu later made explicit in his study of academic life: that physics was also the locus of a competitive struggle. In addition to imagination and technical competence, to be successful in that competition required self-confidence and what Bethe called "conceit." For Bethe this translated into doing only what his off-scale technical capacities and his impressive powers of synthesis would allow him to do. He was very conscious of the fact that his achievements would give him scientific authority, status, recognition, and fame. Already as a *Privatdozent* he understood that it was precisely his *continuing* achievements that would secure him a professorial position.

Before leaving for the United States in 1935, Bethe had always been in situations where the context had determined his status in relation to his superiors. At Cornell he entered an environment where—by virtue of his powers as a scientist and his personality—he gradually became the individual who molded the atmosphere of the department and played a leading role in setting its intellectual agenda. He did so while simultaneously earning the respect and affection of his colleagues.

Hans's father and mother, Albrecht and Anna Bethe, circa 1900. (Courtesy Rose Bethe)

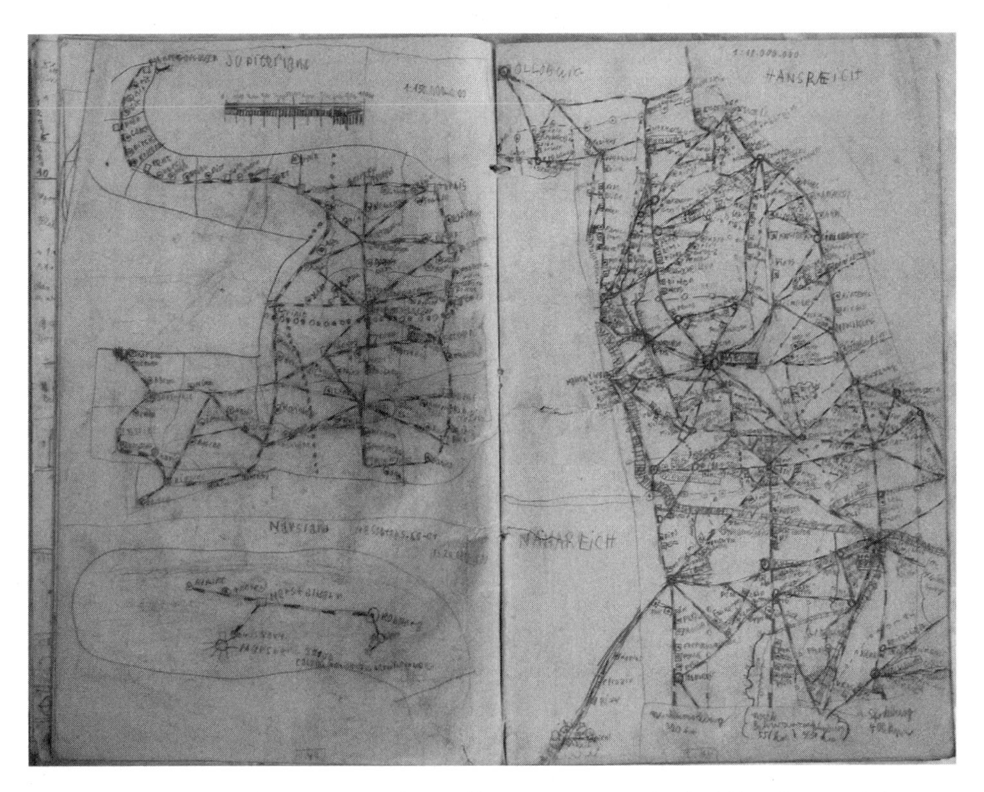

The map of the country the young Hans designed between age six and eight. Taken from an eighty-page booklet of his drawings and writings made between age six and ten. (Courtesy Rose Bethe)

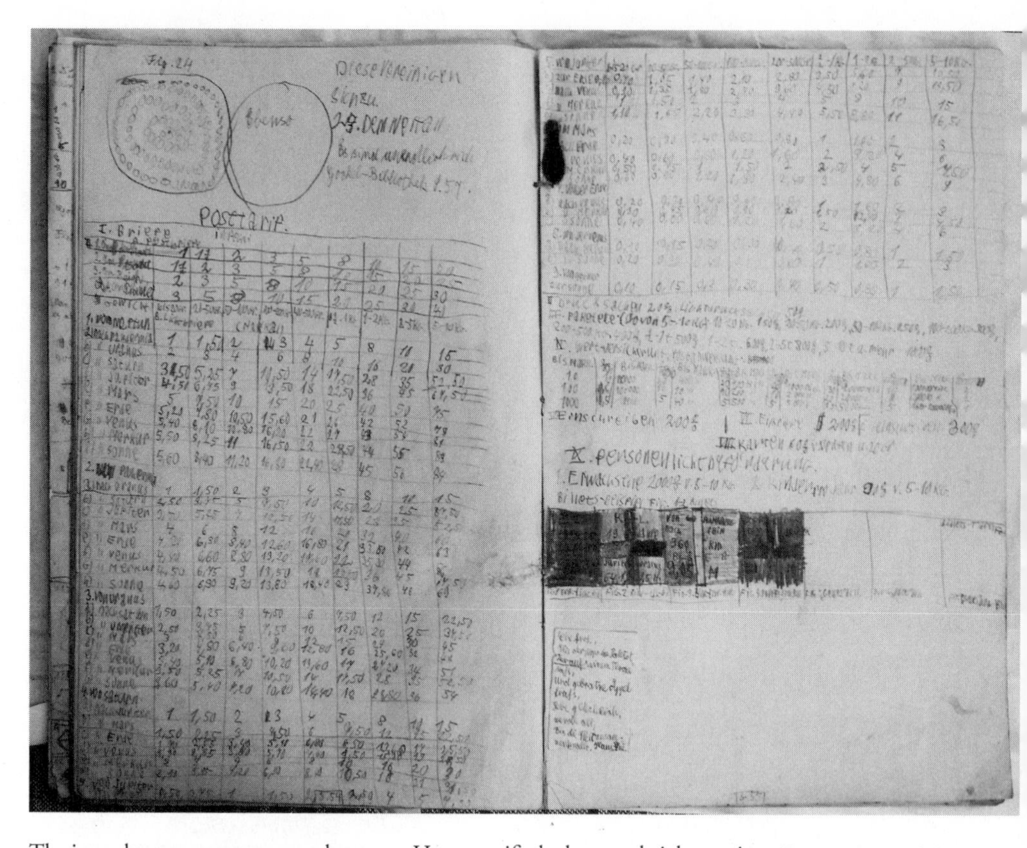

The interplanetary postage rates the young Hans specified when aged eight or nine. (Courtesy Rose Bethe)

Hans, age eight or nine, with his parents in Frankfurt. (Courtesy Rose Bethe)

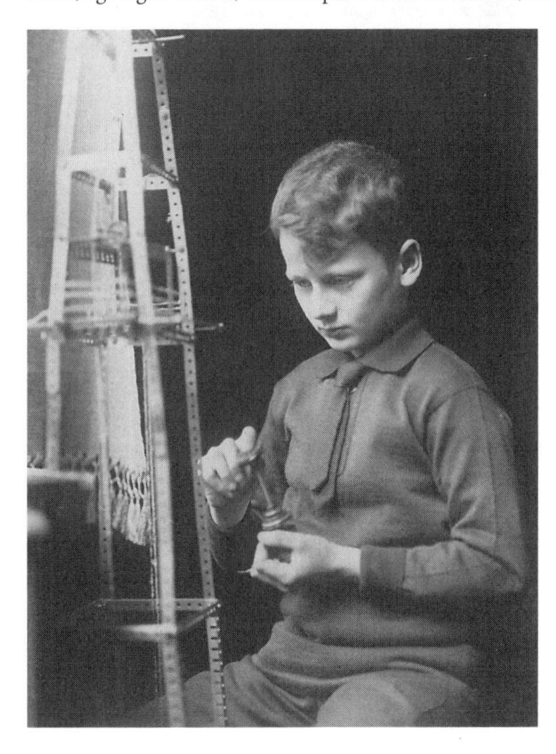

Hans, age ten, working on a construction with his Meccano set. (Courtesy Rose Bethe)

Photo of marriage of Friedrich Levi and Barbara Fitting. (Courtesy of the family of Barbara Levi)

Key to photo of marriage of Friedrich Levi and Barbara Fitting

No.	*Name*	*Née*	*Relationship*
1	Tilla Stein		Friend of Barbara Levi
3	Caroline Voss	Caroline Fitting	Sister of Barbara Fitting
12	Carl Fitting		Husband of Berthe Kuhn & Fanny Levi
13	Anna Bethe	Anna Kuhn	Wife of Albrecht Bethe; mother of Hans Bethe
14	Albrecht Bethe		Husband of Anna Kuhn; father of Hans Bethe
15	Hermann Schmidt		Burgermeister of Oppenheim
16	Eva Wrede	Eva Levi	Daughter of Georg & Emma Levi
17	Carlfritz Fitting		Son of Carl & Fanny Fitting
18	Barbara Levi	Barbara Fitting	Daughter of Carl & Berthe Fitting
19	Hermann Fitting		Son of Carl & Berthe Fitting
20	Friedrich Levi		Son of Georg & Emma Levi
21	Fanny Fitting	Fanny Levi	Wife of Carl Fitting
22	Georg Levi		Son of Simon & Fanni Levi
23	Emma Levi	Emma Blum	Wife of Georg Levi

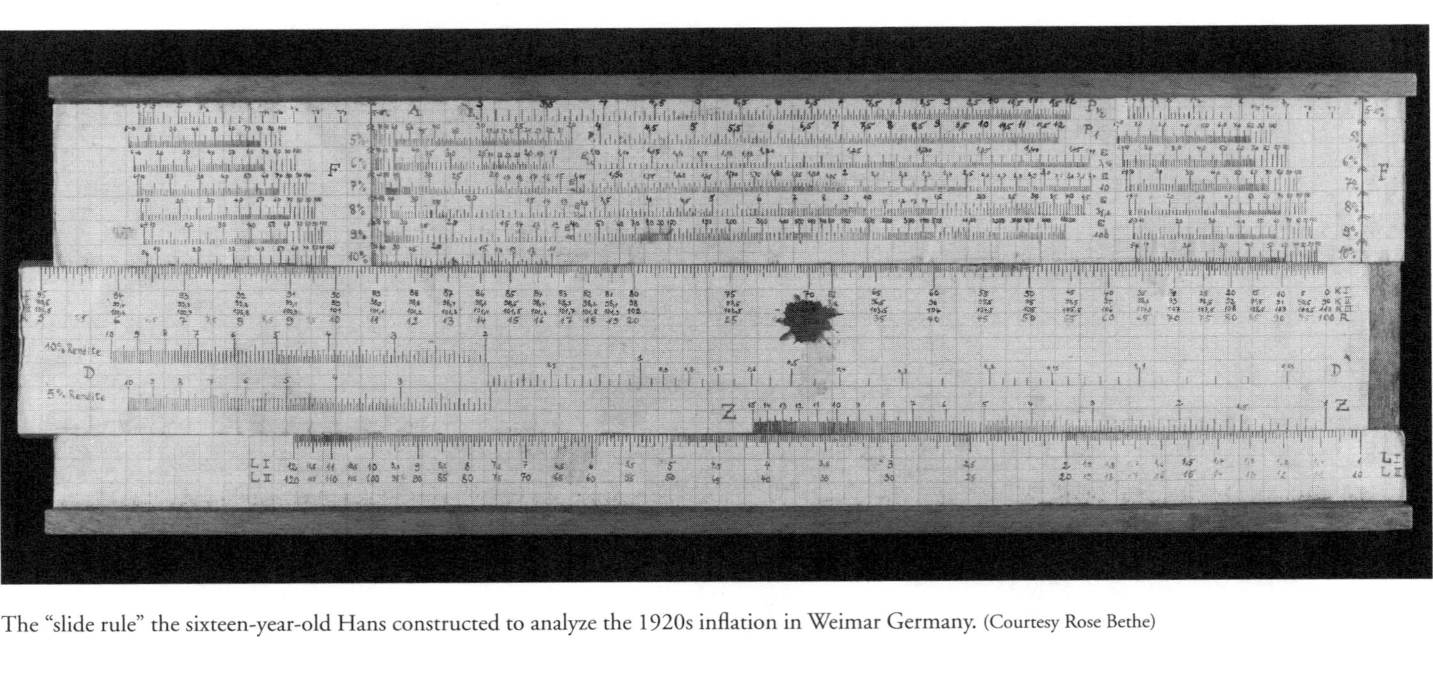

The "slide rule" the sixteen-year-old Hans constructed to analyze the 1920s inflation in Weimar Germany. (Courtesy Rose Bethe)

Hans, in a Napoleonic pose, with his classmates at Gymnasium. (Courtesy Rose Bethe)

Arnold Sommerfeld circa 1926. (Courtesy Rose Bethe)

Left to right: Hans Thorner, Hans Bethe, Rudi Peierls, and Werner Sachs on an excursion to the Austrian Alps in 1927. (Courtesy Rose Bethe)

Franco Rassetti, Enrico Fermi, and Emilio Segré, 1932. (AIP Emilio Segre ıal Archives, E. Recami and Fabio Majorana Collection)

Hilde Levi serving lunch to the conferees of the 1934 Bohr Institute Conference.
(Photograph by Paul Ehrenfest, Jr., courtesy AIP Emilio Segre Visual Archives, Weisskopf Collection)

Hilde Levi. (Photograph by Finn Aaserud. Courtesy Niels Bohr Library, Copenhagen)

Rudolf and Genia Peierls, circa 1934. (Photograph by Francis Simon, courtesy AIP Emilio Segre Visual Archives, Francis Simon Collection)

Hans Bethe with his slide rule, 1935. (Courtesy Rose Bethe)

Physics department, Cornell University, 1936. (Courtesy Rose Bethe)

Participants of Third Washington Theoretical Physics Conference, 1937. (AIP Emilio Segre Visual Archives, George Gamow and Physics Today Collections)

Participants of Fourth Washington Theoretical Physics Conference, 1938. (Special Collections and University Archives, The Gelman Library, The George Washington University)

Rose Ewald, 1938. (Courtesy Rose Bethe)

Hans and Rose Bethe, Santa Cruz, 1940. (Courtesy Rose Bethe)

5

England, 1933–1935

The warm relationship between William Lawrence Bragg and Sommerfeld that made possible Bethe's invitation to Manchester was of long standing. It dated back to the 1910s, following the discovery of X-ray diffraction at Sommerfeld's institute, and the Braggs' subsequent research activities in the field.

William Lawrence was born in Adelaide, Australia, in 1890. His father, William Henry Bragg, was the professor of physics at the decade-old University of Adelaide. In 1895, shortly after the discovery of X-rays by Röntgen, William Henry began to investigate the properties of γ- and X-rays. He received worldwide recognition for his research. The extensive correspondence between Sommerfeld and William Henry was initiated by the latter in 1910. It concerned their differing views of the nature of X-rays: corpuscular for William Henry, wave for Sommerfeld.[1]

William Lawrence obtained his early education in Adelaide. He came to England in 1909 when his father accepted the chair of physics at Leeds University. He then went to Cambridge University and studied physics there. In 1912, a year after reading von Laue's paper on X-ray diffraction, William Lawrence began his own investigations of X-ray diffraction, publishing his first article on the subject that November in *Nature* and a longer one in the *Proceedings of the Cambridge Philosophical Society* in January 1913. From 1912 to 1914 the son collaborated with the father, and their joint work earned them the Nobel Prize for Physics in 1915.[2] William Lawrence was appointed Langworthy Professor of Physics at Manchester University in 1919, succeeding Ernest Rutherford, and held this post until 1937 when he also succeeded Rutherford at the Cavendish. In December 1921 he married Alice Hopkinson, a young woman he had met in Cambridge, though, as David Phillips recounts in his biography of Bragg for the Royal Society, he "had misgivings about bringing a lively 22-year-old wife to grimy Manchester and introducing her to the sober society of middle-aged professors."[3]

Alice was not happy in Manchester. William Lawrence received several offers to move, but he turned them down largely because of his loyalty to Manchester University, "that had given him so much."[4] In 1930 he was offered the professorship of physics at Imperial College, London, and he once again declined the invitation. Phillips notes, "fearful that he had lost his last opportunity to move his family to more attractive surroundings, he broke down." He recovered, "greatly helped by spending the spring of 1931 on leave in Sommerfeld's laboratory in Munich, where he sought to broaden his command of the latest developments in physics, and by the birth in June 1931, of his third child and first daughter."[5] Relations between William Lawrence Bragg and Sommerfeld had become very close and very friendly after they met at the international conference on the determination of crystal structure analysis by X-ray diffraction that Ewald had organized in 1925 in his mother-in-law's house in Holzhausen on the Ammersee, some twenty miles west of Munich.[6]

The Bragg-Sommerfeld connection was an important factor in Bethe being invited to Manchester in 1933. Bragg knew Bethe from his visit to Manchester in 1930–1931, during which Bethe gave a lecture in Bragg's department. Douglas Hartree, who was a member of the Manchester physics department, undoubtedly also gave Bethe a high recommendation. Subsequently, Bragg also played an important role in Bethe's going to Cornell. Bragg spent the spring semester of the academic year 1933–34 as a visiting professor in Cornell University's chemistry department, just when its physics department was looking to hire a young theorist. He strongly recommended the choice of Bethe, having gotten to know him well during the previous fall in Manchester.

The year Bethe spent in Manchester was one of his most productive.

Manchester

Bethe's impression of Manchester was conveyed to Sommerfeld in the first letter he wrote to him after his arrival there.

23 XI 33

. . . Manchester is attractive well beyond my expectations. The city is not even anywhere near as dirty as I had remembered it—it is actually cleaner than London, as is indicated already by the fog which is white and not brown. Out in the suburbs there are even proper landscapes, green meadows and golf courses, creeks and attractive walks. We live in a suburb, Didsbury, where Hartree also has his house. I live together with Peierls and his very high-spirited wife (plus a similar baby) in one of the many identically appearing houses . . .

The best thing in Manchester are the people at the Institute. Bragg is marvelous, humanly and "physically" (physics-wise). He makes life pleasant for everyone at the institute insofar as he can. He introduced us to many nice people (non-physics types), and on various occasions I have been out to his castle, which you also know. I also get much from him scientifically. He makes very interesting experiments about the arrangement of atoms in alloys (superlattices) and I attempt to devise theory for them. It is a pleasure to tell him things: He understands all essential points in the shortest time, while that is mostly very difficult with experimenters . . .

In the past semester I gave lectures to the staff, i.e. the many people who teach in the Institute with Bragg, work here, or are working on their doctorate, at least 25 people. In addition [Michael] Polanyi and his people and various mathematicians came. Theme: *Introduction to wave mechanics.* In the next semester I will then treat specific problems, primarily Hartree's theory of complex atoms and molecules. Finally (during the summer semester) Peierls will do the electron with spin, and if possible, also radiation theory. I think this lecture series is a good idea; in any case the listeners are eager for knowledge and during the final hours even endured matrix mechanics . . .

Most of all I miss the doctoral candidates—I had quite gotten used to them when I was with you, and had enjoyed them. On the other hand the non-existence of doctoral candidates has the advantage that one gets around to one's own work more. A piece of work about the creation of positive electrons is just finished . . .

As is clear from this letter Bethe was very happy in Manchester from the start. He had been in England before, but he had now come as an émigré who had left Germany behind. He came harboring hopes that England would become his permanent home but felt "that in some ways it would be difficult to be fully accepted in England. But I thought that I would want to stay in England permanently, and people were very nice at the lab."[7]

Besides his interactions with Bragg, the reason Bethe was so happy in Manchester was that Rudi Peierls was also there. During the academic year 1932–33 Rudi had held a Rockefeller Foundation Traveling Fellowship with which he followed in Hans's footsteps, but in reverse order. He first went to Rome to work with Fermi and thereafter to Cambridge to work with Dirac. He was in Cambridge when Hitler came to power. Because his parents had been Jewish he, like Bethe, found it impossible to obtain an academic position in Germany. After a long period of uncertainty as to where he might be given a position in England, he finally was told that he had been granted a fellowship from the Academic Assistance Council to go to Manchester.[8]

Since they were all going to be in Manchester, Rudi and Genia Peierls invited Hans to stay with them—it would solve both the problem of housing and of nourishment. In his letter to Bethe of July 24, 1933, in which he suggested the arrangement, Peierls noted, "As you don't like English food and had to suffer so much in Manchester [on his visit there in 1930] . . . I am guessing you might like this solution. Then you would be able to eat 'slowly and plenty.' Regarding the cost, you would of course have to pay simply what you cost us, which would surely be less than any bed and breakfast."[9] Bethe accepted the offer and the Peierlses rented a large, comfortable, two-story house in Didsbury, a southeastern suburb of Manchester. Hans got a room on the second floor, but did most of his work in the big, gas-heated living room on the first floor. "We got along very well indeed." Bethe wrote Sommerfeld that the housing arrangement had "great advantages for the well being of the two physicists and their physics," but since Rudi and Hans spoke German with one another, the living together had "disadvantages for their English."[10]

The Bethe-Peierls Friendship

The evolution of the friendship between Hans and Rudi can be gleaned from their extensive correspondence, which spans the years from 1927 to 1995, the year of Peierls's death.[11] The letters from 1928 to 1933 indicate that their friendship at the time was cemented by their scientific interests and the respect they had for each other's abilities and integrity. They constructively criticized each other's work and egged one another on to greatness. Thus already in June 1928 when Peierls was leaving Munich to go to Leipzig, Bethe wrote him: "I am sending you also my collected scientific works. I hope that you like Leipzig, and will soon become a famous man ≥ Heisenberg!"[12] In these early letters the format is professional: they begin their letters with "Lieber Bethe" or "Lieber Peierls," address one another using the formal "Sie," and sign their letters Hans Bethe and R. Peierls. It is only in 1935, after Bethe had lived with the Peierlses for a year, that they use "du" in their letters.

The letters from 1927 to 1933, which for the most part are predominantly from Bethe to Peierls as many of the Peierls letters from that period have been lost, give valuable glimpses into Bethe's social activities. The letters also detail their scientific activities, and at times give valuable insights into how they arrived at some of their important works during that period: variational principles, the Bethe ansatz, Umklapp processes, and so on. The letters compellingly convey their anxiety about finding suitable positions outside Germany after Hitler's advent to power and the enactment of the racial laws that prevented them from holding university positions in Germany.

To understand the closeness of the friendship that developed between them one must first appreciate the remarkable nature of the marriage of Rudi and Genia Peierls. Every student of Peierls who later spent time in Birmingham under his tutelage commented upon the fact that there weren't any boundaries between the physics department and the Peierls home and that all aspects of life—whether intellectual, emotional, social, or professional—were looked after and helped by the Peierlses. Hans was the first to come under Genia and Rudi's care when he lived with them. He had come to England lacking self-confidence in dealing with people. Genia had told him then that he was "indeed very nice." Hans later revealed to Rudi that her comment had given him "more self-confidence in a personal sense," and that the two of them—Genia and Rudi—had done that throughout their lives. Genia had taught him "that difficulties in external life are problems to be thought about and solved, [and] not to be worried about."[13]

The problems Bethe tackled during his stay in Manchester owed much to discussions with Peierls. A large part of Bethe's research during that year was a collaborative effort with Peierls on problems dealing with order in alloys, phase transitions, nuclear physics, and neutrinos.

> Our interaction was very strong. I think we talked all day long, if we didn't sit down and actually do calculations. We talked physics at the lab, and we talked physics at home, and Genia was quite willing to let us do so—in fact encouraged it.
>
> By the way, the Peierlses had a baby who was born maybe three, four months before I came to Manchester. So Genia had her hands full with the baby, and she took that very easily. She was a very energetic woman . . . There's a baby, one takes care of the baby; there's a house, one takes care of that, [and] buys the necessary food. Genia did all this with her left hand. In the evenings Rudi and I had many conversations about things other than physics in which she would join.

The meaning of the friendship between the Bethes and the Peierlses was movingly conveyed by Rose, Hans's wife, in her letter to Rudi in October 1986 upon hearing of the death of Genia: "She and you—you and she—differently yet inseparably, have been Hans' dearest friends since your common Manchester time more than half a century ago . . . You will be inundated with stories now, of Genia's impact on old and young, of her generosity, of her vitality, sagacity, love of life and people. I am very glad that she was part of Hans' and my life."[14]

And the last words that Bethe wrote to Peierls would have been reciprocated by Peierls: "We traveled together a great deal in physics. You did a lot in physics and you educated countless good physicists, . . . I was greatly impressed at your 80th birthday when they all came to celebrate. You had a full and good life, and I thank you for letting me participate in it."[15]

Manchester Physics

Manchester University is located downtown on a main thoroughfare. It was readily accessible from the Peierlses' house by tram—a thirty-minute ride. Bethe was given a desk in the physics department, located at the south end of the campus, in a room full of glassware. He shared the room with another member of the department, who rarely used it. Although Bethe's large table was by the window, little light filtered through because the window was coated with a thick layer of coal dust most of the fall and winter. In the spring the window got washed, and thereafter Bethe could see the rest of the campus on days when there was no fog. Sometimes the fog was so thick "that you could not see in the daytime people walking two meters ahead."

Bethe's duties as a lecturer were light. They consisted in teaching an introductory quantum mechanics course for advanced undergraduates and beginning graduate students. "I talked a lot about wave-particle duality, connection with classical mechanics, uncertainty principle, and so on, and gave them very little of the working parts [i.e., applications] of quantum mechanics, but occasionally I have met a person or two who was a student, and listened to that, and thought they were very useful." The course was also attended by many faculty members and they "somewhat swamped the students." Even Bragg came quite a few times, and many of the younger staff members came regularly. The course met only two hours a week for a month each semester. This left Bethe with considerable free time, and he "did more research that year than any year before or after." As indicated in his letter to Sommerfeld, he was very impressed by Bragg, with whom he spoke frequently.

> I don't believe that I really inquired very much what went on in the physics department, except for my contact with Bragg. In addition I talked to a man, [Reginald William] James, who was a quite good X-ray crystallographer, and was the right hand of Bragg.[16] They both were very nice men. That was my main impression. Bragg was exceedingly nice, and took both me and Peierls very much into his work, and James was very easy to talk to, in a general way. In addition, there was Hartree. Hartree was part of the physics department. He did however hold his own seminar, occasionally. I don't believe there was very much interaction between him and the experimenters. Hartree was deeply involved in calculating atomic structures. They talked to each other, but it was not a close connection.

During their year in Manchester Bethe and Peierls made it a point to attend regularly the "high energy" London meetings organized by Blackett, who now held a professorship at Birkbeck College. These meetings, when held in London, were at Burlington House. Bethe there heard Walter Heitler report on his calculations of pair production by electrons and γ-rays. This led to Bethe's collaboration with Heitler on problems dealing with *Bremsstrahlung* (the radiation [*Strahlung*] emitted by charged particles when decelerated or slowed down [*Brems*]); pair production of electrons and positrons by γ-rays in the Coulomb field of the nucleus of an atom; and the effect of screening in pair production. These meetings were also the source of much new information concerning cosmic ray showers, superconductivity, and nuclear physics.

Writing to Rudi Peierls on July 6, 1994, shortly before Rudi's death, Bethe commented,

> During the year 1933–1934, you introduced me to nuclear physics. Yes, I think you did—I was too much stuck in my old work, like order in alloys . . ., while you were always open to new problems. So we worked together on the deuteron, neutron-proton scattering, and then we wondered about Wigner and Majorana forces. It was the basis of my work for several years at Cornell, including my three articles in *Reviews of Modern Physics.*
>
> Altogether, my year with you in Manchester was one of my most productive years, and maybe yours, too.[17]

Bethe's time in Manchester was not only extremely productive. His close collaboration with experimentalists and theorists were new ways of doing physics for him. Investigating the frontiers of statistical mechanics, nuclear physics, cosmic ray physics, quantum electrodynamics opened *him* up to new problems. And living with Genia, Rudi, and Gaby, Rudi and Genia's recently born daughter, gave him a new vision of what a marriage and a home meant.

Theory of Superlattices

Shortly before Bethe had come to Manchester, Bragg and Evan James Williams had published a paper in which they pointed out that the arrangements of the atoms in certain alloys depended in a very sensitive manner on the temperature and proposed a mechanism to account for this. For example, in the alloy AlAu, which has a face-centered cubic crystalline structure, the Au and Al atoms are distributed practically at random among the lattice points of the crystal at high temperatures, whereas at low temperature the Al atoms are at the corners, and the Au atoms are at the face centers of the cubic unit cells. This transition from the ordered low-temperature state to the disordered state occurs in a very narrow temperature range and is accompanied by a large increase in the specific heat and the electrical resistance of the alloy, and by similar changes in other properties.

To explain the phenomena—in analogy with Weiss's theory of ferromagnetism—Bragg and Williams assumed that that the "force" tending to produce order at a given lattice point is determined by the *average* state of order throughout the crystal. Doing so, they obtained results that were in qualitative agreement with the observed facts. Their results, however, did not agree with the experiments in details; in particular, the calculated specific heat was too small near the critical point. Bragg asked Bethe to look into the matter, which he did.

In the simplest model of an alloy the energy of a configuration of A and B atoms is expressed in terms of a two-valued variable $\sigma_i : \sigma_i = +1$ if the site i is occupied by an A atom and $\sigma_i = -1$ if occupied by a B atom. The energy E of the configuration is taken to be

$$E = -J \sum_{<i,j>} \sigma_i \sigma_j,$$

with the sum running over pairs of sites that are nearest neighbors. The probability of occurrence of the different configurations is proportional to $\exp(-E/kT)$.

Bethe considered the case of a three-dimensional lattice when the interaction energy between atoms A, between atoms B, and between atoms A and B were all different.[18]

Bethe introduced the concept of *order of neighbors,* now called *short-range order,* characterized by the nearest neighbor correlation $<\sigma_j \sigma_i>$, which is distinct from the long-range order that Bragg and Williams had introduced. By examining very carefully the energy of various configurations Bethe discovered that there is always a short-range order that determines the arrangement of nearest neighbors and second nearest atoms as a continuous function of temperature: perfect arrangement at $T = 0$ K, and total randomness at $T = \infty$. But in addition there is also a long-range order that has a sharp Curie point transition. Below the Curie temperature there is long-range order, but above the Curie point the long-range order vanishes.[19]

> I worked very much at home. I don't really remember why I did that, but I spent many hours, half-days, at home, counting the number of labels, and how many interactions there would be. It was, in a way, a generalization of the Ising model[20] and it impressed Genia Peierls very much that I could sit for hours and make drawings of gold and aluminum atoms, six neighbors, or twelve neighbors, or eight neighbors, and the second ring and so on, and that I had the patience of doing this. It was not very inspired work, but I think it was useful, and it kept me happy for quite awhile.

In the acknowledgment of the paper Bethe wrote on the subject he thanked Bragg for suggesting the problem and for many discussions. He made a special point to thank Rudi "who gave innumerable valuable suggestions" and stressed that "the method of approximation outlined in §5 is entirely due to Peierls, and all improvements of the calculation were done on similar lines." Peierls's suggestion when applied to the Ising model became known as the *Bethe-Peierls approximation* for that model.

Peierls at the time was working on the dependence of phase transitions on the number of dimensions of the system. He proved that contrary to Ising's assertion, the Ising model in two dimensions has a phase transition and that a two-dimensional system with a continuous order parameter could not have a phase transition.[21]

Nuclear Physics

Since the beginning of the nineteenth century, when the search for "ultimate" constituents of matter became empirically grounded, the elucidation of the properties of such entities has had a cyclic history. Each stage—molecular, atomic, nuclear—was characterized by an initial incoherence that gave way to a degree of order through the apprehension of stable patterns and an ensuing classification. Once that order was ascertained, a new level of structure was discovered with the help of new instruments and technologies. Incoherence and confusion again reigned until stable patterns and regularities operating at that level were discerned, classified, and modeled. Four such cycles during the twentieth century can readily be identified: the unraveling of the level of atoms, then that of their nuclei, then that of the constituents of nuclei as described in terms of neutrons, protons, and mesons; and most recently, that of the constituents of these entities: leptons, quarks, and gluons. (*Lepton* is the generic name for electrons, muons, and tau particles, and their associated neutrinos.) This last stage is somewhat different from the previous ones, in that some of the entities that constitute the ontology of that domain—the quarks—cannot exist as isolated, singly experimentally observable particles.

The classification of chemical elements during the nineteenth century is illustrative of the process. As early as 1813 William Prout had suggested that the masses of all the chemical elements were integral multiples of the mass of a hydrogen atom, thus formulating the first "modern" classification of the chemical elements.[22] With Dmitri Mendeleev's periodic table some further measure of intelligibility and simplicity was brought into the classification of the chemical elements.[23] Moreover, by virtue of the gaps that existed in his classificatory scheme, Mendeleev predicted that additional elements should exist with chemical properties that he could specify. These were later discovered. However, the interpretation of the patterning had to await the discovery of the electron and the subsequent model of the nuclear atom formulated by Rutherford. In that model electrons orbited around the minuscule, dense, atomic nucleus in the center. A phenomenological explication of Mendeleev's table, in terms of the quantized orbits that electrons could occupy in their orbital motions around the positively charged core nucleus, was given by Niels Bohr in 1920. The invention of quantum mechanics gave a "foundational" dynamical explanation of the periodic table and much more.[24] Because nuclear radii are 10^5 times smaller than those of atoms and because the nucleus accounts for most of the mass of an atom, the nucleus could as a first approximation be considered as a point charge.[25] Quantum mechanics went on to explain not only the electronic structure of an atom but also its chemical interactions, "in short all the properties of an atom except for very small corrections such as hyperfine structure, isotope shift of spectral lines" that are due to structural properties of the nucleus and require a theory of nuclear structure to account for the numerical values of nuclear properties such as their spin, magnetic moment, quadrupole moment, and so on.[26] Such advances could be made only after the discovery of the neutron by James Chadwick and of the deuteron by Harold Urey in 1932.

Erwin Hiebert succinctly described the attempts at unraveling the nuclear realm before 1932 when the neutron was discovered:[27]

> In trying to establish a roster of fundamental steps in the genesis of nuclear physics, the most conspicuous observation to make, it seems, is the almost total omission of reference to theoretical papers on the part of investigators preoccupied with experimental nuclear phenomena. The theoreticians, by contrast, were swimming rather aimlessly in a sea of novel experimental phenomena about the nucleus that did not fit readily with any of their theoretical moves. There was an ever growing mass of experimental data, but very little that seemed in any way fundamental for the theoretician . . . This state of affairs might well be chalked up to inherent complexities with nuclear phenomena, but also to a situation in which the theoreticians either were too confused to contribute something relevant to the experimenter's craft or too remote from the world of the laboratory, or both.[28]

The history of nuclear physics during the 1930s after the discovery of the neutron was insightfully narrated by the speakers at a symposium that Roger Stuewer organized in 1977.[29] Chadwick was led to his identification of the neutron by reanalyzing the experiments that Walther Bothe and Herbert Becker in 1930 and Irène Curie and Frédéric Joliot in 1932 had carried out in which aluminum, beryllium, and boron were bombarded by α-particles. Curie and Joliot had assumed that the "secondary radiation" emanating from the reaction was gamma rays. However, as Chadwick noted, "The results, and others I have obtained in the course of a week, are very difficult to explain on the assumption that the radiation from beryllium is a quantum radiation, if energy and momentum are to be conserved in the collision. The difficulties disappear, however, if it is assumed that the radiation consists of particles of mass 1 and charge 0, or neutrons."[30]

Within a few months after the discovery of the neutron Heisenberg initiated the development of a theory of nuclear structure. The opening paragraph of Heisenberg's paper on the theory of nuclear structure stated, "Through the experiments of Curie and Joliot, as interpreted by Chadwick, it has emerged that a new fundamental building block, the neutron, plays an important role in building up the nucleus. This result suggests the assumption that the atomic nucleus is built up of protons and neutrons without the assistance of electrons. If this assumption is correct, it represents an extraordinary simplification for the theory of the atomic nucleus."[31] Although Heisenberg later called the papers "relatively trivial," the papers "made a great impression in the scientific world; it was the first attempt to put forward a theory of the nucleus which, although incomplete and imperfect, succeeded in overcoming some of the theoretical difficulties which had so far seemed insurmountable."[32]

To construct his theory Heisenberg assumed first that a nucleus of charge Z and atomic number A could be considered as a bound system of Z protons and $(A - Z)$ neutrons; that a new force of nature was responsible for its cohesion; and, finally, that quantum mechanics was applicable to describe the motion of neutrons and protons in the nucleus.

But whereas in the case of *atomic* structure the forces between electrons and between electrons and the nucleus were known to be electromagnetic, the nature of the forces between the nuclear particles, such as their strength and how they depend on the distance between the particles; the particles' spin and relative velocity, not to mention other properties; and whether there existed three-body forces, was unknown. The nature of the nuclear forces had to be inferred from experimental data.

The task proved difficult. It was clear that nuclear forces were strong since the binding energies of nuclei were of the order of millions of electron volts (eV). They couldn't be electromagnetic in origin since electrical charges whose sum equaled the total charge of the nucleus when distributed within a volume the size of the nucleus could result in energies only of the order of tens of thousands of eVs. Regarding the range of nuclear forces, it had been established by Rutherford in his classic experiment on the passage of charged particles through matter that even in collisions in which a charged particle approaches a nucleus to a distance of the order of a nuclear diameter, the only noticeable force was the electromagnetic one. The range of the nuclear forces must therefore be very small. It was also clear that on balance the nuclear forces had to be attractive since they stably bind together the neutrons and protons that constitute the nucleus. But they could not be entirely attractive because heavy nuclei would then all be of the same size. By 1932 it was known that the average binding energy per particle was roughly the same for all nuclei, and that the volume of nuclei increases with the number of particles in them.

The relation between the size of nuclei and their binding energy supported the assumption that nonrelativistic quantum mechanics could be used to describe nuclear structure. From the comparison of the binding energy of the deuteron (^{2}H) and that of an alpha particle (^{4}He)—2.2 MeV as compared to 28 MeV—Wigner deduced that the nuclear forces had a range of the order of 10^{-13} cm. The fact that the nuclear binding energy is proportional to A, the total number of protons and neutrons in the nucleus (in contrast to atomic binding energies that are proportional to $Z^{7/3}$), indicated that the nuclear forces between the nuclear particles must saturate.[33] From such considerations Heisenberg, Majorana, and others deduced some of the basic properties of nuclear forces.

The success in accounting for some of the properties of the light nuclei, such as the deuteron and the α-particle (^{4}He), led to efforts to model nuclei in general as composites built of neutrons and protons.

At this very same time, John Cockcroft and Ernest Walton, and Ernest Lawrence and M. Stanley Livingston, were building the first nuclear accelerators that could generate protons and deuterons with energies of the order of 1 MeV.[34] These in turn were able to induce nuclear reactions in the lightest elements, providing new data that would have to be taken into account by any new theory of nuclear structure.

A typical experiment in which a nuclear reaction is induced consists in generating a collimated beam of particles, such as protons or α-particles. These are made to strike a stationary target made up of a known element. The products of the resulting reactions are measured. The measuring instruments are designed to determine the mass, charge, energy, and angular distribution of the reaction products. The intent of the theoretical analysis of the data is to obtain information concerning energy levels, spin, and parity of the projectile and the target entities and to test the accuracy of the wave functions used in the description of the nuclear states involved in the reaction. But in order to extract this information one must have some models of nuclear structure and some understanding of the mechanisms involved in the nuclear reactions. Bethe contributed enormously to both of these components of nuclear theory.

Saturation of Nuclear Forces

During his stay in Rome and in Cambridge as a Rockefeller Fellow Peierls had gotten very interested in nuclear physics and was keeping up with the publications in the field. While in Rome he had spoken with Ettore Majorana and had discussed with him the saturation of the nuclear forces, a problem Majorana had studied.

In his first paper on nuclear forces Heisenberg had pointed out that since the binding energy and volume of nuclei are proportional to the mass number $A = Z + N$, where Z is the number of protons and N the number of neutrons of the nucleus, the nuclear forces cannot give rise to equal interactions between all pairs of particles. If this were the case the binding energy would at least be proportional to $A(A + 1)/2$, the number of pairs of nucleons in the nucleus. One way of obtaining binding energies proportional to A is to have the forces "saturate," so that each nucleon interacts only with a limited number of other nucleons. To obtain saturation, Heisenberg proposed that the nuclear forces between a neutron and a proton are "exchange" forces that involve an exchange of their position, so that in an interaction between them the neutron would follow the trajectory that had been the proton's, and vice versa. The "exchange" occurs readily only if the neutron and the proton move in a similar trajectory, or equivalently if they are in the same one-particle state. Since the Pauli principle restricts the number of particles in the same state, an "exchange" force limits the number of particles a given particle interacts with.

The mathematical characterization of such forces was given by Bethe and Peierls in their paper on the "diplon," as the deuteron was then called.[35] For the case of an ordinary (strong short-range) nuclear potential $V(x - \xi)$, the wave equation for the deuteron would be

$$\left(\frac{\hbar^2}{2M}\left(\nabla_x^2+\nabla_\xi^2\right)+E\right)\psi(\boldsymbol{x}s,\boldsymbol{\xi}\sigma)=V(\boldsymbol{x}-\boldsymbol{\xi})\psi(\boldsymbol{x}s,\boldsymbol{\xi}\sigma) \tag{1}$$

where $\boldsymbol{x}$ and s are the space and spin variables of the proton, and $\boldsymbol{\xi}\sigma$ those of the neutron. Wigner had shown that the wave equation (1) when suitably generalized to the case of three and four nuclear particles could correctly describe the behavior of nuclei up to ^{4}He.[36]

For a nuclear force of an exchange type, the wave equation for the deuteron is either of the form

$$\left(\frac{\hbar^2}{2M}\left(\nabla_x^2+\nabla_\xi^2\right)+E\right)\psi(\boldsymbol{x}s,\boldsymbol{\xi}\sigma)=V(\boldsymbol{x}-\boldsymbol{\xi})\psi(\boldsymbol{\xi}s,\boldsymbol{x}\sigma) \tag{2}$$

or

$$\left(\frac{\hbar^2}{2M}\left(\nabla_x^2+\nabla_\xi^2\right)+E\right)\psi(\boldsymbol{x}s,\boldsymbol{\xi}\sigma)=V(\boldsymbol{x}-\boldsymbol{\xi})\psi(\boldsymbol{\xi}\sigma,\boldsymbol{x}s). \tag{3}$$

A force that gives rise to Eq.(2) is called a *Majorana force*; it only exchanges the position of the nucleons and is spin independent.[37] The Heisenberg exchange force in Eq.(3) considers position and spin as intrinsic properties of a nucleon and exchanges both position and spin.[38]

In early spring 1934, Wigner came for a visit to England, and Bethe and Peierls had long conversations with him. They persuaded him that ordinary forces (now called *Wigner forces*) could not be responsible for the binding of nuclei "all the way up." Already at that time Bethe and Peierls had concluded that the nuclear forces must be mostly Majorana forces.

The Diplon

When Peierls was in Cambridge as a Rockefeller fellow he got to know Chadwick and discussed with him the experiments he was doing at the Cavendish. In the spring of 1934 Peierls suggested to Bethe that they go to Cambridge and find out more about Chadwick's current research. Chadwick, with his student Maurice Goldhaber, was at the time carrying out an experiment in which deuterons—then called "diplons"—were photodisintegrated using the $h\nu$ = 2.62 MeV γ-rays emitted by thorium C′:

$$\gamma + \mathrm{D} \rightarrow \mathrm{n} + \mathrm{p}. \tag{1}$$

The importance of their experiment stemmed from the fact that since the deuteron was believed to be a bound state of a neutron and proton, knowing the masses of the neutron and of the proton then allowed one to determine the binding energy of the deuteron. Since neutrons and protons have approximately the same mass, the neutron and the proton coming out of the reaction (1) share equally the excess energy. Energy conservation states that $h\nu + W = 2E$, where W is the binding energy of the deuteron and E the energy of the outgoing proton. The energy, E, of the proton can be obtained by measuring the total ionization it produces or by measuring its range. Chadwick and Goldhaber obtained the value 2.14 ± 0.10 MeV for W $(= h\nu - 2E)$, the binding energy of the deuteron. Once the binding energy is known, the mass of the neutron can be determined. They found to their surprise that the masses of the neutron and proton were different. The neutron had a slightly greater mass. They obtained the value $n^1 = 1.00846$ mass unit.[39]

Chadwick and Goldhaber had also measured the cross section for photodisintegration of deuterons by γ-rays. Their experiment gave a cross section of $(5 \pm 2) \times 10^{-28}$ cm^2. Bethe and Peierls found their photodisintegration experiment and the "numbers" they had obtained for the cross section very interesting.

Bethe recalled that during their visit "Chadwick was most amiable, and contrary to his normal retiring self, talked at some length to us about what they had observed, and then said 'I bet you can't make a theory of that.'"

In the mid-1980s at a conference both Bethe and Goldhaber were attending, Goldhaber reminisced about the Cambridge days and confided to Bethe that Chadwick had been so effusive "in order to stimulate you. He wanted you to undertake making a theory." Bethe and Peierls met the challenge. On the four-hour train ride from Cambridge to Manchester they formulated a theory to account for Chadwick and Goldhaber's data, making use of the short-range character of the nuclear forces that Wigner had established in his analysis of the binding energies of the deuteron and of helium. They solved the deuteron problem assuming that within its range, a, the potential V_0 between neutron and proton is much larger than the binding energy W. For V_0 and a as deduced by Wigner from the binding energy data [$V_0 \approx 50$ MeV, $a \approx 10^{-13}$ cm] there is only one bound state of the neutron-proton system. It has angular momentum $l = 0$, and the wave function for the relative motion given by

$$u_0 = \frac{C}{r} e^{-ar}$$

with

$$C = \sqrt{a/2\pi}$$

$$a = \left(-\frac{1}{ru}\frac{d(ru)}{dr}\right)_{r=0}.$$

The binding energy is then determined to be

$$W = -\frac{\hbar^2}{M}a^2.$$

By analogy with the photoelectric effect of an atom Bethe and Peierls computed the cross section for photodisintegration of a deuteron using the bound state wave function they had derived and the fact that the wavelength of the thorium C′ γ-ray, 75×10^{-13} cm, was large compared to the range of the nuclear forces, 1×10^{-13} cm. Their calculated cross section turned out to be somewhat larger than the one measured by Chadwick.[40]

Upon their return to Manchester, Bethe and Peierls began to write the paper which formalized the preliminary results that they had obtained on the train.[41] For the paper they also considered the scattering of neutrons by low energy protons. By virtue of the finite range of nuclear forces only the $l = 0$, s-wave phase shift contributes to the cross section. They then essentially developed an approximate treatment based on the fact that the logarithmic derivative

$$\left(-\frac{1}{ru}\frac{d(ru)}{dr}\right)_{r=0}$$

of the scattering wave function inside the potential has the same value as that for the deuteron ground state. This allowed them to obtain the following approximate expression for the cross section:

$$\sigma = \frac{4\pi\hbar^2}{M}\frac{1}{E+W}.$$

For thermal neutrons this yielded a cross section of about $4 \times 10^{-24}\,\text{cm}^2$, whereas the experimental value was $\sim 50 \times 10^{-24}\,\text{cm}^2$. When the fact that the protons in these experiments were bound in molecules was taken into account, the cross section of low energy neutrons by free protons would be $20 \times 10^{-24}\,\text{cm}^2$.[42] The reason for the large discrepancy was later explained by Wigner to Bethe in February 1935. Bethe had just come to the United States and was attending the American Physical Society (APS) meeting in New York.

Bethe and Wigner were both taking a noisy subway ride from Columbia University, where the APS meeting took place, to Penn Station, where Wigner would take a train back to Princeton and Bethe one to Ithaca. There being four possible spin states of the neutron-proton system, Wigner asked him, "Why do you assume that the singlet state has the same binding energy as the triplet state? Let's assume a different binding energy and all will be well."[43] And indeed it was. Assuming different binding energies for the singlet and triplet states yielded the following cross section:

$$\sigma = \frac{\pi\hbar^2}{M}\left(\frac{3}{E+W_t}+\frac{1}{E+|W_s|}\right).$$

There the matter stood, for no one knew how to detect the singlet state until Teller suggested that the analysis of the scattering of neutrons by ortho- and para-hydrogen molecules would allow a determination of the binding energy of the singlet state. Teller's suggestion was elaborated in a paper by Julian Schwinger and Teller, and later experimentation corroborated Wigner's insight.

For their paper Bethe and Peierls also computed the cross section for the inverse process to photodisintegration, namely, the capture of a neutron by a proton with the emission of a γ-ray

$$\mathrm{n} + \mathrm{p} \rightarrow \mathrm{D} + \gamma,$$

and obtained a value of the order of 10^{-29} cm^2 for the cross section for low energy neutrons. They concluded that "the capture therefore seems hardly observable." They also computed the cross section for the disintegration of deuterons by electron bombardment and concluded that it was at the limit of observability.

Their paper, which was submitted to the Royal Society on July 14, gave further proof that quantum mechanics could account—semi-quantitatively—for the observed low energy properties and processes involving nuclear particles. In his interview with Weiner, Bethe described it as "quite a nice paper": "And in this paper, in fact, I think we got the first idea of what is now known as the effective range theory—namely, that such matters as the properties of the deuteron do not depend sensitively on the shape of the interaction between neutron and proton but only depend on some overall behavior of the interaction."

The London Meetings

Following the tradition initiated by Kapitza in the early 1920s, the experimentalists and theorists working in Cambridge would get together once a month to discuss recent advances. By 1933 many of them had migrated to professorial positions elsewhere, so they would gather for a day-long meeting only every two months or so. Blackett, then at Birkbeck College in London, and Cockcroft were nominally in charge. The meetings would take place mostly in London, since this was the location most easily accessible by train, although sometimes they were still held in Cambridge. Blackett, Cockcroft, Mott, Herbert W. Skinner, Kapitza, Charles Galton Darwin, and G. P. Thomson were regular attendees. The German refugee physicists who had come to Great Britain—Bethe, Fröhlich, Heitler, Peierls, Teller, and others—were invited to these "very lively and stimulating meetings." It was at one of these gatherings that Bethe and Peierls first heard of the discovery of artificial radioactivity and about Fermi's theory of β-decay.

Beta Decay and Neutrinos

In mid-January 1934, at a session of the Académie des Sciences in Paris, Frédéric Joliot and Irène Curie announced that they had created a new kind of radioactivity by bombarding the light elements with α-particles. They called this radioactivity "artificial" because the radioactive nuclei they had produced in their nuclear reactions emitted positively charged electrons (now called positrons) rather than the negatively charged electrons that are observed in "natural" radioactivity.

The paper by Joliot and Curie was discussed at one of the Cavendish meetings in February 1934.[44] Shortly thereafter, in early April, Bethe and Peierls submitted to *Nature* a paper entitled "The Neutrino" in which they pointed out that Curie and Joliot's findings strongly supported Fermi's β-decay theory. That theory assumed that in the process $p \rightarrow n + e^+ + \nu$ in a nucleus, the positron and the neutrino are not present in the nucleus but are created at the time of emission: this because "one can scarcely assume the existence of positive electrons in the nucleus."[45] Fermi's theory of β-decay also implied that if neutrinos were created in a nuclear decay

$$(A,Z) \rightarrow (A,Z+1) + e^- + \nu$$

then, when energy can be conserved, the neutrino absorption process

$$\nu + (A,Z) \rightarrow (A,Z\pm 1) + e^{\mp}$$

is also possible. In order to ascertain whether such a reaction is observable, and thereby prove the reality of neutrinos, Bethe and Peierls computed its cross section and found that for a 2 MeV neutrino the cross section would be of the order 10^{-44} cm^2. They thus concluded that "it is absolutely impossible to observe processes of this kind with the neutrinos created in nuclear transformations."[46] In a second article dealing with β-decay and the "reality" of neutrinos, submitted to *Nature* on May 5, 1934, Bethe and Peierls considered the possibility of observing the recoil of the nucleus in a (ν, e^+) decay of a light nucleus. In the last part of their paper Bethe and Peierls noted the existence of a new form of β instability in atoms: the capture by one of the protons inside the nucleus of an orbital electron of an atom.

On June 9, 1934 Bethe wrote Sommerfeld: "I recently visited Blackett and persuaded him of the existence of the neutrino. Someone there tried to measure the magnetic moment of the neutrino. Result negative, i.e., magnetic moment $< 10^{-5}$ Bohr magnetons."

Bremsstrahlung and Pair Production by γ-Rays

Walter Heitler was an *Assistent* to Born in Göttingen when Hitler came to power in 1933. When in April 1933 he was dismissed because he was Jewish, Born recommended him to the Academic Assistance Council for a research fellowship to work with Mott at the University of Bristol. While in Bristol he attended the London meetings regularly.

In Germany Heitler had become interested in the calculation of *Bremsstrahlung,* the radiation emitted by a charged particle when deflected by an electric field. The deflection produces a certain amount of acceleration, and the charged particle—according to classical electromagnetic theory—must therefore emit radiation. The quantum electrodynamic description of the process is that there is a certain probability that in the scattering there will be emitted a photon of momentum $\boldsymbol{k}$, the charged particle making a transition from an initial state of energy E_0, momentum $\boldsymbol{p}_0$, to a state of energy $E = E_0 - hck$ and momentum $\boldsymbol{p}$, with $\boldsymbol{p} - \boldsymbol{p}_0 = \hbar \boldsymbol{k}$.

Heitler and Fritz Sauter had calculated the cross section for *Bremsstrahlung* by electrons when scattered by a Coulomb field; and Heitler had calculated the related cross section for pair production by photons in a Coulomb field. *Bremsstrahlung* and pair production were recognized as the most important processes affecting the energy loss of high energy electrons in their passage through matter.

Knowledge of the cross section for *Bremsstrahlung* by electrons when scattered by a Coulomb field and of the cross section for a photon to produce an electron-positron pair in a Coulomb field were necessary in order to explain the multiplicative showers of cosmic radiation that had been observed by Blackett and Occhialini.[47]

In their passage through matter, electrons and positrons of small energy and heavier particles of practically all energies lose their energy primarily in collisions in which the atomic electrons are excited or ejected. High energy electrons and positrons lose most of their energy by radiation—by *Bremsstrahlung*. The energy loss per centimeter path by a high energy electron due to *Bremsstrahlung* is roughly constant up to mc^2 and then increases in proportion to its energy.

The energy lost by radiation is, to a first approximation, uniformly distributed among secondary photons whose energies range from zero energy up to the energy of the primary charged particle. Hence, in the interaction of high energy electrons with matter, high energy photons are produced. These in turn interacting with the electrons and nuclei of atoms produce electron-positron pairs or undergo Compton scattering. Both processes give rise to electrons having an energy comparable to that of the photons that produced them. These new electrons when colliding in turn produce more photons, which in turn produce more electrons through pair production and Compton scattering. At each new step of the process the number of electrons increases, and their average energy decreases. The process continues until the electrons reach the energy at which energy losses by collisions become more important than radiation losses. This phenomenon is called a *multiplicative shower.*

Bethe's interactions with Blackett had gotten him very interested in cosmic rays. In the early 1930s the properties of cosmic rays were little understood. The cosmic rays observed on the surface of the earth were known to consist of a mixture of electrons, positrons, photons, protons, neutrons, and perhaps others. The basic problem was to establish which of these particles were primary and which were produced in the atmosphere as a result of the interaction of the primaries with atoms in the atmosphere. The difficulties in the problem stemmed first from the fact that it was not known whether particles different from the known ones were involved, and second, even if only the known particles were involved, their energy, which could be as high as 10^{11} eV, was so much higher than those manipulated in the laboratory that they might well depart radically from their ordinary behavior and be unrecognizable.[48]

The attraction of cosmic rays to the theorist was that they offered testing grounds for the validity of quantum electrodynamics. As Heitler indicated in his monograph, *Quantum Theory of Radiation,* "The future development of quantum electrodynamics will probably be deeply influenced by the answer to the question whether the present theory will still be reliable at very high energies ($>100mc^2$). For the experimental test of this point the *Bremsstrahlung* will play an important role."[49] The calculations of the cross sections for *Bremsstrahlung* and pair-production, and the comparison of the predicted values with empirical data derived from cosmic-ray experiments, became the focal points of discussions on the limit of validity of quantum electrodynamics. Bethe became deeply involved in these matters.

At one of the meetings in London in late 1933 Heitler reported on his calculations of the cross sections for pair production in a Coulomb field by high energy photons and of *Bremsstrahlung* by high energy electrons in Coulomb scattering. Heitler wanted to use his calculated cross sections to interpret the Blackett and Occhialini data on showers as well as the 1933 experiments of Carl Anderson and colleagues on the energy loss by high energy electrons in passing through lead plates: this to find out whether these experiments indicated a breakdown of quantum electrodynamics.

In his interview with Charles Weiner, Bethe described what had happened after Heitler had made his presentation:

> there were experiments by Anderson which seemed to indicate that electrons could penetrate large thicknesses of material, and Heitler's theory predicted the opposite. It predicted that they would emit radiation at a very great rate in such penetration, and so we did not understand how this could happen. Now, the first idea [I] had, was that maybe this is an effect of the screening of the electric field of the nucleus by the electrons. The first calculation Heitler did . . . was the emission of *Bremsstrahlung* by a fast electron in the field of just one point charge, the charge of a nucleus. He found that this process increases with energy; so that the higher the energy of the electron, the more likely does it become that the electron loses, let's say, half its

energy by radiation in one collision, which is a very unusual way for any particle to behave, or at least so it seemed at the time. So then I set out to see whether [the *Bremsstrahlung* and the pair production cross sections] would be changed if you take into account that the point charge of the nucleus is screened by the electrons around it,[50] and these were fairly complicated calculations but manageable and I found the result—that it didn't really make very much difference. The only difference it made was that at very high energy the *Bremsstrahlung* probability finally became constant—it reached an asymptotic value instead of continuing to increase. And this was definitely in contradiction to the work of Anderson. Well, several years later that work was then explained in terms of Anderson having observed [μ] mesons rather than electrons, and the mesons, because of their much higher mass, do not emit much radiation—the radiation goes inversely as the square of the mass. But at the time we didn't know that, so we pointed out this contradiction, and concluded that electrodynamics *must fail* above a certain energy. In the meantime, by much more experimentation and distinguishing protons, mesons and electrons, we know now that electrodynamics continues to be valid up to much, much higher energies, and we still have not found any place where electrodynamics fails. This is over 30 years later.

Heitler had not initially believed the screening effects. Edward Teller eventually convinced him when the three of them—Bethe, Heitler, and Teller—met in London to resolve the issue.

In the light of their calculations Bethe and Heitler had come to the conclusion that the Anderson experiments indicated a breakdown of quantum electrodynamics.[51] In February 1934 Bethe wrote Sommerfeld—who had made earlier calculations of *Bremsstrahlung* and of the photoelectric effect—to indicate why they had reached that conclusion:[52]

25 II 34

. . . Heitler and I have written up the creation of positive electrons with and without screening, and sent it to the *Proceedings*.[53] For 2.6×10^6 Volt-radiation the calculated "W.Q." for pair creation agrees exactly (within a few percent) with Blackett's latest measurements (for Pb). That, first of all, speaks very much for Dirac's hole theory, which anyway is always confirmed, and secondly it shows surprisingly that Born's collision theory applies to Pb. Thus it is all the more impressive that for the *Bremsstrahlung* of 300×10^6 Volt electrons the theory gives at least 15 times too large a value.

The 1934 London Conference

The opening paragraph of the review in *Nature* of the *Proceedings* of the International Conference on Physics, held in London in the first week of October 1934, noted that the conference was "unique of its kind . . . [C]oncerned with topics among the most important of those about which physical science is conversant, it is small wonder that the conference held for a week the attention of some five or six hundred members, drawn from almost every European country and from America."[54]

Bethe in his interview with Weiner characterized it as "a very nice meeting": "I saw some of the great people for the first time—for instance, Madame Curie and her daughter, Irène, and Joliot. I don't remember who all was there from America, but I think I met several of the American physicists for the first time [Anderson, Robert Millikan, A. H. Compton]. It was an excellent meeting. There were a lot of good papers, and in those days one could still understand all of physics and listen to all the papers and intelligently discuss them."

In his welcoming address the president of the Royal Society, the biologist Sir Frederick G. Hopkins, commented on the dramatic conceptual changes concerning nuclear structure that had taken place: "It seems but yesterday that the atom was thought to be made up of protons and electrons alone; now we hear of the properties of neutrons, positrons, photons, and possibly, I am told, also of neutrinos. More, we have made the acquaintance of artificial radioactive substances and even of atoms new to the universe." And he noted how remarkable it was that the great advances in solid state and nuclear physics had been achieved despite "political unrest, financial stringency" and, alluding to Nazi Germany, despite "anti-intellectual movements in the world."[55]

The conference was divided into two parts. The first addressed issues in nuclear physics; the second concerned solid state physics. Rutherford gave the opening lecture of the nuclear physics sessions, and he presented a masterly, detailed survey. He concluded his address by stating that "At this stage of our knowledge, it is not to be expected that anything but tentative explanations can be given of the observed facts. A complete theory applicable to nuclei in general is probably far distant . . . The development of our knowledge of nuclear physics is now at a most interesting stage and a close collaboration between the theoretical and experimental physicists is important for rapid progress in this most fundamental of problems."[56] Peierls did not attend the conference, but Bethe did, presenting their joint paper on the photodisintegration of the deuteron.

What strikes one in reading the proceedings of the conference is Bethe's participation in it. He made comments in every session of the nuclear physics section, and he was the only one to do so. His remarks indicate how extensive were the calculations he and Peierls had carried out regarding β-decay theory, nuclear structure, and nuclear stability, much of which was never published but which was very important when later Bethe wrote his nuclear physics articles in the *Reviews of Modern Physics.*[57]

Bethe's presence at the cosmic ray session resulted in a joint paper with A. H. Compton in *Nature.*[58]

> [T]he origin of that paper was the following: Compton had somehow come to the conclusion that cosmic rays consisted of protons, and on that he had a terrible fight with Millikan, who maintained that cosmic rays consisted of electromagnetic radiation, which then made electrons. So Compton was very happy about our result—Heitler's and my result—that electrons should suffer very great energy losses when penetrating material—for instance, the atmosphere. And so he said, "The primary cosmic rays cannot be electrons but protons," and we wrote a joint paper in which we gave arguments for this thesis.[59]

Bethe's comments at the London conference following Anderson's presentation of his and Seth Neddermeyer's experimental results on the absorption of high energy electrons by lead merit quotation in full:

> The experiments of Anderson and Neddermeyer on the passage of cosmic-ray electrons through lead are extremely valuable for theoretical physics. They show that a large fraction of the energy loss by electrons in the energy range round 10^8 volts is due to emission of γ-radiation rather than to collisions, but still the radiative energy loss seems far smaller than that predicted by theory. Thus the quantum theory apparently goes wrong for energies of about 10^8 volts, and it would be of special value for any future quantum electrodynamics to know exactly at which energy the present theory begins to fail, in other words to have much more experimental data on the energy loss of fast electrons (energy 10^7 to 5×10^8 volts) passing through matter.

As noted earlier his conclusion that "the quantum theory apparently goes wrong for energies of about 10^8 volts" was erroneous. Anderson was observing

μ-mesons and not electrons. The empirically observed energy loss of these charged particles whose mass was approximately $200m_e$ was in accord with the Bethe-Heitler theory.

Invitation to Cornell

Bethe's appointment in Manchester was for a year only, and thus the question of what would happen the following year came up early. A confluence of events determined Bethe's subsequent life. In the fall of 1933 the physics department of Cornell University started looking for a theorist. On its staff was Lloyd P. Smith, the young theorist who had studied with Bethe in Munich during the academic year 1931–32.[60] Ernest Merritt, who was the head of the Cornell physics department until 1933, had met Bethe in Rome in 1932 when he and Smith were visiting various laboratories in Europe to determine what direction the department of physics at Cornell ought to take to strengthen its research activities. Smith recommended Bethe strongly for the Cornell position. Bragg, who had been visiting Cornell for the spring semester, corroborated Smith's assessment of Bethe.[61] On August 18, 1934, R. C. Gibbs, now head of the physics department, wrote to the dean of arts and sciences to recommend the "appointment of Dr. Hans Bethe as Acting Assistant Professor of Physics for the year 1934–35, at a salary of $3000." In his memorandum Gibbs noted,

> The need for strengthening our work in theoretical physics has been apparent for some time. In taking such a step it was highly important to find a comparatively young man but one with sufficient experience and maturity to insure the likelihood of his assuming a position of leadership in the department. At the same time it was also unusually essential that he be a man capable by his training, temperament and inclination of working in close cooperation with the experimental and theoretical interests of the dept. as a whole. The strong recommendations of Profs. Sommerfeld and Bragg and our intimate knowledge of the admirable way that he exerted his influence in promoting the work in theoretical physics at the University of MUNICH together with his numerous outstanding publications (a list of which is attached) have convinced a large majority of the Faculty in Physics that Dr. Bethe is a most promising candidate in meeting our needs. The situation has been discussed in detail with many members of the department and at a recent meeting with all the Professors and Assistant Profs. in Physics the

> desirability of trying to bring Dr. Bethe here on a temporary appointment was gone over very carefully.[62]

Bethe accepted, but because he had received an offer of a yearlong fellowship in Bristol with Nevill Mott, he asked for and obtained permission from Cornell to assume his duties there in the spring term rather than at the beginning of the academic year.[63] He stayed in Bristol during the fall semester of the academic year 1934–35 and arrived in Ithaca in February 1935.

When interviewing Bethe in 1969, Charles Weiner asked him about his stay in England between 1933 and 1935. Bethe's answers merit quotation.

> WEINER: Let me ask you another question: that is, do you feel that the experience in England changed you in any way—the overall experience? It marked a certain stage. You were by this time about 27 and had already established a certain style in physics, had come from a certain tradition, although you had come in contact with England earlier. But I want to know how you evaluate that entire period in terms of your own career.
>
> BETHE: I think it gave me more contact with other physicists, especially of my own age, and it got me more in touch with the live problems of the day. There was much less of that in Munich and even in Rome, although Fermi of course read very avidly what was going on, it was still very far from the centers of research. So I think it got me much closer to the important problems.
>
> WEINER: And yet one would think the circle around Sommerfeld and the various scientific groups in Germany would be in pretty close touch with important problems.
>
> BETHE: Well, I think they were mostly self-generated problems. In the course of investigating quantum mechanics, you would come to the question of quantum electrodynamics, as Heisenberg and Pauli did, and then you would find in quantum electrodynamics that there are certain divergences, and then you would be interested in investigating the mathematical problem of these divergences. This is the kind of problem that I would have worked on in Germany. Or I worked on the electron scattering problem which was an extension of what I already had done. Then I would go out and look for experiments in which electron scattering was investigated and then use that in the

theory. But it was not so much the style of working on something entirely new like nuclear physics. I doubt very much if I had stayed in Germany whether I would have gone into nuclear physics as soon.

WEINER: In this period, really a brief period, did you become aware of the organized efforts to aid refugees?

BETHE: Yes, I was aware of them. I did not do much about it. I'm embarrassed that I didn't help much, but I knew of the effort. I knew, for instance, Szilard reasonably well. He was deeply engaged in this, and I met, I think, one or two people from the Academic Assistance Council.

Bethe would look back on his year in Manchester as perhaps the most productive of his life. It was indeed productive, but moreover, for Bethe Manchester turned out to be the right place at the right time, with the right collaborator—Rudi Peierls. And Bethe had the right pragmatic orientation and the right tools to solve the problems being posed. Vast new fields at the frontiers of physics were opening: nuclear physics, cosmic rays, quantum electrodynamics. In each, new, accurate, and reliable empirical data were being generated by novel instruments and techniques. And Bethe recognized the opportunities.

In Manchester Bethe buttressed his powers as a modeler. Models, for him, became autonomous in the sense that they could be studied independently of the theories and the data between which they mediated: they could represent and could act like instruments.[64] Bethe constructed models to deepen his understanding of both global and particular features that were incorporated in them. They could be "toy models" like the linear chain of spins for which he derived the ground state wave function; or be more realistic, like the model of alloys he investigated to solve the problem Bragg had posed; or, as in the case of the model of the deuteron he and Peierls advanced to investigate the data Chadwick and Goldhaber had obtained, allow general properties to be deduced that depended only on general and generic features of the model. Thus they formulated a description of the deuteron and of low energy nucleon-nucleon scattering that was insensitive to the details of the short-distance behavior of the short-range nucleon-nucleon potential, i.e., to its particular shape. The scattering cross section they calculated was expressed in terms of two parameters that were to be determined experimentally.

Manchester reinforced Bethe's conception of physics—that it was principally concerned with statements that can be verified by experiments. Knowledge in physics accretes because it is reproducible. The purpose of theory is initially to provide a classification, systematization, and unification of the reproducible experimental results. Theories are mathematical, quantitative, informative compactifications that can "explain" and predict new phenomena. And in view of the enormity of the available reproducible data, in order to grasp and comprehend the regularities discerned in experiments, it is necessary that theories be in some sense simple: they must describe and represent the maximum amount of experimental information with a minimum of concepts.

But Manchester was more than all this. Bethe's collaboration with Peierls deeply affected him. Friendship acquired a new meaning. Also, his interactions with others—Bragg, James, Chadwick, Heitler, and Teller, to name a few—and his participation at the London Conference gave him not only a new perspective on his off-scale abilities as a physicist, but also a perception of what role he could play in the discipline as one of the outstanding young theorists.

During his stay in Manchester Bethe's involvement in physics was total, with very little attention being paid to the plight of other scholars and scientists who had to leave Germany. But it was an involvement that reflected his father's attitude concerning research in the sciences, that satisfaction was to be found in the *search* for knowledge, in the *search* for solutions.

6

Hilde Levi

In the midst of Bethe's professional accomplishments, a personal crisis intruded, one that threatened to disrupt his friendships and his work. Its resolution had enduring consequences: it deeply affected the other person involved, adversely reflected on Bethe in his disciplinary circles, and had professional repercussions. Thus during the 1930s, although Bethe became one of the leading nuclear theorists, he was never invited to the annual conferences hosted by Bohr's institute in Copenhagen, the preeminent international center for theoretical nuclear physics. Nor did he ever visit the institute after coming to the United States in 1935, although he went to Europe every summer to visit his mother and his father.

The *Tanzstunde* Revisited

Hilde Levi was one of the young women in the first *Tanzstunde* that Hans was part of and, like Hans, participated in the second one. Hilde and Hans became very good friends, went together on the skiing trips and summer hikes of the circle, corresponded extensively with one another, and in the summer of 1934 became engaged to be married.[1]

Hilde was born on May 9, 1909, in Frankfurt. Both her parents were thirty-seven years old at the time. An older brother, Edwin, had been born in 1900. Her father, Adolf, was *Verkaufsdirektor* (director of sales) in a large Frankfurt metal company. Her mother, Clara, née Reis, was the daughter of a printer. Books abounded in the Levi home. Almost from the moment that Hilde was born her mother began reading her stories. As a teenager she proofread the galley sheets of the manuscripts her grandfather set. She became a voracious reader with wide interests, and one of her lifetime passions was "books and their content." Another was music. She learned to play the piano at an early age and became a consummate and very gifted pianist. Moritz Wallach, the owner of a very successful retailing business in Munich, was Hilde's cousin on her father's side.[2] Moritz and his wife were generous patrons of the arts and had supported Richard Strauss during the early phases of his career. Their daughter, Ilse, who was six years older than Hilde, was a very fine pianist who had given up a concert career when she married Albert Hesse, a chemist.[3] The Wallachs owned a summer home in Bavaria to which Hilde was invited every summer.[4] Her love of music and her musical talents were nurtured there. Among the guests at the Wallach summer house were many of Germany's outstanding musicians, and Hilde remembered meeting Richard Strauss and Elisabeth Schumann there.[5]

Hilde's parents were liberal and cosmopolitan in outlook. Her father made it a "point of principle" to hire an English "missy" in 1913 so that she and her brother would learn English. Later on, after she graduated from *Oberrealschule,* he sent her to England for half a year. In religious matters her parents were freethinkers and had renounced their membership in the Jewish community. They had a wide circle of friends "whose religion never mattered." Their home was an open, friendly place. Visitors came there frequently and because her father would invite his business contacts to the house, foreign guests were common.

Hilde said the following of her mother's brother's family: "My uncle was—I think—converted, my aunt was Christian and my cousins all married Christians, and made every effort to be 'aufgenordet' and survived the Third Reich in Germany relatively comfortably."[6]

The Levis were close friends of Ludwig and Rosy Fischer, whose son Ernst became Albrecht Bethe's *Assistent* at the university. Although not as "intellectual" as the Fischers, the Levis shared the Fischers' passion for the arts and for music and this had cemented their friendship.[7] The Fischers were a Jewish family of note.[8] Ludwig was born in Breslau in 1860; his wife, Rosy, née Haas, came from a venerable, well-established Frankfurt family. Ludwig was a physician, but poor health made him retire from his practice in Breslau in 1896. After spending a year in Switzerland he and Rosy settled in Frankfurt. In 1905 they started collecting paintings. Being liberal, socially conscious, and independent in their beliefs, they were attracted to the paintings of the German expressionist painters who had banded together around Ernst Kirchner, calling themselves *Die Brücke* (The Bridge). Kirchner's group of painters "sought to serve as a unifying force for all radical and revolutionary artistic elements, to bridge the past with the future, and to bring about social changes in the human condition."[9] From 1905 until Ludwig's death in 1922 the Fischers assembled a magnificent collection of modern paintings, many of them bought from the artists themselves.[10] Except for a few paintings sold for income, half the collection was brought to the United States by Ernst and his wife Anne when they fled Nazi Germany in 1934. The other half, which Ernst's brother Max had inherited, was partly sold, partly seized by the Nazis, and partly lost in transit when Max left Germany in 1935.

Ernst and Max were brought up as freethinkers, "open-minded to everything which was of intellectual, scientific and artistic interest."[11] Ludwig and Rosy gave them great freedom, but also expected them to be responsible for the consequences of their actions. While attending Frankfurt University to obtain his medical degree and his doctorate in physiology with Albrecht Bethe, Ernst became a member of the Nelson circle in Frankfurt. Leonard Nelson, the professor of philosophy at the university, was described by Karl Popper as an "outstanding personality."[12] Ella Philippson had met him in Göttingen when he was habilitating with Hilbert and she was a member of El Bokarebo, a group of influential housemates that included Max Born. He remained a friend of both Ella and Paul Ewald thereafter.[13] Ernst became deeply influenced by Nelson's liberal views and at one time wanted to go into public health as a result of his political convictions.[14] In 1925, after finishing his doctorate, Ernst married Anne Rosenberg. Anne's family had lived in Stuttgart, but shortly after the end of World War I moved to Frankfurt. Her father, too, was a freethinker and, although no Jewish rituals were observed in the house, Jewish ethics were inculcated in the children. Her father was proud to be a "German citizen of the Jewish faith," with an emphasis on the German citizenship.[15] Anne had studied chemistry in Stuttgart and then switched to the humanities when her family moved to Frankfurt. She attended Nelson's seminar and there met Ernst.[16]

After Ludwig Fischer died in 1922, his wife Rosy found that her income from their investments was not sufficient to support herself. They had greatly depreciated in value because of rampant hyperinflation. Hilde's father then came to her assistance and negotiated an agreement with the Staatliche Galerie Moritzburg in Halle, whereby she gave the museum twenty-four paintings by Ernst Kirchner, Erich Heckel, Emil Nolde, and Karl Schmidt-Rottluff, and in return was given a guaranteed annual income of 25,000 DM for a minimum of ten years. That income allowed her to live comfortably.

Looking back on her youth in Frankfurt, Hilde affirmed that her parents and their circle—the Fischers, the Strausses, the Sachses, and the Wallachs—were indeed *Kulturträger* of the Enlightenment ideals of cosmopolitanism and liberalism. And for some of them, the affirmation of these Enlightenment ideals was simultaneously a disavowal of their Jewishness.

Some of Hilde's earliest memories were of World War I. She vividly remembered being on the roof of their house, watching a French aerial bombing of Frankfurt in 1917, and asserting to her father that "a French bomb can't go through a German roof." She quoted her father's reply, "I am afraid that a French bomb *will* go through a German roof" as indicative of his lack of fanaticism. She also remembered her conflicted feelings during the war when asked at school to "pray for victory." "Weren't the French also praying for victory?" Later while attending the Viktoria Schüle, the *Oberreal Gymnasium* for girls in Frankfurt, she too gave up her faith and left the Jewish community. Upon entering the school she was registered as being Jewish and had to attend the obligatory religious instruction, which was part of the curriculum. Sometime during her first year at the *Gymnasium,* she told her father that she was leaving the Jewish community and did not wish to attend the lectures on Judaism since the rabbi who was teaching them was a hypocrite. He had been seen taking a taxi to the back entrance of the synagogue on a Saturday, yet had told the students in his class that one must not ride on the Sabbath. Her father assented to her wishes. Thereafter she consciously rejected everything that "smacks of church. I don't like priests."[17] And as had been the case with her parents, religion was never an issue in the choice of friends. She made a point of stressing the following: the fact that Hans was not Jewish was never a relevant factor in their becoming "best friends" in the *Tanzstunde* or later when they became engaged to be married.

Hilde and Hans

Hans and Hilde became very good friends shortly after they met in the fall of 1925 in the *Tanzstunde.* After Hans went to Munich after Easter 1926 they saw one another whenever he came back to Frankfurt, and they would go together to all the summer and winter activities of the *Tanzstunde* circle. His leaving for Munich also initiated their frequent correspondence. Hans's first letter to Hilde was written in the late spring of 1926 when he was in Heidelberg visiting Werner Sachs, a fellow member of the *Tanzstunde.* Their correspondence continued until December 1934, with some hundred letters exchanged between them. By the end of the summer of 1926 they had become "best friends," and Hans's parents had met her. On November 9, 1926, amid the deep difficulties his parents were experiencing, Hans wrote to them the following about Hilde, who was sick at the time—remarks probably addressed more to his mother than to his father:

> Yes, Hilde. I understand your disappointment. She looks outwardly somewhat different and at the moment looks very bad, so that I too would not think of characterizing her as pretty. At the beginning of the *Tanzstunde* she looked well, and even Torben [one of Hans's friends in the *Tanzstunde*] who is competent in these matters found her pretty . . . From the first moment I never had any lack of material for conversation with Hilde, even when we barely knew each other. Increased liking through long battle and final victory. Strong friendly advances from her side have raised my self-awareness, and only through her have I fully overcome the feeling that "no girl whatsoever likes me." In the first instance she is "lieb" [dear, kind, sweet], and I most of all needed and need a girl who is "lieb," and whose best friend I am—*primo,* and perhaps also *unico loco*—and on whom I can completely rely.[18]

Hans went on to compare Hilde with the other two girls in the circle to whom he was attracted—Fips and Käte—and told his parents that he had come to see that Hilde was the most intellectually mature, the wisest, and the easiest to talk to among them.

Already in *Oberrealschule* Hilde knew that she wanted to become a scientist, and her father encouraged her in this direction. She had a very fine physics teacher in the school, whom she characterized as a "strange fellow." He had been an officer during the war, had been severely wounded, and because of the injuries he had suffered he limped very badly. "He was very hard in his approach. But he knew how to make the physics lessons he gave us very interesting. And therefore, I began to make some experiments with him."[19] The last year of her studies in the *Oberrealschule* was devoted to a yearlong project on colors, spectra, and photography under the supervision of this very supportive physics teacher. The work became her *Oberreal Abiturium.* She graduated in April 1928, the only girl in her class to major in physics. She spent the rest of the spring and part of the summer in London, "learning good manners" in an exchange program that had been recommended to her parents by the "missy" who had been with them before the war and with whom they kept in friendly contact. She did not need to learn manners. She did learn "perfect English," and at the end of her stay was bilingual.

In her last semester in the *Oberrealschule* Hilde decided to study experimental physics at the University of Munich. Hans was not encouraging. When she told him that she wanted to study physics he tried hard to deter her from what Hilde in her *Autobiography* called "this crazy idea." Hans told her "girls should not be in the natural sciences, definitely not in physics—it was not a woman's occupation." A charitable interpretation might view his admonitions to Hilde as stemming from an assessment of the extremely competitive situation in the natural sciences and his trying to make Hilde take into account the realities of the existing strong cultural prejudices against women and against Jews in academia. In Munich Hans had gotten to know Margarete (Margie) Willstätter, the daughter of the chemist Richard Willstätter, and had become friends with her.[20] Margie was studying mathematics and had the reputation of being a mathematical genius. "Not even Margie Willstätter, who is very gifted in mathematics," Hans wrote Hilde, "will have a chance to be able to manage in science." But a more accurate view of the matter is that Hans's advice to Hilde reflected his own prejudices and is to be taken at face value: at the time he did believe that women had no role in the sciences.

Hans's discouraging attitude did not deter Hilde. Illness prevented her from entering Munich University in the fall of 1928, but she started attending lectures in the spring of 1929. By that time Hans had already moved to Stuttgart, yet his disapproving attitude persisted. She became a student in the Institute for Experimental Physics and attended the lectures of Fajans, Wieland, and Wien. She also went to some of Sommerfeld's lectures. She was the only woman pursuing a doctoral degree in the institute. Working in the laboratory she discovered that she had "very skillful fingers," and that she could design and build apparatus that worked and yielded reliable and accurate data; and moreover, that she had good physical intuition. For her dissertation she moved to Berlin, where her father had "managed to get her accepted at the Kaiser Wilhelm Institut für Physikalische Chemie, whose director was Fritz Haber." Her work there was carried out under difficult conditions. The institute was administered in an authoritarian fashion: "Fritz Haber was a very kind man privately, but he ran the institute on a very dictatorial basis with respect to students who didn't have so very much to do with him, and his co-workers, the various scientists working at the institute feared Haber . . . [H]e was the boss and every scientific co-worker had only a very small area to move in independently." For her doctorate Hilde carried out experimental work under Hans Beutler's supervision, studying the spectra of alkali haloids. Beutler, too, was a veteran of the war. "His approaches were . . . almost sadistic . . . If an experiment didn't work he always assumed to a first approximation that it was my fault and that I had done something wrong. And in this way, I really learned to be very critical about my own experimentation."[21] Eventually she successfully completed her doctoral dissertation, despite the obstacles that had been put in her path, and was awarded a doctorate in 1934.

After Hindenburg appointed Hitler chancellor of Germany at the end of January 1933, Hilde knew that her time in Germany was over. Given the Nazi position regarding women and Jews in academia, she saw no future for herself there. With financial support from her father she went to Copenhagen in late April 1933 and became James Franck's assistant in the Bohr Institute.[22]

Hans and Hilde had become close friends by this time, but were not romantically involved. Hans told Hilde about his involvement with Käte Feiss in 1932, when they had traveled through Italy and he had proposed but was rejected. In the summer of 1934, after he had gotten the invitation to join the Cornell physics faculty for the academic year 1934–35, agreeing to come to Ithaca at the beginning of the second semester, Hans went to Copenhagen for a brief visit in September 1934. During that visit he proposed marriage to Hilde.

Undoubtedly, the fact that Hans was anxious about the prospect of going alone to the United States was a factor in asking her to wed him. He admired her, admired her prescience, admired what she had accomplished, and believed that he loved her deeply and that together they would be able to overcome all the adversities they might have to face. Hilde accepted and they became engaged. They had talked about their doubts before Hans left Copenhagen and had written "to each other [independently] a few days later, how happy we were to have cleared [our] doubts and to be quite sure of each other."[23]

Hans then told his parents of his engagement to Hilde. His father was very happy, as he liked Hilde very much and thought very highly of her. His mother's reaction will become clear as the tale unfolds. Extensive preparations were made for the couple's departure to the States and many items were purchased to furnish the home they were going to live in. The Levis bought silverware, towels, and bedding for the couple, and the date for the wedding was set for late December. A few days before the wedding, the engagement was broken.

The Hans Bethe–Hilde Levi Correspondence

During my interview with her, Hilde Levi read to me from some of the letters that Hans had written to her. In the early correspondence Hans was very explicit that he wanted *Kamaraderie,* that he wanted to be a very good friend, possibly a best friend to her, but did not want to get romantically involved. And he criticized her for expecting more of him. But he readily accepted the fact that they were "paired" in the *Tanzstunde* circle and that she clearly was his "girlfriend." He was also unequivocal about his negative views regarding women pursuing scientific careers. Furthermore, he indicated to her that he wanted a wife who would cater to his needs—thus implicitly suggesting to her that he would never marry a woman committed to a scientific career.

Hilde characterized Hans at that stage of his life as "rejecting emotions, not being able to handle emotions." It seems clear that Hans had but a limited awareness of the constraints that his young women friends were experiencing and little understanding of their struggles to achieve a measure of autonomy and professional standing. Nor could he empathize with their emotions in the face of the constraints imposed on them by their cultural and social milieu.

Another trait that becomes apparent from his letters to Hilde is that by the time Hans obtained his doctorate in 1928 he had acquired a very strong ego, was very self-confident, very conceited, and very ambitious. Thus he wrote Hilde in 1928 after finishing his thesis that he was becoming "berühmter und berühmter" (more and more famous). But although Hilde recognized Hans's exceptional abilities, she was not intimidated by him.

The letters Hans wrote to Hilde before and after his parents had gotten divorced are particularly revealing. In them he described how "schrecklich" the situation was at home. He told her of his mother's emotional instability and difficult character, and of the fact that his parents' marriage existed only as a legal entity. He divulged that his father, in order to find emotional solace, had had affairs with other women. He depicted the turbulent, "roller-coaster"–like interactions between his parents and explained that this had been the reason that, except on one occasion, the *Tanzstunde* never met at his house. He told her that when his father had recently come down with a bout of phlebitis, his mother had nursed him back to health only to find out that he was involved with a young woman he wanted to marry, and that he wanted a divorce. Hans then told Hilde that his mother, unable to face that possibility, had had a nervous breakdown and had gone to a sanatorium, from which she had been recently discharged; that she had finally consented to a divorce and was presently living in Baden-Baden. In a letter dated August 1927, while his mother was in the sanatorium, Hans described her as deeply depressed, despondent, and "lebensmüde" (tired of life). Hans believed his mother had considered suicide when faced with the possibility of divorce. Hans's letters to Hilde from that period also reveal the hostility he bore toward his father at that time.

Turning to some of the later correspondence, Hilde corroborated the fact that Hans had been startled to be identified as a non-Aryan and having Jewish roots at the time of his dismissal from Tübingen in April 1933. Hilde also knew that Hans's mother remained in denial of her Jewish roots and was quite sure that the National Socialist regime and the new laws would not affect her status as Frau Geheimrat Professor Bethe. It took *Kristallnacht* to convince Mama otherwise.

From the letters that Hans wrote to Hilde it is clear that at least until the summer of 1934 Hilde was much more interested in and attracted to Hans than he was to her and that she was much more deeply involved emotionally in the relationship than he was. Mama became the reason for breaking the engagement. From the moment that Hans had told her that he and Hilde were engaged to be married, Mama relentlessly badgered him, stated her objections, magnified his doubts and apprehensions, and finally convinced him that given who Hilde was and what she was committed to, Hilde would not be able to meet his needs, nurture his gifts, and allow him to fulfill his promise. According to Hilde, there was another reason for Mama's vehemence: Hilde was Jewish.

Loschka Michel, in whose nursing home in Queens Hans's mother stayed from 1941 until 1960, and who got to know Mama well, at ninety wrote the following of her:

> Of course she made plenty of remarks that she would have loved not to be born as a jewish girl. But she was absolute "antisemitisch!" if it would have been possible to akt [*sic*] as one!
>
> Of course she was "antisemitic." A number of people like to convince themselves that they are not jewish (if they could help it!)[24]

Hans had initially believed that the distance between Ithaca and Mama would attenuate her influence on him and moderate the vehemence of her ill feelings toward Hilde. But her continual battering eventually wore him down, and a few days before the wedding he wrote to Hilde to break off the engagement.

The letters between Hilde and Hans's father following the breakup are particularly moving. It was the first time they addressed one another as "du." Albrecht was aware of the enormous hurt that Hans had inflicted on Hilde. He spoke of Hans's difficult character and bemoaned the fact that she was not Hans's "Glück."

In my interview with Hilde, she was still irate over what she considered Hans's deception and dishonesty. "How could he write me these effusive letters from September to December professing his love and then write a few days before the wedding that he wasn't sure that he loved me and breaking the engagement?" And with even greater emphasis, she wrote, "either those [effusive] love letters [he wrote when we were engaged] were dishonest, i.e., he lied to me, or the statement in his farewell letter: that he did not love me enough during the months we were engaged is dishonest, i.e., he lied at this time."[25]

Hilde did overcome the trauma of her involvement with Hans "thanks to the unique support of all my Cph. [Copenhagen] friends."[26] She told me that after the breakup and over the years she deliberately tried to forget and to block out everything about that painful event and her relationship with Hans, for many reasons. "My deliberate efforts to forget were not directed towards HB, but due to the fact that I was German, I belonged to this terrible people who shook the world with their crimes. I was terribly ashamed of being taken as one of them. I made a great effort to learn Danish and was unhappy when people recognized my German accent. This wish to forget became still more forceful after the occupation of Denmark (April 1940) when it became really painful—even dangerous—to be mistaken as a German."[27]

Hilde had reread the letters Hans wrote her just before my visit. A tone of bitterness often crept into her narration. When referring to herself during the late 1920s and the 1930s, she spoke of herself as "that girl Hilde" and as "little Hilde."

A glimpse into her status at the Bohr Institute during the early 1930s is revealed in photographs from the period. In one, she serves meals to the participants at one of the annual conferences; in another, she pours coffee for Pauli and Peierls, even though she had her doctorate and ran Georg von Hevesy's laboratory. These photos reflect a culture that thought it natural for young women to serve men. But her *Autobiography* also reveals that she became good friends with many of the visitors to the institute in the 1930s: Victor Weisskopf, Edward Teller and his wife Mici, Otto Frisch, George Placzek, Leon Rosenfeld, Christian Møller, and Rudolf Peierls, among others. She was considered and treated as a peer by them.

Instead of becoming Hans's companion and partner, Hilde became a loyal and highly competent laboratory assistant to James Franck and to Georg von Hevesy. Hevesy made use of her extensive talents.[28] In effect, she was his collaborator, responsible for much that got accomplished. During my interview with her she deprecated her contributions to the activities in Hevesy's laboratory and focused on her vulnerability, her dependence—"for political and economic reasons"—and her luck.[29] But it is clear that it was not only out of their sense of responsibility and their caring that Bohr and Franck recommended her to be Hevesy's assistant. If Franck and Bohr had not thought highly of Hilde's technical and intellectual abilities, they would not have recommended her, no matter what their feelings. In her *Autobiography,* written a few years after I interviewed her, she did state that in 1933, after she had applied to join the institute, "Professor Bohr asked Franck whether he knew me and whether he would like to have me as a kind of assistant. And Franck replied that he didn't know me but he did know my thesis. And that was a fairly good thesis. So he wouldn't mind to have me as his assistant when he came to Copenhagen." When Franck decided to leave Denmark for the United States and Bohr asked Hilde to become Hevesy's assistant, she had proven herself to be a most valuable and highly competent experimentalist in the laboratory she had helped Franck set up. Even had other paths been open to her, Hilde would have accepted the offer, because it was impossible to refuse Bohr's "fatherly way of asking."[30]

During my interview with her, Hilde regarded her activities during the

1930s and early 1940s, when she had worked with Franck, run Hevesy's laboratory, written joint papers with them, and gotten close to the Bohr family, as the high points of her life. This because "I came very close to the greatest personalities I ever got to know, and I experienced the difference in spirit between academics in Germany—as I had known them—and that of the Bohr Institute. This was much more essential for my 'growing up' and becoming a mature person than was my scientific development in later years." Arguably, it is her scientific accomplishments from 1946 on that stand out in terms of her professional career. What she accomplished after World War II when she became an independent researcher, making use of the techniques of radiocarbon dating and of autoradiography that she had learned from Hevesy in the 1930s and from Willard Libby in Chicago after World War II, was outstanding. Her papers from 1946 on are clear proof of her powers and creativity as a scientist.

Instead of becoming the caring and dutiful wife that Hans had wanted, Hilde became a loving and devoted daughter to James Franck and to Niels Bohr. When in late September 1943 the Jewish population of Denmark was taken by boat across the Øresund to Sweden, and thus saved from forced deportation and being murdered by the Nazis, Hilde was on the same small boat as Sophie Hellman, Bohr's trusted secretary, looking after the oldest and youngest of the Bohr sons. Hilde had cemented a very special relationship with all the Bohr sons as children. And the closeness of their bond continued when they became adults. After she retired, she collected all of Hevesy's personal and scientific correspondence and wrote his biography. She was then entrusted with looking after the Bohr family personal papers in the Bohr Archive.

Hilde's ties to the Bohrs were similar to her bond to James Franck. She herself characterized her relationship to Franck as "more a father-child relationship than the relationship of a teacher to his pupil."[31] In her *Autobiography* she noted this about coming to Copenhagen: "For Franck just as well as for me this was a turning point in our lives. He had left his great school of physics in Göttingen, I was just a beginner. My first year in Copenhagen was shaped by James Franck." Franck had left Göttingen in protest. Having served in the army during World War I he was not, at least for the time being, affected by the April 1933 racial laws that led to the dismissal of all Jewish civil servants in Germany. But he resigned shortly after the promulgation of the edict to protest the dismissals, believing—naively—that this would contribute to altering the governmental anti-Semitic policy. In a letter to the rector he gave the reasons for his resignation:[32]

> We Germans of Jewish descent are being treated as aliens and enemies of our homeland. [The edict] require[s] that our children grow up with the knowledge that they will never be allowed to prove themselves as Germans. Those who fought in the war are supposed to have permission to continue to serve the State. I refuse to avail myself of this privilege, even though I understand the position of those who consider it their duty to remain at their post.

When he wrote this letter on April 16, Franck indicated that although he was resigning, he intended to try to continue his scientific work in Germany. But shortly thereafter, realizing the futility of his resignation and the bleak future he faced in Nazi Germany, he left for Copenhagen.

Besides being Franck's laboratory assistant in the new researches he had undertaken on photosynthesis, Hilde "also sort of played the role of his secretary and helper."[33] Franck's wife was often ill, and he was quite lonely, so Hilde and he became close friends and would, weather permitting and the situation in the Franck home allowing, go cycling at night in the Dyrehaven or some of the other large parks of Copenhagen. But as Hilde commented in her *Autobiography*, "that was . . . in the summer of 1934 and the summer days in Denmark are very long. It is light very late so we really went on quite extensive trips and talked and had a very wonderful time together."[34]

Franck left Copenhagen in the fall of 1934 to accept a position at Johns Hopkins University in Baltimore. He had a great deal of difficulty adjusting to his new environment and was "fairly unhappy and sad."[35] During his first years in the United States he wrote Hilde a large number of "very moving letters," telling her how he missed all that had been and that he had left behind in Germany.[36] It is testimony to Hilde's judiciousness, sensitivity, and maturity that such a close bond could be cemented between them.

Life became happier for James Franck after he met Hertha Sponer again and later married her. Sponer had been a student of his and had become a distinguished experimental physicist and a colleague of his in Göttingen.[37] She was dismissed from her position when Hitler came to power due to Nazi intolerance of women in academia. She emigrated to the United States in 1936 after accepting a full professorship at Duke University in Durham,

North Carolina. Rose Ewald was Sponer's "housekeeper" after she came to the United States in 1936. James Franck had gotten Rose the position.

The Aftereffects

In her *Autobiography* Hilde refers tersely to her involvement with Hans:

> The only one of the many gifted emigré physicists who appear in Mrs. Schultz's [one of the institute's secretaries] guestbook only once for a very brief time in the mid-thirties is Hans Bethe. During his visit, we decided to get married, but in the last moment he gave in to his mother's wish and broke off our relationship. This brought him a storm of disapproval from friends and colleagues and a lasting reservation on Bohr's part. Maybe it triggered Bohr's fatherly support of my endeavors to build a home and a career in Denmark. In the years to come, I have been ever so grateful for my rich life within the circle around the Bohr Institute and the biological laboratories where I worked later.

Both Bohr and Franck were shocked by what Bethe had done in breaking the engagement a few days before the wedding. They disapproved greatly and because of its callousness and cruelty found it morally wrong. Bethe was not invited to visit the Bohr Institute from 1935 until after World War II.

Bethe had first met Bohr in Rome when he came there on the first leg of his Rockefeller fellowship. Bohr had come to Rome to attend the September 1931 international conference on nuclear physics that Fermi had organized. Bethe did not attend the conference because he thought nuclear physics was "witchcraft" at the time but clearly met Bohr.[38] The sociology of the theoretical physics community was such that in June 1932 Bohr could write Bethe and tell him that in view of the fact that the proceedings of the conference were to be published soon he had taken the liberty of "suggesting" to the editor of the proceedings, Vice-Chancellor Bruers, that in order to prevent delays and expedite matters he—Bruers—should send the proofs to Bethe, so that Bethe could look at the additions that Bohr had made to the text of his comments at the conference. In his letter to Bethe he indicated that he hoped "that you (*Sie*) Bethe would perhaps undertake this assignment out of friendship, in which case I would be very much in your debt."[39]

The exchanges that make up the second set of letters in the Bethe-Bohr correspondence of the early 1930s were written after Bethe had been dismissed from his Tübingen assistant professorship in April 1933 and was seeking a position for the academic year 1933–34. After much confusion and some delay Bohr answered Bethe on August 10, 1933, stating that there were no funds available to support him for the coming academic year; that he was glad that he had obtained an appointment in Bragg's department in Manchester; and that he hoped that Bethe would be able to come to Copenhagen for the academic year 1934–35. In addition, Bohr urged Bethe to attend the forthcoming Theoretical Physics Conference to be held September 14–16 at the institute, and informed him that he would send him an official invitation to it. Bethe wrote back to thank him, but indicated that he would be unable to attend the conference.

In May 1934 Bethe wrote Bohr that Nevill Mott had offered him a position in Bristol for the coming academic year, a position that might be extended, and that therefore he would not consider coming to Copenhagen. Bohr quickly answered him, telling him that he was pleased about Mott's invitation. Evidently, Hilde Levi had told Bohr that Bethe would be coming to Copenhagen to see her in early September, so Bohr invited him to give a lecture while in Copenhagen and offered to have the institute defray the cost of his journey. On August 24 Bethe wrote Bohr to thank him for the invitation, although the primary reason for his letter was to extend his condolences to Bohr, who had just lost his eldest son in a tragic accident at sea.

Bethe's reaction to meeting Bohr in September—as he remembered it in the early 1990s—was the following: "I was of course in great awe of Bohr. I was not persuaded to stay there longer than this one month. His style of doing physics was so different from mine, very vague. Bohr would talk, and would really talk around things, and not directly about things. It was obvious that he always had something [i.e., was on to something]." Bohr's style of thinking and his approach to physics were so different that there was no meeting of minds between the two physicists.

Stimulated by the experiments that Fermi had been carrying out in Rome, Bethe gave a lecture at the institute on the theory he had formulated concerning nuclear reactions induced by neutrons. In Bethe's model, constructed in analogy with the quantum mechanical theory of the scattering of electrons by atoms, the passage of a neutron through a nucleus was described as if the neutron were moving in a (suitably averaged) potential created by the neutrons and protons that constitute the nucleus. It is said that Bohr went away from Bethe's talk agitated, shaking his head, and saying: "Well, I am quite sure this is wrong. It cannot go this way."[40] This talk by Bethe has been said to have influenced Bohr's subsequent introduction and use of the concept of the compound nucleus.

The next exchange between the two was a note from Bethe to Bohr acknowledging the receipt of 400 Kr to cover the expenses of his visit, "in connection with scientific discussions at the institute."[41]

There were no interactions between Bohr and Bethe from October 1934 until November 1936, reflecting Bohr's disapproval of Bethe's breaking his engagement with Hilde in December. In November 1936 Bethe sent Bohr a partial manuscript of his work with George Placzek that presented their version of the compound nucleus model, asking for his reaction to what they had done.[42] Bohr answered the letter (in English) two weeks later, commenting that his perusal of the manuscript had made him very happy since there was a convergence of their respective views regarding nuclear reactions. He concluded his letter by informing Bethe that "Kalckar and I intend to come to America in the early spring, where we not least [i.e., greatly] look forward to meet you at the conference in Washington which Gamow is arranging in February, and to the opportunity for a closer discussion of all the problems mentioned."[43] Physics had trumped all other considerations!

After November 1936 all traces of the memories of the events of December 1934 disappeared in the professional contacts between Bethe and Bohr. They were together at the 1937 Washington Theoretical Physics Conference.[44] They met again at the end of April at the annual national meeting of the American Physical Society in Washington. And the two of them were the speakers at the colloquium on nuclear theory that Hertha Sponer organized in early May at Duke University. On that occasion, Bethe once again met Rose Ewald. Bethe invited Bohr to lecture at Cornell in the spring of 1939, when Bohr was visiting the Institute for Advanced Study. Bohr accepted but had to cancel because of other duties.

Bethe and Bohr were also together at Los Alamos, and this is how Aage Bohr, Niels Bohr's son, described their interactions there:

> The first time I met Hans was in Los Alamos, when I was together with my father. We had many discussions with Hans, and I remember vividly how much my father appreciated these discussions and admired Hans's abilities. There were, of course, many brilliant stars in the theoretical division at Los Alamos, contributing to the solution of the technical problems and provid-

ing new ideas, but Hans seemed to be the person who held it all together, and he would always be able to give brilliantly clear and concise account of the current situation . . .

We had quite a bit of personal contact with Hans and Rose at Los Alamos, and my father enjoyed talks with Hans on many different matters. They talked about the future perspectives of the project that deeply occupied my father and also about some of the personal problems that had created difficulties at Los Alamos. My father had considerable confidence in Hans's straightforwardness and fairness in such matters.

. . . I do not believe that my father's relations to Hans at that time was at all affected by Hans's earlier relationship to Hilde Levi. As a matter of fact, my father never mentioned the episode to me, and I only learned about it at a much later time, long after my father's death.[45]

Let me conclude this chapter with Bethe's reaction to my telling him of my visit with Hilde Levi. I was deeply impressed by his willingness to talk candidly and unequivocally about this painful episode in his life. He was—and had been—conscious of the fact that he had deeply hurt and scarred Hilde and commented that "I hope that I never hurt anyone else like that." He admitted that he was somewhat dishonest in his letters to Hilde in the fall of 1934, that he should have been more forthright and should have told her of his apprehensions and doubts. But he "had wanted it to work." He was genuinely pleased—and comforted—to find out about the closeness of her relationships with James Franck and with the Bohrs, and to learn of her scientific accomplishments and of her full and rewarding professional life.[46]

I should also mention that before I went to Copenhagen Rose Bethe asked me to tell Hilde how lucky she was not to have had to deal with Mama. I believe there was more to the message. I believe that Rose was also trying to tell Hilde that marrying Hans would have curtailed the full professional life that she led.

Rose made a further comment after she read a draft of this chapter: Hans—for all his hesitations and apprehensions—was ready to go through with his marriage to Hilde. It was Mama who had forced the issue. Mama had sent a telegram to Hilde stating that the engagement was being broken. Hans's letter to Hilde came thereafter.

7

Cornell University

Cornell University owes its existence to the unlikely association of two men of very different background: Ezra Cornell (1807–1874), a reticent, self-made man of humble background, a Quaker and tough-minded idealist; and Andrew Dickson White (1832–1918), a voluble, urbane, highly educated scion of a wealthy patrician family. Ezra Cornell was a successful inventor-industrialist in the emergent field of telegraphy who made his fortune by being the largest stockholder in the Western Union Company.[1] Carl Becker, in his lectures on the founding of Cornell University, offered a portrait of both these men. He depicted Ezra Cornell as "large, slow moving, self-contained . . . , a bit dour and austere in appearance, as well weathered as a hickory knot by fifty-seven years of harsh experience in the world of men and affairs, knowing much, saying little." He portrayed White, whom he had met personally in 1917, as "slight, nervously active, buoyant and vital . . . , [an] intellectual out of the academic world, fully equipped and armored with ideas newly polished and pointed by the battle of the books, . . . expatiating and expounding at length with friendly confidence and persuasive facility."[2] Becker characterized him as "essentially a crusader, by profession a promoter of good causes, [who was] primarily interested in changing the world rather than understanding it."[3] From his youth on, White's dream had been to build a "truly great" secular university, and he molded himself to become an educational entrepreneur. His father had wanted him to attend Hobart College and become an Episcopalian minister, but White insisted on attending Yale to obtain a broader education. After earning his bachelor's degree at Yale in 1854 he went to the Sorbonne in Paris and thereafter to the University of Berlin, where he earned a master's degree in history. In 1857 the twenty-five-year-old White became professor of history at the University of Michigan. Its president, Henry Philip Tappan, was a forerunner of the new breed of university presidents exemplified by Charles William Eliot of Harvard, Daniel Coit Gilman of Johns Hopkins, William Rainey Harper of Chicago, and White himself at Cornell. Tappan was an admirer of the German research university. He had undertaken to build the University of Michigan as a nonsectarian institution in which students would have considerable freedom in choosing their curricula. Tappan had recognized the increased importance of science and engineering and strongly supported research and laboratory studies. White was deeply impressed by Tappan. However, Tappan's vision for the university was zealously opposed by religious and educational traditionalists, and his overbearing demeanor grated on the members of the Board of Trustees. The combination brought about his dismissal without cause in 1863.

Witnessing liberal education under attack at Ann Arbor galvanized White to action. At the height of the Civil War, White outlined his conception of an institution of learning in a letter to a potential donor. Its mission would be

> First, to secure a place where the most highly prized instruction may be afforded to all—regardless of sex or color.[4] Secondly, to turn the current of mercantile morality which has so long swept through this land. Thirdly, to temper and restrain the current of military passion which is to sweep through the land hereafter. Fourthly, to afford an asylum for Science—where truth shall be sought for truth's sake, where it shall not be the main purpose of the Faculty to stretch or cut science exactly to fit "Revealed Religion" . . .

And in concluding he noted that such an institution should "afford a nucleus around which liberally-minded men of learning—men scattered throughout the land, comparatively purposeless and powerless—could cluster, making this institution a center from which ideas and men shall go forth to bless the nation during ages."[5]

Cornell and White first met in January 1864 in the chambers of the New

York Senate, to which they both had been elected. Cornell, White's senior by some twenty-six years, was by then "a rich but honest man who could make a case of conscience out of the prosaic fact that he had five hundred thousand dollars more than he thought his family would ever need."[6] White's vision of a university resonated with Cornell's hard-nosed idealism, and Cornell agreed to endow such an institution but with the stipulation that the State of New York match his contribution. White and Cornell succeeded in having the state legislature fulfill this condition, and the university, which was to bear the name of its endower and be located in Ithaca, was incorporated in 1865.

The following year Andrew White became its first president, and the university welcomed its first class in October 1868.[7] In a brief address during the opening ceremonies White elaborated in clear and plain terms the thesis that "The individual is better, society is better, and the state is better, for the culture of its citizens; therefore we desire to extend the means for the culture of all." The university became shaped by White's enlightened liberalism.[8] His task was made easier by Ezra Cornell's Quaker conscience and his insistence that the university he was endowing have no ties with any religious denomination. The charter of the university stipulated that "Persons of every religious denomination, or of no religious denomination, shall be equally eligible to all offices and appointment." At the time this requirement was exceptional among American colleges and universities, all of them having some denominational affiliation.[9] To this day Cornell does not have a school of divinity.

The university grew under White's able governance, but its growth was not without difficulties, stemming in part from the fact that it was widely believed that White could not be trusted in matters of religion. Deeply influenced by his readings of Herbert Spencer and Charles Darwin, White considered himself an evolutionary deist, and though he called himself a Christian he did not identify with any denomination, and rejected becoming a member of even as liberal a fellowship as the First Unitarian Society of Ithaca, the congregation that counted Ezra Cornell as one of its most steadfast participants. White's stand was at odds with most Ithacans, and with the vast majority of the population of the state of New York and of the United States.[10]

The events following the appointment of Felix Adler as nonresident professor of Hebrew and Oriental literature in the spring of 1874 are indicative of White's and Cornell's stand on religion.[11] Adler had given a series of lectures on the history of religions in Lyric Hall in New York in the fall and winter of 1873 and early 1874. These lectures attracted a great deal of attention in both intellectual circles and the popular press. White attended several of them and came away deeply impressed and ready to explore the feasibility of having Adler join the Cornell faculty. This became a reality when some of Adler's patrons, who had underwritten his Lyric Hall lectures, agreed to fund a visiting professorship in Hebrew and Oriental literature. Although he considered Cornell a "nonsectarian Christian institution," White harbored the hope that it would become truly nonsectarian. As he indicated in a letter to Joseph Seligman, one of Adler's patrons, "Cornell is one of the first institutions established on the basis of complete equality between men of every shade of religious ideas and of complete liberty in the formation and expression of thought. Dr. Adler if I am not mistaken is a man calculated by his lectures as well as by his influence in other ways to promote those studies calculated to break down all unfortunate barriers of creed which so long afflicted mankind. And this it was that led me at the outset to take a real interest in the case."[12]

Evidently Adler lectured not on what had been expected of him—the creed of Judaism—but on the topics closest to his heart: the comparative history of religions, the recent philological studies of the Old and New Testament, the challenges of Darwinism to religion, and on the lessons to be drawn from these studies for contemporary religion. The majority of Cornell's professors and the residents of conservative Ithaca, many of whom had attended the public lectures Adler had delivered in town, found his presentations heretical and believed that they undermined the Christian faith.[13] Adler thus became embroiled in the impassioned, and at times bitter, religious controversy then raging on American campuses, brought about by the confrontation between orthodox beliefs and Darwinism and higher criticism. When Adler's contract expired in 1876 it was not renewed, even though Seligman had indicated that he was willing to continue underwriting the appointment. Adler's liberal religious views and the vehemence of the reactions against them had made his presence at Cornell injurious to the best interests of the institution. But in rejecting Seligman's offer Cornell advanced a principle that became of great importance in guaranteeing academic freedom at American universities.[14] The Executive Committee of Cornell's Board of Trustees wrote Seligman that it "object[s] to having professors nominated to them from the outside. Views, often independent of talents and singleness of purpose on the part of the candidate, influence appointments and of such none but those considered in the management of educational institutions can judge. The Executive Committee, therefore, will not again accept any propositions for endowment where the choice of the incumbent is not left without restrictions to the Trustees."[15]

Having been established as both a private and a public institution, Cornell acquired a distinctive character of its own. It is a successful graft of a liberal arts college and graduate school, which reflected White's vision of a great university, with a land grant component consisting of an engineering and an agricultural school, which embodied Ezra Cornell's practical idealism.[16] The dual mission of the university was also impressed on the science departments of the arts and science division. To this day, the ideals of pure research and the advancement of knowledge, commitments Cornell shared with other private liberal arts universities molded on the German pattern, live in harmony with the American land grant tradition of fostering applied science and encouraging usefulness to the state and its citizenry.[17]

By the end of the nineteenth century roughly 2,000 undergraduates and 200 graduate students were on campus. Of the 2,000 undergraduates, some 650 were enrolled in the arts, humanities, and the sciences, some 685 in civil, mechanical, and electrical engineering, but only 85 in agriculture. Being a land grant college and mindful of its ties with agriculture and "the mechanic arts" and of its responsibilities to the state and to the nation, Cornell invested heavily in the growth and strength of that component of the university. By the outbreak of World War I the number of students enrolled in the School of Agriculture had risen to 1,700, and some 950 were electing to major in engineering. Meanwhile, the number of liberal arts and science students had grown to 1,700.

Cornell's expansion during that period reflected the rapid growth of the American population and changes within it, both its rising per capita income and its increasingly urban nature. The railroads had expanded markets dramatically with a consequent finer division of labor, transforming the American economy and its institutions. Specialized expertise became ever more relevant, and a larger fraction of the population was able to take advantage of these opportunities. The growth of the economy, its demands, and the new channels it opened were accompanied by the formations of national specialized disciplinary associations: the American Chemical Society (1876), the American Historical Association (1884), the American Economic Association (1885), the Modern Language Association (1883), American Physical Society (1899), and the American Sociological Association (1905). The concurrent expansion of the undergraduate and graduate student body was accompanied by a transformation of university faculties: they became reorganized into departments that were identified by their disciplinary commitments. To secure promotions and tenure it thereafter became essential for faculty members to obtain national and international reputations by virtue of scholarly and research activities in their discipline. Very early in Cornell's history, departments were set up, and their members taught and carried out research in specialized areas such as mathematics, chemistry, physics, botany, zoology, history, literature, and philosophy.

Physics at Cornell before World War II

The story of the buildup of the physics department at Cornell before World War II is similar to that of the dozen or so American universities that were considered the premier institutions of higher learning. This growth, at Cornell and elsewhere, was facilitated by the great power in the hands of the president and the heads of departments. At Cornell the president was the chief administrative and the chief academic officer. One of the characteristic features of the physics department at Cornell was that until the 1930s most of its faculty members had attended Cornell themselves, most as undergraduates but many as doctoral candidates. Another feature of the physics department was that from 1887 until 1946 only three people occupied the position of head of department: Edward L. Nichols (1887–1919), Ernest G. Merritt (1919–1934), and P. Clifton Gibbs (1934–1946). The fact that in 1946 Lloyd P. Smith became the chair of the department, rather than its head, is indicative of the changes brought about by Edmund Day, who assumed the presidency of Cornell in June 1937. Also of note is the fact that from 1909 when the position was first created until 1939, two of the three deans of the graduate school were physicists: Merritt (1909–1914) and Floyd K. Richtmyer (1929–1939).

Physics has always been strong at Cornell.[18] Edward Nichols, the founder and first editor of the *Physical Review,* came there in 1887. He brought Merritt, Frederick Bedell, and Richtmyer to the department and made it one of the foremost research centers in spectroscopy. By the mid-1920s it became apparent that quantum mechanics was altering the character of physics. To keep abreast of the advances the department invited H. A. Lorentz to spend the spring semester of 1926 in Ithaca.[19] That same year Merritt made an offer of a full professorship to Max Born at the very attractive salary of $8,000 a year.

After Born turned down the invitation, attempts were made to have Peter Debye and James Franck come. The sensitivity of the department to the changes in physics can also be inferred from its course offerings. The first edition of Richtmyer's *Introduction to Modern Physics,* which was based on a very successful course that he was teaching to the undergraduate physics majors, was published in 1928.[20]

Quantum mechanics was brought into the graduate curriculum in 1928 when Earle Hesse Kennard started teaching a course on wave mechanics.[21] In 1927 Gibbs initiated a program that brought distinguished physicists to the campus each summer.[22] The department also encouraged its young theorists to go abroad. Kennard spent his sabbatical year in 1926 and 1927 with Born and Bohr. Lloyd P. Smith spent a year in 1930 at Caltech and the following year went to Germany to study with Sommerfeld on a National Research Council postdoctoral fellowship.[23] As Sommerfeld was too busy, he was assigned to work with Bethe, who was then in Munich as a young *Privatdozent.*[24]

By the early 1930s it had become clear that the principal areas of research within the department—atomic and molecular spectroscopy, electron and ion physics—no longer were frontier fields. Intense discussions were held to decide which new fields of research ought to be supported. Merritt went to Leyden in 1931 to visit the low-temperature laboratory there, and to Göttingen to acquaint himself with work carried on in Franck's laboratory. He also went to Berlin to confer with von Laue and others,[25] and to Rome to the first international conference on nuclear physics. In Rome Merritt met Bethe when he visited Fermi's institute. In 1933, Merritt recommended to the Cornell administration that the department expand its activities and go into nuclear physics, thus setting it on its modern course. Not everyone supported Merritt's recommendation. In particular, Richtmyer, the most successful and best known of the experimental physicists at Cornell and dean of its graduate school, fearing that nuclear physics would drain support from his X-ray laboratory, opposed the move. Nonetheless, the department went ahead and asked M. Stanley Livingston and Bethe to join it. In retrospect, this was undoubtedly the most important decision taken by the department. Livingston, who had helped Lawrence build his cyclotron at Berkeley and, according to Bethe, "is generally credited with having made it actually run after Lawrence had the idea for it," came in the fall of 1934.[26] He immediately began building Cornell's first accelerator, the first to be built outside of Berkeley. In early February 1935 Bethe arrived, and in September 1935 Robert Bacher joined the department. Gibbs, who had become head of the department in 1934, explained to Bethe when he arrived that the department "was changing from one in which research was done to provide thesis topics for graduate students to one in which graduate students could participate in ongoing research." Gibbs added that "Not everyone agreed with this new emphasis on research, and there was some disagreement on which fields to expand into. It was the progressives versus the conservatives. The progressives had won the fight and now had the backing of the administration." The appointment of Bethe—a "theoretical nuclear man, and a foreigner to boot"—was one of the signs of change.[27]

Bethe arrived in the United States at a time when the American physics community was undergoing enormous growth. The number of physics doctorates awarded had increased from an annual rate of around ninety in 1930 to over 170 by 1940, with almost all the degrees earned during the decade awarded by just fifteen universities. Whereas in the early 1930s Heisenberg, Pauli, and Ehrenfest had four or five students at most working for a university degree at any one time, Harvard, Illinois, Berkeley, Princeton, and Wisconsin were *annually* accepting that number of graduate students interested in theory.

When the refugees from Nazi Germany began to arrive in the United States they strengthened measurably the theoretical activities at some of the less well-developed centers—Bethe at Cornell, Bloch at Stanford, Teller and Gamow at George Washington University, Lothar Nordheim at Purdue, Fritz London at Duke, and later in the decade Weisskopf at the University of Rochester. They greatly enriched all physics activities in the United States.[28] Conversely, however, the American experience likewise enriched these "illustrious immigrants" intellectually by virtue of their closeness to experimental activities and results. They altered their style to suit their new environment, particularly so where research in nuclear physics was being actively pursued. In 1937 Lawrence's thirty-seven-inch cyclotron was the largest operating in the United States, and similar machines were under construction at Chicago, Columbia, Cornell, Illinois, Michigan, Pennsylvania, Princeton, Purdue, and Rochester. A little later, cyclotrons were being built at Harvard, Indiana, and Washington University. At all these places experimentalists depended on theorists to help them interpret their data and to explain the theory that had been advanced to account for the observed phenomena.

The presence of the new immigrants was readily apparent to their colleagues. Livingston, who built the first cyclotrons at Berkeley and Cornell and became both a colleague and a collaborator of Bethe at Cornell, commented that Bethe "gave me a feeling for the fundamentals of physics, and what was going on in nuclear physics. I learned of many new concepts like magnetic moments and quantum aspects, that I had never heard while with Lawrence. It was a different environment. I was now following a scholar and was really impressed."[29]

And Emil Konopinski, who was a National Research Council postdoctoral fellow with Bethe from 1937 till 1939, gave the following description of his supervisor: "Bethe is really an amazing man. I had no idea that a human being could think so quickly, and thoroughly at the same time, have such a tremendous capacity for getting things done, and possess such a mine of information. He is especially a godsend to experimenters. One day I sat in his office while, in succession, he advised persons on nuclear physics, spectroscopy, X-rays, crystal structure and surface phenomena. In every case he instantly had positive, detailed information to give."[30]

The influence of the émigré scientists was particularly noticeable at the many theoretical conferences that were being organized to assimilate the insights that quantum mechanics was giving to many fields, especially molecular physics and the emerging field of nuclear physics. The Washington Conferences on Theoretical Physics, jointly sponsored by the Carnegie Institution and George Washington University, and held annually from 1935 until 1942, were paradigmatic of such meetings. They were the creation of George Gamow, who in 1934 had insisted on the establishment of this series as a condition of accepting his appointment at George Washington University. The sponsors were John Fleming, the director of the Carnegie Institution of Washington's department of terrestrial magnetism, and Cloyd H. Marvin, the president of George Washington University. Their intellectual agenda was set by Merle Tuve of the Carnegie Institution, and by Gamow and Teller. Their purpose was "to evolve in the United States something similar to the Copenhagen Conferences, in which a small number of theoretical physicists working on related problems assemble to discuss in an informal way difficulties met in their researches." The conferences proved to be extremely influential and seminal, partly because they were restricted to theory and partly because their size was strictly regulated so they could remain "working" conferences.[31] The unanimous opinion conveyed by the members of the third conference, that this type of "working meeting [was] much more effective in furthering their investigations than any other scientific meetings for physicists now organized in this country," surely also expressed the sentiments of those attending the later conferences. Bethe attended all the conferences until 1940. After attending the third Washington conference Bethe wrote Tuve

> [once again] how much I appreciate the opportunity for free discussions among theoretical physicists afforded by the Washington Conferences. These conferences are something so entirely different from other meetings in their informality, that I am looking forward to them the whole year, and consider them the only meetings at which we can really do valuable work together. If I had to, I would gladly miss all the meetings of the American Physical Society in order to attend the Theoretical Conferences in Washington.[32]

The 1930s

The threat posed by totalitarianism deeply worried Livingston Farrand and Edmund Ezra Day, Cornell's presidents during the 1930s.[33] Farrand, Cornell's fourth president, upheld its liberal tradition. In the early 1920s, at a time when a wave of conservatism and isolationism was sweeping the country at large, Farrand was giving speeches deploring "the outbreak of intolerance and the reign of prejudice which seems to have seized the American people." He was deeply disturbed by

> The right of the majority to rule [being] conceived of as a charge to translate a prevalent, and often ill formed, opinion into a code of conduct to which all must conform . . . So called patriotism insist[ing] upon teaching a history which shall titillate national complacency even though inconsistent with fact. Traditional conservatism demand[ing] that natural science shall suppress the implications which the patient search for truth has made logically inevitable. Reactionary privilege see[ing] anarchy in unbiased discussion of the uncertain field of economic relations. Education [becoming] the theatre of attempts to enforce uniformity of method at the very time when latitude is the great desideratum.

When new discriminatory and restrictive legislation was being considered in Congress in 1923 and 1924 to curb immigration into the United States, Farrand called for liberalizing immigration policy. He asserted that "Immigration has served immeasurably to increase our population and our wealth, has, in a word, made us what we are." He recognized that "while injecting from year to year an unceasing stream of vitality and strength it has just as unceasingly held active and irritant the fact of divergent and often, in the first generation, irreconcilable racial traits and traditions." But sharply curtailing immigration was not the way to ensure prosperity and growth, nor the way to strengthen democracy. He was not discouraged by the fact that success had not yet been achieved in overcoming ethnic and racial prejudice. Rejecting despotism and anarchy as viable alternatives, he envisioned "Democracy's task" as this: "To arouse a sense of collective interest, to establish common ideals of the common good, to enlist the forces of humanity and justice and knowledge for the improvement of the general welfare and this with the preservation of individual initiative and the safeguarding of individual opportunity."[34]

An assessment of Farrand's character can be gleaned from the note the eminent historian Carl Becker wrote him in 1929: "Your unfailing frankness and courtesy in all matters and to all persons, and your cordial support of the intangibles that make Cornell what it is, are not the least of those contributing factors that make for my great contentment."[35] In 1933 after Hitler took power in Germany and dismissed all civil servants of Jewish descent, including university professors, Farrand helped organize the Emergency Committee in Aid of Displaced German Scholars, and was an active member of its executive committee.[36] During 1933 and 1934 he often spent several days a month in New York to direct the activities of the Emergency Committee.

Farrand strongly supported the sciences at Cornell. For him the scientific method was a guideline for the conduct of life. It resulted in informed belief and "respect for expert as opposed to uninformed opinion . . . Certainly an application of the habits of the laboratory to, say, the problems of legislation would produce a marked improvement in the annual grist of statutes which emanate from our halls of legislation, federal and state."[37] He considered the sciences—in particular, physics and chemistry—the foundations for engineering and for medicine. The curriculum of the Cornell Medical School, which he helped establish during his presidency, reflected these views. He advocated the revamping of the engineering curriculum during the 1920s and encouraged the physics department to expand its activities into nuclear physics during the early 1930s.

Shortly after assuming the presidency of Cornell in 1937, Day urged a comprehensive program of national defense. Day was a social scientist, an economist and statistician by training. During the two years before the United States entered World War II, Day lectured extensively on the topic: "What really threatens American Democracy?" Four of these addresses were collected together in a book published in 1941, *The Defense of Freedom.* In a lecture delivered in February 1939 he also outlined these threats. Although he didn't believe that Germany, Italy, and Japan seriously threatened the United States by direct attack, nonetheless they constituted a menace in two important ways. First, they might "in their program of imperial expansion" precipitate a general European war that would certainly become a world war. The propaganda the Axis powers were spreading—in which democratic ideals became objects of contempt and scorn, and democracy branded as an outmoded form of government—constituted a second threat to American democracy. But the chief danger to democracy, Day emphasized, did not come from without but from within. The Depression and the changes in the economic situation since World War I were posing a challenge that was not easily met. The disappearance of economic security had eroded public confidence, and Day believed that this loss posed the greatest danger to democracy in the United States: "Starvation in the midst of plenty, idleness in the face of need, unemployment despite a desire and a capacity for work, these are poisons no body politic can long withstand. If they cannot be substantially eliminated under democracy, an ultimate change in the social order is inescapable."[38]

Even in June of 1940, after the fall of France, when Day had seen "what devastating powers of destruction can be developed when a great power, completely regimented under a despotic and ruthless government, employs the full arsenal of science and technology," and was no longer so confident that the oceans that separated the United States from Europe and Japan would give it the protection they "once so surely afforded," and was urging a program of full mobilization for national defense, he still felt that the chief threat came from within, from attacks that might break the national unity.[39] "Let us not forget that the most insidious dangers of these next few years will be those which threaten our solidarity as Americans." These attacks "will play upon our prejudices, our fears, our dislikes, our hates. They will play upon our loyalties, and our unthinking patriotism. They will exploit our discontent and failure; the disillusionment and frustration we have suffered. Suspicion and mistrust will be fomented. Liberals will be called communists . . . Colleges and universities will be said to be centers of 'red' activities."

Day was convinced that an essential defense against such attacks was "the better operation of our economic system," for "we have taken all the economic punishment we can stand without revolutionary changes in our government, if not our society . . . [Our] traditional system of regulated private enterprise . . . must be made to work . . . The enforced idleness of millions of young people cannot be longer tolerated . . . If we fail to put our economic house in order America is headed for revolutionary changes whether or not there be attack from without."[40]

The loss of economic security and of equality of opportunity for the many, the undue concentration of economic power in the hands of a few, had created attitudes that threatened the operation of democratic procedures. Quoting Bertrand Russell, Day characterized the true signs of civilization as: "thought for the poor and suffering, . . . the frank recognition of human brotherhood irrespective of race or color or nation, the narrowing of the domain of mere force as a governing factor in the world, the love of organized freedom, abhorrence of what is mean and cruel and vile, ceaseless devotion to the claims of justice." To make his point, Day contrasted these words of Lord Russell with words of Mussolini: "Words are beautiful things. Machine guns, ships, aeroplanes are still more beautiful."[41] He ended his address with some deeply felt exhortations: "It is in the humble actions of our daily living that some of our most important defenses must be built . . . We must seek out the facts—if possible find the truth. We must combat fraud and greed and sheer selfishness. We must strive to act fairly and justly. We must not entertain suspicions without supporting evidence. We must be quick to respond to the needs of our fellows . . . We must be tolerant of honest differences of opinion if these not be seditious. We must seek the common good."[42] For Day, universities were the bastions of civilization, places

> in which there are relatively undisturbed opportunities to live with ideas. Much of life is otherwise engaged. But on campuses such as [Cornell] thoughtful men and women, of faculty and student body alike, should be led to seek out all sort of ideas: . . . ideas that have more recently opened the doors of new knowledge of nature and man; ideas that afford the foundation of our system of law and order, of justice and liberty . . . Men and women on a campus like this should learn how knowledge is gained and wisdom won . . . They should . . . through their common interests and activities . . . come to know what is really meant by the intellectual life.[43]

In his inaugural address Day had noted some of the forces that made it difficult for universities to maintain the primacy of the intellectual function. One of these was "the current eclipse of the liberal tradition," the tradition that

Andrew White had imparted to Cornell. Universities can develop only in an atmosphere that permits the independence of the communities of scholars.[44] Universities have the duty to guarantee and protect the freedom of inquiry of teachers and students and cannot allow activities within their walls that would threaten that freedom.[45] Echoing White, Day noted that "the love of money" is another external force of "great potency" that makes it difficult for universities to cultivate intellectual interests. "To a dangerous degree," Day warned, "we have come to regard the accumulation of wealth as the hallmark of individual success . . . [It] has dominated our social psychology to such an extent as to make the intellectual life appear to many pale and academic." And closely affiliated with love of money, though certainly not identical with it, is "the widely prevalent insistence upon vocational results in American education." The danger stemmed from the fact that "vocational interests are in many instances narrow in outlook and distressingly shortsighted." Since Cornell was engaged in vocational education—in engineering, architecture, agriculture, and medicine—Day insisted that it must do so "in ways becoming an institution of higher learning devoted basically to the intellectual life": vocational education must be essentially professional in character. This, Day indicated, involved recognition of at least three governing principles: an "emphasis upon fundamental disciplines" rather than "narrowly conceived" techniques; "a sustained pursuit, through scholarly and scientific research, of new knowledge within the field of the vocational art"; and a "steadfast recognition of the broader implications and social obligations of the vocation." What was done within universities and how these activities were supported from without must be consistent with their ideals and goals. Even though universities were no longer ivory towers, they were not meant to be entrepreneurial businesses nor were they meant to be places where secret research was carried out for the armed forces. "It is for the long pull that our universities exist."[46]

Under Day, Cornell maintained its position in the front ranks of American universities and remained faithful to its ideals.

Bethe and Physics at Cornell

At the end of January 1935 Bethe sailed from Bremerhaven to New York. In the morning of his first day in New York, he took a subway to the Battery at the southern tip of Manhattan, and then walked all the way to Columbia University on 120th Street. He recalled being deeply impressed by what he saw on his ten miles, especially Rockefeller Center, in the final stages of construction, whose majestic RCA Building scraped the sky over a plaza with arresting sculptures and over the adjacent Radio City Music Hall.

The next day Bethe took a train to Ithaca, where he was met by Lloyd P. Smith. He stayed with the Smiths for two weeks, during which time Mrs. Smith found him a nice furnished room in a house within walking distance of the university. It was located on Kelvin Place, off Wait Avenue, in the part of Ithaca that abuts Cayuga Heights. Breakfast came with the room.

Bethe vividly recollected his welcome during his first few days at Cornell: he was received "with open arms" by the physics department. After being shown his large office on the first floor of Rockefeller Hall, the department's home building, he was familiarized with the workings of the department. Soon after he arrived he could assert, "I felt very much at home, and, except for Richtmyer who wasn't there anyway, everybody seemed to embrace me, much more than even in the Manchester and Bristol departments. There it had been individuals, but here it was the whole department." Gibbs and Richtmyer, both of the same generation, were the dominant personalities in the department. Floyd Richtmyer was a remarkable man. He had obtained his doctoral degree from Cornell in 1910 and there he remained, except for the two years he spent at the Drexel Institute in Philadelphia just after obtaining his degree. He had helped organize the American Association of Physics Teachers and the Optical Society of America, and he was an active member of some fifteen other professional societies. He had been at one time or another president of four of them: the American Physical Society, the Optical Society, the American Association of Physics Teachers, and Sigma Xi. He was fully engaged outside of work as well: a life trustee of the National Geographic Society, he had organized some of its trips to observe solar eclipses. He was married and an active participant in his church and a couple of local clubs. When Bethe came to Cornell, Richtmyer, besides running a busy and productive X-ray laboratory, was the dean of the graduate school and the editor of *Reviews of Scientific Instruments* and the *Journal of the Optical Society.*[47] Richtmyer was much better known as a physicist than Gibbs, by virtue of his extensive X-ray research, his authorship of a widely used textbook, and his numerous professional activities.[48]

Gibbs headed the Cornell physics department, from which he had obtained his doctorate in 1911 and had stayed on. He was a very fine spectroscopist whose work turned out to be quite important, although it was not recognized as such at the time.[49] He had strongly supported Merritt's ambitious plans to expand into nuclear physics. The department had recruited Stanley Livingston in 1934, and Gibbs had gotten him a grant of $1,000 from private sources to enable him to build a small cyclotron. He had also obtained approval from the Cornell administration to appoint a second experimentalist in nuclear physics. After he became head of the department Gibbs didn't do much research because he took his administrative responsibilities very seriously. He would spend a great deal of time discussing with every member of the department his plans for the department and informing them of the issues he was going to bring up at the departmental meetings.

Relations between Gibbs and Richtmyer had always been strained. At the time Bethe arrived, their most recent disagreement had stemmed from the fact that Richtmyer wanted the research activities within the department to be entirely devoted to X-rays, his area of expertise. He strongly opposed the department going into nuclear physics, a move that he realized would take emphasis and positions away from his laboratory. He had been against the appointment of Livingston, and had been particularly unhappy when Gibbs was appointed head in 1934. He essentially took no notice of Bethe when he joined the department.

Bethe recalled that at the end of his first semester at Cornell he was supposed to submit a report to the graduate school about the graduate students on whose Ph.D. committees he was a member. Although Bethe had only been at Cornell a few months, he already had five or six such students. He didn't know about the regulations, however, so he didn't submit his evaluations on time. Richtmyer, as dean, immediately wrote him a letter taking him to task.

In fact, when Bethe joined the department Gibbs and Richtmyer were not on speaking terms. Richtmyer never came to the departmental meetings. He did supervise a number of doctoral theses, however, and so a graduate conference was established, chaired by Kennard, to oversee matters dealing with graduate studies. There were about forty graduate students working toward a Ph.D. when Bethe came in 1935. Both Richtmyer and Gibbs were comfortable talking to Kennard, therefore Richtmyer would come to the graduate conference meetings, as did all the faculty members who had responsibilities in the department's graduate program. Since Bethe was teaching a graduate course, he also attended the graduate studies meetings. He recalled that at one of them Richtmyer made a statement that Gibbs contradicted, whereupon

Richtmyer told Kennard, "Mr. Chairman, would you please inform Mr. Gibbs that such and such is so and so."

Bethe characterized both Kennard and Richtmyer as "old-fashioned." Thus Kennard, though very knowledgeable about quantum mechanics, "saw no purpose for the quantum theory of the solid state." And Richtmyer was of the opinion that internal developments within physics and the interests of professors—not societal needs beyond the training of competent scientists—should determine the fields of research within the department. The preamble of a request for funding from the Graduate Conference that Richtmyer wrote in 1938 encapsulates the ethos of the "old order" that Gibbs had mentioned to Bethe when he arrived. At the time Richtmyer was still the dean of the Cornell Graduate School, and thus an influential member of the Cornell faculty and administration. I quote Richtmyer at some length because of his aristocratic vision of the professoriate. Undoubtedly, one of the reasons for his composing the document was to protect his research operation. But the document was also meant to defend his vision of the university. He saw its mission to be the teaching and training of its undergraduate and graduate students, with research to be carried out in order to implement these functions. In the following the remarks within parentheses are Bethe's reaction to Richtmyer's statements:

Report to Physics Faculty on Present Status and Needs of Research in X-Ray Spectroscopy. F. K. Richtmyer

I. The primary function of a university is to train men at both the graduate and undergraduate level. The primary purpose of research in a university is to contribute to that training. The scientific results of such research, however important they may be scientifically, are secondary to their main purpose and are to be regarded as a by-product.(!)

II. A more important purpose of research by faculty members is the providing of suitable subjects for Ph.D. theses. Subjects for such theses must come from research programs actively carried on by faculty members. Ph.D. theses subjects cannot be "found in books." (yes)

III. In the logical development of an integrated program for any department, one should start first with the interest in research of the individual professors and others who make up the permanent staff of the department. In order that each department member may contribute most effectively to the department's work of training men, his research interests should be promoted and supported to the extent of the department's resources. (additions to staff)

IV. Each of the several fields of research in which the various members of the department are interested should be carried out along scientifically logical lines. This will occasionally mean a major expenditure for assistants, apparatus or both, in order to pass some hurdle. Some of the research problems in a given field are too complex to be done by graduate students. Other problems involve work which is too much of a routine nature. Special assistants should be provided for such problems.

V. It . . . should be kept in mind at all times that the primary unit upon which a comprehensive research program should be built is the individual's interest.

VI. Since in the last analysis, the primary purpose of any university is the training of men—that is to say, teaching—and since in reality our several research programs in physics are contributory to teaching, there should be an adequate balance between attention given to developing the teaching program proper and the research program.

Work Schedule

Bethe quickly settled into his new life within the Cornell physics department. He rose at around 7 AM, breakfasted at home, and arrived at his office by 9 AM. He would eat lunch in the nearby Home Economics cafeteria, usually with another young faculty member and with graduate students. Bethe thought the food served there was "quite good," although he would not say that it was "good food." Yet it was "very convenient." He dined by himself in a small restaurant on Wait Avenue that served, in his estimation, "very good food," not far from his house. He arrived as late as possible, just ten minutes before they closed at 7:30 PM. After supper he would go home and work, often until quite late. This was his routine six days a week.

In the fall of 1935, after Lloyd Smith had taught him how to drive, Bethe bought himself a Plymouth. He would then drive to one of the nearby state parks on Sundays and go for long walks, mostly by himself.

Bethe's teaching duties required him to teach two courses per semester: a theoretical physics lecture course for advanced undergraduates and first-year graduate students and a graduate course on some aspect of quantum theory—solid state theory, collision theory, quantum theory of radiation, or relativistic quantum mechanics.[50] There were approximately twenty students attending the undergraduate lectures and some ten in the graduate courses. The undergraduates in the theoretical physics course were less advanced than the students Sommerfeld taught in Munich, and the graduate students knew less physics and mathematics than those in the Sommerfeld seminar. But on balance, Bethe found the graduate students "quite equivalent" both in ability and knowledge to those in Germany.

The Cornell physics department was fairly large. There were some fifteen regular faculty members and ten recent Ph.D.s without jobs because of the Depression, who were being kept on as instructors.[51] Until Robert Bacher joined the department in the fall of 1936, Smith and Livingston were Bethe's closest colleagues, with Livingston's interests closest to Bethe's. Smith's office was next to Bethe's and they talked a great deal with one another: about Smith's researches in electron physics; about the Cornell summer conferences on solid state that Smith was organizing; about the course Smith was teaching on mathematical methods of physics that was similar to Sommerfeld's lectures on differential equations; and about the theory courses being given in the department. When discussing the latter they were often joined by Kennard.

Bethe also spoke frequently with Kennard, who was quite knowledgeable in quantum mechanics and its interpretations. Since they were the three theorists in the department, Bethe, Kennard, and Smith ran the weekly theoretical seminar, at which their contrasting views of physics were often made evident: Smith was more "applied," Kennard more "pure," and Bethe more "pragmatic."

When Bethe first arrived at Cornell, he spent many hours discussing nuclear theory with Livingston, who had joined the department as an instructor in the fall of 1934. With the assistance of three graduate students Livingston was building a small cyclotron, two feet in diameter. He was a superb instrument builder with a natural aptitude for making machinery work. Although a very good experimentalist, he knew little theory. Livingston had a big card file, two or three boxes with references to the literature on nuclear physics. The card file became very important to Bethe. He read many of the papers noted in the file and discussed their content with Livingston. "We talked a great deal, both about the cyclotron, and about experiments." As a result of his interaction with Livingston it occurred to Bethe that it might be a good idea to write down what he was telling him. A visit to the University of Minnesota presented him with the opportunity to do so.

Meetings and Lectures

Shortly after his arrival at Cornell Bethe and three other colleagues drove with Smith to New York to attend the February 22–23 meeting of the American Physical Society (APS). He remembered the drive on very icy roads as "quite adventurous" and was very impressed by Smith's "prowess" as a driver.

The APS meeting was held at Columbia University. Gibbs had evidently nominated Bethe as a fellow of the Physical Society, for the minutes of the meeting show that on Friday, February 22, the council of the APS elected Bethe a fellow. That Bethe was elected a fellow immediately upon becoming a member of the APS indicates that his accomplishments and his stature were already widely recognized in the United States. Bethe, most likely at Gibbs's request, gave a talk, "The capture and scattering of neutrons."

At the meeting Bethe encountered some of the young American physicists working in nuclear theory—in particular, Eugene Feenberg and David Inglis; and some of those doing research in cosmic rays—Richard Woodward, Jabez Street, and Donald J. Montgomery. At the meeting he also got reacquainted with some of the theorists he had met in Germany and England, such as Eugene Wigner and Lothar Nordheim, and met some of the senior American theorists: Edward Condon, Otto Halpern, and Gregory Breit. Perhaps the most consequential reacquaintance was with Isidor Rabi, whom he had first met in Munich in 1927.

Rabi in 1935 was the leading physicist at Columbia, making molecular beams yield wide-ranging precise information regarding the electromagnetic structure of nuclei. Bethe visited Rabi at Columbia while he was in New York. He there met Robert Bacher, who was an instructor in the department and spent a great deal of time with him. Some fifty years later Bacher still vividly remembered that meeting and how deeply impressed he was by Bethe's knowledge and command of so many different subjects. When later he mentioned this to Rabi, "he just shook his head in disbelief that Hans could know so much about so many areas of physics."[52]

Rabi invited Bethe to visit him during the semester, which Bethe did several times. On each of these occasions Rabi welcomed him very warmly. "We talked at great length. He talked to me about his experiments, and their interpretations . . . That was a very satisfying connection, and I met a few of the older people also, in particular, [J. R.] Dunning, who was doing experiments with slow neutrons."

During one of his visits Rabi gave Bethe the draft of a paper that a very young sophomore student at the City College of New York (CCNY) had written as a freshman.[53] That student, the seventeen-year-old Julian Schwinger, was failing some of his courses in the humanities because he spent most of his time in the library reading advanced physics and mathematical textbooks, and rarely went to classes. Schwinger's grades reflected his erratic class attendance. Although he had no difficulty getting A's in his mathematics and physics courses, even though he rarely attended classes, the same was not so in his other courses. The matter got serious enough for Lloyd Motz, one of Schwinger's instructors in physics at CCNY, to bring Schwinger's problems to Rabi's attention.

Rabi vividly remembered the episode and told it many times:

> There was this paper of Einstein, Podolsky and Rosen, and I was reading it. And one of my ways of trying to understand something was to call in a student, explain it to him and then argue about it. And [in this case] this was Lloyd Motz, who was also at that time I think an instructor at City College, we talked about it. And then he said there was somebody waiting for him outside, so I said call him in. So he called him in and there was a young boy there. [I] asked him to sit down and we continued. And then at one point there was a bit of an impasse and this kid spoke up and used the completeness theorem [of orthogonal functions] to settle an argument . . . I was startled.

After hearing of Schwinger's problems at CCNY, Rabi suggested to him that he transfer to Columbia, where he would be able to take graduate courses as an undergraduate. To justify Schwinger's admission to Columbia given some of his failing grades at CCNY, Rabi needed strong letters of recommendation regarding Schwinger's abilities in physics. Rabi gave Bethe Schwinger's manuscript on quantum electrodynamics and asked him to meet Schwinger. Thereafter Rabi requested Bethe to write a letter of recommendation for him to be admitted to Columbia. Bethe did so. The twenty-nine-year-old Bethe wrote the following remarkable letter about the seventeen-year-old Schwinger to Rabi:[54]

10 July 1935
Dear Rabi,

Thank you very much for giving me the opportunity to talk to Mr. Schwinger. I was greatly impressed by him.

When discussing with him his problem, I entirely forgot that he was a sophomore 17 years of age. I spoke to him just as to any of the leading theoretical physicists. His knowledge of quantum electrodynamics is certainly equal to my own, and I can hardly understand how he could acquire that knowledge in less than two years and almost all by himself.

He is not the frequent type of man who just "knows" without being able to make his knowledge useful. On the contrary, his main interest consists in doing research, and doing it exactly at the point where it is most needed at present. That is shown by his choice of his problem: When studying quantum electrodynamics, he found that an important point had been left out in a paper of mine concerning the radiation emitted by fast electrons. That radiation is at present one of the most crucial points of quantum theory. It has been found to disagree with experiment. It is quite conceivable that the error which Mr. Schwinger found in my paper might bring about agreement between theory and experiment which would be of fundamental importance for the further development of quantum electrodynamics. I might add that the mistake has not only escaped my own detection but also that of all the other theoretical physicists although the problem has been at the centre of discussions last year.

The way in which Schwinger treated his problem is that of an accomplished theoretical physicist. He has the ability to arrange lengthy and complicated calculations in such a way that they appear simple and can be carried out without any great danger of errors. This gift is, I believe, the most essential requirement for a first-class theoretical physicist besides a thorough understanding of physics.

His handling of quantum theory is so perfect that I am sure he knows practically everything in physics. If there are points he doesn't know, he will certainly be able to acquire all the necessary knowledge in a very short time by reading. It would just be a waste of time if he continued listening to the ordinary physics course, 90% of whose subject he knows already while he could learn the remaining 10% in a few days. I feel that nobody could assume the responsibility of forcing him to hear any more undergraduate (or even ordinary graduate) physics courses.

He needs, of course, some courses in minor subjects, principally mathematics and chemistry and a small amount of physical laboratory work. In physics the only thing he has to learn is teaching physics, i.e., to explain himself very simply—an art which can be learnt only by experience. He will learn that art automatically if he works at a great institution with other students of similar caliber.

I do not need to emphasize that Schwinger's personality is very attractive.

I feel quite convinced that Schwinger will develop into one of the world's foremost theoretical physicist[s] if properly guided, i.e., if his curriculum is largely left to his own free choice.

With kind regards,
I am
Sincerely yours,
H. A. Bethe

Visits

In addition to Rabi's invitation to visit him at Columbia, Bethe received several others, including one from Edward Condon to visit Princeton and one from Lee DuBridge to visit Rochester. And early in the semester he received several more invitations to lecture on nuclear physics: from Karl Lark-Horovitz at Purdue, from John Torrence Tate at the University of Minnesota, from Francis Wheeler Loomis at the University of Illinois at Urbana-Champaign, and from John van Vleck at Harvard.[55] Bethe went to all these campuses during the spring of 1935.

In Princeton he spoke at length with Wigner about nuclear physics. The Institute for Advanced Study at the time was located on the campus of Princeton University in Fine Hall, the mathematics building adjacent to Palmer Laboratory, which housed the physics department. Wigner suggested that Bethe and he invite Einstein to go with them for lunch. Unfortunately Einstein was not on campus that day.

The physics departments of Cornell and Rochester had close ties. DuBridge at Rochester was building a slightly larger cyclotron than the Cornell one. There thus were constant visits both ways. Also DuBridge was very eager to learn nuclear physics; hence the invitation to Bethe.

To visit Purdue, Bethe drove to Buffalo and from there flew by airplane to Chicago. It was Bethe's second airplane ride, but the first by any member of the Cornell physics faculty. The fact that he had flown made quite a splash in the department upon his return to Ithaca. In Chicago he visited Arthur Compton, whom he had met at the London conference the previous year.

Bethe's visits to Urbana-Champaign and Minneapolis were consequential. His visit to the University of Illinois resulted in being reinvited there in 1937 for a more extended stay. And in Minnesota John Torrence Tate, the editor of *Reviews of Modern Physics,* asked him to write articles on nuclear physics for the periodical. His lectures at Rochester, Purdue, and Harvard had made him aware how little American physicists knew about nuclear theory. Livingston had made that evident at Cornell. He therefore accepted Tate's offer. Bethe's expertise, abilities, and stature were such that he accrued most of the credit that came from writing the three long *Reviews of Modern Physics* articles on nuclear physics, one of which he co-authored with Bacher and one with Livingston.

The invitations to lecture that Bethe had received had also made clear to Gibbs that Bethe's status within the department had to be secured. In early May Gibbs wrote to Robert Ogden, the dean of the faculty, recommending that Bethe be appointed a regular assistant professor starting in the fall of 1936. The recommendation was accepted. In addition, Gibbs made clear to Bethe that promotion to a full professorship, which included tenure, would certainly be forthcoming in time. Shortly thereafter Bethe acquired a Ph.D. student, Ralph Myers, who would write his doctoral thesis in solid state under Bethe's supervision.

When Bethe went back to Germany that summer to visit his mother, he was convinced that the United States would be a good place to stay. The Cornell physics faculty was being extremely nice to him, he felt "quite at home" there, and he had been given the strong impression "that probably [he] would remain here for a long, long time."

Robert Bacher

In order to strengthen the nuclear physics activities of the department Gibbs wanted to have a second experimental nuclear physicist join the department and consulted Bethe on whom to invite. The two of them analyzed the credentials of various applicants. Both of them, as well as Livingston, decided that Robert Bacher was the best man.[56] He had previously done mostly work in spectroscopy that included hyperfine structure, and thus had become acquainted with nuclear properties. He was at the time an instructor at Columbia. Bacher was offered the job, but not until Gibbs had asked him whether he was Jewish. For all its liberality Cornell, like the other Ivy League schools, discriminated against Jews, Catholics, Blacks and other minorities.[57]

At many American universities the appointment of German refugee scientists who had been dismissed because of their Jewish roots met with opposition. Thus, when the Cornell physics department in 1934 was considering appointing Bethe it was also examining the possibility of appointing Frederick Seitz, who had just completed an exceptional dissertation under Eugene Wigner at Princeton on the quantum theory of metals. Edward Condon wrote Gibbs that he was very concerned about American universities appointing German Jews instead of Americans. However, the next year, he sent another letter to Gibbs saying, "I was wrong about Bethe. He's clearly absolutely outstanding, and you were right."

Similarly, when the appointment of Teller was being considered at George Washington University Gregory Breit wrote a letter objecting to Teller and suggested they appoint Eugene Wigner, who did not have tenure at Princeton. The issue here was not anti-Semitism, but the fact that the appointment of foreigners prevented the appointment of bright young Americans who were looking for jobs, or in this case, a brilliant physicist like Wigner who had come to the United States before 1933.

In the fall of 1935 Bacher came to Cornell and became Bethe's closest friend in the physics department.

Bacher was born in Loudonville, Ohio. His family moved to Ann Arbor, Michigan, when he was five. When he was in junior high school his family bought a summer place at a lake near Ann Arbor. Professor Harrison Randall, the head of the physics department at the University of Michigan, owned the adjacent house. In high school, Bacher became aware that he was interested in the physical sciences. Randall helped him channel his interests into atomic physics. After graduating from high school he went on to the University of Michigan, and in 1926 earned a bachelor of science degree in physics. He began his graduate studies at Harvard. During his year there he took graduate physics courses with Edwin Kemble, John Slater, and George Pierce, as well as two graduate courses in mathematics. Toward the end of that year, his father took ill, and Bacher returned to Ann Arbor and enrolled in the University of Michigan.

During the academic year 1926–27, when Bacher was at Harvard, Otto Laporte came to Michigan as the first of a series of theoretical physics professors. The Michigan summer school in physics was started after his arrival. In the fall of 1927 Samuel Goudsmit and George Uhlenbeck joined the physics department at Ann Arbor as theorists. They had been trained by Ehrenfest and Lorentz and were "wonderful teachers." In 1929, after Goudsmit, Uhlenbeck, Laporte, and David Dennison were all in residence, the summer school began to exert its full impact and to become very well known. Heisenberg, Dirac, Fermi, and Oppenheimer were among its first lecturers. In his interview with Charles Weiner, Bacher noted the large initial impact that these summer sessions had on physics graduate students and faculty members who came to them from nearby universities. Many of them had had relatively little knowledge of developments stemming from the new quantum theory. Since it was relatively difficult and costly to travel at the time, Bacher stressed that these summer schools were invaluable to those in Ann Arbor and the whole Midwest.

Bacher became Goudsmit's first graduate student at Michigan. In his thesis he worked out the theory of the Zeeman effect in relatively simple atoms for magnetic fields of intermediate strengths. He married in May 1930 and, after obtaining his doctorate in June of that year, went to Caltech for the academic year 1930–31 and to MIT in 1931–32 as a National Research Council fellow. In 1932 he and Goudsmit published their *Atomic Energy States, as Derived from the Analyses of Optical Spectra.* While at Caltech he did some collaborative experimental work with Stuart Campbell on the hyperfine structure of some simple atoms. In 1931–32 at MIT he worked with Slater and with Condon, who was visiting in the summer of 1932. While at MIT he also spent some time at Princeton and at Columbia. Rabi was then starting his research with molecular beams to measure nuclear magnetic moments. Bacher was working on hyperfine structure, so he was very interested in Rabi's work and very much impressed by it. The following year he went back to Michigan as a Lloyd postdoctoral fellow for a year, and then spent a year in Michigan without a job. It was the height of the Depression and very difficult for physicists to find an academic position. Supported by his family, he continued doing research at the University of Michigan and deepened his knowledge of nuclear physics. In 1934 he obtained a job at Columbia, and in the fall of

1935 he joined Cornell's physics department. One of the reasons he wanted to go to Cornell was that it offered him the opportunity to get into experimental nuclear physics.

Bethe, Bacher, and Livingston made Cornell into one of the outstanding centers of nuclear research. The Bethe Bible, as the three articles Bethe wrote with Bacher and Livingston in the *Reviews of Modern Physics* of 1936 and 1937 came to be known, played an important role in the explanation of Cornell's reputation in nuclear physics. In these three articles everything that was then known about nuclear physics was presented in a coherent and accessible manner. How the collaboration with Bacher and Livingston came about merits a few words. This is how Bacher put it in his interview with Charles Weiner in June 1966:

> Well, of course, it was Bethe's idea. I would say that practically all the credit for those review articles ought to go to Bethe because while some of the rest of us worked at it, they never would have been done without Bethe. And it was Bethe who, by his extraordinary force of intellect, ingenuity, and tremendous energy, managed to pull all of this stuff together, rework it where it had holes in it, and get it out in the form of a *Review* article. . . It involved looking into all sorts of things, including the re-interpretation of experiments. And when Bethe looked at experimental work and found that they didn't seem to have made certain corrections, we went right back to the raw data and started working it up again. Or, if he went to a theory and he got into it and found out that certain things in the theory had been neglected, he reworked the whole theory [with the help of his postdocs Rose and later Konopinski]. So that parts of these articles, particularly the things that Bethe worked on in them, are really simply very much more than simply summarizing what's in the literature. My contributions were, I would say, somewhat minimal, compared to Hans Bethe's.
>
> . . . The way it started was that Bethe laid out the plan of it, and at first it was going to be a very much shorter thing than it subsequently turned out to be . . . He first, sort of laid out what the first part would be, and he had certain parts of the first volume that he thought I might work with him on, particularly the part on nuclear moments, and I think some other parts, but essentially the part on nuclear moments. Then he had all the second volume that had to do with various problems about dynamics and so on, and then the third part which was all experimental. Well, he got Livingston to work with him on the third part, that had to do with the experimental things, and I worked with him on the first part, and he did the second one

all by himself. I think it's fair to say that, at least in the parts with which I'm familiar, the bulk of the credit for that goes to Hans Bethe.

It took a little over a year and a half to write the three articles.

Victor Weisskopf, when interviewed by Charles Weiner and Gloria Lubkin in December 1966, gave the following assessment of the Bethe Bible:

> Bethe's way of writing review articles is something unique. He doesn't only review—he also reworks the papers, so there is not a single conclusion or statement that he hasn't calculated from beginning to end, including the evaluation of the experiments, so that he reshapes the work done by others in his own style, and that makes these Bethe reviews tremendously valuable . . . [They were], I think, perhaps more important than a book. More important because it came at a time when perhaps one would not have been able to write a book, although these three articles would make a perfect book—in fact, a better book than many books that have been written. But the fact that it came so quickly at a time when this field was completely new and in the center of interest made these articles the most important piece of literature in this field.

Actually in 1937 Gamow published his *Structure of Atomic Nuclei and Nuclear Transformations.* Though he called it a second edition of his *Constitution of Atomic Nuclei and Radioactivity,* it was a completely new book by virtue of the advances in nuclear physics since 1931. A comparison between Bethe's and Gamow's presentations corroborates what Weisskopf had asserted. In the preface of his book Gamow indicated that he had made no attempt "to go rigorously through all the mathematical calculations concerning different nuclear processes, the main aim [of the book] being to consider questions of principle concerning nuclear structure, and to understand the different nuclear processes from the point of view of the present quantum theory."[58] But if, in contrast to Bethe, Gamow did not attempt to formulate a more rigorous mathematical basis to quantitative explanations of nuclear processes, he did offer numerous insights and suggestions for qualitatively modeling nuclear phenomena. And in the last chapter of the book he outlined what was known about the relative cosmic abundance of the chemical elements and speculated on the mechanisms by which they are produced in stars. The nuclear processes responsible for nucleosynthesis and for energy generation in stars would become the focus of Bethe's research in 1938—and this by virtue of Gamow's interest in the subject!

More Nuclear Physics

Although Cornell's cyclotron only produced 1.2 MeV deuterons, Livingston and his associates developed an arc source that transformed it into a particularly useful tool for neutron research. Early in the fall of 1935 Bethe was induced to participate in an experiment with Livingston and his student, D. F. Weekes, on a method to determine the selective absorption regions of slow neutrons and actually "spent at least one long evening taking counts with Mr. Weekes, and observing how the counts changed, with different absorbers."[59]

Bethe "moved among the experimental physicists as a source of brilliant conjecture and practical aid."[60] He not only provided suggestions for experiments and the theory for their interpretation, but he was intimately involved with their design and the analysis of the data.[61] At Cornell, like at the other centers where nuclear physics was being cultivated, theorists and experimenters worked intimately together. But, as emphasized by Wigner in his article on the development of the compound nucleus model, experimental work constituted the most important factor in the growth and expansion of the theoretical understanding of nuclear phenomena.[62] Bethe was fully aware of this fact. Indeed all of his papers in nuclear physics were stimulated by experiments. They all were intended to clarify empirical data and to model nuclear structure and nuclear reaction mechanisms that could be tested experimentally.

Bethe's Social World

During his first term at Cornell in the spring of 1935, Bethe was mostly occupied with physics and with departmental matters and did not get to meet many people outside the physics department. He did get to know a mathematician,

> a very nice man, also an assistant professor, William [Flexner] who was the nephew of the Flexner who established the Institute for Advanced Study . . . Both he and his wife were very nice to me. He explained to me lots of mathematics, including Gödel's theorem. In the meantime, I totally forgot how that was proved, but Bill Flexner made it clear to me. He was a very free spirit, and so was his wife. I saw a lot of the Flexners socially. I saw more of them than any of the physicists . . . I met some of the other mathematicians, Ralph Agnew, who later was department chairman, who was a somewhat straitlaced gentleman. Him I met mostly at examinations of doc-

toral candidates, there had to be a mathematician on the committee, and I met also Hurwitz. Hurwitz I think took his Ph.D. at Cornell, and stayed at Cornell until he retired. And he was terribly popular with the students, because he was full of fun, in his lectures, probably was the most popular teacher, of either physics or mathematics.

In 1937 George Placzek came to Cornell, supported by Farrand's Emergency Committee funds. The friendship between Bethe and Placzek was thus resumed. The extent of the friendship can be gauged by the fact that Placzek joined Bethe when he was visiting his mother in Switzerland in the summer of 1938, and stayed in the same hotel for the month Bethe and his mother were together.

Contact with chemists came early, because John Kirkwood was an assistant professor in the department of chemistry.[63] Bethe had first met Kirkwood in the spring of 1932 when Kirkwood was a Rockefeller International Research Fellow. He first worked with Debye in Leipzig and thereafter spent time in Sommerfeld's institute. Kirkwood worked closely with Bethe when he was in Munich, and at Cornell regularly attended the theoretical physics seminar that Bethe, Kennard, and Smith ran weekly. Kirkwood left Cornell in 1937. When he rejoined the department of chemistry in 1938 as the Todd Professor of Chemistry the interaction between Bethe and Kirkwood resumed and became more intense, and they collaborated on several papers. Even before Kirkwood had left in 1937 they had collaborated on a paper dealing with the statistical mechanics of cooperative phenomena.[64] Kirkwood had presented its content to the July 1937 Cornell symposium on the structure of metallic phases that Bethe had helped organize.

Professional Standing

Upon his arrival in the United States Bethe was immediately recognized as an outstanding physicist by the physics community. He was invited as one of the lecturers to the 1936 Michigan summer school and received many more lecture invitations than he could handle.

At the beginning of the academic year 1936–37 Bethe confessed to Sommerfeld that although he had gone to Cornell with mixed feelings,

> like a missionary going to the darkest parts of Africa in order to spread there the true faith . . . already half a year later I no longer held this opinion

> and today I hardly would return to Europe even if I would be offered the same amount of dollars as at Cornell.
>
> The characteristic trait of physics in America is team work. Working together within the large institutes, [and] in every proper one everything that physics encompasses is being done, the experimentalist constantly discusses his problems with the theorist, the nuclear physicist with the spectroscopist. By virtue of this cooperation many of the problems are immediately disposed of, [whereas] that would take months in a specialized institute. *More team work* [in English in the original]: the frequent conferences of the American Physical Society . . .
>
> What presently one is interested in, is of course nuclear physics. With the result that 90% of all work in this area is done in America.

He concluded his letter with an insight into what made possible the success of Cornell's physics department: "Gibbs has earned the greatest praise for the development of Cornell. Though not a very good physicist, he is a fabulous institute director. He is busy the whole day to obtain money for the institute and to make sure that everybody has all he needs for his work."[65]

In the spring of 1937 Bethe was invited back to the University of Illinois at Urbana-Champaign to lecture there for three days. At the end of the third day Wheeler Loomis, the head of the department, offered him a full professorship at the considerable salary of $6,000. At the strong recommendation of Gibbs, Farrand matched the offer, and Bethe stayed at Cornell.

The department thereby maintained its standing as one of the best in the country by strongly supporting nuclear physics. The 1938 proposal submitted by the department to the Cornell administration included plans for a new and larger cyclotron, the design of which was worked out the following year. Its contemplated cost was approximately $50,000.[66]

But while strongly supporting nuclear physics, the department made sure that its other research activities—in spectroscopy, X-ray and electron diffraction, optics, and electronics, and particularly those that were of importance in training students who would seek jobs in industry—would not suffer. To make itself better known the department in 1935 started sponsoring a series of annual summer symposia.[67] Edmund Day recognized the importance of the sciences in the mission of the university. On July 1, 1937, in one of his first public appearances as president of the university, he gave the welcoming address to the Symposium on Metallic Phases that the department of physics had organized for that year.

By 1938 Bethe would write Peierls, "There are terribly many *(schrecklich viele)* people here these days and Bacher speaks of my production line: 5 people working on doctoral dissertations, 3 more who are likely to be doctoral candidates with me, in addition to Groenblom, Rose and Ludloff [postdoctoral fellows]. It's becoming a little too much for me, as I don't get to do my own work any more." No photograph better illustrates Bethe's status as a theorist than the group photograph of the 1937 Washington Theoretical Physics Conference on nuclear physics, in which he is at the center of a group of like-minded colleagues.

8

The Happy Thirties

In the spring of 1976, Roger Stuewer, an eminent historian of physics at the University of Minnesota, in discussions with his colleagues, formulated the idea of holding a symposium "to chart key developments in the history of nuclear physics, principally in the 1930s, by inviting eight distinguished nuclear physicists to lecture on and discuss these developments."[1] These lectures were to be based on their personal knowledge and experiences, and to include, if possible, information from their personal correspondence or other published materials. With funding from the University of Minnesota the conference was held May 18–21, 1977. The speakers included Bethe, Segrè, Frisch, Goldhaber, Edwin McMillan, Eugene Wigner, Peierls, and John A. Wheeler. The title of Bethe's talk was "The Happy Thirties."[2]

The distinguished experimental physicist Alfred O. Nier was the chair of the opening session of the conference. In his introduction of Bethe he indicated that it was "most appropriate that we officially start our symposium with a speaker who might be called 'Mr. Nuclear Physics.'" He commented that when he was studying physics in the mid-1930s Bethe's name was already "virtually a household name among physicists" and that the articles, totaling 487 pages, that he had written with Bacher and Livingston in the 1936–1937 *Reviews of Modern Physics* "amounted essentially to a compendium of everything known in nuclear physics at the time." Everyone at the conference echoed these sentiments. Thus, Wigner in the opening remarks of his lecture on the neutron indicated that he would "compliment highly the article by Bethe and Bacher" when he would come to it in his lecture.[3]

Before delivering the historical part of his lecture, Bethe commented that it was certainly a "little unusual" to have called his lecture "The Happy Thirties," given that "politically the thirties were anything but happy. Many of us in this room . . . emigrated from Germany and Italy because of the dictatorships prevailing in these countries, and it was only by great good fortune and by the wonderful hospitality of this country that we were then able to lead happy and productive lives. However, in nuclear physics certainly the thirties were a very happy period."[4]

In this chapter I review some of Bethe's articles concerning the theory of nuclear reactions, contributions that made the 1930s "a very happy period" for him. This research was stimulated by his intense collaboration with Livingston and Bacher in writing the Bethe Bible, and by his active participation in the experiments they were carrying out. His mastery of the theory of nuclear reactions—as then formulated—and of the contemporaneous version of β-decay theory is what made possible his crucial contributions to the elucidation of the stellar energy problem at the 1938 Washington Theoretical Physics Conference. That conference and the research that led to it contributed substantially to Bethe's happiness in the 1930s. The career of George Gamow, whose research in nuclear physics paralleled Bethe's, is also detailed in this chapter. In fact, it was Gamow's interest in nucleosynthesis and stellar evolution that was the reason for the 1938 Washington Theoretical Physics conference dealing with these subjects. Others, in particular Bengt Strömgren and Carl von Weizsäcker, were likewise addressing these problems at the time.[5] I have concentrated on stellar energy generation because Bethe felt that it was his *decoupling* of the problems of energy generation and nucleosynthesis that allowed him to accomplish what he did.

During the 1930s Bethe formed several professional associations and friendships with colleagues, in particular Robert Marshak and Victor Weisskopf. Marshak was Bethe's most brilliant student during the 1930s, and under Bethe's guidance made important contributions to the theory of white dwarf stars and to the understanding of stellar evolution. Victor Weisskopf, who joined the Rochester physics department in 1937, became Bethe's closest professional friend and was very influential in enlarging the scope of Bethe's atomic and nuclear investigations. The chapter concludes with an examination of Bethe's field theoretical investigation of the origin of nuclear forces.

Nuclear Reactions

Bethe's 1930 *Habilitationsschrift,* "On the passage of fast charged particles through matter," was a wide-ranging paper in which he had calculated in Born approximation the transition amplitudes for the elastic and inelastic scattering of an incident charged particle with an atom and had computed the excitation of X-ray and optical levels of the atom, the number of primary and secondary ions produced, the velocity distribution of the secondary electrons, and the energy loss by the incident charged particle. As important as the quantitative results that Bethe had obtained were his the qualitative insights.

In the interaction of an electron of moderate energy (keV to MeV) with an atom, the electron passes through the atom essentially undeflected. In such an atomic collision the interaction between the incident electron and the individual electrons of the atom is small, and the electron seldom imparts energy to an atomic electron. It primarily undergoes elastic collisions. Also, since the time spent by the projectile in the atom is of the order of an atomic dimension divided by the velocity of the incident particle, it is very improbable that radiation is emitted. However, because of the long range of the Coulomb force between charged particles there will be numerous *inelastic* collisions for impact parameters larger than atomic sizes. In such collisions the electron can be regarded as interacting with the average field of the atom because the interaction with an individual electron is much smaller than the average interaction with all the electrons. And, because the energy of such an incident electron is much larger than the interaction energy between it and the bound electrons, treating the problem in Born approximation is justified.

Initially Bethe used these same concepts to describe nuclear reactions, and in particular to calculate cross sections for the neutron reactions with different nuclei and compare them with the cross sections yielded by the experiments published in 1935 by Fermi and Edoardo Amaldi, Philip Burton Moon and J. R. Tillman, T. Bjerge and C. H. Westcott, and others.[6] Fermi's experiments had indicated that for heavier elements the reaction $n + {}^{A}Z$ proceeds principally by radiative capture, i.e., $n + {}^{A}Z \rightarrow {}^{A+1}Z + \gamma$, and furthermore, that slow neutrons had a larger cross section for capture than the fast neutrons coming directly from the source. These developments stimulated the theoretical analysis of the scattering and absorption of slow neutrons by nuclei in papers published in 1935 and 1936 by Bethe, Francis Perrin, Perrin and Elsasser, and Amaldi and Fermi. Bethe presented his theory both in Copenhagen on his visit there in September 1934 and at the January 1935 New York meeting of the American Physical Society. In 1935 Bethe published a long paper that described his theory of the elastic and inelastic scattering of slow neutrons by a nucleus.[7] His model was based on the assumption that the interaction between neutron and nucleus could be described by an average nuclear potential represented by a potential well.[8] This implied that the interaction between the neutron and the nucleus takes place only when the neutron is inside the nucleus or very close to its boundary. Although the rate of change of the nuclear potential at the boundary was important for the quantitative results, the qualitative results were independent of it. The large disintegration cross sections resulted from two factors. First, the cross section is inversely proportional to the velocity of the neutron, $\sigma_c = \frac{k}{v}$, because slow neutrons stay longer in the nucleus. The second factor (reflected in the k in the cross section) depended on the phase of the neutron wave function at the nuclear boundary, which could not be predicted theoretically, but could, under reasonable assumptions, account for the large difference in cross sections of different elements in the experimental data available at that time. Gregory Breit and Wigner thereafter pointed out that quantum mechanics predicted a large scattering cross section whenever there was a large capture cross section; and experiments by Dunning, Fink, Pegram, and Mitchell (1935) at Columbia showed that the scattering cross section of neutrons on cadmium (Cd) was at most 1 percent of the capture cross section. Furthermore, the experiments by Tillman and Moon had indicated that some nuclei had selective narrow energy absorption bands, contrary to the theoretical expectation of a smooth energy dependence of the cross section. The sharpness of the absorption bands implied that the time spent by the slow neutrons in the nucleus was at least an order of magnitude longer than the estimate given by *nuclear size/velocity of neutron* = 10^{-13} cm/10^{9} cm/sec = 10^{-22} sec.

Upon hearing these results Bohr immediately recognized that the theoretical description of the interaction of neutrons with nuclei required a radical revision of the nuclear dynamics formulated by Bethe and others. The experiments implied that models and calculations mimicking the atomic model were wrong. Instead Bohr proposed in 1936 the compound nucleus model that focused on the many body features and stressed the strong interactions of the different degrees of freedom of the nuclear system.[9] Rutherford, upon reading Bohr's ideas on the subject in *Nature,* succinctly summarized them in a letter to Born: "The main idea is an old one of Bohr's, viz. that it is impossible to consider the movements of the individual particles of the nucleus as a conservative field, but that it must be regarded as a "mush" of particles of unknown kind, the vibrations of which can in general be deduced on quantum ideas. He considers, as I always have thought likely, that a particle on entering the nucleus remains long enough to share its energy with the other particles."[10]

The conceptual basis for the Bohr model was the following: the short range of the nuclear forces and the fact that they saturate implies that in the interaction of a nucleon with a nucleus the nucleon will interact with but a small number of the nucleons constituting the nucleus. These individual interactions will be of the same importance as the interaction with the average total interaction and it is not permissible to model the interaction by an average potential. Any nuclear projectile that hits a nucleus will be caught and a new nucleus will be formed which Bohr called the *compound nucleus.* This "compound nucleus" is long lived compared to the natural nuclear time. This because the excited state will subsist until the excitation energy that had been imparted to the compound nucleus is concentrated again on one "particle" and allows it to escape. But it seldom happens that enough kinetic energy is concentrated on one neutron or proton to enable it to escape from the nucleus. The more probable process is a transition of the compound nucleus to a stable state with the emission of a photon—thus explaining why radiative capture occurs more frequently. And most importantly: the lifetime of the compound nucleus is such that complete statistical equilibrium is established —and it essentially "forgets" how it was formed. The final break-up is therefore independent of the mode of formation of the compound state.[11]

Bohr had a nice model of the process. The nucleus was represented as a shallow dish with little balls in it.[12] A nuclear reaction was then pictured as letting an extra ball with considerable energy come into the dish. Its energy would very soon thereafter be distributed among the various balls in the dish.

Bethe took up the idea and wrote extensively on it in the Bethe Bible. But Bethe was to say later, "When I wrote my *Reviews of Modern Physics* article I was too much convinced by the authority of Bohr, so I said that the compound nucleus theory applies to everything . . . This was wrong."[13] Bohr's compound nucleus had many of the properties of the stable nuclei, in particular, fairly well-defined energy levels. The general characteristics of the energy levels of a nucleus were believed to be the following: With the exception of the ground state, they have a finite lifetime because of the possibility of transition to lower states by the emission of electromagnetic radiation or by the ejection of a neutron (or a proton, α-particle, or more complex nuclear entity). An excited state i thus has a finite width Γ_i that is connected to its lifetime τ_i by $\Gamma_i = \hbar/\tau_i$. If the lifetime is due to different emissions, the total width Γ_i is the sum of partial widths:

$$\Gamma_i = \sum_j \Gamma_i^{\,j} + \Gamma_i^{\,r},$$

where Γ_i^j is the partial width corresponding to the emission of a "particle" j (which can be a neutron, a proton, an α-particle, and so on), and Γ_i^r is the radiation width. The emission probability per second of a particle j by the excited state i is $\Gamma_i^j/\hbar$. Mathematically, one can conceive the energy of the jth level to be described by the complex quantity $E_j - \frac{i\Gamma_j}{2}$, so that the time dependence of the probability of finding the nucleus in the state j is given $\exp(-\Gamma_j t)$.[14]

In the Bohr model a nuclear reaction thus consists of two steps:

1. The formation of the compound nucleus, with the probability of its formation being very small unless the energy of the incident projectile and that of the target nucleus coincide fairly closely with one of the energy levels of the compound nucleus. As a function of the energy of the incident particle the cross section for the formation of the compound nucleus therefore shows sharp maxima, but very small values between the maxima.
2. The disintegration of the compound nucleus. The important feature of the Bohr model is that probability of a particular mode of disintegration is *independent* of the mode of formation of the compound nucleus. However, the probabilities for the different modes of disintegration of the compound nucleus are energy dependent, sometimes quite critically.

The experiments indicating that the cross sections for nuclear reactions induced by slow neutrons were characterized by high maxima and by low values between the maxima had led Bohr to formulate his compound nucleus model. For Wigner, who had worked with the physical chemist Michael Polanyi in the mid-1920s on the theory of chemical reactions,[15] these features reminded him of the compound molecule theory of chemical reactions and of the behavior of the cross sections of the scattering of light by atoms, which manifested strong maxima for those frequencies of the light that coincided with an energy level of the atom. In 1930 Weisskopf and Wigner had formulated the quantum electrodynamical description of these processes. It too was a two-step process: the absorption of a photon by the atom induced a transition to an excited state. This state subsequently decayed and emitted a photon. In the nuclear context the excited state was assumed to correspond to a state of the compound nucleus, and the decay to the disintegration of the compound nuclear state.

In 1936, Breit and Wigner thus arrived at a model very similar to that of Bohr and also developed a two-stage theory of neutron-induced nuclear reactions. They wanted to explain the anomalous behavior of the absorption of slow neutrons in cadmium and formulated the simplest possible version of the model. They assumed a single (bound) compound state and posited that the continuum compound states do not couple with one another. If the energy of the neutron is close to the resonance energy, they deduced the following expression for the cross section:

$$4\pi\lambda\!\!\!^{-}\lambda\!\!\!^{-}_r \frac{\Gamma_n \Gamma}{(E - E_r)^2 + \frac{1}{4}\Gamma^2}.$$

This has become known as the Breit-Wigner cross section; $\lambda\!\!\!^{-}$ is the de Broglie wavelength divided by 2π of the neutron of energy E, and $\lambda\!\!\!^{-}_r$ the same quantity for the case in which the neutron has the resonance energy E_r; Γ is the width of the resonance, and Γ_n is that part of the resonance that results from neutron emission.

In an extensive paper with Placzek, Bethe in 1937 extended the work of Breit and Wigner for neutron-induced reactions to take into account the spin of the neutron and that of the target nucleus. They generalized it to allow an arbitrary number of compound nuclear energy levels, but only if the widths of the various resonances are small compared with the distance between them. Their paper also contained a detailed analysis of all the experimental results available until the writing of their paper.[16]

Nuclear structure and nuclear reactions were discussed in Bethe's singly authored article in the Bethe Bible. In it he critically reviewed all the different models of nuclear structure.[17] The fact that the density of energy levels of medium and heavy nuclei was much larger than previously thought had led him to consider a nuclear model that assumed that the interaction between pairs of nucleons was small compared to their kinetic energy so that the nucleus could be treated as a gas of nucleons.[18] This in turn led to the introduction of the concepts of the temperature and entropy of a nucleus. In his *Reviews of Modern Physics* article Bethe illustrated this approach by deriving a number of general relations, such as the relation between the total excitation energy and the nuclear temperature, and that between the nuclear entropy and the density of nuclear energy levels.

But the compound nucleus model clearly contradicted the assumptions that justify a gaseous model since to a first approximation it modeled the nucleus as a liquid drop in which the interactions between the nucleons were large compared to their kinetic energy.[19] To justify the physical picture of the independence of the formation and decay of the compound nucleus, Weisskopf in 1937 extended the liquid drop model and proposed an evaporation model that could be mathematically formulated.[20]

How to reconcile the various models advanced to account for the different properties of nuclei and nuclear reactions became a central problem for nuclear theorists. In the *Reviews of Modern Physics* article Bethe was very much concerned with the problem, and the article was written so that the issues involved were highlighted. He continued being concerned with this fundamental problem throughout the late 1930s. Thus in a 1940 paper entitled "The Continuum Theory of the Compound Nucleus" Bethe tried to answer the following question: "How is it possible that a proton or a neutron is so strongly absorbed by the nucleus?" Granting that there is a compound nucleus to explain Fermi's experiments, there was still the problem of how the projectile particle gets in and gets caught in the target nucleus. In that paper Bethe indicated that it was important first to have a complex potential, $V_0 + iV_1$, act on single particles, and second for this potential to have a gradual surface rather than a sharp boundary. The size of the imaginary part of the potential was to be a measure of how much of the incident beam is absorbed by the target nucleus.[21] After the war the model was elaborated by Barschall (1952) and by Feshbach, Porter, and Weisskopf (1954) in their optical model of nuclear scattering.

It would take us too far afield to give a detailed exposition of Bethe's further elaboration of nuclear theory before World War II.[22] Suffice it to say that the Bethe Bible gives proof of his mastery of everything that was known about scattering theory, many-body physics, statistical mechanics, electrodynamics, β-decay theory, nuclear systematics, and nuclear reactions. His mastery of these topics allowed him in 1938 to successfully solve the problem of energy generation in stars.

Robert Marshak, who came to Cornell as a graduate student in the fall of 1937, played an important role in the solution of the stellar energy problem. He was one of Bethe's most distinguished students, and after working with Bethe on the sources of stellar energy, later extended the theory to explain energy generation in white dwarfs.

Robert Marshak

Working with Hans Bethe as a graduate student was, for Robert Marshak, an opportunity made available through years of hard work and self-direction. Robert Marshak's father was born in Minsk, Russia, and came from a long line of rabbis famous as Talmud scholars. As a child he was educated in a *Cheder* (religious elementary school) and never went to public schools. Yiddish was the only language Marshak's father spoke throughout his life. As a teenager he experienced the virulent discrimination against Jews that was endemic to tsarist Russia and came under the influence of the Jewish Enlightenment, and though he remained observant he became sympathetic to the liberal political activities then taking place in Russia. He lived through the pogroms that swept Russia on the heels of the Kishinev pogroms of 1903 and 1905 and was deeply shaken by them. After Russia lost the Russo-Japanese War and in the aftermath of the 1905 Revolution he had expected that the creation of the Duma in 1906 would lead to reforms. But it soon became apparent that these hopes were unfounded. Many Jews left Russia, and Marshak's father was among them. He came to the United States in 1907.

Marshak's mother was born in a little town not far from Minsk and emigrated to the United States in 1910. His parents met in New York and were married in January 1916. Robert was born in the Bronx on October 11, 1916. In his interview with Charles Weiner, Marshak stressed that his parents "were always broke." The family, which consisted of Robert, his parents, and two younger sisters, lived in a six-room apartment, but in order to make ends meet they rented out two rooms to two boarders. Their economic situation became worse during the Depression, and his mother began working as a seamstress because his father could find no employment. He "was reduced to fruit peddling, taking care of the house and doing some of the cooking . . . [He] always had very great respect for learning and education; and always encouraged me to work hard at being a student." Marshak indicated that his father, during the time that he was a fruit peddler, would get up fifteen minutes earlier than necessary, at four o'clock in the morning, "in order to polish my shoes, because he argued that I would then have more time for my studies." When Marshak was a teenager his father didn't let him work during the summers to earn extra money, "because he argued that I should be a student. In that sense I had very strong encouragement from my parents as far as school was concerned."[23]

Marshak rewarded his parents by being an outstanding student. His mathematical talents became clear in elementary school. He attended James Monroe High School in the Bronx, then one of the best public high schools in New York City, and became the captain of the school's mathematics team and the president of the school's Service League.[24] He was valedictorian of his class and won all the gold medals the high school awarded at his graduation: the medal for scholarship because he had the highest grade point average of the graduating seniors; the medal in service; and a third gold medal for "good citizenship." Normally the latter medal was given to the president of the senior class, which Marshak was not. However, the faculty adviser to the senior class was Marshak's history teacher. He felt that in view of what Marshak had done for the school and for him—Marshak used to grade some of his class assignments—that he should get the gold medal for citizenship.

In his interview with Weiner, Marshak related that a young man by the name of Hippocrates Apostle, who had graduated two years earlier from his high school, had influenced him deeply. Hippocrates, who came from an immigrant Greek family, had also excelled in mathematics at Monroe and had likewise become the captain of the math team. After graduating he remained interested in the math team and was an assistant coach. Hippocrates had won a Pulitzer Scholarship to attend Columbia College, and urged Marshak to do the same. At Columbia he had become interested in philosophy and used to tell Marshak "about the great joys of philosophy and the exciting ideas to which he was exposed." He introduced Marshak to the first philosophy books he ever read, John Dewey's *How We Think* and John Herman Randall's *Mak-*

ing of the Modern Mind, a stimulating and very informative presentation of Western intellectual history.

Marshak graduated in January 1932. Columbia admitted students only for the academic year. Likewise, the Pulitzer Scholarship was only awarded annually, in September. The question was therefore what to do between February and September. Even though Marshak's family was in serious financial trouble, his father thought it essential for him to go to college in the interval. So Marshak went to City College in February 1932 and supported himself with the Regents' Scholarship he had won, worth $100 a year. He also began tutoring high school students in mathematics. But he also applied for admission to Columbia and, separately, sent in his application for a Pulitzer Scholarship.[25] In May 1932 he was informed that he had been rejected for admission by Columbia College. He was told that if interested, he could be admitted to Columbia's Seth Low Junior College, which was located in one of the office buildings in Borough Hall, in Brooklyn. And if he did well there he might, after a few years, be allowed to transfer to the main college. This was Columbia's mechanism for handling the applications of bright Jewish students. Marshak replied that he wasn't interested in going to Seth Low Junior College. He then was in a quandary as to whether he should take the College Entrance Examinations in June, the results of which were to be submitted with the application for a Pulitzer Scholarship. According to the will of Joseph Pulitzer, who was Jewish, ten scholarships were to be awarded each year to graduates of New York City public high schools. Thirty applicants who obtained the highest scores on the college entrance examinations would be interviewed by the Pulitzer Scholarship Committee, and ten scholarship recipients would be chosen from among them. In 1932, there were 500 applicants for the scholarships. Marshak was one of the thirty students interviewed by the committee. Dr. Adam Leroy Jones, a reasonably well-known philosopher, was the chair of the Pulitzer Scholarship Committee. Jones, in his role as director of admissions, had rejected Marshak's application to Columbia. Marshak was therefore not optimistic about the outcome of his application. He told himself: "if Dr. Jones, as Director of Admissions, had rejected a poor boy from the Bronx with all those gold medals, what chance would he have to be awarded a Pulitzer Scholarship by Dr. Jones as Chairman of that committee."[26]

Then "a curious thing happened." Marshak, after a personal interview with Jones, was awarded a scholarship. Marshak thought that this happened because Dr. John L. Tildsley, the associate superintendent for the New York high schools, was on the Pulitzer committee. Tildsley had been commencement speaker at Marshak's graduation, and on that occasion had met him personally. After that meeting, Tildsley invited Marshak to visit him at his Spuyten Duyvil home. Tildsley wanted to find out what Marshak thought of the teachers at James Monroe. Evidently, Marshak was "a little naive and talked frankly about the performance of various teachers, their teaching style," and made some mildly critical comments about the principal.[27] Marshak's guess was that Dr. Tildsley was chiefly responsible for the favorable outcome of the Pulitzer decision. Of the ten students who received Pulitzer Scholarships that year, Marshak was the only Jewish recipient. Marshak commented: "This is very strange because I knew quite a few Jewish boys who were among the first 30."

Since he had been offered the Pulitzer Scholarship but had not been admitted to Columbia College, Marshak wrote to Dr. Jones that he would stay at City College and just collect the Pulitzer money. A few days later he received a reply from Jones stating that a place had opened up for him at Columbia College. He decided to go to Columbia starting in September 1932 "because of the great opportunities in a university setup."[28]

Although he took the introductory physics course during his freshman year, a course in relativity theory as a sophomore, and one in classical mechanics based on the Lagrangian and Hamiltonian approaches in which the mathematical aspects were emphasized as a junior, the majority of his undergraduate courses were in mathematics and in philosophy. In fact, he graduated in 1936 as a mathematics and philosophy major. It was his philosophical interests that had led him to study the theory of relativity. In the relativity course he came in contact with some of the undergraduate physics majors: Herbert Anderson, Morton Hamermesh, Henry Primakoff, and Norman Ramsey. Later he got to know Julian Schwinger, who transferred to Columbia from City College in the fall of 1935. The five of them organized an undergraduate physics club at which they would make presentations to one another on various topics in physics. Through his interactions with these outstanding young people Marshak got to know what was going on in the Pupin labs, the site of the physics department, and what doing physics was like. During the second semester of his senior year he took Rabi's course on quantum mechanics, which convinced him, at that late stage of his college career, to become a physicist. Although Rabi was not a good lecturer—he was then very busy working on the experiments for which he would earn the Nobel Prize and didn't prepare for his lectures—he nonetheless communicated a sense of excitement about quantum mechanics. Although he got to know Rabi fairly well, Rabi didn't take Marshak's conversion to physics seriously because he thought he was still really interested in philosophy.

Since the decision to switch to physics had come so late, Marshak hadn't applied to any graduate program in physics. Moreover, there was no one in the Columbia department of physics he could ask to write a letter of recommendation for him. In any case, applications to graduate schools had been due in February, so the only option open to him for the 1937 academic year was to inquire whether Columbia would accept him as a graduate student in physics. It did, and offered him a scholarship to cover his tuition.[29]

Marshak characterized the year he stayed at Columbia as a graduate student as "quite unspectacular." He took the usual first-year graduate courses in physics: electromagnetic theory, statistical mechanics, and quantum mechanics. He convinced Rabi he was serious about becoming a physicist, but he didn't think that Rabi believed he was "particularly promising as a theoretical physicist" because "the norm of comparison all the time was Julian Schwinger, which was a little rough."

Toward the end of that first year Marshak began worrying about what to do the next year. He knew he wanted to be a theorist, but Rabi was the only theorist at Columbia. Rabi, devoting all his attention to molecular beams, was becoming an experimentalist, so Marshak felt he should consider going elsewhere.

Then a chance comment transformed his life. In late June 1937, Sidney Siegel, a graduate student at Columbia who had just completed his doctorate in solid state physics, was looking for a university job as an instructor. He decided to drive to Ithaca to attend the symposium on the structure of metallic phases that Bethe and Smith had organized and talk to people there. Siegel asked some of the graduate students whether they would like to accompany him. Marshak accepted the invitation. During the conference, Siegel asked Gibbs whether there was an opening in the department. Gibbs answered that there weren't any for instructors but that there was still an opening for a graduate assistantship. On the way back to New York Siegel mentioned this to Marshak, and although Marshak didn't think he had much of a chance, he decided to apply. Rabi must have written a reasonable letter on his behalf, because his application was taken seriously. At first Marshak was turned down.

But a few days later a telegram from Gibbs arrived, saying that if Marshak were still interested, he could have the assistantship. Apparently the recipient of the remaining assistantship had withdrawn to accept a Harvard assistantship. Marshak was delighted and accepted the Cornell offer.

"A completely new world opened up" to him because Bethe took an interest in Marshak from the start:

> He was a friend all through my graduate student years . . . he sort of took a paternal interest in me and encouraged me to come around and talk to him about physics whenever I wished . . . He was not married at that time, and so during the two years I was at Cornell we went to lunch together on an almost daily basis. At lunch we would carry on scientific conversations, and we were joined by two mathematicians named Walker and Curtis. And then when Placzek came a year later, Placzek joined the group.
>
> [Bethe had] two post-doctoral fellows at that time, M. E. Rose and Emil Konopinski, who used to join the luncheons, too. They were . . . post-docs, but I was the only graduate student. There were several other graduate students studying theory with Bethe but he didn't invite them along. I guess we worked a little more intensely together.
>
> I guess I had the good fortune as a graduate student to be accepted at small faculty luncheons; and that, of course, opened up horizons. So in many ways, like in Grecian days, I was apprenticed to Socrates or something like that.
>
> I guess one reason Bethe accepted me was that he . . . knew little about philosophy or the arts or the humanities and, of course, American culture. Occasionally these matters were discussed, and I guess he found the conversations interesting. I think I gave him something in return, not much but a little.
>
> I even recall several instances where I took a girl out to dinner and would invite Bethe along, because I thought he would enjoy it. He was pretty shy as far as women were concerned, and maybe he also appreciated that. There is a ten-year age difference between us. I was, say, 21—and he was 31. He was still a pretty young man.
>
> I was at Cornell for only two years, because I was able to finish up with the help of a President White Fellowship during the second year.[30]

In their interview, Weiner asked Marshak to characterize Bethe's style. Marshak answered as follows: "I think that he [then had] a very logical mind and a very rigorous mind, although, in later years, he didn't want to have very much to do with the very formal rigor of modern theoretical physics . . . ;

and in trying to understand what you were telling him, he would make you sharpen up your statements." And he found Bethe very precise and very serious.

Energy Generation in Stars

> We do not argue with the critic who urges that the stars are not hot enough for this process [the formation of helium from hydrogen]; we tell him to go and find *a hotter place.*
>
> A. S. Eddington 1926, p. 301

Rational descriptions of stellar evolution and of cosmogenesis demand some notion of conservation laws, such as the conservation of energy, and an appreciation of the immense extent of time. Similarly, reasonably well-grounded *scientific* conjectures concerning the evolution of the cosmos require reliable observational data. Such conjectures could be advanced only after telescopes had become standard scientific instruments, and the data they collected standardized and widely disseminated. Furthermore, mathematical models of the entities the telescopes revealed and of their dynamics could be constructed only after Newton had laid the foundations of dynamics and formulated the law of universal gravitation.[31]

Natural explanations for the formation of our planetary system were advanced during the eighteenth century, and by the beginning of the nineteenth century the idea of the evolution of stellar systems and that of our own planetary system was familiar to astronomers by virtue of the nebular hypotheses that Immanuel Kant, William Herschel, and Pierre-Simon Laplace had advanced.[32] But evolutionary concerns came to occupy a more central and accepted position only after the publication of Charles Darwin's *Origin of Species.*

Once again it was empirical methods that transformed astronomy and made it possible to advance testable hypotheses regarding stellar and planetary evolution. In 1908 the American astronomer George Ellery Hale noted: "In astronomy, the introduction of physical methods has revolutionized the observatory, transforming it from a simple observing station to a laboratory, where the most diverse means are employed in the solution of cosmical problems."[33] What Hale was referring to was the introduction of spectroscopic methods in astronomy and the greatly increased power of telescopes through the use of photography.[34] The new data had revealed that stars differ greatly among themselves in some properties and little in others. Their luminosity—the rate at which they radiate energy—ranges roughly from one million times the sun's luminosity to a thousandth of it; their diameter from a thousand times that of the sun to one-fortieth of it; their masses from a hundred times that of the sun to one-tenth of it. On the other hand, the surface temperature of stars varies in a much narrower range: from ten times that of the sun to one-quarter of it. Although there were obvious instrumental limitations in detecting and measuring the properties of faint, of small, and of cool stars, there was no such limitation on the side of great brightness. There must therefore be a natural limit on the brightness attainable by a star in its steady state, which would explain why stars of very large radius are never found to be very hot.

The new physical methods also revealed that stars could be classified into different populations characterized by distinctive peculiarities in their luminosity, which marked "definite stages in an orderly process of development."[35] William Huggins had established the gaseous nature of the nebulae and had suggested that stars had their origin in these filmy masses of luminous gas, "taking form after long periods of condensation, through processes regarding which our ideas are still vague and ill defined."[36] The new techniques also validated the notion that "the materials of the sun, of the stars, and of the nebulae are essentially the elements of which our own earth is formed, and with which chemists had already become well acquainted."[37] The experiments that had established these facts—those of Joseph von Fraunhofer, Robert Bunsen, and Gustav Robert Kirchhoff—provided further evidence for believing that the physical laws that hold terrestrially are valid in the sun, the stars, and the nebulae. Their groundwork gave additional support to Newton's vision of the possibility of simple, universal laws of nature.

Just as Newton's *Principia* in 1686 marked a turning point in the history of the physical sciences, so did the formulation of the two laws of thermodynamics in the middle of the nineteenth century. Their impact on what became known as "astrophysics" later in the century was quick. In particular, the source of the sun's energy was immediately recognized as a critical problem after the formulation of the law of conservation of energy. Both William

Thomson and Hermann von Helmholtz, who played key roles in establishing the laws of thermodynamics, addressed the problem and concluded that gravitational contraction offered the only reasonable explanation. In a popular lecture delivered in Königsberg in 1854 on the occasion of the centennial anniversary of the publication of Kant's cosmogonic hypothesis, Helmholtz estimated that the sun could radiate at its present rate for millions of years if its diameter contracted some two hundred meters a year.

Some years later Thomson expanded on this suggestion and estimated that the sun's age was probably not more than 100 million years.[38] This timescale was much shorter than the age of the earth advanced by geologists and much shorter than the time Darwin required for evolution to produce the diversity of flora and fauna found on earth.[39] The conflict between physicists and geologists stayed unresolved for the rest of the nineteenth century. Physicists and astrophysicists generally continued believing that a slow, secular gravitational contraction was the explanation for the sun's enormous radiant output, and that the only sources of terrestrial energy responsible for the temperature distribution inside the earth were the internal heat that remained from its molten origin and the radiant energy it received from the sun.[40] Thus in 1892, Thomson, now Lord Kelvin, could still assert: "Within a finite period of time past the earth must have been, and within a finite period of time to come must again be, unfit for habitation of man as presently constituted, unless operations have been and are being performed which are impossible under the laws going on at present in the material world."[41]

The resolution of the dilemma came with discovery of radioactivity in the 1890s. Rutherford, the towering figure in unraveling "subatomic" radiation, convinced himself that the decay of radioactive elements, and of radium in particular, was an important source of the earth's internal heat, thus invalidating Thomson's assumptions and his conclusion regarding the age of the earth. By measuring the thorium-lead ratios in radioactive terrestrial rocks Rutherford determined the minimum age of the earth as between 1 and 2 billion years and inferred that the sun must have existed in essentially its present state for at least this length of time. The geological record thus disproved the Helmholtz-Kelvin contraction hypothesis for the sun and therefore presumably also for similar stars. Rutherford expounded these ideas in a Friday evening discourse at the Royal Institution in May 1904. He later gave the following account of his lecture:

> I came into the room, which was half dark, and presently spotted Lord Kelvin in the audience and realized that I was in trouble at the last part of my speech dealing with the age of the earth, where my views conflicted with his. To my relief Kelvin fell fast asleep but as I came to the important point, I saw the old bird sit up, open an eye and cock a baleful glance at me. Then a sudden inspiration came and I said Lord Kelvin had limited the age of the earth *provided* no new source was *discovered.* That prophetic utterance refers to what we are considering tonight, radium! Behold! the old boy beamed upon me.[42]

The much longer age of the earth and sun—a few billion years—which became accepted at the beginning of the twentieth century, created new problems. Gravitational contraction could no longer be invoked to explain the energy generation of the sun, but neither could radioactivity. On the one hand, spectroscopic evidence indicated that only very tiny amounts of uranium, thorium, and radium were present in the sun and hence could not account for the present rate of energy generation; on the other hand, and more important, radioactivity takes place at a rate that is not influenced by conditions external to the nucleus, such as the external temperature, and could therefore not explain the variation in observed luminosities. By the 1920s it was recognized that any physical mechanism that was to explain the present observed rate of energy production had also to meet the requirement that the supply of energy last for a time of the order of a few billion years.[43] As James Hopwood Jeans stressed, the problem of stellar energy is not only one of intensity but also one of timescale. By then, it was also generally appreciated that, in Arthur Eddington's words, "No source of energy is of any avail unless it liberates energy in the deep interior of the star."[44]

Special relativity—and Francis William Aston's precise measurements of atomic masses using his mass spectrograph—made possible a new explanation for the source of stellar energy: the transformation of matter into energy. The idea seems first to have been suggested by Jeans, and a summary of the conditions that such sources of energy must satisfy was given by Henry Norris Russell in 1919.[45] Eddington, too, became committed to a "subatomic" source for the energy generation in the stars.[46] The epigram quoted at the beginning of this section was first delivered by Eddington in an address to the British Association for the Advancement of Science at Cardiff in 1920. Incidentally, that same lecture contained the following prophetic statement: "If, indeed, the sub-atomic energy in the stars is being freely used to maintain their great furnaces, it seems to bring a little nearer to fulfillment our dream of controlling this latent power for the well being of the human race—or for its suicide." Eddington thoroughly reviewed the question of energy generation in stars in 1926 in chapter 11 of his monograph, *The Internal Constitution of the Stars.* He there noted that the formation of helium from hydrogen involved a loss of 0.8 percent of the mass and that this energy must have been set free in the reaction. He found this process attractive and a very probable source for energy production in stars, but found the process so mysterious that he mistrusted all predictions as to the conditions under which it occurs: "How the necessary materials of 4 mutually repelling protons and 2 electrons can be gathered together in one spot, baffles imagination. One cannot help thinking that this is one of the problems in which the macroscopic conception of space has ceased to be adequate, and that the material need not be at the same place (macroscopically regarded) though it is linked by a relation of proximity more fundamental than the spatial relation."[47] Recall that Eddington was writing in 1925, before the existence of the neutron was known and at a time when electrons were assumed to reside in the nucleus. These difficulties led him to take seriously his earlier proposal that energy generation in stars originated in electron-proton collisions in which both particles were "mutually cancelled," or in other words, annihilated: $p + e^- \rightarrow \gamma$-rays.

One consequence of this mechanism would be that stars would diminish in mass as time went on, with a timescale of the order of 10^{11} to 10^{12} years. This Eddington found very attractive, for he believed that the Hertzsprung-Russell (H-R) diagram represented the life history of a star. Electron-proton annihilation would thus explain the fundamental difference of stars descending the main sequence, namely, that they had decreasing mass. Eddington was primarily interested in the problem of energy generation in order to explain the evolution of stars rather than to shed light on their internal constitution. Even though he could not calculate the rate for the $p - e^-$ annihilation process, Eddington's stature was such that $p - e^-$ annihilation was taken seriously until the end of the decade and made fashionable a timescale of a trillion years. The situation changed in 1929 when Edwin Hubble reported evidence that most galaxies were receding from the Milky Way with velocities proportional to their distance, with the proportionality constant indicating a lifetime of the universe of the order of a few billion, not trillion, years. Therefore, stars were much younger than the annihilation process suggested.

At this very same time, a new avenue for the explanation of energy generation in stars opened up. On the heels of Gamow's theory of radioactive decay and his exploration of the inverse reaction, Robert d'Escourt Atkinson and Fritz Houtermans indicated that the central temperature of the sun was such that nuclear processes involving protons and light nuclei in which the light nuclei were transformed into heavier ones could occur in its interior, and that in such nuclear reactions energy would be liberated. However, the problem of energy generation in stars could not really be addressed until more empirical data had been obtained about nuclear structure and nuclear cross sections, and a better theoretical understanding of nuclear reactions and beta decay was available. In other words, the problem could be satisfactorily resolved only after the construction of the first "high energy" machines, such as Cockcroft and Walton's 500 keV accelerator and Lawrence's cyclotron, the discovery of the neutron by Chadwick in 1932, the formulation of a theory of beta decay by Fermi in 1933, and a quantum mechanical description of nuclear structure and nuclear reactions as given by Heisenberg, Bethe, Peierls, Bohr, and others.

On the other hand, the modeling of stellar constitution to explain the diversity and varied properties of stars had made important advances since the initial work of Homer Lane in 1869. The fact that the first mathematical model of stellar structure combined gravitation and thermodynamics is another illustration of how fruitful research in thermodynamics had been. Lane suggested that stars were gaseous spheres and inquired about the conditions for convective equilibrium. His aim was to investigate the internal temperature and pressure distribution in such models in order to see whether the observed surface conditions of the sun could be reproduced.[48] Others, especially August Ritter, developed the subject further, and the theory of such polytropic spheres was given its definitive formulation in Robert Emden's book, *Gaskugeln,* published in 1907.[49]

The *Gaskugel* model assumed that the stellar material is a perfect gas of uniform composition.[50] Thus at any point r the pressure is related to the density at that point by the gas law:

$$P(r) = \frac{K}{\mu}\rho(r)T(r) \tag{1}$$

where $P(r)$, $\rho(r)$, $T(r)$ are the pressure, density, and temperature at position r, K is the gas constant, and μ the mean molecular weight. The value of μ appropriate to the conditions of stellar interiors has an interesting history and will be discussed further. The conjecture that stars are in a state of convective equilibrium translated into the polytropic assumption that the P, T, and ρ relations were

$$(2) \qquad \frac{\rho}{\rho_c} = \left(\frac{T}{T_c}\right)^n \quad ; \quad \frac{P}{P_c} = \left(\frac{T}{T_c}\right)^{n+1} .$$

where the suffix c refers to the center of the sphere and n is given by $\gamma = 1 + 1/n$ where γ is the ratio of the specific heats of the gas. In addition to Equations (1) and (2), two differential equations connect P and ρ with $M(r)$, the mass enclosed in a sphere of radius r. These are

$$(3) \qquad \frac{dP(r)}{dr} = -\frac{GM(r)\rho(r)}{r^2},$$

which states that the internal pressure supports the weight of the gas above it (G is the gravitational constant, equal to 6.67×10^{-8} dyne cm^2/gr^2), and

$$(4) \qquad \frac{dM(r)}{dr} = 4\pi\rho(r)r^2,$$

which is the equation specifying the mass distribution: $dM(r)$ is the mass enclosed in a spherical shell of radius r and thickness dr; $dM(r) = 4\pi\rho(r)r^2dr$. However, by the end of the nineteenth century astronomical observations had made it clear that stars were much more complicated objects than what was being modeled by the Emden models.

During the first two decades of the twentieth century observational astronomy yielded a whole new set of data. The seminal representation of the new data for the problem of stellar structure was given by the Hertzsprung-Russell diagram, which exhibited the relation between the luminosity and the surface temperature of stars.[51] Initially the Russell magnitude-type diagram—the name by which the H-R diagram was first called—was thought to represent the life history of stars. Stars were thought to start out as cool red giants with low density and enormous bulk. As they contracted, in accordance with Lane's model, they got hotter; that is, they moved to the left on the H-R diagram to near the top of the main sequence—the region of the H-R diagram in which most of the stars are located. As they mature they move down the main sequence, becoming cooler and redder and finally ceasing to radiate.

By the 1920s, the advances in the theoretical understanding of atomic physics stemming from the old quantum theory found applications in astrophysics.[52] Eddington's discovery of the great importance of radiative equilibrium as compared to the former theory of purely convective equilibrium, both in the interior and in the outer layers of a star; Meghnad Saha's explanation of the spectral classes;[53] and Milne's analysis of the degree of ionization of atoms under stellar conditions were probably the most fundamental contributions. They indicated the transition to more realistic stellar models in which atomic physics had the major role. In fact, the research of Svein Rosseland, Russell, and others on ionization and opacity using the old quantum theory had yielded a fairly accurate knowledge of the physical properties of matter in the interior of a star.[54] The state of knowledge just prior to the introduction of the new quantum mechanics was masterfully and excitingly summarized in Eddington's classic *The Internal Constitution of the Stars.*[55]

Eddington had opened a new era in the modeling of stellar structure in 1917 when he noted that radiation pressure must play an important role in maintaining the equilibrium inside the star. He assumed that the energy transport over most of the star is by radiation and replaced Lane's hypothesis of convective equilibrium by radiative equilibrium. Radiation pressure is proportional to the fourth power of the temperature and amounts to some 25.5×10^6 atmospheres at 10×10^6 K, the central temperature of a typical Lane-Ritter model of a star. Eddington compared the outward flowing radiation "to a wind blowing through the star and helping to distend it against gravity."[56] Eddington still assumed that the perfect gas law described stellar matter, but this assumption was now more credible than in Ritter's days because inside stars the atoms were ionized, and high ionization makes the effective size of atoms very small. Eddington also assumed a constant opacity throughout the star, which he believed to be a convenient assumption to make in a first attack on the problem. In addition, he needed to stipulate the distribution of the sources of energy through the star in order to deduce the flux of energy at each point. This led Eddington to investigate various stellar "models," the most important of which were the "standard model" and the "point source model." In the standard model the sources are assumed to be distributed uniformly throughout (most of) the star; in the point source model essentially all the energy is produced near the center of the star.

Eddington was able to account for the fact that there is a close relation between the mass of a star and its luminosity. It had been observationally established that over almost all the available range the total heat radiation of a star is nearly proportional to the fourth power of its mass, and that for a given mass, the luminosity essentially does not depend on the radius of the star. Eddington's explanation of this marked the greatest success of modern atomic physics in interpreting the stars.

There was, however, one problem with the model. The virial theorem allows an estimate to be made of the central temperature of a star, or equivalently of the thermal energy per particle at the center of the star, and yields the result that $kT_c = \alpha\mu\ GHM/R$ where α is constant, depending on the specific model of the star, but is usually of the order of 1 for main sequence stars, and H is the mass of the hydrogen atom.[57] For the sun Eddington arrived at a value of about 40 million degrees (when $\mu = 2.1$). If the theoretical value of the opacity that Hans Kramers had derived, $\kappa = \kappa_0\rho\ T^{-7/2}$, is used, one obtains a contradiction with the equation of radiation flow. Or alternatively stated, if one accepted the 40 million degree central temperature, there was an order of magnitude discrepancy between the observed and the Kramers prediction for the opacity of stellar matter. Eddington had noted that the discrepancy could be removed if one assumed a much higher hydrogen abundance in the stars, but in *The Internal Constitution of the Stars* he rejected that possibility. By the end of the decade, however, the analysis of the line shapes in solar and stellar spectra by Albrecht Unsöld and Russell indicated that hydrogen was extremely abundant in the atmosphere of the sun and other stars. Unsöld had used the observed line shapes of resonance lines to determine the number of atoms involved in the production of the line and had ascertained that hydrogen was a million times more abundant than the metals in the solar atmosphere. Russell in 1929 had determined quantitatively the relative abundance of fifty-six elements in the solar atmosphere and uncovered the "almost incredibly great" cosmic abundance of hydrogen. His paper established the fact that the material of the universe is a mixture of rather pure hydrogen and helium, with all other elements present as slight impurities.[58] In the early thirties Eddington and Strömgren independently produced strong arguments that hydrogen was also abundant in stellar interiors. This settled the matter, and thereafter astrophysicists regarded hydrogen as the predominant constituent not only *on* but *in* the stars obeying the mass-luminosity relation.[59] Belief in the universality of the observed abundance of elements—that it was true of all stars—introduced another problem, however.

The stage for a successful attack on the problem of energy generation in stars was set during the first half of the 1930s. It required the question of the abundance of hydrogen in stars to be settled. It also required the new insights that quantum mechanics made possible into the dynamics of collisions in general, and of nuclear reactions in particular. George Gamow was an important contributor to these developments, and it is to his work that I turn next.

George Gamow

George Gamow's life and work present an interesting contrast to Bethe's, and their intellectual biographies parallel each other in many ways. During the 1930s their interests overlapped. Gamow's explanation of α-particle decay in 1928 was the first application of wave mechanics to nuclear physics, and it suggested that nuclear physics might also be understood in terms of the new quantum mechanics. In 1930, he wrote the first monograph on nuclear physics that attempted to account for the properties of nuclei using quantum mechanics.[60] Starting in 1928 Gamow began applying the lessons being learned in nuclear physics to astrophysical problems. After the discovery of the neutron and subsequently of artificial radioactivity, he was one of the first to address the problem of nucleosynthesis in stars. A man of wide interests and great imagination, he seems to have been more interested in addressing "big questions" and offering plausible answers than to carefully working out the details of his suggested answers. The big questions ranged from the birth and death of the sun, the genesis of the chemical elements, and the origin of the planetary system to cosmogenesis and the alphabet of the genetic code.

In 1937 he was ideally positioned to solve the problem of energy production in stars. He recognized the interrelation of nucleosynthesis and energy production and, together with Edward Teller, his colleague at George Washington University, fashioned the tools to solve the problem. Perhaps because of his fascination with problems of origins and genesis he came to regard nucleosynthesis as *the* all-important problem, and the explanation of the relative abundances of the elements the criterion by which the theory would be tested. He was unable to see that energy generation and nucleosynthesis need not be addressed simultaneously. The pragmatic Bethe—always the theoretician who based his work on firm empirical data and sound phenomenological knowledge—decoupled the two aspects of the problem and was thus able to give the definitive answer to the problem of the energy generation in stars.

George Gamow was born in Odessa on March 4, 1904.[61] His father was a teacher of Russian literature at one of Odessa's private *lycées* for boys.[62] His mother's family stemmed from a long line of Ukrainian clergymen. She died when Gamow was nine. As a boy Gamow was precocious and talented in many different areas. At age seven his mother was reading him Jules Verne, and he dreamt of going to the moon. He went to high school during the 1917 revolution and the subsequent civil war. For a while he thought of being a mathematician after he had become fascinated by number theory, topology, and set theory. But physics won out. He enrolled in the physico-mathematical faculty of Novorossiya University of Odessa in 1922 but left after a year to go to Petrograd (later called Leningrad). Petrograd was the scientific capital of the Soviet Union, and the Academy of Sciences and its main academic institutes were located there. The physical sciences, and physics in particular, were being nurtured in Petrograd and had recovered from the upheaval of the revolution. Petrograd had two institutions of higher education, the Polytechnical Institute and the university, and both offered a sound training in physics. Gamow enrolled at the university. He there came under the influence of Alexander Alexandrovich Friedmann.[63] Friedmann was trained as a mathematician and was deeply interested in physical problems. In 1922, after studying general relativity, he discovered a mistake in Einstein's treatment of the cosmological problem and indicated that the equations of general relativity admit nonstatic solutions corresponding to both open and closed universes. In 1923–24 Friedmann taught a course at the university on the mathematical foundations of Einstein's theory of general relativity that Gamow attended. This was Gamow's introduction to both general relativity and to the concept of an expanding universe. Fellow students in the class were Lev (Dau) Landau and Dmitry (Dim) Ivanenko, and the three of them became inseparable friends. This brilliant trio became known as the "three musketeers." Genia Kannegiesser, who later became Rudolf Peierls's wife, was another member of this social circle.

Gamow's talents were recognized by his teachers, and in 1928 he was sent to study with Bohr at his Theoretical Physics Institute at the University of Copenhagen. The fellowship gave him the opportunity to visit the centers of theoretical physics in Germany and England. Léon Rosenfeld, a close associate of Niels Bohr who became a distinguished nuclear physicist, recalled that his experience with nuclear physics started in the summer of 1928 with the sudden appearance one morning of Gamow in the library of the physics institute at Göttingen. Rosenfeld described Gamow as "a fair-headed giant with shortsighted, half-shut eyes behind his spectacles."[64] Gamow explained to Rosenfeld that he was working on the quantum mechanical description of α-radioactivity in heavy nuclei. At this time essentially all that was known about nuclei was that they were very small, had an electric charge, and had a spin. It had become clear to Gamow that one must take into account the fact that the α-particle was quasi-bound inside the nucleus before it came out. Thus the α-particle must experience an attractive potential inside the nucleus, which combined with the repulsive potential outside to form a "potential barrier" through which it could "tunnel." The decay of radioactive elements is a purely quantum mechanical process in which the α-particle "leaks through" the nuclear potential. Gamow constructed a solution of the Schrödinger equation for this type of potential, which satisfied the boundary condition that there was an outgoing wave at large distances (corresponding to the escaping α-particle). The quantum mechanical formula he obtained for the "transparency" of the nuclear walls allowed him to derive the Geiger-Nuttall relation between the lifetime of the nucleus and the energy of the emitted α-particle. Gamow's calculation was the first application of quantum mechanics to nuclear physics.[65]

That summer Gamow met Fritz Houtermans in Göttingen.[66] Houtermans and Gamow were kindred spirits: cosmopolitan, bon vivants, somewhat reckless, carriers of a vast store of jokes they loved to tell, and displaying at times curious manners. They became very good friends. Hendrik Casimir noted in his autobiography that Houtermans "was so colorful a person that one is liable to forget that he was also a very good physicist."[67] Houtermans had obtained his doctorate working with James Franck on resonance fluorescence in mercury. After he heard Gamow expound his theory of nuclear α-decay he discussed with Gamow refinements of the theory and its application to specific nuclei. A joint paper ensued that was submitted for publication to the *Zeitschrift für Physik* in October 1928, by which time Gamow had gone to Copenhagen to visit Bohr and his institute, and Houtermans had moved on to Berlin. Houtermans not only had fully mastered Gamow's theoretical description of α-decay, but also realized that the quantum mechanical tunneling effect made possible the inverse process in which a nucleus absorbs an α-particle—and subsequently gives rise to various reaction products. All nuclei are positively charged and, in order to react, they must penetrate their mutual Coulomb barrier. Gamow's theory indicated that in general the effective cross section for nuclear reactions is given essentially by the product of:

1) the collision cross section. Houtermans took it to be equal to $\pi r_0^{\,2}$, where r_0 is the nuclear radius. Wave mechanically, using plane waves

for the particles, this cross section for head-on collisions is $\lambda^2/4\pi$ where λ is the de Broglie wavelength of the incident particle with respect to the center of mass of the combined system.

2) the transparency of the Coulomb barrier for the colliding particles. If two nuclei of charge Z_1e and Z_2e and masses M_1 and M_2 collide with a kinetic energy of relative motion E, the probability that the nuclei will penetrate each other is given by

$$W = \exp\left\{-2\left(\frac{2M_1M_2}{M_1+M_2}\right)^{1/2}\frac{\pi^2 Z_1Z_2e^2}{hE^{1/2}}\right\}.$$

3) the probability of a disintegration when the nuclei are together.

On the assumption that the value of the third item is essentially 1, and that the collision cross section is of the order of $\pi r_0^{\ 2}$, Houtermans determined that the cross sections for such reactions were too small to be easily experimentally detected at the kilovolt energy range, which was the magnitude of the energy then available in laboratory experiments.[68] In discussions with Atkinson, a young English astronomer who had attended Eddington's lectures on the internal constitution of stars in Cambridge and who was visiting Göttingen in the summer of 1928, it became clear that stellar interiors were hot enough for *sustained* nuclear reactions initiated by protons to take place. Atkinson and Houtermans realized that in the calculation of the transparency of the nuclear barriers for the thermal protons in the interior of a star one should not take for the proton energy the mean energy 3/2 kT, since according to the Maxwellian distribution there were always present fast protons in small number, and these are very much more efficient in their disintegration effect. With some help from Gamow, Atkinson and Houtermans wrote a joint paper on "thermonuclear" reactions in which they obtained an expression for the dependence of the reaction rate on the temperature of the mixture and on the atomic numbers of the colliding nuclei. In their paper Atkinson and Houtermans gave a schematic representation of the number of particles $N(E)$ and of the cross section for disintegration $\sigma(E)$ as functions of the energy of the particles in order to highlight the fact that the total number of disintegrations, which is essentially given by the product $N(E)\sigma(E)$, reaches a maximum for an energy well above the average energy corresponding to a given temperature. In order to calculate the total number of disintegrations one has to integrate $N(E)\sigma(E)$ over all energies from 0 to ∞; it is clear that most of the contributions to the integral will come from particles with the optimum energy. They then computed the probability per unit time of nuclear reactions between nuclei of charge Z and mass A and a proton gas with a density of 10^{23} particles/cm^3 and Maxwellian velocity distribution corresponding to a temperature of 6×10^7 K. They estimated the periods of time necessary for 50 percent transformation of different substances mixed up with hydrogen under conditions holding inside the sun on the assumption that each penetration gives rise to disintegration.[69] Their calculations indicated that for heavy elements the probability of processes of this type was exceedingly small, but for light elements—assuming a collision radius of 4×10^{-13} cm—they obtained half lives ranging from 8 seconds in ^{4}He, to 34 minutes for Li, 14 years for B, 10^9 years for ^{20}Ne, and 10^{61} years for Pb.[70]

Writing at a time when the only information available was based on α-particle bombardment—no accelerator existed and thus no experimental data on proton reactions was available—they concluded that the only reactions suitable for explaining energy production in the sun are those between protons and the nuclei of some light elements between lithium and neon in the periodic table. They suggested that the process was a cyclic one in which four protons are captured consecutively by a nucleus and then ejected in the form of an α-particle. The tentative title of the paper Atkinson and Houtermans submitted to the *Zeitschrift für Physik* was "Wie kann man ein Helium Kern in einem potential Topf kochen?" ("How can one cook a helium nucleus in a potential pot?")[71] Houtermans later recalled, "That evening, after we finished our essay, I went for a walk with a pretty girl. As soon as it grew dark the stars came out, one after another, in all their splendour. 'Don't they shine beautifully?' cried my companion. But I simply stuck out my chest and said proudly: 'I've known since yesterday why it is that they shine.'"

Later that year, after his discussions with Houtermans on nuclear reactions, Gamow visited the Cavendish Laboratory and expounded his findings. He stressed the fact that from his theory the transparency of nuclear barriers is much larger if the bombarding particles are protons. Due to the smaller charge and mass of a proton, the probability of penetration of a proton into a given nucleus is the same as that of an α-particle with sixteen times larger energy. Since Rutherford had observed nuclear transformations only for α-particles of several million electron volts one could conclude that in the case of proton bombardment one could expect observable results for energies of a few hundred kilovolts. These considerations played a crucial role in the decision to have Cockcroft and Walton build at the Cavendish a high voltage apparatus that could produce a 500 keV beam of protons. Bombarding a target of lithium with these protons, Cockcroft was able to detect the reaction

$$p + {}_3Li^7 \rightarrow {}_2He^4 + {}_2He^4$$

with the liberation of 18 MeV, a large amount of energy.

The discovery of the neutron in 1932 by Chadwick and the subsequent systematic analysis of neutron-induced nuclear reactions by Fermi initiated a new stage in the study of the "thermal transformations of elements in stars." Henceforth the subject matter of nucleosynthesis could be addressed empirically. Gamow became fascinated by the subject. At a symposium, "The Nucleus of the Atom and Its Structure," in 1935 he gave a talk entitled "Nuclear transformation and the origin of the chemical elements," and in the lecture he noted "that there are many different phenomena inside a star giving rise to formation and transformation of different elements and the hope may be justified that further investigation will clarify the relative importance of various processes and lead to a complete explanation of the relative abundance of different elements in the universe."[72]

Nor was Gamow alone in his interest in this problem. In the fall of 1936, Rutherford delivered the Henry Sidgwick Memorial Lecture, in which he reviewed the developments in the physicists' understanding of the structure of nuclei since the beginning of the century and described the important new findings since the discovery of the neutron by Chadwick in 1932.[73] He indicated that during "the last few years . . . almost all elements had been shown to be capable of transformation by suitable agency," in particular by bombardment by fast neutrons, and concluded his lectures with the observation that:

> The information we have gained on transformation processes may prove of great service too in another direction. In the interior of a hot star like our sun, where the temperature is very high, it is clear that the protons, neutrons, and other light particles must have thermal velocities sufficiently high to produce transformation in the material of the sun. Under this unceasing bombardment, there must be a continuous process of building up new atoms and of disintegrating others, and a stage at any rate of temporary equilibrium would soon be reached. From a knowledge of the abundance of our earth, we are able to form a good idea of the average constitution of the sun

> at the time 3000 million years ago when the earth separated from sun. When our knowledge of transformations is more advanced, we may be able to understand the reason of the relative abundance of different elements in our earth . . . We thus see how the progress of modern alchemy will not only add greatly to our knowledge of the elements, but also of their relative abundance in our universe.[74]

What Rutherford had in mind was not merely the relative abundance of the elements, but also the relative abundance of the isotopes of a given element, that is of nuclei with the same number of protons but differing number of neutrons.[75] Isotopes behave alike in the chemical and the physical processes responsible for geological change. Only nuclear processes differentiate isotopic differences. Hence isotopes reveal the nuclear history of the cosmos.

In 1936, while writing the second edition of his monograph on nuclear physics, Gamow reviewed the data on the relative abundances of the elements that Aston and others had obtained and was struck by "the remarkable constancy in isotopic constitution of the elements found on the earth [and in the universe]." It was generally believed that the data represented a "universal" or "cosmic" abundance. For Gamow, explaining theoretically this uniformity of the relative abundance of the different nuclei in the universe became *the* interesting puzzle, and his investigations of "the creation and transformation of elements in different part of the universe" had as their objective finding a solution to that problem. Assuming that protons, neutrons, and electrons were the fundamental building blocks, he suggested that "by some process taking place inside a star . . . a considerable amount of neutrons can be produced." When these neutrons were slowed down they would be captured by nuclei, especially heavy nuclei, in reactions such as

$$_{Z}X^{A} + {}_{0}n^{1} \rightarrow {}_{Z}X^{A+1} + h\nu,$$

forming heavier isotopes, which by the process of β-decay,

$$_{Z}X^{A+1} \rightarrow {}_{Z+1}X^{A+1} + e^{-} + \text{neutrino},$$

may give rise to elements located higher in the periodic table. He believed that such reactions might be effective throughout the whole range of elements and could lead to an understanding of the relative amounts of the various elements in the universe. He concluded his section with this statement: "We may also notice at this point that reaction-chains of the [above] types, corresponding to the building up of heavy nuclei, must in general involve the emission of large amounts of energy—probably enough, in fact to provide a source for the radiation of the stars, which for a long time presented an inexplicable puzzle to astrophysicists."[76]

Thus, explaining energy production was an addendum to explaining nucleosynthesis. In 1937, Gamow gave to his graduate student Charles Critchfield the problem of calculating the cross section for the reaction $p + p \rightarrow d + e^+ + \nu$ under stellar conditions in order to see whether the cyclic reactions that Carl Friedrich von Weizsäcker had proposed in 1937—in which reaction $p + p \rightarrow d + e^+ + \nu$ is the first step in the conversion of four protons into an α-particle—could account for the luminosity of the sun. But there was another reason for Gamow to investigate the reaction, which he divulged in the lectures he delivered at the Institut Henri Poincaré during the summer of 1938 after Bethe had proposed the carbon cycle: "The energy liberated suffices to explain the radiation from stars, but it does not produce free neutrons. If we want to explain the formation of the heavy elements, we must envisage another reaction that produces neutrons in great quantity."[77]

That is why the reaction $p + p \rightarrow d + e^+ + \nu$ was so important to Gamow. Once deuterons are produced they can react as $d + d \rightarrow {}_2He^3 + n$ to copiously produce neutrons, which in turn helps generate the heavy elements through neutron-capture reactions.

The 1938 Washington Conference of Theoretical Physics

The annual Washington Conferences on Theoretical Physics had been initiated in 1935. They were sponsored by the Carnegie Institution and George Washington University, and their intellectual agenda was set by Gamow and by Teller. Bethe had attended the 1935, 1936, and 1937 conferences, but when invited to the 1938 meeting he first declined because he was not interested in the topic for that year, the problem of stellar energy generation. The topic had been chosen by Gamow. It was only after Teller's repeated urgings that Bethe agreed to come to the conference.

The conference opened on March 17, 1938. Some thirty people had been invited, among them the physicists Bethe (Cornell), Gregory Breit (University of Wisconsin), Merle Tuve, and Lawrence Hafstad (Carnegie Institution of Washington); the astronomers Bengt Strömgren, Subrahmanyan Chandrasekhar (Yerkes Observatory, University of Chicago), Donald Menzel, and Theodore Sterne (Harvard University); and John von Neumann (Princeton Institute for Advanced Study).[78]

According to Chandrasekhar's recollection in the opening session Gamow gave a talk on the theory of nuclear reactions and presented the content of the paper he and Teller had just written on how to calculate the cross section for nuclear reactions in stellar interiors.[79] Chandrasekhar then gave a talk on white dwarfs. The most extensive lecture during that session and the next was given by the Danish astrophysicist Bengt Strömgren, who had just written a seventy-page paper that critically reviewed all that was known about stellar structure and the evolution of stars.[80] In his presentation Strömgren focused on the problem of the temperature and density distribution in the interior of stable stars, for which there is an equilibrium between the amount of energy generated in the star and the amount radiated. He indicated that spectroscopic data indicated that the most reasonable model for the sun was for hydrogen to be the prevalent element in its composition. He noted that a model of the sun with a central temperature of 19 million degrees, a central density of 76 g/cm^3, and a hydrogen content of 35 percent by weight would result in its generating 2 erg/g sec. The challenge he posed to the physicists in the audience was to find the thermonuclear reactions that would give rise to the observed luminosities of the sun and other main sequence stars. In the subsequent discussion Bethe was critical of any theory, such as Gamow's and von Weizsäcker's, both published in 1937, that proposed a chain of nuclear reactions that would simultaneously generate energy and account for the building up of heavy elements in stars; this because of the instability of He^5 and of Be^6.[81] Thus there was no obvious way to create elements heavier than He. It seemed more likely to Bethe, given the great abundance of hydrogen in the sun, that von Weizsäcker's suggestion that the reaction in which two protons fused to form a deuteron,

$$p + p \rightarrow D + e^+ + \nu, \quad (1)$$

which was the first step in proton reactions that led to the formation of helium, was a more likely source of stellar energy. Moreover, since the reaction proceeds via the β-decay interaction, the timescale of the reaction inside the sun would be of the order of 10^{10} years.

Upon hearing of Critchfield's interest in the reaction, Bethe proposed that they collaborate. Before the conference's end Bethe was able to report that the reaction (1) together with the chain of reactions

(2) $$D + p \rightarrow He^3 + \gamma,$$

(3) $$He^3 + He^3 \rightarrow He^4 + p + p,$$

(4) $$He^3 + He^4 \rightarrow Be^7 + \gamma,$$

(5) $$Be^7 \rightarrow Li^7 + + e^+ + \nu,$$

(6) $$Li^7 + p \rightarrow 2\, He^4$$

could account for energy production in the sun. Effectively, the chain of reactions combined four protons into one α-particle and released a large amount of energy.

Bethe explained the details of the reactions: The neutrino produced in the pp reaction (1) would escape from the star because it interacts so weakly with matter. The positron produced in the reaction would soon collide with an electron, and both would be annihilated, producing photons. The deuterons that are produced in the pp reaction (1) disappear very rapidly as they interact with protons to produce He^3 and more radiation. Thus there are no deuterons around to have collisions with He^3. The He^3 produced in reaction (2) only undergoes elastic collisions with protons. As a result the He^3 builds up in abundance until pairs of He^3 interact to form He^4 and also make possible the reactions (4)–(6).

Bethe asserted that the pp cycle would account for the energy production of the sun, but not of heavier bright stars. Eddington earlier had shown that the central temperature of a star increases but slowly with the mass of the star. Astronomical observations had corroborated this, but had also indicated that the amount of observed radiation—their luminosity—increases very rapidly with increasing mass. The pp reaction, which has a moderate T^4 dependence, could therefore not account for the luminosity of stars heavier than the sun. Processes with a much stronger temperature (T) dependence were required. Since the T dependence is determined by the Coulomb barrier Bethe inferred the necessity of involving heavier nuclei in the energy-generating processes. At the talk in which he announced these results Bethe also reported that he had discovered another cyclic process that converted four protons into an α-particle: namely, his carbon cycle.

In his interview with Spencer Weart in 1977, Chandrasekhar described these events as follows:

> CHANDRASEKHAR: Bethe got very much interested in the nuclear aspects, and Bethe one morning said that he had thought about Weizsäcker's cycles and that he had alternative ideas, and he gave a talk right at the meeting on his carbon cycle.
>
> WEART: I see—was that the carbon cycle then already?
>
> CHANDRASEKHAR: That's right, he gave the carbon cycle, exactly as he formulated it later when he wrote the paper.
>
> WEART: I see. Because in the paper you published with Gamow and Tuve on the conference, you don't mention the carbon cycle specifically. You simply talk about the idea of helium, and two protons getting together, and the possibility of the proton-proton cycle—
>
> CHANDRASEKHAR: That was also mentioned by Bethe, you see.
>
> WEART: Oh, that was Bethe who mentioned the proton-proton cycle?
>
> CHANDRASEKHAR: Critchfield was there, and then of course, there was the question of the Gamow-Teller selection rules, or the alternative selection rules. Bethe proposed that hypothesis, then.
>
> WEART: —and the possibility of resonance levels was brought up—even something of the idea of shell burning—
>
> CHANDRASEKHAR: Yes.
>
> WEART: Were some of these ideas new to you, this nuclear physics? Had you had much contact with nuclear physics before?
>
> CHANDRASEKHAR: It wasn't new to me in the sense that I had not read about these things. In fact, I gave at Yerkes that year a course on nuclear physics, based upon Rasetti's book which had just come out. I tried to keep abreast of what was going on at that time.[82]

After the Washington Conference Chandrasekhar, Gamow, and Tuve had issued a brief report on the meeting in *Nature*.[83] In it they reported that "some interesting conclusions had been reached" concerning nuclear transformations as the source of energy production in stars. They noted that the so-called *Aufbauhypothese,* the suggestion that the production of the heavier elements from hydrogen is continually taking place in the interiors of stars and that in the process energy is liberated, had run into difficulty. Recent experiments had indicated that ^{5}Li and ^{5}He were both unstable. Thus the reaction $^{1}H + {}^{4}He \rightarrow {}^{5}Li + {}^{5}He$ could not be invoked, which in turn invalidated von Weizsäcker's proposal for the buildup of the heavier elements. Similarly, the other possible chain of reactions for the synthesis of heavier elements from hydrogen would require the stability of ^{8}Be, but recent experimental evidence indicated that this was not the case.

Chandrasekhar, Gamow, and Tuve reported that therefore "the only course not excluded by present evidence on the binding energies of the light nuclei is the formation of ^{6}Be in triple collisions involving an α-particle and two protons," a possibility that Bethe had suggested at the conference. "Although the existence of ^{6}Be had not been established experimentally," they reported, "there are some indications that it might be a stable nucleus." Evidently Bethe had also suggested triple α-particle collisions to form C^{12}. Chandrasekhar, Gamow, and Tuve had estimated the probability of such reactions involving triple collisions and had concluded that "under the conditions in the stellar interiors, the rate of liberation of energy will be sufficient to account for the radiation of the stars."

They reported "another possibility," the reaction $^{1}H + {}^{1}H \rightarrow {}^{2}H + \beta^{+} + \nu$, which had also been proposed, and they commented, "It seems that the rate of such a reaction under the conditions in stellar interiors would be just enough to account for the radiation of the sun, though for stars much brighter than the sun other more effective sources of energy are required." They corroborated that by the end of the conference Bethe and Critchfield had reported on the calculations the two of them had made. In addition, Chandrasekhar, Gamow, and Tuve mentioned that the consequences of resonance effects in nuclear reactions for stellar models had been discussed at the conference.

The *Nature* report indicates that in the minds of most of the participants the problem of nucleosynthesis was conflated with the problem of energy generation. This was certainly the case for Gamow and von Weizsäcker. Bethe separated the two problems. They would later once again be brought together, but at that stage the separation of the two problems was an important step for the resolution of both.

Upon his return to Cornell from the conference Bethe started investigating quantitatively reactions involving heavier nuclei that would explain energy production in massive stars. Lithium, beryllium, and boron could be ruled out because of their comparative scarcity in stellar interiors. The next element was carbon. The detailed investigation of the cyclic reaction of carbon with protons, which he had put forth at the Washington Conference, yielded a positive result:

$$C^{12} + p \rightarrow Ni^{13} + \gamma, \tag{7}$$

$$Ni^{13} \rightarrow C^{13} + e^{+} + \nu, \tag{8}$$

$$C^{13} + p \rightarrow Ni^{14} + \gamma, \tag{9}$$

$$Ni^{14} + p \rightarrow O^{15} + \gamma, \tag{10}$$

$$O^{15} \rightarrow Ni^{15} + e^{+} + \nu, \tag{11}$$

$$Ni^{15} + p \rightarrow C^{12} + He^{4}. \tag{12}$$

At the end of the cycle the C^{12} nucleus is recovered and four protons have been combined into an α-particle. The C^{12} nucleus thus acts as a catalyst for the reaction and hence the relatively low abundance of carbon nuclei could still allow the reaction to proceed frequently. Under the same condition as the rates calculated for the pp cycle, and with a concentration of N^{14} of 10 percent, Bethe calculated an energy production of about 25 ergs/g sec.[84] The reaction has a temperature dependence of T^{18}, and thus accounted for the sharp increase in luminosity with slight increases in core temperature. Bethe could exclude on various grounds almost all other reactions besides the pp and carbon-nitrogen cycles by detailed investigations of their properties.[85] He was thus able to explain why both stars like the sun and heavier ones burn for billions of years at the rate that they do.

Karl Hufbauer noted that Bethe was so busy during the month following the Washington Conference that he could not have given more than intermittent attention to the stellar energy problem.

> In addition to a short spring vacation and his teaching, he made summer plans for participating in the University of Michigan's Symposium on Theoretical Physics and for seeing friends and family in Europe, hosted visits by colleagues on two successive weekends, sent a note on Li's magnetic moment to *Physical Review,* and prepared a paper on nuclear forces for the Washington meeting of the *American Physical Society* . . . [But] by the last week of April, if not before, he was arranging to give colloquia on "Energy

> Production in Stars" to the physics departments at Cornell and Harvard on 2 and 16 May respectively. Moreover, while at the Washington APS meeting of 28–30 April, he began telling people about the results of his inquiry. He took the opportunity afforded by Critchfield's delivery of some calculations on the proton-proton reaction to present his latest ideas about the energy-generating reactions . . . Bethe even told Laurence of the *New York Times* about his new interest in the stellar-energy problem during an interview about his APS paper on nuclear forces.[86]

By mid-May Bethe had written up the paper he was going to publish with Critchfield on the pp cycle and wrote Critchfield,

> In writing up the paper, I found that it would be rather inhomogeneous if the proton process and the discussion of the other reactions in stars were put into the same paper. The proton calculations are very explicit and quantitative, the rest is only qualitative and partly speculative. It would be much easier to write it as two separate papers. In this case, I think I should not make you share the responsibility for my speculations about the other processes, so I would suggest publishing part II alone whereas you would, of course, have to be the senior author on the proton paper. Is this alright with you? Please let me know soon because the writing up of the papers depends on your decision. Perhaps you [should] discuss this question with Teller, I am sending him a copy of this letter.

Critchfield acquiesced but insisted on having "Bethe and Critchfield" as the order of the names of the authors of the paper.

In mid-July Bethe gave his manuscript on the carbon-nitrogen-oxygen (CNO) cycle to his postdoctoral fellow Emil Konopinski for a critical review of its calculations and asked him to oversee its typing as he was leaving for Europe for his annual visit to see his mother and his father. He first went to England, where he saw Rudi Peierls, who asked him to visit his parents in Berlin and try to persuade them to leave Germany. During Bethe's brief stay in Germany he saw his mother in Baden-Baden and did go to Berlin, but failed to convince Rudi's parents to leave.[87] In early September, while Bethe and his mother were still vacationing in the Swiss Alps, his paper on the CNO cycle was mailed to the *Physical Review.* Hufbauer observed that

> Its length and sweep seem to have dismayed John Tate, the editor of the *Physical Review,* for he wrote a month later to suggest that the *Reviews of Modern Physics* would be a more appropriate venue. Bethe, who was back

> teaching at Cornell by this time, invoked the paper's high originality quotient to defend its appearance [in] *Physical Review.* Still, having recently learned from his student Robert Marshak that the New York Academy of Sciences was offering a $500 prize for the best unpublished study of "solar and stellar energy,"[88] he welcomed the opportunity that Tate's letter gave him to request his paper's return for the interim. As things worked out, Bethe secured his typed copy, won the prize in mid-December, got out a one-page summary in the *Physical Review*'s issue for 1 January 1939, published the full paper with several minor revisions in the issue for 1 March, and used $250 of the prize money to liberate his mother's furniture from the Nazi authorities later that spring.[89]

The conclusions of the paper were rapidly accepted by the astrophysical community, helped greatly by Russell's strong support.

In 1967 Bethe was awarded the Nobel Prize for this work. Oskar Klein, in the presentation of the Nobel award to Bethe, stressed that

> Even when Bethe started his work on the energy generation in stars there were important gaps in the knowledge about nuclei which made the solution very difficult. And it is by a remarkable combination of underdeveloped theory and incomplete experimental evidence, under repeated comparison of his conclusions with their astronomical consequences, that he succeeded in establishing the mechanism of energy generation in the sun and similar stars so well that only minor corrections were needed when many years later the required experimental knowledge had made considerable progress and when, moreover, electronic computers had become available for the numerical work.[90]

Marshak, Revisited

After Bethe came back from the Washington Conference and was working out the quantitative features of the carbon cycle, he asked Marshak to do some of the calculations for him. Bethe acknowledged his help in his *Physical Review* article. He also asked Marshak to investigate what was known about the quantitative aspects of stellar evolution and to give a seminar on it. In preparing for his talk Marshak became aware of the existence of white dwarfs, stellar objects much denser than the sun. After his talk, Marshak suggested to Bethe that a possible thesis problem would be an investigation of the source of energy generation in white dwarfs and the mechanism by which the energy is transported to the surface of the star given its much greater density. Bethe agreed.

Marshak was awarded a White fellowship for the academic year 1938–39, which freed him from any teaching responsibilities, so he was able to complete his research and write his doctoral thesis by the spring of 1939. Its first part consisted in further explorations of the proton-proton reactions and of those of the carbon cycle when applied to main sequence stars. The second and "most interesting part" of his thesis dealt with white dwarfs.

Chandrasekhar had earlier studied these stars but had only considered the case of a white dwarf at zero temperature. Under these circumstances the pressure of the degenerate electron gas is responsible for the stability, and Chandrasekhar had computed what mass could be sustained for a given central density; or conversely, if the mass of the star is given, he could derive the radius and the central density. There were actually two parts to Chandrasekhar's calculations. One calculation yielded the radius as a function of mass, and Chandrasekhar proved that there was a critical mass above which the white dwarf would not be stable; this limit is called the *Chandrasekhar limit.* This limiting mass depends critically on the amount of hydrogen in the star. If only hydrogen is present, the critical mass is 5.6 solar masses; if no hydrogen is present, the critical mass is 1.4 solar masses.

In a second calculation Chandrasekhar predicted the radius of the white dwarf Sirius B as a function of its hydrogen content, assuming a completely degenerate electron gas (in other words, zero temperature). But in order to know the hydrogen content, he had to know the internal temperature, because the cross section for the proton-proton reaction that Bethe and Critchfield had calculated depended sensitively on this temperature. But Chandrasekhar had not calculated how a finite temperature inside a white dwarf star would change the dependence of the radius on the mass.

In his thesis Marshak examined the conditions inside white dwarf stars in much greater detail than Chandrasekhar had. He noted that the mechanism responsible for the opacity of a white dwarf was different than in main sequence stars: rather than radiative processes being responsible, in white dwarfs the electrons transferred energy through elastic collisions with nuclei.[91] Having obtained a value for the opacity, Marshak could then derive the internal temperature of a white dwarf like Sirius B. He obtained a value of ten million degrees for the central temperature of Sirius B (whose mass had been determined empirically). But under those circumstances the proton-proton reaction is so copious that it yields too much luminosity for Sirius B unless its hydrogen content is essentially zero. With the knowledge of the central temperature, and that the amount of hydrogen present was very small, Marshak could calculate the effect of a finite temperature on Chandrasekhar's prediction of the radius from the observed mass and make a precise prediction for the radius of Sirius B. Measurements in the 1970s confirmed Marshak's prediction.

Marshak's thesis won the A. Cressy Morrison Prize of 1940, the same New York Academy prize that Bethe had won two years earlier. Reciprocating Bethe's gesture of having given him 10 percent of the 1938 Morrison prize as a finder's fee, Marshak gave 20 percent of the 1940 prize money to Bethe.

After Marshak had finished his dissertation Bethe received an invitation to write a review article on stellar energy for the *Reports of Progress in Physics.* He made the same offer to Marshak as Sommerfeld had made to him regarding the writing of the *Handbuch* article on *Elektronentheorie der Metalle.* He, Bethe, would be its author, and would be paid for it—but Marshak would write most of the article. It would be published under both of their names and Bethe would give Marshak all the money he would receive for it. "This because the original invitation had come to the professor and not to the student."[92] And indeed this is what happened. The honorarium for the article was eight guineas, which was about forty dollars at that time. Marshak spent most of the summer of 1939 writing that article. Bethe read it very carefully and made some additions and a few changes. Published in 1940, their review article is a very informative, detailed account of the state of astrophysical knowledge concerning stellar evolution.

Victor Weisskopf

In the fall of 1937 Victor Weisskopf joined the University of Rochester's department of physics.[93] Bethe had met Weisskopf in Copenhagen during his visit there in the summer of 1934. He admired Weisskopf's subsequent research in quantum electrodynamics and in nuclear physics, and Weisskopf's arrival in Rochester was a turning point in Bethe's scientific career. Bethe thereafter became deeply interested in quantum field theory, in particular in the higher-order perturbative aspects of quantum electrodynamics and in the field theoretic description of nuclear forces.

Weisskopf was born in 1908 in Vienna. His mother's family had lived there since the beginning of the nineteenth century. His great-grandmother, Charlotte Cohn, known as "die Urmama," was born in 1828 and was a very shrewd businesswoman who turned a failing bank into a very successful financial institution, becoming very rich. One of her daughters, Weisskopf's grandmother, married a Hungarian businessman named Alexander Gut. Weisskopf's only recollection of them was the "nice presents they always gave us at Christmas." They had but one child, Weisskopf's mother, who was provided "with the easy, comfortable life of a . . . girl of the upper middle class . . . sheltered and protected with the usual nannies and servants to smooth the way for her."[94]

Weisskopf's father, Emil, came from a very different background. He was born in Sucice, a small town in Czechoslovakia then part of the Austro-Hungarian Empire, the son of the ghetto's *schochet,* the ritual slaughterer of the community. Yiddish was his mother tongue. The rabbi of Sucice, recognizing Emil's unusual intellectual abilities, got the community to provide the necessary funds to send him to a *Gymnasium.* He did outstandingly there but was denied the full scholarship to study at a university usually awarded to the top student in the class because he was Jewish. He nonetheless went to Vienna and enrolled in the university, supporting himself by tutoring. After receiving his law degree he became a district judge in a suburb of Vienna. He met Martha, and they fell in love. When they decided to marry, "die Urmama was not pleased. Her assimilated, nonobservant Jewish family prided itself on its enlightened outlook. They were Austrian[s] who happened to be Jewish. That Emil was a district judge was not as important as his background. The fact remained that he was a poor young man from a tiny *shtetl.* His father was a butcher—and a kosher butcher at that. It wouldn't do at all."[95] But they married anyway. They had three children: a son who was older than Victor, Victor, and a younger daughter; and all three had very happy and carefree childhoods despite the Great War.[96]

Weisskopf went to *Gymnasium,* and after passing his *Abitur* became a student at the University of Vienna. He had always been deeply interested in science and elected to study physics there. Hans Thirring, his physics teacher at the university, recognized his great gifts and sent him to Göttingen to study with Max Born. His doctoral thesis, on the quantum electrodynamic theory of the widths of atomic spectral lines, was submitted in 1930 and resulted in an important paper with Wigner.[97] After obtaining his doctorate he went to

Leipzig to work with Heisenberg and to Berlin to work with Schrödinger. In 1932 he was awarded a Rockefeller Fellowship and spent the fall semester of the academic year 1932–33 in Copenhagen with Bohr and the spring semester in Cambridge with Dirac.[98]

The two-and-a-half years he was in residence at the ETH in Zurich as Pauli's *Assistent* were formative in molding him as a theoretical physicist. While there he published two very important papers. In the first he calculated the self energy of an electron when the electron is assumed to obey the Dirac equation interpreted hole theoretically and its interaction with the quantized electromagnetic field likewise treated quantum mechanically. Weisskopf found that in contrast to the linear divergence obtained when the electron is described non–hole theoretically, and in contrast to the linear divergence found in classical theory, the self energy of an electron in hole theory diverged only logarithmically with decreasing size of the electron's charge distribution. The second paper, which was co-authored with Pauli, formulated the quantum field theoretic description of charged spin 0 particles. Weisskopf continued his quantum field theoretic investigations while a postdoctoral fellow in Copenhagen during the academic year 1936–37. He there made a crucial calculation in which the concept of charge renormalization in quantum electrodynamics was elucidated.

In 1936, recognizing the precarious position in Europe of scientists of Jewish descent after Hitler had come to power, he decided to do research in nuclear physics in order to enhance his chances of obtaining a position. The shift proved wise, for on Bohr's recommendation he was offered a lectureship at the University of Rochester, beginning in the fall of 1937. He accepted the invitation.

Weisskopf's arrival in Rochester, only about ninety miles or a two-hour drive from Ithaca, marked the beginning of a close friendship between Bethe and him. In his interview with Weiner and Gloria Lubkin, Weisskopf recalled that they saw one another every month:

> when I had a new idea or got stuck with an idea, I just went over to Bethe to have simply somebody with whom I could talk. Talking always helps. I think I'm not wrong in saying I was there every month once, let's say, for a day. You asked me also what kind of collaboration it was. I do not remember that we ever wrote a paper together. In fact, I don't think there exists a paper "Weisskopf and Bethe" as far as I know. However, we were working in the same field [of nuclear physics], and there's hardly a paper in that period, in the pre-war period, that I wrote which wasn't thoroughly discussed

with Bethe; and most of the Bethe papers he discussed with me—in particular, in the field of nuclear physics, but I think also in other fields.[99]

It was Weisskopf who got Bethe deeply interested in meson theory and in the divergences encountered in perturbation theory in quantum electrodynamics. Weisskopf was working on these problems as well as on the nature of the nuclear forces that various meson theories predicted.

Meson Theory

In September 1940 the University of Pennsylvania celebrated its bicentennial anniversary with a weeklong series of conferences. The theme of the physics conference was "Nuclear Physics" and the speakers included Fermi, Breit, Rabi, Wigner, Oppenheimer, and John van Vleck.[100] The talks surveyed the then-current state of knowledge in the field and highlighted the problems to be solved. Coming a year after the outbreak of World War II and a few months after the fall of France, it was the last time for many years these physicists would address problems in "pure" physics. In fact, all of them were already deeply immersed in work that would culminate in the assembly of an atomic bomb.

At the conference, Fermi spoke about the experiments he was performing at Columbia on the reactions produced by the neutron bombardment of heavy elements and, in particular, on the analysis of neutron-induced fission in uranium. In his talk Breit reviewed what had been learned about the two-body nuclear forces from neutron-proton and proton-proton scattering. For his contribution to the symposium Wigner updated the paper he had published in 1937 on nuclear binding energies in light of "the new experimental data that had greatly enriched the field over the past four years." And Rabi in his presentation described the experiments that were being carried out at Columbia using molecular beams to measure the spin and magnetic moment of the lighter nuclei and reported on the most recent data for the quadrupole moment of the deuteron, 2.73×10^{-27} cm^2. He also reviewed the implication that the existence of the deuteron quadrupole moment had for the neutron-proton force and noted that the 1938 paper of Herbert Fröhlich, Walter Heitler, and Nicholas Kemmer, and a more recent one by Bethe, had indicated that spin 1 meson theories could explain not only the deuteron's binding energy and its level structure, but also its quadrupole moment.[101]

Oppenheimer talked about the mesotron and the quantum theory of fields and undertook to assess the current state of knowledge and use of meson theory. It was a natural task for him to undertake, for at the time Oppenheimer was at the center of meson theory research in the United States.

In 1937, shortly after Carl Anderson and Seth Neddermeyer had provided evidence for the existence of a new type of particle in the penetrating component of cosmic rays, Oppenheimer and Robert Serber published a short note in the *Physical Review* pointing out that the mass of the newly discovered particle specified a length, which they connected with the range of the nuclear forces. They referred to Hideki Yukawa's 1935 paper in which he had proposed a field theoretical model in which the neutron-proton force was mediated by the exchange of a scalar particle whose mass was adjusted so as to yield a reasonable range for the nuclear forces. Oppenheimer and Serber's note was responsible for drawing the attention of American physicists to the "meson theories" of nuclear forces that Yukawa, Ernst Stueckelberg, and Gregor Wentzel had advanced.[102]

The direct observation in a cloud chamber by J. C. Street and E. C. Stevenson of this "heavy electron" in 1937 authenticated its existence. They determined that it existed in both a positive and a negative variety, estimated its mass to be 150 to 220 electron masses, and estimated its lifetime to be about 10^{-6} sec.[103] By 1939 Bethe could assert that "it was natural to identify these cosmic ray particles with the particles in Yukawa's theory of nuclear forces."

From 1937 until the attack on Pearl Harbor in December 1941, Oppenheimer and his students were actively engaged in investigating possible descriptions of the newly discovered cosmic ray meson and its connection to nuclear forces. Two of Oppenheimer's students—Willis Lamb and Leonard Schiff—found in 1938 that Yukawa's theory in its original form (charged mesons of zero spin) gave the wrong sign for the interaction potential in the deuteron, in other words, a repulsion instead of an attraction in the ^3S state. This could be remedied if it was assumed that the meson had spin 1—and Yukawa, Shoichi Sakata, Mituo Taketani, and Minoru Kobayashi in Japan, and Fröhlich, Heitler, and Kemmer, and Homi Bhabha in Great Britain, developed this approach. A spin 1 vector meson theory gives rise to a spin dependent tensor force. However, any theory involving only charged mesons will give no force between like nuclear particles in lowest approximation (that is, with the emission and absorption of a single meson). To second order, Fröhlich, Heitler, and Kemmer found a strong repulsive force at small distances. This contradicted the experiments of Tuve, N. P. Heydenburg, and

L. R. Hafstad in 1936 at the Carnegie Institution in Washington. Their experiments indicated that the force between two protons is strongly attractive and very nearly equal to that between neutron and proton in the singlet state. A force between two like particles can only be obtained with neutral mesons. Kemmer developed a theory of nuclear forces in which neutral and charged mesons occur in a symmetrical way and thus explained in a natural way the equality of the forces between like and unlike nuclear particles.[104] A characteristic of the symmetrical theory is the specific exchange character of the nuclear forces, which made it possible to explain saturation of the nuclear forces simply, in a manner analogous to the homopolar chemical bond.

An alternative way of explaining charge independent forces is to assume interactions with neutral particles only. Then the charge of the nuclear particle—whether it is a neutron or a proton—becomes irrelevant, and the equality of the forces follows immediately. This alternative was explored by Bethe and found to be successful when applied to the two-body problem.[105] It suffered, however, from the disadvantage that must appear in any purely neutral meson theory, namely that there was no connection with the cosmic ray mesons that had been observed and no explanation of beta decay, nor of the anomalous magnetic moments of neutron and proton.[106]

The "mesotron" that had been identified by Neddermeyer and Anderson, and by Street and Stevenson, exhibited weak nuclear interactions in its passage through the atmosphere. Field theorists were thus trying to find a meson theory that yielded strong nucleon-nucleon forces yet a small meson-nucleon cross section.[107]

It is with this background in mind that Oppenheimer's paper at the Pennsylvania bicentennial celebration should be read. Oppenheimer judged his report as "negative," for his purpose was to explain why "the theory of the mesotron, and more generally, the quantum theory of fields, has failed so completely to deepen our understanding of nuclear forces and processes, has left almost totally unanswered the many questions earlier speakers have put to it; and to try to explain, too, why in spite of this the quantum theory of fields still seems to me a subject worth reporting on at all."[108]

He noted that the discovery of the mesotron, although it sharpened all the difficulties of field theory, had also "given us some confidence that the fundamental ideas of field theory are right, for the mesotron was a prediction, very general and qualitative it is true, of this theory."[109] He concluded his presentation with the following somewhat despairing but accurate assessment:

"The largeness of the couplings, and Yukawa's relation between nuclear range and mesotron mass, seem to us about the only fairly sure conclusions that we have."[110] The war interrupted all pure physics activities in the United States. Except by Pauli, who spent the war years at the Institute for Advanced Study in Princeton, and some of his associates who were foreigners, essentially no meson theoretic work was carried out in the United States during the war.[111]

But Bethe's involvement with meson theory and quantum electrodynamics—a result of his interactions with Weisskopf—made possible his quantum electrodynamic explanation of the Lamb shift in 1947.

9

Rose Ewald Bethe

Rose Ewald is the woman Hans Bethe loved all his adult life. She raised their children, and was the guardian of his home life and the person greatly responsible for his moral growth and stature in the face of his ever deeper involvement in weaponry and politics.

Rose was born on March 20, 1917, in Munich and was baptized shortly thereafter. Her father, Paul Peter Ewald, was then in the German army in charge of an X-ray unit in the rear of the eastern front. As noted earlier, her father had been one of Sommerfeld's first *Doktoranden,* and after having obtained his doctorate in 1912 he became an *Assistent* to Hilbert in Göttingen. He there met Ella Philippson, who was studying medicine at the university. They fell in love and married in May 1913.

Rose's mother, Elise Berta (Ella) Philippson (1891–1993) was born in Magdeburg. The Philippsons were a distinguished German Jewish family that had adopted the ideals of the Enlightenment and of *Bildung* at the beginning of the nineteenth century. Throughout that century Philippsons became eminent physicians, scholars, and theologians. Ludwig Philippson, who considered himself a "historical Jew, neither Orthodox nor reformed," was one of the most influential Reform rabbis in Germany. For him it was the writings of the Prophets that characterized the essence of Judaism: "The prophets changed from the hope of a personal Messiah to that of a future Kingdom of peace and justice for all mankind." He wanted Jews to adopt the universalistic vision of the Prophets as their creed and to commit themselves to their moral principles. Several of the Philippson men converted and several of the Philippson women married Christians—as in the case of Rose's mother and of Ewald's mother, who was also a Philippson. Yet even as the nation as a whole veered to the right politically after the rise of Bismarck to power, in a home with an adult who had Philippson roots liberal ideals were fostered. A belief in progress, in justice, in internationalism, in the Kantian notions of universal peace and other Enlightenment ideals, in *Bildung,* in the dignity of man and in human freedom permeated the cultural values of such a family. At the end of the nineteenth century many of the Philippson women were attending university and were becoming professionals.[1]

Ella's family was Jewish, assimilated, liberal, and well off. Ella's mother was born and educated in Frankfurt. Her father was a stockbroker, the representative in Magdeburg of a Berlin brokerage firm. He was "a happy man who loved his wife and children."[2] Ella, in her old age, looked back on her childhood as a golden time that ended abruptly when her father died at age fifty-six. Ella was seventeen at the time, her mother forty-four, and her youngest sister had just been born. Ella's mother was the one who looked after the well being of the household. She wanted her four daughters to have the best education possible and the independence this would give them. When Ella was growing up there was no *Gymnasium* for girls in Magdeburg. So together with like-minded women, Ella's mother established a private school where girls would obtain a similar education. Ella thus became part of a small private circle of girls organized by their feminist-inspired mothers, and many of these girls went on to become professionals of various kinds. Ella went to Freiburg University in 1910 to study medicine, but in 1911 became ill and had to withdraw from her studies for a year. In 1912 she enrolled in Göttingen University to continue her medical studies.[3] She lived there in a small pension whose other boarders were the physicists Max Born and Theodore von Kármán, and two other young academics, Hans Bolza and Albrecht Renner.[4] The landlady of the boarding house, who longed for female company, happily accepted Ella, who had answered her advertisement. Ella, Born, von Kármán, Bolza, and Renner became good friends, and the five of them became known as *El Bokarebo*: "El" for Ella, "Bo" for Born and Bolza, "Ka" for von Kármán, and "Re" for Renner. There were others who came to the pension for their midday meal, among them Paul Ewald and Richard Courant. Paul Ewald had come to Göttingen in 1912 to be Hilbert's "physics tutor." Eating at the pension is how he met Ella.

Rose characterized her home as that of a "professorial family," with her father acting very much the *Herr Professor* and her mother attending to his needs and managing the household. Ella, though assimilated, never converted. By virtue of Ella's presence, "righteousness" and "probity" held sway in the Ewald home. It was a very "warm" home, even though it was very much a "professorial" household.

To better understand Paul Ewald a few words about his upbringing are in order. Paul Peter Ewald was born in Berlin in 1888 and was named after his father, Paul, who had died three months before his birth. His great grandfather Anton had been a banker, and the family owed their fortune to his efforts. Late in life he changed his name from Cohn to Ewald. Paul Peter's paternal grandfather, Arnold Ewald, studied philosophy but then became a painter; Arnold's two brothers were scholars, one a mathematician and the other a geologist. Arnold had four children. His daughter Johanna married the director of the Berlin Museum of Arts and Crafts; Anton became a professor of medicine at Berlin University and the director of the Augusta Victoria Hospital there; Richard became professor of physiology at Strassburg University (and the professor with whom Albrecht Bethe got his *Habilitation*); and Paul Sr., at the time of his death at age thirty-six, was a respected historian and a *Privatdozent* at the University of Berlin.

Rose characterized her grandmother Clara Philippson as having "a very independent nature." She was a distant relative of Ella Philippson. Clara's father, a Philippson, had married a non-Jewish woman, née Kapp. Clara told Rose of having suffered from anti-Semitism when going to school in Bonn. She went to Lutheran churches all her life, and her son Paul Peter considered himself a Protestant. As a young woman Clara was fluent in English, French, and Italian, in addition to German, her mother tongue. She had wanted to study medicine, but medicine, and for that matter "any other serious study," was not considered ladylike by her social milieu, and so "she developed her talents for painting seriously." After her parents moved to Berlin, she studied painting there. In 1885 she rented an apartment and opened a professional studio in the house where Paul Sr. lived with his parents. Paul Sr. courted her and "after much hesitation on her part they were married in 1886."[5] While in Berlin she had painted two very professional portraits, the first of many to follow.[6] She was twenty-eight when Paul Sr. died, and Paul Peter became the focus of her life. Painting became the means by which she supported herself and maintained her independence from her husband's family. But her livelihood as a portrait painter required extended travelling, and Paul Peter thus experienced a nomadic upbringing, first in Paris, then in Cambridge, Berlin, and Potsdam.

Paul Peter's first three years of schooling were in a *Realgymnasium* in Berlin and the next few years in a humanistic *Gymnasium* there. After Clara's father suffered a stroke, she moved to Potsdam to help. Paul Peter finished *Gymnasium* in Potsdam and obtained his *Abitur* in the spring of 1905. Greek had been his favorite subject, but he also developed an interest in the sciences. In the fall of 1905 he enrolled in Caius College, University of Cambridge, to study chemistry. But the calculus, physical chemistry, and physics as taught in Cambridge did not resonate with him, and in the spring of 1906 he enrolled in Göttingen. During the academic year 1906–07 he became enthralled by Hilbert's lectures on calculus. Later, he told Constance Reid, Hilbert's biographer: "It was wonderful." When introduced to derivation and integration in the course, "we felt as if we had actually seen Hilbert *create* the new concept."[7] Although he was not an advanced student, Paul Peter nonetheless became the official note taker for the course and got Ernst Hellinger, one of Hilbert's *Assistenten,* to give him extra instruction to be able to do so. Paul Peter claimed that he learned almost all the analysis he ever needed in Hilbert's calculus class and in the sessions with Hellinger.

To obtain a more rigorous, more "epsilontics" based, and less intuitive approach to mathematics Paul Peter went to Munich in the fall of 1907 to study with Alfred Pringsheim. By chance, a friend of his took him to the lecture course by Sommerfeld on hydrodynamics, and Paul Peter became captivated by the harmony between mathematical formalism and physical phenomena. According to Bethe, "From then on he considered himself a theoretical physicist."[8] In 1910 he became a doctoral student in Sommerfeld's seminar. His thesis was an explanation of the double refraction of light in its passage through an anisotropic crystal. He successfully solved the mathematical problem of adding the responses of all the atoms—modeled as dipole resonators—to the incoming light beam and obtained qualitative agreement with the observed double refraction for one type of crystal. His work gave strong support to the idea that crystals consisted of regular geometric arrangements of atoms or molecules. His thesis was the first step in a formulation of a general dynamical theory of the behavior of electromagnetic radiation in crystals, the latter modeled as regular periodic arrays of dipoles. It led to Paul Peter's dynamical theory of X-ray propagation "which kept [him] busy, in some form or other, all his life."[9]

Paul Peter Ewald obtained his doctorate in 1912, and became Hilbert's *Assistent.* Hilbert at the time was devoting most of his efforts to physics. Paul Peter liked and admired Hilbert, but found him "a bit of an arrested juvenile," doing things that he thought would be shocking to more conventional citizens.[10] In 1913, after his year as Hilbert's "physics tutor," Ewald moved back to Munich, becoming Sommerfeld's *Assistent.* Ewald's mother, Clara, had moved to Munich in 1909 when he had first gone there and had become a close friend of the Sommerfelds, so that Ewald's children called the Sommerfelds "aunt" and "uncle."

In 1911 Clara built a house in Holzhausen, a small village on the Ammersee, a large lake some twenty-five miles west of Munich. Except for her professional extended trips, a good part of the year and all summers were spent there, and Paul Peter and his wife, Ella, visited often. Shortly after the outbreak of World War I in September 1914, Lux, the first of Ella and Paul Peter's four children, was born. Clara early on in the war recognized that it would be lengthy and that severe food shortages would develop, and so she began keeping animals in Holzhausen; first two goats so that there would be enough milk, even near the end of the war in 1917 when Rose was born, then two pigs and turkeys, ducks, and geese, and finally a cow. The two other Ewald children, Linde and Arnold, were born in 1919 and 1921, respectively.

In 1921 Paul Peter Ewald received an offer of an associate professorship from the Technical University of Stuttgart and so the family moved there. As there was some difficulty in finding suitable lodging in Stuttgart, Ella and the children went to live with Grandma Clara in Holzhausen. In 1922 a beautiful second floor apartment became available in a large three-story house that had a magnificent view of all of Stuttgart. The house was surrounded by an enormous garden in which the children could play. So the family moved to Stuttgart. As the apartment had only four rooms, and Ewald's study was "sacrosanct," all four children shared the same room, which included the bathroom. But the family was well off enough to employ a general maid.

Ewald eventually became the director of the theoretical physics department at the Stuttgart Technical University and became an internationally recognized leader in X-ray crystallography. In 1932 he was elected *Rektor,* and remained in that post even after the Nazis came to power in January 1933. He resigned in April 1933 because university life had become so politically polarized that remaining in that the position had become unbearable for a liberal with a Jewish wife. The fact that he had served in the army during World War I prevented his dismissal as a professor. However, by the end of 1936 the situation had become intolerable. When Paul stopped teaching, he was pleasantly surprised to find himself pensioned. Lawrence Bragg secured him a research grant in Cambridge, and he left Germany in the fall of 1937. The rest of the family joined him in the spring of 1938, and Clara followed in the fall of 1938.

Growing Up

Rose Ewald grew up aware that there were tensions in the house. Her father was a disciplinarian and at times had difficulties maintaining an even disposition: he could be "playful" one instant, and "harsh" the next. Her grandmother, Clara, developed a very severe case of shingles in the fall of 1922 that essentially paralyzed her and necessitated around-the-clock nursing care for almost a year. She was cared for in the Ewald apartment, a bed having been placed in Paul Peter's study. She recovered, became once again her demanding self, and remained in the house. Relations between Ella and Clara were evidently somewhat difficult. However, Ella never complained to her daughters about her mother-in-law. "She took it silently and patiently." When Ewald was under pressure, relations between Ella and him were tense. The hyperinflation of the mid-1920s was one such extended period, and Rose remembered "there not being any money in the house."

The household dynamics while she was growing up became etched in Rose's mind and became her model of a "professorial life." Contrary to the impression of "outsiders" there wasn't much social life in the Ewald home, and except for Ewald's *Assistenten* coming to the house occasionally there were very few informal gatherings. The formal ones were very "formal."[11] For Rose, the chief characteristic of the professorial life was that "mother was totally devoted to father." But both parents invested a great deal of time and effort to stimulate the children by reading to them, having them read books on many diverse subjects, and by discussing with them current political and cultural issues. It was Rose's mother who was more open, easier to talk to, and more demonstratively loving; and it was her mother who saw to it that the interactions among siblings were affectionate, considerate, and friendly.

Rose started going to a *Volkschule* at age seven, and stayed there for a year and a half. For three months thereafter she attended a private school and then was sent to a *Mädchengymnasium* at age ten. The *Gymnasium* was not very stimulating, and "she did what she was asked to do." She read a lot, mostly the classics of German poetry and nineteenth- and twentieth-century novels by German writers.

Rose didn't remember much of Hans's visit to Stuttgart in the spring of 1928. She did remember more of his stay when he was an *Assistent* to her father in 1929, and in particular, her walk with Lux and him in the park, an occasion Hans described as "babysitting." When in Tübingen Hans made several visits to Stuttgart. She recalled that on those occasions Hans would visit the house and stay for dinner, and her mother would tell the maid to "add another pound of spaghetti." Werner Ehrenberg, who was an *Assistent* to Erich Regener, the professor of experimental physics in the institute, later told Rose that whenever Hans came, there would be a long line of people waiting to see him to get advice on physics problems. Rose also remembered very clearly the visit of Hans's mother in 1930 or 1931 when one of her children's plays was performed in Stuttgart. She came to dinner and "performed" there.

Rose began feeling the economic depression in 1930, becoming aware of the huge number of unemployed workers. Watching the clashes between communists and Nazis that could be seen from her house—with the authorities doing nothing to restore calm—brought home the political consequences of unemployment. Ewald was *Rektor* during the academic year 1932–33, and thus was responsible for hosting the annual *Rektor*'s ball. The ball was held on January 31, 1933, the day Hitler came to power. Rose, who attended the ball, remembered people saying that "the center would eventually come back." The burning of the Reichstag was a "turning point" for her. No one in her circle of friends believed what was officially said, namely, that the fire had been set by the communists. And the illegal, forceful detention of a considerable number of opponents to the regime thereafter made it evident that she was witnessing the end of democracy and the rule of law in Germany.

It was clear to Rose that "life had become uncertain and unpredictable." Two of her mother's closest friends, the Adlers, were Jewish. Their daughter was Rose's friend. The father was a lawyer, the mother a physician. On April 1, 1933, Stuttgart experienced the first organized destruction of Jewish property and businesses, including the Adler law office. Rose retained a vivid memory of the weekend she spent at the Adlers' after the office had been wrecked by Nazi SA storm troopers wearing brown shirts.[12] That episode left an indelible mark.

When Rose went to school at the beginning of the new academic year in

April 1933, she found that "school had changed. The teachers were talking differently; the teaching of history became twisted; even the tone of the teacher of religion had changed." Fellow students teased her for being half-Jewish, even though she had been baptized and confirmed, and didn't feel Jewish. She wanted to leave Germany and told her mother so. Whenever visitors from outside Germany came to visit her parents she would tell them, "take me along when you go back."

At the end of the academic year in 1934, after six years of *Gymnasium* studies, Rose quit school, went to England, and worked in a school for wayward children. After she came back to Stuttgart she attended a private secretarial school for three months. At that time she was not yet concerned with "money matters" nor with the fact that she "should be responsible for herself."

One of the visitors to the Ewald home in early 1935 was Hans, who had come to Stuttgart to pick up his visa to go to the United States. Rose remembers Hans telling her mother, whom he had gotten to like very much and to feel close to, his "sad song" of "going alone to America." At which point the eighteen-year-old Rose interjected, "Why don't you take me with you to the United States? I'll marry you—if it doesn't work out we can get divorced in Ithaca." Ella glowered as a result of the "cold-blooded proposal," and the matter was not explored further.

In the fall of 1935, Max von Laue, a good friend of Ewald,[13] visited and asked whether he could do anything for Ewald on his trip to the United States. Ewald said no. But Rose piped up and said, "I want to emigrate to America." Von Laue spoke with two families while in the States, the Courants and the Weyls, regarding the possibility of Rose becoming an *au pair* in their house. The Courants welcomed the idea.[14] Rose then applied for an immigration visa to the United States, and Richard Courant, Hermann Weyl, Rudolf Ladenburg, and Otto Stern—all friends of her parents who had emigrated to the United States—were contacted to support her application. However, they had all sponsored too many people already and could not do so. Her parents then turned to the Landés, close friends of the Ewalds. Mrs. Landé was Ella's friend—one of the circle of girls who had been privately educated in Magdeburg so that they would be able to obtain their *Abitur.* In addition, both Alfred Landé and Ewald had completed their doctorates with Sommerfeld. Ewald had obtained his degree one year earlier than Landé. Landé had become Hilbert's "physics tutor" in 1913 after Ewald left to go to Munich. The

Landés emigrated to the United States in 1931, when Alfred accepted a position at Ohio State University in Columbus, an institution he had visited twice before during the 1920s.[15] The Landés did sponsor Rose.

Rose worked in France until her visa came through in April 1936. She returned to Stuttgart, packed her belongings, and at the end of the month sailed for New York, where the Courants met her. She went on to Columbus in early May and had a few lessons on how to drive a car while waiting to become an *au pair* with the Courants. Richard Courant had been a close associate of Hilbert, and Ewald and Ella had gotten to know him well during their stay in Göttingen in 1912, remaining in close touch with him and his wife thereafter.[16] Rose eventually met the Courants in Buffalo and stayed with them at their house in New Rochelle, New York, looking after the Courant children. She thereafter went to Durham, North Carolina, to become Hertha Sponer's housekeeper, a position James Franck had gotten for her.

Rose was struck by the restrictive, constraining atmosphere in North Carolina and the fact that she could not go to Durham without wearing gloves, a hat, and stockings. She couldn't take walks there freely because of the "negro and white problem." She was able to complete her household duties quickly, so she became a part-time student at Duke, taking courses in remedial English, English poetry, and French history and government. She began to read extensively in English, particularly Jane Austen's novels. And much like Elizabeth in *Pride and Prejudice,* she felt she was learning social grace in her new environment.

Rose and Hans

Rose met Hans again in late February 1937. By then Hans had been recognized nationally as an outstanding theorist, was being courted by several universities, and had been promoted to a full professorship by Cornell. Bohr and Bethe had been invited to talk about nuclear theory at Duke following the Washington Conference on Theoretical Physics. Rose induced Hertha Sponer to take her to the banquet after the lecture. Hertha had doubted that Hans would speak to Rose. At the dinner she asked Rose whether she thought that he would. Rose's answer was "no," but as they were ready to leave, Hans approached Rose.

Shortly after the Duke visit Hans wrote his mother: "There also lives in Durham Ewald's daughter Rose, who has suddenly become an adult and does the housekeeping for the woman professor of physics there. She is also studying something. She has become an exceptionally nice girl; she was always very nice when 8 years ago I played baby-sitter at the Ewalds. But it is always surprising to see how such a girl suddenly becomes an adult. Furthermore she has become very pretty as one might expect from her father but not from her mother. And she is very untouched by life far away from her parents, and as natural as ever. It is worth keeping her in mind." A few weeks later in another letter to his mother Hans commented: "Your letter reminds me again how attractive I found Rose Ewald. Is she too young? 20, I think. Like a schoolboy, I don't know what I should initiate in this connection."

For her part, after she had met Hans at the Duke banquet Rose found that he "was still awkward and shy, but self-confident." Rose was right: he indeed was self-confident. At precisely this time Hans wrote to his mother, "I think I am about the leading theoretician in America. That does not mean the best. Wigner is certainly better and Oppenheimer and Teller probably just as good. But I do more and talk more and that counts too." As far as Rose was concerned, not much more than making contact again had taken place at the banquet. "Perhaps all the happy days in Stuttgart came back to him. Perhaps he was searching for a wife, being tired of bachelor quarters and of living alone. Perhaps he was struck by my beauty" are the reasons Rose gave for Hans writing her a letter a few weeks after the meeting in which he proposed marriage.

After his visit to Duke, Hans had spoken with the Tellers about including Rose in the car trip to the Rocky Mountains the Tellers and he were planning, "provided," as Hans wrote his mother, "the Tellers like her sufficiently to have around for four weeks." In that same letter to his mother in early April 1937 he informed her that "The Tellers are at the moment in Durham and are taking the opportunity to view said girl (Mädchen). Of course the question is still, whether she has the time and wants to come. In any case it would be nice, it is only too bad that it is not for a longer time."

According to Rose, "in 1937 Erwin Strauss was Hans's closest friend, and Edward [Teller] his closest scientist friend."[17] Hans and Teller had seen a good deal of one another in England in 1934, when Teller was in London and Hans in Manchester. After they both had emigrated to the United States in 1935, they met every year in Washington: first at the annual February George Washington Theoretical Physics conference that George Gamow, Edward, and Merle Tuve organized, and then again at the April national meeting of the

American Physical Society.[18] Bethe stayed with the Tellers during these extended visits to Washington. Rose knew Teller from Stuttgart. Teller had studied with Ewald in Karlsruhe when Ewald was lecturing there in 1927 and had kept in contact with Ewald thereafter, visiting him in Stuttgart on a number of occasions.

Rose did not accept Hans's marriage proposal at the time. But she said "yes" when he asked her to travel west with him, with the Tellers as chaperones, and agreed to being considered engaged to him. The car trip west lasted from mid-June until mid-July. They visited Estes National Park in Colorado, the Tetons, Yellowstone, Glacier National Park, Seattle, Mt. Rainier, the Olympic Peninsula, and down the Pacific coast to Crater Lake. From there they traveled to Berkeley and Stanford. Fermi was lecturing at the Stanford summer school and went with them to Yosemite.

Rose was a fearless driver. She had learned how to drive in Columbus. Teller, who was a timid driver, was "appalled" by Rose's driving, and Rose found Fermi "annoying" as he constantly commented on her gear shifting while she was at the wheel of the car. The most anxious moments of the trip, however, were when Fermi drove the car from Tuolumne Meadows to the Yosemite Valley, speeding into the tight curves. The road is extremely narrow, with a steep, straight precipice on the right side. Rose commented, "All four of us were scared to death."

The Tellers, Hans, and Rose proceeded to travel to Reno, where Hans took a train east. Before parting for the rest of the summer—Hans to sail to Europe to visit his parents and Rose to housesit for the Courants in New Rochelle—they talked about marriage. What transpired in the conversation was conveyed by Hans to his mother in a letter written July 9, 1937. Rose had told him that she loved him, but did not want to marry. Hans in his letter commented that Rose was "A modern girl. Too modern for me, because in this case for the first time, I want permanence. 'Why legalize it,' she says, and 'I need to develop myself more.' She has her will in this and most other things. And curiously, it doesn't matter. She is terribly smart, and understands everything that I feel without my saying it. And I am happy for the first time really happy, and would be totally so if it were for ever . . . It is all like a dream, and I still cannot believe it, that again a girl who I love, loves me. And everything can be like one dreamt it."

In the fall of 1937 Rose enrolled in Smith College, the liberal arts college for women located in Northampton, Massachusetts, and thus within driving distance from Ithaca. Hans helped by giving money to her parents to make this possible. She was admitted as a sophomore and elected to major in social work. Her two years at Smith were "good years." Smith was "rational" compared to Duke. Her teachers were excellent: they were committed to making the courses they taught relevant to their students and to prepare them to live useful and rewarding lives. Since colleges at the time considered themselves *in loco parentis* as far as the personal life of their students was concerned, especially in the case of women's colleges, leaving the Smith campus was restricted. Students had to sign out, indicate where they were going, and come back and sign in at a stipulated time. When Hans and Rose went off together on weekends, the warden at Smith, also the Dean of Women, considered Hans Rose's "guardian."

Hans wanted very much to get married. In November 1937 he wrote a long letter to his mother expressing his "sweet torment" ("süsse Qual") stemming from Rose's unwillingness to marry. Rose was sure she did not want to get married as yet. She was happy at Smith studying for a career in social work. She didn't think of the financial burden involved. She knew that finishing school would give her some independence and autonomy and that this was crucially important to her.

In a letter to his mother in December 1937 Hans told of his happiness whenever he expected a phone call from Rose and of the pleasure of making up after a small quarrel between them. Also, that he had indulged in the fantasy of the three of them—Rose, his mother, and he—spending Christmas together. He concluded that particular letter with the statement that since he could be totally open with her: "My dear Mamuzzelchen, I must tell you again what I already said many years ago: You are my best friend" ("meine beste Freundin").

Rose, for her part, was firm in her resolve to obtain a professional degree. Moreover, as Hans reported in a letter to his mother on January 6, 1938, she had told him: "You do everything to win me, but once you have got me what then?" And Hans admitted to his mother: "I could not say anything in reply, and I fear she is right. With all men and more with me than with many others the attraction disappears to a large degree with fulfillment. I could only say that I believe *her* charm for me would never cease."

Rose had countered with this argument: "But what would give me a justification for trusting you? Perhaps you will one day do something which will give me confidence. I don't know what. For the time being it is not there."

And Hans thereafter confessed to his mother, "They sound like hard words. But they did not hurt me because they are true. They did not hurt me because they were said with an undertone 'I would like it very much if that moment came soon.' They did not hurt because they are the true reason why she doesn't want to. A good reason, which shows that this girl is really extraordinary, as I always thought and more."

The summer of 1938 was an exciting one for Hans. He had finished his paper on energy generation in stars. "My best work of the last three years," he wrote his mother in April 1938. In May he had lectured at Harvard with "much success." Henry Norris Russell, the dean of American astrophysicists, commended him: "It is nice when an important problem is finally solved." In June Hans had driven to Ann Arbor to see Weisskopf and Placzek at the summer school there, and then for a week went to lecture at Purdue and at Ohio State; and then back to New York City to celebrate his birthday with Rose and her parents. Rose's parents had come to the United States because Rabi had invited Ewald to teach in Columbia University's summer school.[19] Rose spent most of that summer with them in New York. After his birthday, Hans wrote his mother that Rose had revived his trust in her, "the certainty that she indeed loves me. That was the most beautiful birthday present."

In late July Hans went to Switzerland to meet his mother and his father. It had become dangerous for him to go to Germany for an extended stay. Before leaving he met Rose in New York. On that occasion it became painfully clear to him that Rose's love was circumscribed. Rose had earlier revealed to him that there was someone else in her life, but that the relationship was hopeless.

Although seeing her again was "wonderful," he wrote his mother, "In the moment when I thought I had lost her she has very genuinely come to me. In spite of that she will not marry, but she is a true friend. And if I ever need someone, she will be there for me. She says so, and I believe it too. She is the only person except you who tells me of my mistakes. And like you, she takes the sting out of these conversations."[20]

The extant letters from Rose to Hans during 1938 confirm the picture drawn by Hans in his letters to his mother. Soon after the 1938 George Washington University Conference Hans had told Rose what he had accomplished in his solution of the problem of energy generation in stars. Sometime before he left for Europe Rose sent him a little poem she had written. If not great poetry, the lines are charged with emotion:

O thanks for the memories
Of stars & coal lit sun
Of neutrons on the run
Hydros splitting
Helios fitting
Each replacing one by one
How lovely it's been
Thanks for the memories
Of collisions in the dark
Of light created with a spark
Sunlight rising
Earth devising
Methods for the mark.
Ed . . . mitations
O thank you so much [etc]
My poor dear

During his trip to Europe Hans went briefly to Germany. He was trying to get his mother out of Germany and went there to make inquiries whether the government pension she was getting would be transferable to the United States should she emigrate. His father assisted him in these matters. Hans also spoke to him about Rose, of his frustrations with the relationship, and intimated to him that he was thinking of breaking off the relationship. On October 13, 1938, Albrecht wrote him: "I do not know very much about her [Rose], but according to the little that I do, I have the impression that your decision is the right one. Love which is so much stronger on one side than on the other rarely leads to happiness."

Later that summer, Genia Peierls likewise wrote Hans a letter concerning Rose. Hans had spent some time with the Peierlses in Cornwall before returning to Cornell in September and had talked to Genia about Rose. Her advice to him was: "Hans, you must marry her as soon as possible, or much that is beautiful can be wrecked." She suggested that probably Rose was waiting for deeper feelings to develop, "but on its own it never come[s]. From a girl (*Mädchen*) one must make a woman." So she advised him to take Rose somewhere for a couple of weeks without the Tellers, without Weisskopf, and without Placzek. "A woman begins to form with long evenings of body warmth and unhurried lips." In addition, Genia gave him advice on birth control and stressed its importance. Well meaning, but not very realistic advice! Given

Hans's great professional success and the deep frustration in his personal life, his relationship with Rose was becoming problematic.

On November 26, 1938, Hans wrote his mother, "I am no longer in love, but I love her [Rose] just as much. She is just as attractive as ever, but no longer mysterious. The 'problem' is solved and for that reason there is much more friendship than before. The torment is past. And now I could even imagine that on some occasion I might fall in love with someone else. For example, Mrs. Winter, if she were not Mrs. She is very charming. Very pretty, unusual face, dark eyes and very slim figure. Very intelligent."

Anne Winter

Annie and George Winter had arrived in Ithaca in the late summer of 1938. George Winter was a close friend of Victor Weisskopf.[21] They had grown up together in Vienna, went to *Gymnasium* together, co-authored a scientific paper when they were seventeen, and shared the same interests in music and literature. Annie Singer had been a common friend of both of them since their *Gymnasium* days, and both George and Victor had fallen in love with her. George graduated from the Munich Technical University in 1930 with a degree in construction engineering. But 1930 was the beginning of the Great Depression, and George was lucky to obtain through family connections an engineering position for the design of a seventeen-story building—which until 1999 was *the* skyscraper of Vienna. Annie had intended to study art, "very much influenced . . . by George and his great interest in it," but the observation by the teacher of her exercise class that she would be a very good dance instructor made her get a diploma in that field. After George and she married in July 1931 she gave dance lessons in their spacious apartment.

When in January 1932 the firm George was working for went bankrupt, he found himself without a job and with no prospect of one in Austria. The only opportunities open to him were in Argentina and the Soviet Union. His socialist views made him accept the job in the Soviet Union. In April 1932, Annie and George departed for what turned out to be a six-year stay in Miasochlastroy, a small development town near Sverdlovsk in the Urals. Victor Weisskopf, who had spent a half-year in 1932 in Kharkov with Lev Landau, kept in close touch with them throughout their stay in the USSR. In September 1933 he visited them on his way to Leningrad to attend a physics conference he had helped organize.[22] In the summer of 1937 Annie and George went to Vienna to introduce their son Peter to his grandparents. They there met Weisskopf, who had followed closely the August 1936 and January 1937 trials of Grigory Zinoviev, Lev Kamenev, Karl Radek, and the others who were accused of plotting with Leon Trotsky to undermine the security of the Soviet state. As forcefully as he could, Weisskopf urged Annie and George to leave the Soviet Union because of the purges that Stalin was carrying out, the show trials, and the imprisonment of many foreigners. It was no longer safe for them to be there. After several harrowing experiences the Winters succeeded in leaving the Soviet Union in February 1938. A month after they arrived in Vienna Austria was annexed by Germany and, "because they were Jewish, Austria was even more dangerous for them than Russia."[23] Weisskopf helped them obtain an American immigration visa, and after some hardships they arrived in the United States in mid-September 1938.

Weisskopf, with the help of Bethe, had arranged for George to be given a research assistantship in the Cornell school of engineering were he to enroll as a graduate student for the spring term of the academic year 1937–38. Even though he arrived a semester late, George was admitted and given a scholarship. He proved himself to be an outstanding student, earned his doctorate in two years, and was asked to stay on as a postdoctoral fellow. He eventually joined the faculty of the school of engineering, becoming the head of Cornell's civil engineering department and one of the world's experts on the properties of light-gauge steel.

Annie became known as Anne and gave private lessons in physical exercises. Their son Peter went to the Cornell nursery school. Being very adept at making friends, Anne created a lively social circle that included Hans. Hans became very attracted to, very fond of, and very close to Anne. At one stage he even thought about marrying her. Knowing Anne Winter affected his relationship with Rose. In the face of Rose's insistence that *for the time being* she could not marry him and that she was "not a bit jealous" of his growing friendship with Anne Winter, Hans wrote his mother that he was consoling himself by believing that since "our days together are so beautiful . . . and what stands between us is so small as it is now, then it is just as good or better than marrying."[24]

But in a somewhat later letter to his mother on New Year's Day, 1939, after he had spent two days in New York with Rose, who was there because a friend of hers had suffered a deep tragedy, he noted that the time spent with her had been "beautiful and clarified many things," and that he had come to the conclusion that "it would really be better if I did not marry Rose . . . There is much love, and warmth and trust . . . She says: 'Stay with me, because you are as if I still had my father and mother with me.' I am for her home and a foothold and warmth in a cold world—and she is the same for me. We talk to each other like two old people, like two young animals that crawl next to each other because they give warmth to one another." Later that month Hans broke his engagement to Rose. But they remained in close touch and kept seeing one another.

In addition to the loving attention of two women, Hans was then also receiving much public acclaim for having solved the problem of energy production in stars. In a letter to his mother in February 1939 he described the novel situation he found himself in: "At present it is so beautiful and I am living through as much in one week as formerly in a year. The experience is both personal and scientific. And in both cases it is success. I am famous and can hardly help myself in the face of articles in newspapers and periodicals. *Time,* the most widely read weekly, had my picture and a nice article. The *New York Times* meanwhile published two long articles on the front page . . . all on the sun and the stars."

Regarding his personal life, he commented that

> It is peculiar. A little while ago there was no one, and now there are two. With Annie, that is Mrs. Winter . . . it has become a great friendship. We have gone on walks—with Peter, the child, and without him. I have been there in the evenings and it was very homey and we all got along very well. The husband is an engineer and intelligent. The boy is still more intelligent! (for age 4) and not an engineer. With him it is science that connects us, with her I talk more about other things. Everything—books, people, landscapes, philosophy and very personal things.
>
> Last week we went to New York to a physics meeting. The meeting was irrelevant. But in the evening I went out with Annie, and she told me for the first time what went on. She said "You were everywhere in my thoughts. I often wrote you letters but I never sent them . . . Now it is better, you are no longer inside me all the time." . . .
>
> She is wonderful, the most complete "woman" that I know. Compared to her Rose is very male and very young. Annie is very mature and at the same time has the childish joy in play, which I too still have. She is very impulsive and clever. Too bad she has a husband and son. That naturally means to forego . . . It means forgoing much. And still it is wonderful to be

> loved by such a woman, fully and completely—hundred percent says Rose. And to love her.

Rose had come to New York and had met Anne, and Hans could tell his mother that the meeting went very well. "Annie likes Rose. Rose is fond of Annie." He concluded his letter by noting that "Life is wonderful since January 1, 1939. Everyday brings something new, surprising. There is love and work. Much work and beautiful work. For the first time in my life I am completely happy and fulfilled . . . No plans for the future, only the now." The "now" meant working hard with his Ph.D. student, Robert Marshak, on the problem of stellar evolution and of white dwarfs and beginning new research in the field theoretic description of nuclear forces.

That spring Hans's relations to Rose changed once again. On May 20 Rose sent him a letter consisting entirely of drawings, as in cartoons or comics. Hans was identified as a rabbit with long ears. Some of the drawings conveyed what she was doing; others indicated when she hoped he would telephone her and that she was looking forward to being with him for four days in June. The letter's farewell greeting was a big heart.

Mama Again

During Hans's visit in the summer of 1938 his mother indicated to him that she was "determined to go to America. 'I won't be leaving anything behind and will be happier with a room next to Hans than a house in Baden-Baden.'" In fact they had surprised one another: Hans had believed that he would have to convince her and she thought that Hans was against her emigration.[25]

Kristallnacht persuaded Hans's mother that it was no longer safe for her to stay in Germany and that it had become urgent for her to leave. Hans had taken the necessary steps for her to obtain an immigration visa to the United States in the fall of 1938, but it was only in late April 1939 that she finally got both her passport and her visa.[26] She then booked a cabin for her transatlantic voyage on a boat that was to leave on May 26.

Although Hans was very concerned about whether she would be able to manage everything and face her departure without breaking down, she got all the packing and shipping done, attended to various financial matters, overcame the emptiness and sadness she felt after receiving her visa, and got herself to the boat. In his last letter to her in Germany Hans tried to cheer her up with the prospect that she would enjoy the spring which had arrived in the last few days. "Everywhere things are flowering, whole streets are yellow with forsythia, and red almond trees and white fruit trees stand among them. And the green is wonderful and fresh and the air cool and the sky blue."[27]

Mama came to the United States in the beginning of June 1939. Hans met her in New York and took her for a drive through New England. He thereafter brought her to Concord, New Hampshire, where he had rented a room in a hotel for her. She had told Hans that she didn't want to meet Rose on that trip. They did stop in Ithaca, where she met the Winters. Her reaction to Anne was negative, and the same was true for Anne.

Somewhat later that summer Rose went to England to be with her parents. While she was there Hans wrote her and once again proposed marriage. This time she accepted, feeling that Hans needed her—both for himself and to help him cope with his mother. She also felt that she needed him, that she wasn't sure that she "could make it on her own." Moreover, she had grown accustomed to and valued highly her interactions—both personal and intellectual—with him. Her refusal in the past had stemmed from her desire to obtain a professional degree and her realization that Hans wanted her to be there for "him," and what that implied should she marry him. Although in the summer of 1938 Rose had spoken "spontaneously" about marrying Hans and had told him that the thought was giving her pleasure, at that time she was not ready to give up the possibility of a career of her own. There was thus a "contractual" element when she agreed to marry Hans in the summer of 1939. Rose recognized this, and accepted that she, like her mother, would be assuming the role of a Frau Professor.

Hans met Rose at the dock when she returned from England on September 4. Together they went to City Hall in Manhattan to get married, but since they had not taken a blood test, and Rose could not provide an address for her domicile, the official couldn't marry them. Having taken care of these details, they were married a week later, on September 14, in Westchester County with the Tellers and Rose's housemother from Smith College as witnesses. The Tellers gave Rose as a wedding present a large serving plate with the inscription "He who hesitates is lost."

Thereafter Rose and Hans went off for a brief honeymoon to Vermont, after which they drove to Ithaca. Hans had rented a house large enough to accommodate them and his mother. The furniture Rose and Hans had bought in New York arrived gradually, as did that of Hans's mother. They furnished the house with them and were ready to pick up Mama and bring her to Ithaca.

Rose was apprehensive about having Mama live with them. Hans had told her how difficult and crotchety a person she was. Actually, Mama had approved the marriage, believing that she could mold Rose into becoming her maid. When Rose was in England that summer she had bought Mama a tartan blanket. When Hans went to New Hampshire to pick up Mama he gave her the blanket. When Mama came to Ithaca she had the blanket in hand, and on meeting Rose handed her the blanket, saying, "I don't need this."

That fall Rose enrolled as an undergraduate at Cornell in order to complete her bachelor's degree. She was accepted very happily by the physics department. Bethe recalled that "In those days, there were teas of the physics department, to which the wives were invited and everybody embraced her with open arms, the Smiths and the Bachers particularly so. As did Placzek. He was very much a guest in our house, very frequently so."

Rose was not so enchanted with Placzek. He made life difficult for her when visiting, frequently disparaging the social sciences with Hans joining in and supporting him, and often breaking some of Mama's fine dishes when drying them. And life with Mama was demanding. Rose made breakfast for her and at first came home every day to prepare lunch for her. Eventually Mama learned to scramble some eggs for herself for lunch.

Bethe described what happened after he and Rose agreed to have Mama live with them:

> My mother stayed with us in Ithaca from October 1939 to January of 1941, a year and a half. The first year, or, academic year, went quite well. She had her own living room, and kept sorting her many letters and so on. She was troubled by Rose and I having too much social life, especially with the Winters and Placzek, neither of whom she liked very well . . . We occasionally had colloquium guests, including Fermi, James Franck, and Felix Bloch. She liked Felix Bloch, who has been described as a very aristocratic person. Well, he had quite an aristocratic bearing, and, so he was the only one who got a really strong favorable comment from my mother. She had nothing against Franck and Fermi, but it was too much excitement for her. As Rose remembers, we had a fair amount of contact with Kennard and Mrs. Kennard. Kennard spoke German, and was willing to talk about my mother's interests, so that was favorably received. Generally things went quite well that first year, in fact, in retrospect, surprisingly well. Then we went on our

usual summer holiday—well, I should say summer travels, which were quite extensive. We must have left, in June, fairly early June, and returned probably in early September.

This was the summer of 40, and we asked my former landlady, Mrs. Dean, to take her as a boarder. That was a catastrophe. My mother considered Mrs. and Ms. Dean as her servants, treated them like she had treated her servant in Baden-Baden, and she had many wishes, many demands, and she gossiped very freely about our life, which also was not very good. So, she was very unwelcome by the end at the Deans, and I guess it had reacted on her also. She felt unwelcome, and so, this was for her the start of a depression. So, the fall of 1940 was a very bad time, her depression got worse and worse. She got a doctor who had come from Germany just then. In fact, he had come from Stuttgart. He was really a pediatrician, and she made him do whatever she wanted.[28] He gave her lots of medicines which were very bad for her, especially a lot of sleeping pills containing bromides, which I think deepened the depression. And at that time she learned of the internment of her aunt in a French concentration camp, and that was partly on her mind, and the war for the first time entered her consciousness. She spent practically all her time in bed, feeling "paralyzed."

When the situation had become intolerable, Hans and Rose asked Hans's cousin Charlotte to transfer from Goucher College to Cornell and to help look after Mama. "We were hoping that she would help with my mother, and help in the household. She was a cousin from my mother's side whom my mother had liked very much earlier. She was a little younger than Rose. And that didn't go well at all." Mama made life difficult for Charlotte by constantly speaking of her resentment of Rose and suggesting that Hans should have married Charlotte rather than Rose. Charlotte, finding life intolerable in Ithaca, returned to Goucher in January 1941. Mama's antagonism to Rose became ever more explicit, and she would harangue Hans every night, criticizing her.

So, I was between my mother and my wife, so it got more and more intolerable. Rose remembers that I came back from work, and said, "I can't work any more with this strain."

So then, Rose began searching for places where my mother could stay, and through "Self-help for German Émigrés," to which I had contributed and where Rose had worked, she found a place in Flushing, Long Island, owned by Mrs. Loschka Michel, who spoke German. We remained in contact with the Michels ever since. Her sons were in their teens when my mother was there, and one became a very prosperous engineer in New York. And with him, Henry Michel, we have stayed in contact all the time. Mrs. Michel is still alive, she is, I think, maybe three years older than I.

We brought Mama to that place, driving to New York, and delivered her. She was in pretty bad shape. In fact, Mrs. Michel and I had to carry her into the house, but within a month she recovered, as she now was being medically taken care of by a very good nurse. This nurse was very sensible and didn't give her her bromides so she very quickly became human again —and was happy there. Happily, at the same time, I went for a sabbatical to Columbia, and Rose went with me.

The 1930s in Retrospect

World War II broke out on September 1, 1939. The war, marriage to Rose, cutting the umbilical cord to Mama, and becoming an American citizen in March 1941 marked the beginning of a new life for Hans Bethe.

When the war broke out Bethe felt "at home" in the United States. He was recognized internationally as one of the outstanding theorists of his generation, had earned the admiration and affection of his colleagues at Cornell, and had married the woman he had fallen in love with. His own perspective on what had happened to him was movingly conveyed to Sommerfeld after the war, upon being offered the chair in theoretical physics in Munich.

20 May 1947

I am very gratified and very honored that you have thought of me as your successor. If everything since 1933 could be undone, I would be very happy to accept this offer. It would be lovely to return to the place where I learned physics from you, and learned to solve problems carefully. And where subsequently as your *Assistent* and as *Privatdozent* I had perhaps the most fruitful period of my life as a scientist. It would be lovely to try to continue your work and to teach the Munich students in the same sense as you have always done: With you one was certain to always hear of the latest developments in physics, and simultaneously learn mathematical exactness, which so many theoretical physicists neglect today.

Unfortunately it is not possible to extinguish the last fourteen years. My father-in-law has written to you about that already—and I believe he expressed my feelings very well. For us who were expelled from our positions in Germany, it is not possible to forget. The students of 1933 did not want to hear theoretical physics from me (and it was a large group of students, perhaps even a majority), and even if the students of 1947 think differently, I cannot trust them. What I hear about the nationalistic orientation of students at many universities starting up again, and about many other Germans as well, is not encouraging.

Perhaps still more important than my negative memories of Germany, is my positive attitude toward America. It occurs to me (already since many years ago) that I am much more at home in America than I ever was in Germany. As if I was born in Germany only by mistake, and only came to my true homeland at 28. Americans (nearly all of them) are friendly, not stiff or reserved, nor have a brusque attitude as most Germans do. It is natural here to approach all other people in a friendly way. Professors and students relate in a comradely way without any artificially erected barrier. Scientific research is mostly cooperative, and one does not see competitive envy between researchers anywhere. Politically most professors and students are liberal and reflect about the world outside—that was a revelation to me, because in Germany it was customary to be reactionary (long before the Nazis) and to parrot the slogans of the German National *(Deutschnationaler)* party. In brief, I find it far more congenial to live with Americans than with my German *Volksgenossen.*[29]

On top of that America has treated me very well. I came here under circumstances which did not permit me to be very choosy. In a very short time I had a full professorship, probably more quickly than I would have gotten it in Germany if Hitler had not come. Although a fairly recent immigrant, I was allowed to work and have a prominent position in military laboratories. Now, after the war, Cornell has built a large new nuclear physics laboratory essentially "around me." And 2 or 3 of the best American universities have made me tempting offers.

I hardly need mention the material side, insofar as my own salary is concerned and also the equipment for the Institute. And I hope, dear Mr. Sommerfeld, that you will understand: Understand what I love in America and that I owe America much gratitude (disregarding the fact that I like it here). Understand, what shadows lie between myself and Germany. And most of all understand, that in spite of my "no" I am very grateful to you for thinking of me.

We can identify three fairly well-delineated stages in Bethe's life until the mid-1950s. In the period from 1906 until the early 1930s, German culture and German institutions molded him. The two *Handbuch der Physik* articles are the fruition of this first stage. The molding of his personality in the period from the early 1930s until 1940 reflects his interactions with Sommerfeld, Ewald, Fermi, with the physicists at Cambridge, Manchester, and Bristol, with Peierls, and especially his contacts with the physics community in the United States and the sense of belonging that Cornell, the American Physical Society, and the Washington Conferences gave him. The "Bethe Bible" and his solution of the problem of energy generation in stars epitomize the capacities of the mature scientist, of the scientist who had a dominant voice in shaping nuclear physics, the new field of subatomic phenomena. Bethe's scientific output during the 1930s is overwhelming. More than half of the papers that were particularly meaningful to Bethe and included in his *Selected Works* were from the 1930s.

The third period, which began with the outbreak of World War II, saw Bethe acquiring new powers at the Radiation Laboratory at MIT and at Los Alamos. The postwar years from 1946 to 1955 constituted one of the most exhilarating phases of his life, both scientifically and professionally. The stage of his activities became national and international. He was at the center of important new developments in quantum electrodynamics and meson theory. He helped Cornell become one of the outstanding universities in the world. He was a much sought-after and highly valued consultant to the private industries trying to develop atomic energy for peaceful purposes. He was deeply involved and exerted great influence in issues concerning national security. He was happily married and the proud father of two very bright children. But the demands from his activities outside Cornell were enormous and the pace grueling. They were exacting a heavy toll both at home and in his scientific research. In 1955 Bethe went to Cambridge University to spend a sabbatical year there. It was a year of taking stock and of narrowing the focus of his scientific research.

Hannah Arendt, in an article entitled "We Refugees," described her experiences as a refugee, first in France, and then in the United States following the rise of Hitler to power: "We lost our homes, which means the familiarity of daily life. We lost our occupation, which means the confidence that we are of some use in the world. We lost our language, which means the naturalness of reactions, the simplicity of gestures, the unaffected expression of feelings."[30] Bethe's experience was almost the opposite of Arendt's. As a "refugee" he did not lose the familiarity of daily life. On the contrary he became less isolated in his daily life, and life in general became more intense, more rewarding, and more fulfilling. Nor did he lose his occupation: in fact, he obtained a temporary position that very quickly became permanent, and which allowed him to grow, and to meet and surmount new challenges on a timescale much shorter than a similar career path in Germany. In addition, he became much more creative and productive in America by virtue of the collective efforts in which he became engaged. Until he left for England, he was the sole author of his research publications. The prank with Beck and Riezler was his first collaborative effort, and a paper with Fermi his first true scientific collaboration. Thereafter many of his publications were joint efforts—with Peierls, with Livingston, with Bacher, with his graduate students and postdoctoral fellows. Nor did Bethe lose his language, nor did he suffer the consequences of its loss. He had secured his command of English during his first visit to England in 1930, and his stay in Manchester and Bristol had made him a native speaker—except for a slight accent. Furthermore, the Anglo-American context had allowed him to give expression to some of his feelings. At the symposium on the history of nuclear physics during the 1930s Bethe had entitled his talk "The Happy Thirties." Although he was referring primarily to developments in nuclear physics, his personal and professional life could likewise be characterized as "The Happy Thirties"—despite the fact that "politically the thirties were anything but happy."

Conclusion

Past and Future

Hans Bethe was an unusually gifted individual. In visualizing the past for this biography, I explored his roots, his family, his friendships, his emotional maturation, and his development as a theoretical physicist. I have highlighted the differences between the physics communities he was associated with in the 1920s and 1930s—Frankfurt University, Sommerfeld's seminar in Munich, Ewald's institute in Stuttgart, the Cavendish, Fermi's institute, Bragg's department in Manchester, and Cornell's physics department—and indicated how these differing contexts were reflected in the physics produced in them, in Bethe's persona, in his presentation of self, and in his physics, and how they enabled him to become the off-scale scientist that he was.

Writing this biography required me to comprehend the various intertwined components that made up the young Bethe's cultural, political, social, and professional world.

The Jewish Moral Element

Although Hans Bethe never considered himself to be Jewish, Judaism certainly played a role in his moral development. The majority of his friends in his youth came from highly assimilated Jewish families or families who had recently converted to Christianity from Judaism. Both Hans and his father chose wives of Jewish descent. His mother was of Jewish descent, though she did not like to admit it, and opposed World War I. Rose, the daughter of a Jewish mother who never converted, had a great influence over Hans's moral evolution. The intensification of Hans's moral concerns is due to her presence in his life. Rose's mother was a relative of Ludwig Philippson, one of the central figures in shaping Reform Judaism.

Bildung—with its belief in individualism and the potential of human reason—and the collectivist Liberal (Reform) Judaism movement both influenced Hans's moral worldview. They help explain the remarkable contributions of Jews, or individuals of Jewish descent, to German culture until the demise of the Weimar Republic, until Hitler and National Socialism seized power in 1933. I came to understand Liberal Judaism in mid-nineteenth-century Germany as the attempt to separate in Judaism the "peoplehood" component from its "religious" ethical component. Rejecting orthodox, Halacha-based Judaism, with its emphasis on commandments, on the "chosen people" aspect of its beliefs, on prayers for the return to the Holy Land and the rebuilding of the Temple there, Liberal Judaism made the vision of the Prophets and their universalism the core of its tenets and made it a religion similar to Protestantism. As a member of the Liberal Judaism community one could be a German of the Jewish faith, without any "peoplehood" strings attached to it. Reform Judaism and *Bildung* instilled a liberal outlook in many of its adherents.

Daughters in Reform Jewish families became better educated than other Jewish women and than the general population, many becoming distinguished professionals. These women also became desirable partners to members of the *Külturburgertum,* irrespective of the latter's religion. Reform Judaism of course also became a way station for eventual conversion. But in the process it fostered skepticism, autonomy, and independence, important qualities when trying to understand the creativity of individuals of Jewish descent in the sciences and in the arts in Wilhelmine and Weimar Germany.

In pointing to the "Jewish" component in Bethe's moral development I, of course, do not wish to minimize the liberal outlook of Bethe's father that resonated with many of these "Jewish" values, and deeply influenced Bethe. And the same is true for Rose's father.

Personality and Institutions

Trying to understand why Bethe characterized himself as "very conceited" when young led me to recognize the deeply hierarchical structure of German universities and the intense competition that existed among them and within them. The necessity of proving oneself outstandingly excellent could, in some circumstances, lead to asocial behavior. To aspire to a professorship at any of the leading universities demanded enormous productivity. Bethe *was* remarkably productive during the 1930s. Rabi told this story about Bethe: he was once asked by a lady in England, when he was introduced as Dr. Bethe: "Are you the Dr. Bethe who writes the *Physical Review*?" And Rabi went on to comment "that at one time it did look like Bethe wrote the *Physical Review*, and if that wasn't enough, he also wrote the *Reviews of Modern Physics*."[1]

Following Bethe's peregrinations from Munich to Stuttgart, Cambridge, Rome, Manchester, and Ithaca made clear to me certain facets of the social dimensions of theoretical physics. Theoretical physics did not have—nor does it have—a predetermined relationship to experimental physics. It is a form of practice whose toolkit of mathematics, theories, models, and metaphors is to the theorist much like the instruments and pieces of apparatus are to the experimentalist. Jeff Hughes stated the matter succinctly in his article on the practice of theory at the Cavendish in the 1920s and 1930s: theoretical physics "is a historically variable, context-specific set of practices whose relevance, meaning and relationship to other forms of practice have to be negotiated in a succession of local instances."[2]

Bohr's Institute for Theoretical Physics in Copenhagen, Born's institute in Göttingen, the Kaiser Wilhelm Institute in Berlin, and Sommerfeld's seminar in Munich had given theoretical physics an institutional framework for the application of the new mathematical technologies and for the training of new theorists. These institutions gave theorists a sense of collective identity, and visits to them fostered a collective image among the practitioners of what the enterprise of theoretical physics was about. The success of quantum mechanics gave new standing to theorists and their work, and new importance to these institutions.[3] But Munich, where Sommerfeld was the *Herr Geheimrat Professor*, was different from Göttingen. And Copenhagen differed sharply from Göttingen, Munich, and the Cavendish. Nevill Mott put it thus: "Theoretical physics at the Bohr Institute was a 'social phenomenon,' while in Cambridge it was a solitary activity." But the "social phenomenon" at the Bohr Institute manifested a characteristic that was common to them all: it was very elitist. Visitors to the Bohr Institute would either be "in" or "out." Weisskopf in his interview with Kuhn and Heilbron in 1965 in Copenhagen put it thus:

> I was "in" right away due to my old friends who were already there. [Max] Delbrück had been there a year or so and [Felix] Bloch also, . . . good friends from before. It is very difficult to get into Copenhagen; I have seen cruel things happen if you come and cannot get through the 'Guard.' Bohr was surrounded by five or six, maybe even more, of his disciples, who were a very arrogant crowd. If you were not accepted by them you would have a very difficult time with him. That was always so, and I can give you a few examples. Rabi is one; a number of Americans had a very bad time here . . . [The "outs"] might even not have noticed it, fortunately; but some did, the more intelligent ones. They would see Bohr very little because we watched it. I know because I was one of those disciples; we were not nice. Well we did it out of tremendous enthusiasm, to keep the level high. We stayed together in the evenings and discussed things. We went to the movies together. A new fellow who was not well known had a very hard time.[4]

The elitist component was transformed as theoretical physics evolved as a discipline and transplanted itself from Europe to the United States, but elitism remained an element of the social dimension of the theoretical physics community.

Scientific Style

Bethe's stay in Sommerfeld's and Fermi's institutes made him into a very special kind of theoretical physicist: a pragmatic, highly mathematically skilled problem solver. But it is also well to remember Nietzsche's insight: "Educators and formative teachers reveal . . . to you what the true basic material of your being is, something in itself ineducable and in any case hard to realize . . . : your educators can only be your liberators."[5] Sommerfeld and Fermi were superb expositors and lecturers and Bethe followed in their footsteps, honing his skills in communication. Bethe's lectures became masterpieces of organization and exposition. Like Fermi and Sommerfeld Bethe would not make any concession to expediency when lecturing on a difficult subject. He presented it in a clear, concise, and insightful manner. However, it would at times be difficult to reconstruct his lectures, for what was clear and simple to him was not necessarily so to his audiences.[6]

As a physicist Bethe molded himself into a theorist who explained experimental data and suggested new experiments to verify the predictions of particular theories, assumptions, and models. And as I previously noted, in many ways Bethe was a modeler. For him models could be autonomous in the sense that they could be studied independently of the theories and the data they mediated between.[7] For him, models, like theories, had the ability to represent and acted like instruments.

And very early on, as a result of his father's mentoring of his scientific interests, Bethe recognized that common theories and models were a potent force in structuring research communities, configuring the research articles being written and tying them to each other. As Bazerman noted, "acceptance of a common theory and of common models creates common interests among practitioners, and connects and unites wide ranges of empirical research gathered by many different people at different places and different times. Furthermore, commitment to a common theory makes possible suggestions that certain kinds of experiments be performed, indicates what kind of data ought to be collected, facilitates proposals for interpreting data, and allows the data to be harmonized with the results and ideas of others."[8] Bethe wrote his *Handbuch* and his *Reviews of Modern Physics* articles with this in mind.

Bethe also appreciated intuitively what Howard Stein called the exact-approximate duality, the general method by which some of the mathematically describable physical sciences have proceeded since Newton's time.[9] Observations are first described by a *model,* which simulates the observations. The model in turn gives rise to a first approximate *theory.* This theory in turn confronts the data that in the meantime may have become more accurate, which in turn gives rise to refinements to the first approximate theory. The discrepancy between theory and the observational or experimental data thus may become smaller through successive approximation, and the confrontation of the theory with the empirical data tests not only its domain of validity but also its correctness. The discrepancy may in fact become corroboration for a subsequent theory, as in the case of the perihelion motion of Mercury, and the replacement of Newton's gravitational theory by Einstein's general relativity. For its success the method requires that the underlying theory (for example, Newton's theory of universal gravitation and his laws of motion) have some "robustness," which can be translated into the statement that in the chain of successive approximations of the theory or in the chain of theories, each approximation or each theory is a limiting case of its successor. In the case of theories, the replaced theory will be derivable from its successor when the latter is restricted to a certain natural domain—as in the case of classical mechanics being able to be derived from the special theory of relativity in the limit that all the particles involved have velocities that are much smaller than the speed of light.

Conversely, Bethe had appreciated the *novelty* that emerged from the synthesis of quantum mechanics and special relativity. Their melding implies that when describing particles of mass m a new length scale appeared, the Compton wavelength of the particle h/mc, which vanishes in the classical limit, $h \rightarrow 0$, and in the nonrelativistic limit, $c \rightarrow \infty$. Heisenberg's uncertainty principle then implied that probing distances much smaller than the Compton wavelength entailed involving momenta much greater than mc and hence energies much greater than mc^2, and thus the possibility of particle creation. Establishing the limits of validity of theories in which particle creation and annihilation are intrinsic features of the theory, such as quantum electrodynamics, became an important component of Bethe's research program until the mid-1950s. These theories could only be analyzed by perturbation theory, and exhibited divergences in higher orders of perturbation theory. Very early on it was recognized that these divergences were connected with the short distance structure of the theory. Bethe became deeply interested in these problems in the late 1930s. The war shifted the focus of his research to problems of armor penetrations, shock waves, wave guides, neutron diffusion, and atomic bombs. It is only during the academic year 1946–47 that Bethe returned to quantum field theoretical problems when teaching a course in advanced quantum mechanics.

Visualizing the Future

Being one of the outstanding theorists in the United States, Bethe was invited to the June 1947 Shelter Island Conference. While attending that conference Bethe saw a way to get rid of the infinity that appears in the perturbative treatment of quantum electrodynamics when the dynamics of the charged particles are described nonrelativistically. The episode is important because the Shelter Island Conference turned out to be one of the most seminal conferences held right after the end of World War II—a conference whose impact was comparable to that of the Solvay Congress of 1911. Just as the Solvay Congress of 1911 set the stage for all the subsequent developments in quantum theory, similarly Shelter Island provided the initial stimulus for the post–World War II developments in quantum field theory: effective, relativistically invariant, computational methods; Feynman diagrams; renormalization theory. The conference was also responsible for the elucidation of the structure of the mesonic component of cosmic rays. Richard Feynman many years later recalled: "There have been many conferences in the world since, but I've never felt any to be as important as this."

The conference ushered in a new stage in Bethe's life. His participation at the conference demonstrates once again his most impressive abilities as a physicist. His seminal contribution stemming from his attendance can be seen as the fruition of his labors during the 1930s. But the episode also reveals certain aspects of his character that remained a legacy of his German university education.

Though my biography is concerned with Bethe's life before 1940 or so, I shall conclude with a glimpse of the Bethe that emerged from World War II.

Bethe after World War II

The Bethe who came to the Shelter Island Conference had directed the Theoretical Division of the Los Alamos Laboratory and had made essential contributions to the success of the project. And when going back to Cornell after the war he had restructured physics there. A word about that transformation is in order.[10]

In late 1944 it had become evident that the tide of the war had turned and that the defeat of the Axis powers was in sight. At Cornell as elsewhere extensive plans were being drawn for the expansion of physics activities once the war was over.[11] The Cornell administration had become aware that the character of the physics enterprise had dramatically changed. Many of the leading members of its physics department who were working on wartime projects in government and industrial laboratories were being wooed with attractive positions elsewhere. Bethe and Bacher received lucrative offers from the University of Rochester in 1944.[12] Extensive discussions between Bethe and Bacher took place thereafter at Los Alamos, resulting in a memorandum by Bacher to Gibbs outlining "Plans for Nuclear Physics at Cornell University." Bethe and

Bacher's conclusion was that "the most important plan of post war research will be the high energy particle physics using a betatron source."[13] Edmund Ezra Day, Cornell's president, took up the challenge. He committed himself to make university funds available to build an accelerator.

Lloyd Smith had received an offer from the RCA laboratories in Princeton to head a new division that was to devote itself exclusively to "fundamental pure" research, and thereafter a tug of war developed over the direction that the Cornell physics department was to take. Should it concentrate on "fundamental" or on "applied" physics? These differences were settled in the fall of 1945 by the adoption of a "Proposal of Organization of Physics at Cornell University" that Bethe drafted. The document proposed that

> The entire development in physics shall be divided in two parts, (1) the Physics Department and (2) the Physics Research Laboratory. The latter is at present designed to develop high energy particle physics including nuclear physics, cosmic rays, and researches with the help of a high energy electron accelerator. The Physics Department will retain all other research and teaching functions. There will be a Chairman of the Physics department whose functions will be essentially the same as those of the present chairman. There will be a Director of the Research Laboratory. In addition there will be a Coordinating Committee whose purpose it is to correlate the activities of the department and the research laboratory. The Committee will consist of the Chairman of the Department, the Director of the Laboratory, and a third person who will be the Chairman of the Coordinating Committee.[14]

The proposal, circulated to the faculty members of the physics department, was accepted by them and thereafter by the administration. Thus, the power struggle was resolved by having a triumvirate govern the physics operations, with Smith as chairman of the department responsible for faculty and teaching, Bacher in charge of the semi-autonomous Laboratory for Nuclear Studies, and Bethe the coordinator and moderator between the two. The three were referred to as the father, the son, and the holy ghost, with Bethe clearly the dominant personality in the trinity and the ultimate arbiter.

In September 1946, the Cornell Board of Trustees unanimously voted to appropriate $1,200,000 for the "specific support of the Nuclear Studies Project." Writing to Bacher to inform him of the board's action, Day asserted that

"No decision of the Board of Trustees during my term of office has been of greater importance to this institution and to the prospects of scientific research in this country."[15]

Bethe and the Shelter Island Conference

Bethe thus came to the Shelter Island Conference in June 1947 as one of the most admired, respected, and influential theorists in the United States. The conference turned out to be the first of three small post–World War II conferences on theoretical physics sponsored by the National Academy of Sciences.[16] The physicists invited to the conference were primarily theoretical physicists, most of whom had been leaders at the highly successful wartime laboratories: the MIT Radiation Laboratory, the Chicago Metallurgical Laboratory, and Los Alamos. Also in attendance were several experimentalists: Lamb and Rabi, who reported on experiments on the spectrum of hydrogen they had recently performed at Columbia University; and Rossi, who reported on the results of research carried out in Rome on the absorption of cosmic rays in the atmosphere.

The three conferences—Shelter Island, Pocono, and Oldstone—were small, closed, and elitist. In a sense they marked the postponed end of an era, that of the 1930s, and its characteristic style of doing physics: small groups and small budgets. None cost more than $1,500.[17] Coming after World War II, these conferences reasserted the value of pure research and helped to purify and revitalize the theoretical physics community. They also made evident the new social reality implied by the newly acquired power of the theoreticians and helped integrate the most outstanding of the younger theoreticians into the elite: Richard Feynman, Julian Schwinger, Robert Marshak, David Bohm, and Abraham Pais at Shelter Island, and Freeman Dyson at Oldstone.

The date of the Shelter Island Conference was set to be June 2, 3, and 4, 1947—Monday, Tuesday, and Wednesday—to accommodate the presence of Oppenheimer, who was to write a five-hundred-word paper to outline subjects for discussion at the conference. The other two persons who were asked to prepare papers for the conference were Victor Weisskopf and Hendrik Kramers.[18] In his paper Kramers reviewed the difficulties encountered in quantum electrodynamics since its inception in 1927 and indicated one way out of these problems by referring to his own work and that of his students, J. Serpe and W. Opechowski. In these papers, which had been published in

1940, a model had been elaborated in which all contributions to the description of the interaction of charged particles with the electromagnetic field that could be characterized as structural effects had been eliminated. The model described "how an electron with *experimental mass* behaves in its interaction with the electromagnetic field."

By the end of May 1947 the preliminary results of Lamb and Retherford's experiment—that the $2s_{1/2}$ level of hydrogen lies 1,000 megacycles above the $2p_{1/2}$ level—had been widely circulated. Schwinger and Weisskopf discussed the theoretical implication of that finding on their train ride from Boston to New York to attend the Shelter Island Conference and agreed that the effect was very likely quantum electrodynamic in origin.

The conferees gathered in New York on Sunday, June 1, 1947, at the headquarters of the American Institute of Physics on East 55th Street. From there they were taken "on an old and shaky" bus to Greenport at the tip of Long Island. On the final phase of the trip they were accompanied by a police motorcycle escort, and their bus didn't stop at any traffic lights. As they passed each county line a new police escort would meet them. In Greenport the conferees were wined and dined as guests of the Chamber of Commerce. The dinner, although officially given by the Chamber of Commerce, was actually paid for by John C. White, its president. In an after-dinner speech he told the audience that he had been a Marine in the Pacific during the war and that he —and many like him—would not be alive were it not for the atomic bomb.

Karl Darrow chaired the conference, but Oppenheimer was the dominant personality and "in absolute charge."[19] On the first day of the conference Lamb presented the most recent results from his experiment with Retherford on the fine structure of hydrogen. In the discussion which followed Oppenheimer pointed out that if one calculated the *difference* in the 2p 2s energy levels using hole theory, a finite answer might be obtained given that the divergences encountered were logarithmically divergent. Lamb's talk was followed by Rabi, who presented the data John Nafe, Edward Nelson, and he had obtained on the hyperfine structure of H and D. That afternoon Rossi reported on the experiment of M. Conversi, E. Pancini, and Oreste Piccioni on the absorption of mesons in the atmosphere.

The next morning Kramers presented his version of the Lorentz theory of an extended charge in which structural effects had been encapsulated in the experimental mass of the particle. He concentrated on the classical version of his nonrelativistic theory, but the extensive notes that Bethe took of Kramers's presentation make clear that in the last part of his talk he indicated what quantizing the theory would do.

After the conference was over, Bethe performed his famous Lamb shift calculation on the train ride from New York to Schenectady. Bethe's paper that proved that the level shift would be accounted for quantum electrodynamically was completed by June 9, and he circulated it to the participants of the conference.[20] Recalling his behavior in Germany as a young theorist, Bethe did not acknowledge Kramers's talk in his paper, even though Kramers's presentation had been crucial by stressing the importance of expressing observables in terms of the *experimental* mass of the electron.[21]

Why this was so is probably because Bethe thought he had a much simpler way to incorporate Kramers's insight than Kramers had found. Bethe had noted that the quantum electrodynamically calculated self-energy of a free nonrelativistic electron could be ascribed to an electromagnetic mass of the electron, and though divergent, had to be added to the mechanical mass of the electron. The only meaningful statements of the theory involve the combination of electromagnetic and mechanical masses, which is the experimental mass of a free electron. In contrast to Kramers, Bethe's was a *model independent* formulation of mass renormalization that did not assume an extended charge distribution to the electron. And in contrast to Schwinger and Weisskopf's initial insight that the finiteness of a hole theoretic calculation could be achieved by computing the difference between the energies of two levels, Bethe's mass renormalization allowed computing the energy of each level and gave an unambiguous formulation of mass renormalization in the nonrelativistic case. Moreover, he knew what was required to give an analogous relativistic prescription. Weisskopf and Schwinger, though emphasizing Kramers's insight, could not do so at Shelter Island.[22]

Max Dresden, in his biography of Kramers, suggested that Kramers did not receive adequate credit for his contributions at Shelter Island.[23] The recently discovered extensive notes that Bethe had taken at Shelter Island indicate that Dresden was right. To Bethe, the pragmatist for whom numbers were always the criterion of good physics, and who had just been so deeply and successfully involved in the war effort calculating numbers that translated into physical effects and measurable empirical data, the challenge was to get the numbers out and account for the magnitude of the 2S-2P shift in hydrogen and for the new values of the hyperfine splitting in H and D that Nafe, Nelson, and Rabi had measured. Accounting for the empirical data would be explaining the data. Kramers's approach was too model dependent, too theoretical, and too far removed from calculating numbers. For Bethe, the value of a novel idea was gauged by whether it could help you calculate numbers that could be compared with empirical data.[24]

The calculation was a "crucial calculation." By introducing the concept of mass renormalization and its associated consequences so simply, Bethe's calculation gave a new perspective on how to address quantum electrodynamic calculations.[25] It was crucial also because Bethe could convincingly justify the cut-offs he had to introduce in his nonrelativistic calculation and obtain a logarithmic expression for the Lamb shift that agreed with the observed level shift.[26] It should be added that only Bethe could have evaluated the logarithmic contribution as quickly as he did. He had encountered similar logarithmic expressions when calculating quantum mechanically the energy loss of a fast charged particle traversing matter in his *Habilitationsschrift* in 1929!

Schwinger at the time made another crucial calculation—perhaps a more important one. He calculated the quantum electrodynamic contribution to the magnetic moment of the Dirac electron. It was crucial because it was the first *field theoretical* calculation of quantum electrodynamic radiative corrections. Up to this point all calculations had been based on hole theory. Schwinger's calculation asserted that a quantum *field theoretic* description of both electrons and radiation is the generative and effective way to describe atomic and subatomic processes.

The Shelter Island, Pocono, and Oldstone conferences were the precursors of the Rochester Conferences on High Energy Physics.[27] They differed from these later conferences in important ways. While the Shelter Island, Pocono, and Oldstone conferences reflected the style of an earlier era, the Rochester conferences were more professional and democratic in outlook and had the imprint of the new era: the large group efforts and the large budgets involved in machine physics.[28]

Indeed a new era had begun; not only in physics and the role of physics but in national and international politics, in American economic growth, and in American hegemony. For Bethe it meant entering a new world with new challenges and rewards. At the personal level, he enjoyed a happy home life and saw his two children grow up; in his professorial life, he was surrounded by outstanding colleagues such as Richard Feynman and Robert Wilson. He also collaborated with outstanding postdoctoral fellows such as Edwin Salpeter and mentored outstanding students such as Freeman Dyson. In his professional life outside the university, he was deeply involved in trying to make nuclear power a viable peaceful option. He became an influential consultant to Los Alamos and a trusted adviser to the Pentagon. It was an exhilarating period in his life that his talents and upbringing had made possible—but a phase that could not last.

Appendixes

A. The Bethe Family Genealogy
B. Courses Taken at Frankfurt University
C. A Brief History of the Genesis of Quantum Mechanics
D. Courses Taken at Munich University
E. Bethe's Doctoral Thesis
F. The *Habilitationsschrift* Defense

Notes

References

Acknowledgments

Index

APPENDIX A

The Bethe Family Genealogy

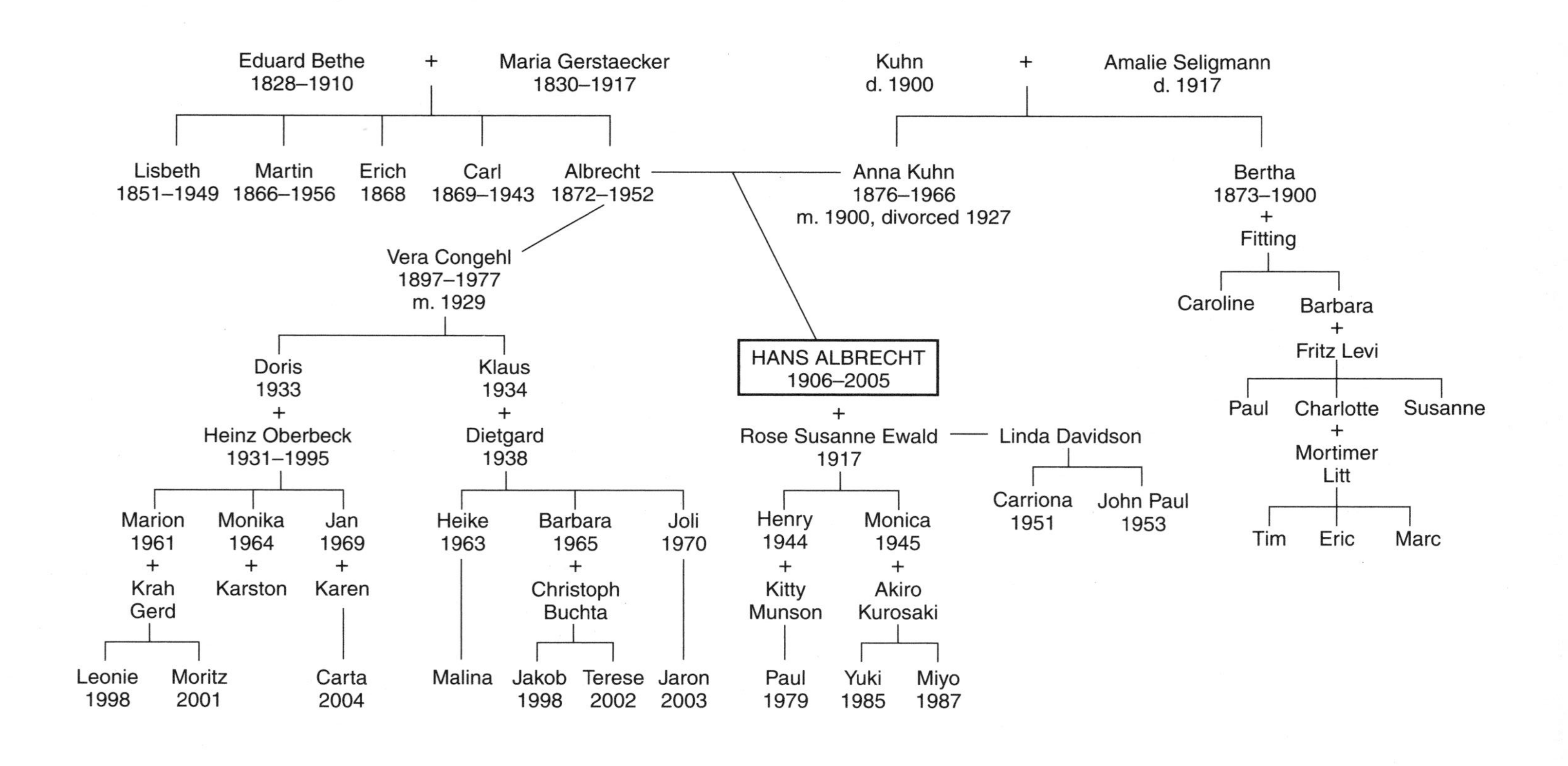

Eduard Bethe 1828–1910
+
Maria Gerstaecker 1830–1917
Kuhn d. 1900
+
Amalie Seligmann d. 1917
Lisbeth 1851–1949
Martin 1866–1956
Erich 1868
Carl 1869–1943
Albrecht 1872–1952
Anna Kuhn 1876–1966 m. 1900, divorced 1927
Bertha 1873–1900
+
Fitting
Vera Congehl 1897–1977 m. 1929
Caroline
Barbara
+
Fritz Levi
HANS ALBRECHT 1906–2005
+
Rose Susanne Ewald 1917
Linda Davidson
Paul
Charlotte
+
Mortimer Litt
Susanne
Doris 1933
+
Heinz Oberbeck 1931–1995
Klaus 1934
+
Dietgard 1938
Carriona 1951
John Paul 1953
Tim
Eric
Marc
Marion 1961
+
Krah Gerd
Monika 1964
+
Karston
Jan 1969
+
Karen
Heike 1963
Barbara 1965
+
Christoph Buchta
Joli 1970
Henry 1944
+
Kitty Munson
Monica 1945
+
Akiro Kurosaki
Leonie 1998
Moritz 2001
Carta 2004
Malina
Jakob 1998
Terese 2002
Jaron 2003
Paul 1979
Yuki 1985
Miyo 1987

APPENDIX B

Courses Taken at Frankfurt University

Summer semester 1924

Szász	Differential and Integral Calculus I with Exercises
Siegel	Higher Arithmetic
Siegel	Exercises for Higher Arithmetic
Wachsmuth	Experimental Physics I
Gerlach	The Experimental Foundations of Atomistics
Sieverts	General Chemistry I (Inorganic Chemistry)
Kalverlam	Accounting with Exercises

Winter semester 1924–25

Dehn	Mechanics
Dehn	Exercises in Mechanics
Wachsmuth	Experimental Physics II
Lorenz	General Physical Chemistry
von Braun	General Chemistry II (Inorganic)
F. Hahn	Seminar on Inorganic Chemistry
von Braun, Sieverts, F. Hahn	Inorganic Chemistry Laboratory
Gerlach	The Mechanical and Electrical Properties of Matter
Gerlach	The Foundations of Modern Physics

Summer semester 1925

Frobenius	Organization of People and States
Dehn	Differential Geometry
Madelung	Special Topics in Theoretical Physics (Quantum Theory, Theory of Relativity)
Meissner	Modern Spectroscopy
Wachsmuth	Physics Laboratory for Beginners
Lorenz	Electrochemistry
Magnus	Chemical Thermodynamics
F. Hahn	Seminar on Inorganic Chemistry
von Braun, Sieverts, Hahn	Chemical Laboratory in Inorganic Chemistry

Winter semester 1925–26

Dehn	Differential Equations in the Real Domain
Dehn	Exercises in Differential Equations
Madelung	Mechanics
Meissner	Advanced Experimental Physics I: Mechanics, Acoustics, Heat
Meissner	Exercises in Advanced Experimental Physics
Meissner	Universal Physical Constants and the Methods of Their Determination
Madelung	Exercises in Theoretical Physics
Wachsmuth	Physical Laboratory for Advanced Students
Magnus	General Chemistry II: Atomics
von Braun, Sieverts, Hahn	Chemical Laboratory in Inorganic Chemistry

APPENDIX C

A Brief History of the Genesis of Quantum Mechanics

This appendix conveys the state of affairs in the development of quantum mechanics when Bethe was admitted to Sommerfeld's seminar. Schrödinger's contribution seems magnified because Willy Wien and Sommerfeld, as editors of the *Annalen der Physik,* received his papers before anyone else. Schrödinger's approach was carefully analyzed in the seminar and resonated with Sommerfeld's outlook.

By the spring of 1925 the theoretical models and representations that had been patched together and formed what subsequently was called the "old quantum theory" was a rather tangled web of successes and failures. Pauli captured the tenor of the times when in May 1925 he wrote Ralph Kronig, "Physics is at the moment very muddled again, it is much too hard for me anyway, and I wish I were a movie comedian or something like that and had never heard anything about physics."[1]

When Pauli was writing his letter, the following were accepted experimental "facts":

1) The Einstein-Bohr relation between atomic energy levels E_n and the frequencies ω_{nm} of the radiation emitted or absorbed in a transition

from n to m: $E_n - E_m = \hbar\omega_{nm}$; as well as the Ritz combination principle, $\omega_{nm} + \omega_{mr} = \omega_{nr}$;
2) The Pauli exclusion principle and the attribution of a two-valued quantum number to an electron stemming from its spin;
3) The particle-wave duality for light that had been advocated by Einstein since 1905 and rendered uncontroversial with the discovery of the Compton effect in 1923.

In addition, there was a consensus about the following two useful, efficacious, and widely used theoretical elements:

1) Bohr's correspondence principle, which stated that in the limit of large quantum numbers the results of the quantum theory reduce to classical theory, and
2) The all-important Einstein relation between the *probability* for spontaneous emission, A_{nm}, and the *probability* of stimulated emission, B_{nm}, between the states n and m, and more generally his relations between the rates for absorption and spontaneous and stimulated emission. This is where probability entered the quantum theory.

The situation changed dramatically during the summer of 1925. Heisenberg took the next crucial step in a remarkable paper submitted to *Zeitschrift für Physik* at the end of July in which he "fabricated" quantum mechanics. He there noted that "The Einstein-Bohr frequency condition already represents such a complete departure from classical mechanics, or rather from the kinematics underlying this mechanics, that even for the simplest quantum-theoretical problems the validity of classical mechanics cannot be affirmed." The guiding principles that Heisenberg adopted were:

1) The new mechanics must satisfy the Bohr correspondence principle; i.e., that in the limit of large quantum numbers the theory should yield the classical results;
2) The difficulties encountered in the "old quantum theory" were not due mainly to a deviation from classical dynamics, but rather a breakdown of the kinematics underlying this dynamics;

3) To establish a quantum theoretical mechanics one must base oneself entirely upon relations between observable quantities.[2]

Heisenberg believed that the position of an electron in a Bohr orbit is not observable. The question he had to answer was therefore the following: Assuming that the equation of motion of an electron in a quantum state n,

$$\frac{d^2x(n,t)}{dt^2} + f(x(n,t),t) = 0, \tag{C1}$$

could be retained and that only the kinematical interpretation of the quantity $x(n,t)$ as the position had to be abandoned, what kind of quantities should replace the $x(n,t)$ that appears in the classical equation of motion? The solution Heisenberg gave was to associate with $x(n,t)$ an "ensemble of quantities" X_{mn} that depended on two quantum states, m and n. He replaced the Fourier series representation of $x(n,t)$ in the case of classical *periodic* motion

$$x(n,t) = \sum_{m=-\infty}^{m=+\infty} X_m(n)e^{im\omega t} \tag{C2}$$

by a new kind of expansion, wherein the Fourier coefficients *and* the frequencies depended on a quantum number m:

$$x(n,t) = \sum_{m=-\infty}^{m=+\infty} X_{nm}e^{i\omega(m,n)t}. \tag{C3}$$

The reason for this was that in previous work on dispersion theory, Kramers and Heisenberg had shown that the transition frequency corresponding to the classical $m\omega(n)$ is in general not a simple multiple of the fundamental frequency but is given by the Bohr-Einstein relation

$$\hbar\omega(n,n-m) = [E_n - E_{n-m}]. \tag{C4}$$

Hence Eq.(C3).

Now according to classical theory the power (the energy per unit time) emitted in a transition corresponding to the mth harmonic $m\omega$ is:

$$-\left(\frac{dE}{dt}\right) = \frac{4e^2}{3c^3}[m\omega(n)]^4 |X_n(m)|^2. \tag{C5}$$

In the quantum theory the lefthand side of (C5) has to be replaced by the product of the Einstein transition probability per unit time, $P(n,n-m)$, and the emitted energy, $\hbar\omega(n,n-m)$. Equation (C5) then becomes

$$P(n,n-m)=\frac{4e^2}{3c^3}[\omega(n,n-m)]^3\left|X_{n,n-m}(m)\right|^2. \tag{C6}$$

Heisenberg then posed the question of how the classical quantity $|x(t)|^2$ was to be represented in order to yield Eq.(C6). Classically,

$$|x(n,t)|^2=\sum_r Y_r(n)e^{ir\omega(n)t}. \tag{C7}$$

where

$$Y_r(n)=\sum_s X_s(n)X_{r-s}(n) \tag{C8}$$

In order to obtain (C6) and simultaneously satisfy the Ritz combination principle, $\omega_{nm}+\omega_{mr}=\omega_{nr}$, Heisenberg had to combine the quantum amplitudes so that

$$Y_{n,n-m}e^{i\omega(n,n-m)t}=\sum_r X_{n,n-r}e^{i\omega(n,n-r)t}X_{n-r,n-m}e^{i\omega(n-r,n-m)t}$$

or

$$Y_{n,m}=\sum_r X_{n,r}X_{r,m}, \tag{C9}$$

which is Heisenberg's rule for multiplying transition amplitudes.

This new approach was quickly developed into an elaborate mathematical formalism by Born, Jordan, and Heisenberg. Born had received from Heisenberg a copy of his paper on a quantum mechanical reinterpretation of kinematic and dynamical relations at the time of its submission to *Zeitschrift für Physik.* Upon pondering the meaning of the rule that Heisenberg had given for the multiplication of two quantum mechanical transition amplitudes, Born came to the conclusion that Heisenberg's symbolic multiplication was nothing but the matrix calculus, a calculus that was well known to him.[3] Like Kramers and Heisenberg, Born and Jordan had considered such symbolic multiplications of transition amplitudes during their discussions when writing a paper on the absorption of radiation.[4] However, they had not realized the implications of such symbolic multiplications, whereas Heisenberg had.

Born once again asked Jordan to collaborate with him, and within two months Born and Jordan had laid the foundations of the new matrix mechanics.[5] They showed that by starting with the basic premises given by Heisenberg, it is possible to build a closed mathematical theory that displays strikingly close analogies with classical mechanics, but at the same time preserves the characteristic features of quantum phenomena. They completed this work without knowing how to impose quantum rules for systems with more than one degree of freedom. Heisenberg, in early September 1925, formulated such rules after becoming acquainted with Born and Jordan's matrix formulation of quantum mechanics. The famous *Dreimänner* paper extended Born and Jordan's work to systems with an arbitrary (but finite) number of degrees of freedom.[6] For the paper Born and Heisenberg developed a perturbation theory that took into account the possible degeneracy of the energy spectrum of such a system. The paper also included a derivation of the conservation laws for energy and angular momentum. And in a chapter entitled "Physical Applications of the Theory," as a result of Jordan and Heisenberg's work, the commutation rules for the components of the angular momentum operator $\boldsymbol{M}$,

$$[M_x, M_y] = \frac{h}{2\pi i} M_z, \quad \text{(C10)}$$

and cyclic permutations were derived and a proof given that in the representation in which M_z and $\boldsymbol{M}^2$ are diagonal, the eigenvalues of M_z are $m(h/2\pi)$, with m integer or half-integer, and those of $\boldsymbol{M}^2$ are $j(j+1)(h/2\pi)^2$ with j integer or half-integer. Using these results, formulae for the intensities and polarization of atomic transitions were derived.

Heisenberg had given a seminar in Cambridge in mid-July 1925, right after having finished the calculations that went into the paper he was to submit to *Zeitschrift für Physik* later that month. He mentioned this work only briefly in his talk but amplified on it in a subsequent discussion with Ralph Fowler. Fowler asked him to send on the proof sheets of his paper when they became available. When these arrived in late August, Fowler sent them to Dirac with the question, "What do you think of this? I shall be glad to hear."

Dirac thought hard about what he read and came to the conclusion that it was the noncommutativity of the quantum variables which Heisenberg had introduced that "would be the key to understand[ing] the atom." In early October he had worked out his formulation of quantum mechanics. "It is not the equations of classical mechanics that are in any way at fault but the mathematical operations by which physical results are deduced from them that require modification. All the information supplied by the classical physics can thus be made use of." He then established a relation between the commutator xy − yx of two "quantum quantities," what he later called *q*-numbers, and the Poisson bracket in the correspondence limit. Generalizing this, he postulated the commutation relations to be satisfied by the dynamical variables q_i and p_i, corresponding to position and momentum

$$\text{(C11)} \qquad [q_l, p_m] = i\hbar\delta_{lm} \quad ; \quad [q_l, q_m] = 0 \quad ; \quad [p_l, p_m] = 0,$$

and derived the equation of motion for a dynamical variable $Q(t)$,

$$\text{(C12)} \qquad i\hbar\frac{dQ(t)}{dt} = Q(t)H - HQ(t)$$

where H is the Hamiltonian of the system. He thus had independently obtained all the results that Born, Jordan, and Heisenberg had published somewhat earlier, but of which he had been completely unaware.

Although the *Dreimännerarbeit* had laid the foundations of a consistent quantum theory, it did so by requiring that the theory deal only with relations among observable quantities and by relinquishing the possibility of giving a physical, visualizable picture of the processes it could calculate. Thus, it was not clear how to formulate the description of scattering processes with it. Heisenberg's matrix mechanics was seen by many as a kind of *phenomenological* theory precisely because he had taken his problem as establishing relationships between *observable* quantities. Furthermore, matrix mechanics appeared strange and alien. This was one of the reasons that the wave mechanics of Schrödinger was so welcome by Planck, Einstein, and Lorentz.[7] Wave mechanics seemed to avoid the unconventional features of Heisenberg's formulation and was based on more traditional foundations: variational principles, differential equations, and the properties of waves. And although it operated with new conceptions and abstractions, some of which were in principle inaccessible to observations, its claims of being foundational resonated with the outlook of the older generation.

Schrödinger's wave mechanics was an outgrowth of insights and suggestions by de Broglie and by Einstein. De Broglie in 1923 had an idea that was the obverse of Einstein's attribution of particle properties to wave radiation: de Broglie endowed discrete matter with wave properties. To a particle of (rest) mass m_0, energy E, and velocity v, de Broglie associated a wave of frequency ν with

(C13)
$$E = h\nu = \frac{m_0 c^2}{\sqrt{1 - \left(\frac{v^2}{c^2}\right)}}$$

and a phase velocity V given by

(C14)
$$V = c^2/v.$$

Equivalently, this implies attributing to the particle a wavelength $\lambda = V/\nu$ with

(C15)
$$1/\lambda = p/h$$

where p is the momentum of the particle

(C16)
$$p = \frac{m_0 v}{\sqrt{1 - \left(\frac{v^2}{c^2}\right)}}.$$

Thus,

(C17)
$$\nu = E/h; \lambda = h/p.$$

These relations permitted de Broglie to give a novel geometrical interpretation of the Bohr-Sommerfeld quantization condition $\oint pdq = nh$:

(C18)
$$\left(\frac{1}{h}\right)\oint p\,dq = \oint dq / \lambda = (\text{length of orbit}) / \lambda = n,$$

i.e., the Bohr quantization condition could be translated into the requirement that an integral number of wavelengths fit the perimeter of the orbit. De Broglie amplified these ideas in his thesis, which was subsequently published in *Annales de Physique* in early 1925.[8]

The particle-wave duality of radiation was thus extended to matter. Einstein in his paper on Bose statistics endorsed de Broglie's idea of attaching a wave field to matter and indicated that "This wave field, whose physical nature is still obscure for the time being, must in principle be demonstrated by the diffraction phenomena corresponding to it. A beam of gas molecules which passes through an aperture must, then, undergo a diffraction analogous to that of a light ray." Einstein was aware that the diffraction angle would be extremely small for physically realizable slits and, therefore, that a straightforward experimental verification of the hypothesis might prove difficult.

Gregor Wentzel, who was a student in Sommerfeld's seminar in the early twenties, recalled in his interview with Thomas Kuhn that Sommerfeld's seminar was "considerably" impressed by de Broglie's paper.[9] Sommerfeld had heard of it from Einstein, and Einstein's great reputation had been responsible for the seminar studying it. However, the members of the seminar did not do much with de Broglie's paper, except for Wentzel's favorable review of it in *Physikalische Berichte.*[10] A student of Max Born, Walter Elsasser, did follow up Einstein's comment and observed that electrons with energies below 25 eV would be particularly well suited to test the "assumption that to every translational motion of a particle one must associate a wavefield which determine[s] the kinematics of the particle." Elsasser furthermore pointed out that the existing experimental results of Carl Ramsauer, Clinton Davisson, and Charles Kunsman already seemed to give evidence of diffraction and interference of matter waves. Bethe would take up this problem after the advent of wave mechanics and make it his own.

But it was left to Erwin Schrödinger to extend de Broglie's and Einstein's ideas to a dynamical scheme.

Schrödinger

In the fall of 1925, Felix Bloch was a young graduate student in physics at the ETH and had started attending the joint physics colloquium of the ETH and the University of Zurich. Peter Debye, who was a professor at the ETH, and who was acknowledged as a physicist of the first rank, ran the colloquium with firm authority. Schrödinger had been the professor of theoretical physics at the university since 1922. Bloch later recalled that

> Once at the end of a colloquium I heard Debye saying something like: "Schrödinger, you are not working right now on very important problems anyway. Why don't you tell us sometime about that thesis of de Broglie, which seems to have attracted some attention." So, in one of the next colloquia, Schrödinger gave a beautifully clear account of how de Broglie associated a wave with a particle and how he could obtain the quantization rules [of Niels] Bohr and Sommerfeld by demanding that an integer number of waves be fitted along a stationary orbit. When he had finished Debye casually remarked that he thought this way of talking was rather childish. As a student of Sommerfeld he had learned that, to deal properly with waves, one had to have a wave equation. It sounded quite trivial and did

not seem to make a great impression, but Schrödinger evidently thought a bit more about the idea afterwards.[11]

In mid-January 1926 Schrödinger gave another colloquium, which he began by saying: "My colleague Debye suggested that one should have a wave equation; well I have found one." Between the two lectures Schrödinger had worked with singular concentration and had obtained his equation.

As Schrödinger was developing his "wave mechanics" he kept in close contact with both Willy Wien and Arnold Sommerfeld. Thus on December 27, 1925, shortly after he had obtained his nonrelativistic wave equation, Schrödinger had written to Willy Wien, "I am very optimistic, and expect that . . . it will be *very* beautiful . . . I can specify a vibrating system that has as eigenfrequencies the hydrogen *term* frequencies—and in a relatively natural way, not through ad hoc assumptions."[12] Schrödinger had first become acquainted with Wien in the spring of 1925 when, as one of the two editors of the *Handbuch der Experimentalphysik,* Wien had asked him to contribute an article to the *Handbuch* on color theory and the physiological processes connected with vision, a subject that Schrödinger had previously worked on. Wien, as one of the editors of the prestigious *Annalen der Physik,* had subsequently requested that both Schrödinger and Debye, the two leading physicists in Zurich, submit papers to the *Annalen.* Schrödinger complied with the invitation and sent all his papers on wave mechanics to Wien for publication in the *Annalen.* Since Wien asked Sommerfeld to look at all the theoretical papers submitted to the *Annalen* and comment on their suitability for publication, Wien and Sommerfeld were among the first physicists to study Schrödinger's papers.

Shortly after Schrödinger had submitted his first paper on January 27, 1926, *Quantisierung als Eigenwertproblem,* Wien wrote him to ask how one would derive Planck's radiation law from his theory and to tell him that his paper, "which I have passed on—following your request—to Sommerfeld, has raised great interest here. Of course it is not yet possible to oversee all consequences."[13] At the end of January, Schrödinger sent Sommerfeld a letter to tell him the further results he had obtained regarding the wave mechanical description of the one-dimensional harmonic oscillator, that of the rigid rotator in three dimensions and that of a free particle using his wave equation, results that would form the second half of the second article that Schrödinger submitted to *Annalen.* Sommerfeld promptly replied, telling Schrödinger that

"What you write in the [first] paper, as well as in the letter is terribly interesting." Sommerfeld expressed the view that his new method is "a substitute for the new quantum mechanics of Heisenberg, Born and Dirac; [it is,] in particular, a simplified method, so to speak, an analytic resolution of the analytic problem stated there, because your results fully agree with theirs."[14] A few days later, Sommerfeld wrote Pauli, informing him that Schrödinger had obtained exactly the same result as Heisenberg for the harmonic oscillator and as Pauli had for the hydrogen atom "but in a quite different, totally crazy way, not by matrix algebra but by solving boundary value problems." He ended his letter by commenting that "Surely, in some way or another, something reasonable and final will soon emerge from all this."[15]

Schrödinger kept Sommerfeld informed of the progress he was making in developing his "wave mechanics," an appellation he first used in a letter to Sommerfeld in February 1926. In that letter he described his calculation of the Stark effect in hydrogen using his newly formulated perturbation theory. Shortly thereafter he apprised Sommerfeld of the fact that he had been able to show the equivalence of the Heisenberg-Born-Jordan quantum mechanics with his own. This last paper was particularly influential, for it unified what previously had been seen as distinct approaches. Equally important, it paved the way to the probabilistic interpretation of the ψ function. It had always been explicit in Heisenberg's approach that his x_{nm} were related to the Einstein A and B coefficients. A probabilistic element was thus intrinsic to this approach and its elaboration by Born and Jordan. With the equivalence of Schrödinger's approach to that of Born, Jordan, and Heisenberg—and that of Dirac—it become immediately obvious to several people that similarly the ψ function must have a probabilistic connection.

In his first communication on "Quantization as an Eigenvalue Problem," Schrödinger indicated the procedure by which he had arrived at his equation for a one particle system, in particular for an electron in a hydrogen atom.[16] He had re-expressed the dynamics of the system as described by the Hamilton-Jacobi equation

$$H\left(q, \frac{\partial S}{\partial q}\right) = E, \tag{C19}$$

where H is the Hamiltonian of the system,

$$H(p, q) = \frac{p^2}{2m} + V(q), \tag{C20}$$

by introducing a function ψ defined by

$$\psi = \exp\,[S/K] \tag{C21}$$

where the constant K had to have the dimension of action, i.e., energy × time. ψ thus obeys the equation

$$H\left(q, \frac{K}{\psi}\frac{\partial\psi}{\partial q}\right) = E. \tag{C22}$$

Explicitly written out in Cartesian coordinates Eq.(C22) becomes

$$\left(\frac{\partial\psi}{\partial x}\right)^2 + \left(\frac{\partial\psi}{\partial y}\right)^2 + \left(\frac{\partial\psi}{\partial z}\right)^2 - \frac{2m}{K^2}(E - V(r))\psi^2 = 0. \tag{C23}$$

Schrödinger then required that the integral of the quadratic form on the left side of Eq.(C23) be an extremum. This variational condition replaced the quantum conditions and implied that ψ satisfied the equation

$$\frac{K^2}{2m}\nabla^2\psi - \frac{2m}{K^2}(E - V(r))\psi = 0. \tag{C24}$$

Schrödinger then proceeded to show that with suitable boundary conditions, Eq.(C24) yielded the Bohr energy levels for $E < 0$ provided K is given the value $h/2\pi$.[17] In his next communication Schrödinger gave a more elaborate exposition of the genesis of wave mechanics based on Hamilton's analogy between ordinary mechanics and geometrical optics. But he there also indicated an approach based on Debye's suggestion that the "phase wave" should satisfy a wave equation. He combined de Broglie's idea of associating wave properties with a particle and Einstein's remark that, as a consequence, "it is possible to associate a scalar wave field" with a Bose gas. He took the phase wave to satisfy the equation

$$\nabla^2\psi = \frac{1}{V^2}\frac{\partial^2\psi}{\partial t^2} \tag{C25}$$

where V is the phase velocity.

The wave field attached to a particle of definite energy, and hence of definite frequency, has a time dependence given by $\exp(2\pi i\nu t)$, and for this situation ψ satisfies the equation

$$\nabla^2\psi = \frac{4\pi^2\nu^2}{V^2}\psi. \tag{C26}$$

For a nonrelativistic particle,[18] de Broglie's relations imply that

$$\frac{v^2}{V^2} = \frac{p^2}{h^2} = \frac{2m}{h^2}(E - V(r)), \tag{C27}$$

and thus Eq.(C26) becomes

$$\frac{h^2}{8\pi^2 m}\nabla^2\psi - (E - V(r))\psi = 0. \tag{C28}$$

In his fourth communication, Schrödinger came to the conclusion that the "simplest" time-dependent equation that yielded Eq.(C28) for stationary processes was

$$i\hbar\frac{\partial\psi}{\partial t} = -\frac{\hbar^2}{2m}\nabla^2\psi + V(r)\psi. \tag{C29}$$

The success of Schrödinger's program and the simplicity of his approach—the quantization rules followed from the imposition of simple boundary and continuity conditions on the "wave function" that obeyed his wave equation, very much like the situation that existed for the amplitude of a classical vibrating system—resulted in wave mechanics enjoying almost universal acceptance. Schrödinger was invited to talk everywhere. In July he accepted an invitation to come to Berlin, where he was enthusiastically congratulated by Planck, Einstein, and Nernst. On the return trip to Zurich he stopped in Munich to deliver a lecture in the Wednesday Wien and Sommerfeld colloquium.[19] Heisenberg, who was a lecturer in Copenhagen at the time, had studied Schrödinger's papers during a holiday in Norway, and had arranged to spend the rest of his summer vacation in Munich with his parents to be on hand when Schrödinger gave his lecture and to be able to "discuss his theory with him in person."[20] Heisenberg was very much taken by the beauty of the mathematical part of Schrödinger's presentation but was "unconvinced" by Schrödinger's interpretation of wave mechanics. During the discussion period Heisenberg asked Schrödinger how he hoped to explain quantized processes such as the photoelectric effect on the basis of his continuum model and "pointed out that Schrödinger's conception would not even help explain Planck's radiation law."[21]

For this he was taken sharply to task by Willy Wien, who told him "that while he understood [Heisenberg's] grief that [his] quantum mechanics was finished, and with it all such nonsense as quantum jumps, etc. the difficulty

[Heisenberg] had mentioned would undoubtedly be solved by Schrödinger in the very near future."[22]

Schrödinger was not as emphatic in his answer, but he too felt confident that the problems would be resolved in due time. Sommerfeld was less sanguine. He wrote Pauli a few days later: "My general impression is that wave mechanics is indeed a marvelous micromechanics, but that the basic quantum riddles have in no way been solved by it. I cease to believe Schrödinger the moment he starts to calculate with his c_{bk}." Indeed, Schrödinger's interpretation of his "wave function" was problematic because he tried to give his wave function a realistic interpretation. In his first three papers Schrödinger had spoken of the ψ-vibrations "as though of something quite real." He had suggested representing a mass point as a wave packet, the approximate position and velocity of the packet corresponding to the position and velocity of the particle. In his second paper he had indicated that for the harmonic oscillator a wave packet built by an appropriate superposition of eigenfunctions has the property that it "holds *permanently together,* [and] does *not* expand over an ever greater domain in the course of time." This result led him to believe that "it could be anticipated with certainty" that the same would be true for suitable superpositions of hydrogenic wave functions describing an electron in large n orbits. While his calculation for the harmonic oscillator was correct for the harmonic oscillator, his anticipation for the general case turned out to be fallacious. As pointed out to him by Lorentz most wave packets disperse, and therefore his dream to have a quasi-classical, continuum wave description of particles was unrealizable. In the fourth of his papers Schrödinger backtracked somewhat. In addition to Lorentz's criticism he had come to realize that ψ is in general complex, and that for an n-particle system ψ is a function of the coordinates of all the particles and is thus defined on the 3n dimensional configuration space and "cannot and may not be interpreted directly in terms of ordinary three-dimensional space." He suggested that $\psi\psi^*$ has physical meaning, and that from it the actual electric charge distribution of a given particle can be recovered (by suitable integration over the other particles' coordinates). Schrödinger thus still regarded waves as the fundamental reality, with the particles as derivative constructs. In fact, Schrödinger never gave up his belief that "he had accomplished a return to classical thinking" and from 1926 until his death he argued "that the idea of particles and of quantum jumps be given up altogether." According to Born "he never faltered in this conviction."[23]

Born on Collisions

Schrödinger considered the electron not a particle but a charge distribution whose density is given by the square of the wave function. In a short paper sent to *Zeitschrift für Physik* in late June of 1926 Born explicitly rejected Schrödinger's continuum viewpoint and proposed a probabilistic interpretation for the ψ function.[24] The article, some five pages long, had the title "Zur Quantummechanik der Stossvorgänge" [On the quantum mechanics of collision processes]. Even though he had been one of the founders of matrix mechanics Born was deeply impressed by Schrödinger's approach. He had come to the conclusion that "among the various formulation of the new quantum theory only Schrödinger's has proven itself appropriate for the description of collision phenomena; and for this reason Born was "inclined to regard it as the most profound formulation of the quantum laws."[25] Yet he could not accept Schrödinger's interpretation of the ψ function as describing a real material wave. In his autobiography written late in life, Born remarked that: "Every experiment by Franck and his assistants on electron collisions appeared to me as a new proof of the corpuscular nature of the electron."[26] In 1926 Born was the director of the theoretical physics institute at Göttingen University, which was housed in the same building as James Franck's institute for experimental physics. Franck had been awarded the Nobel Prize for his beautiful experiments on the inelastic scattering of electrons by hydrogen that had demonstrated the quantized nature of the energy levels of hydrogen.

The brief June 1926 communication by Born constituted a preliminary report on the contents of two subsequent papers in which he gave a full discussion of the quantum mechanical description of collision processes.[27] In the first of these he reiterated his belief that Schrödinger's formalism, being equivalent to that of Dirac and that of Born, Jordan, and Heisenberg, was to be welcomed. However, Schrödinger's interpretation of his ψ function, which attempted to describe the motion of an electron by a classical continuum theory, could not be upheld. And precisely because his formalism was retained it was necessary to give it new physical content. To formulate the new picture Born drew on an unpublished "observation" that Einstein had made in the early 1920s concerning the relation between the Maxwell field and light quanta. Einstein had suggested that the fields are present "only to show the corpuscular light quanta the way, . . . in the sense of a 'ghost field.' The [guiding field] determines the probability that a light quantum, the bearer of energy and momentum, takes a certain path; however the field itself has no energy and no momentum."[28] Embracing the full analogy between a light quantum and an electron, Born suggested regarding the de Broglie–Schrödinger waves that obey Eq.(C20) as the "ghost field" or the "guiding field" for the electron, and stipulated that the wave function $\psi(x,t)$ determines the probability of finding the electron at the position x at time t. A wave packet thus does not represent the electron, but only determines the probability of the position of the electron—which is essentially pointlike. He was thus led to conclude, "somewhat paradoxically," that "the motion of the particles follows laws of probability, but the probability itself propagates in harmony with causal laws." To validate his interpretation, Born made use of Schrödinger's proof of the equivalence of matrix and wave mechanics for the hydrogen atom, and proceeded as follows:

If ψ_n are the normalized eigenfunctions of the energy operator H,

$$H\psi_n = E_n\psi_n \text{ with } (\psi_n,\psi_m) = \delta_{mn}, \tag{C30}$$

the expansion of a normalized wave function $\psi(q)$ in terms of the $\psi_n(q)$ yields

$$\psi(q) = \sum_n c_n \psi_n(q)$$

with

$$\sum_n |c_n|^2 = 1. \tag{C31}$$

Born then interpreted $|c_n|^2$ as the probability of the particle with wave function ψ having energy E_n, since Eq.(C31) then expresses the fact that the sum of all the (disjoint) probabilities must add up to 1.

The interpretation was bolstered by the fact that the expectation value of the Hamiltonian, i.e., of the energy, is given by

$$\int \bar{\psi}(q) H \psi(q) dq = \sum_n E_n |c_n|^2, \tag{C32}$$

with the righthand side being the possible energies averaged with the probabilities $|c_n|^2$. Born then extended the analysis to the case where the "observable" has a continuous spectrum, such as the momentum operator, and indicated that a parallel procedure existed in that case with an analogous interpretation.

APPENDIX D

Courses Taken at Munich University

The list of courses taken also gives the fee paid per course, indicated within parentheses in Reichsmark, with one Reichsmark being worth approximately twenty-one cents. In addition Bethe paid thirty marks per semester for health insurance.

Summer semester 1926

Carathéodory	Continuum Mechanics	(14)
Fajans	Physical Chemistry II	(10.50)
Fajans	Chemical Forces and Constitution	(4.50)
Sommerfeld	Partial Differential Equations of Physics	(14)
Sommerfeld	Exercises—Partial Differential Equations of Physics	(2)
Sommerfeld	Theory of Magnetism	(3.50)
Wentzel	Theory of Band Spectra	(3.50)
Wentzel	Dispersion of Light and X-rays	(3.50)
Wien	Physics Exercises for Advanced Students	(87)

Winter semester 1926–27

Bechert	Psychology	(14)
van Calker	Introduction to Politics	(3.50)
Fajans	Electrochemistry	(9)
Kirchner	Experiments in Electricity and Light	(4.50)

Sommerfeld	Theoretical Physics Seminar	—
Sommerfeld	Exercises—Mechanics	(2)
Sommerfeld	Selected Problems in Quantum Theory	(7)
Wien	Experimental Physics for Advanced Students	(87)

Summer semester 1927

van Calker	The Political Parties of the Reichstag	(1)
Fajans	Colloid Chemistry	(1.40)
Ott	Experimental Methods, Properties, and the Theory of Crystal Lattices	(2)
Sommerfeld	Continuum Mechanics	(4)
Sommerfeld	Exercises—Continuum Mechanics	(2)
Sommerfeld	Structure of Matter	(2)
Sommerfeld	Theoretical Physics Seminar	—

Winter semester 1927–28

Sommerfeld	Electrodynamics	(4)
Sommerfeld	Exercises—Electrodynamics	(2)
Sommerfeld	Quantum Mechanics	(2)
Sommerfeld	Theoretical Physics Seminar	(2)

Summer semester 1928

Emden	Physics of the Sun	—
Sommerfeld	Optics	(4)
Sommerfeld	Wave Mechanics	(2)
Sommerfeld	Theoretical Physics Seminar	—

APPENDIX E

Bethe's Doctoral Thesis

In William Bragg's explanation of the observed peaks in X-ray diffraction by a crystal, the crystal is considered made up of parallel planes of atoms spaced a distance *d* apart. The condition for a sharp maximum in the intensity of the scattered radiation is that the various planes reflect the radiation incident upon them specularly (i.e., with the angle of incidence of the radiation to it equal to the angle of reflection from it) and that the reflected rays from successive planes interfere constructively. The condition for this is that the difference in the path length travelled by two rays reflected from two neighboring planes be an integral multiple, n, of the wavelength of the radiation, which translates into the Bragg formula, $n\lambda = 2d\sin\theta$, with θ the angle of incidence.

In von Laue's explanation the crystal is considered to be made up of oscillators (ions or atoms) located at positions $\boldsymbol{l} = l_1\boldsymbol{a}_1 + l_2\boldsymbol{a}_2 + l_3\boldsymbol{a}_3$ (l_1, l_2, l_3 integers) of a Bravais lattice, with $\boldsymbol{a}_1, \boldsymbol{a}_2, \boldsymbol{a}_3$ the primitive vectors that generate the lattice. These oscillators reradiate the incident radiation in all directions at the same frequency. The observed sharp peaks occur at those energies and directions for which the radiation reemitted by all the oscillators interferes constructively.

The condition for constructive interference is most easily stated in terms of the primitive vectors g_1, g_2, g_3 of the reciprocal lattice that Ewald had introduced in his 1912 doctoral thesis in which he also analyzed the geometry of the reciprocal lattice. These basis vectors g_1, g_2, g_3 are defined by

(E1) $$g \cdot l = 2\pi \times \text{integer}.$$

For given basis vectors a_1, a_2, a_3, the basis vectors g_1, g_2, g_3 of the reciprocal lattice can be readily and explicitly constructed:

(E2) $$g_1 = \frac{a_2 \times a_3}{a_1 \cdot (a_2 \times a_3)}, \quad g_2 = \frac{a_3 \times a_1}{a_1 \cdot (a_2 \times a_3)}, \quad g_3 = \frac{a_1 \times a_2}{a_1 \cdot (a_2 \times a_3)},$$

If $\mathbf{n}$ and $\mathbf{n}'$ are unit vectors in the direction of the incident and scattered radiation, the condition for constructive interference is that

(E3) $$k' - k = g$$

where $k' = \mathbf{n}'k$ and $k = \mathbf{n}k$ with $k = 1/\lambda = \omega/c$, λ is the wavelength of the radiation, and $\mathbf{g}$ is a vector in the reciprocal lattice

(E4) $$g = h_1 g_1 + h_2 g_2 + h_3 g_3 \; (h_1, h_2, h_3 \text{ integers}).$$

Multiplying the equation $k' - k = g$ by g yields the standard form of the von Laue equation:

(E5) $$g \cdot (\mathbf{n} - \mathbf{n}') = m\lambda \; (m \text{ an integer}).$$

For a given g a maximum does not occur for arbitrary direction and frequency of the incident radiation. The strongest maxima occur when the von Laue equation is satisfied exactly. Since $k' = k + g$ and $k^2 = k'^2$,

(E6) $$g \cdot k = -\frac{1}{2} g^2.$$

This equation determines the values of the wave vector k for which maxima occur with the given value of g. Since $|k' - k| = 2k \sin\theta$ where θ is the angle of incidence, the equation $k' - k = g$ becomes $2k \sin\theta = g$, which determines the angle of diffraction at a maximum, and is Bragg's statement of the condition for a maximum.

The equivalence of the von Laue and the Bragg approaches was demonstrated by Ewald in the three wartime papers in which he presented his dynamical theory.[1] He represented the crystal as an (idealized) regular three-dimensional array of atoms located at the points $l = l_1 a_1 + l_2 a_2 + l_3 a_3$ (l_1, l_2, l_3 integers) of a Bravais lattice (with a_1, a_2, a_3 the primitive vectors that generate the lattice). As Lorentz had done earlier, the dynamics of the vibrating electrons of an atom was modeled by harmonic oscillators whose frequencies corresponded to the frequency of the light that the atom emitted.

The response of an electron of charge e (represented by an oscillator of natural frequency ν_0) to the electric field $E\cos 2\pi\nu t$ of the X-rays incident (in the x direction) is to be set in an additional vibrating motion and to be displaced from its equilibrium position by an amount x:

(E7) $$x = \frac{e}{4\pi^2 m}\frac{E\cos 2\pi\nu t}{\nu_0^2 - \nu^2}.$$

It thus acquires a dipole moment equal to ex, and emits dipole radiation of the same frequency as the initial radiation. If there are n_1 electrons in the atoms with natural frequency ν_1, n_2 with natural frequency ν_2, . . . the total dipole moment[2] is

(E8) $$ex = \frac{e^2}{4\pi^2 m}\sum_i n_i \frac{E\cos 2\pi\nu t}{\nu_i^2 - \nu^2}.$$

A given dipole in the lattice is affected by the electric field of the other dipoles in its neighborhood. Ewald took this into account by recognizing that this interaction results in a slight change in the natural frequency ν_0. A given dipole also responds to the radiation emitted by the other dipoles. Moreover, the radiation emitted at time t by oscillator i located at r_i affects an oscillator located at r_j only at time $t + \frac{|r_i - r_j|}{c}$ later, due to the fact that electromagnetic radiation travels with speed c. It is this retardation effect that is responsible for a macroscopic description attributing a different velocity of propagation to the beam travelling inside the crystal than in free space.[3] Ewald was able to sum the effects of the incident beam on the oscillators and that of the response of the oscillators to the radiation emitted by them in a *self-consistent* manner and to show that in this model the incident wave was "extinguished" inside the crystal, and that the superposition of the radiation emitted by the dipoles together with the incident radiation gave rise, by virtue of the periodicity of the lattice, to reflected radiation at angles seen in the von Laue–Bragg X-ray diffraction patterns and to a refracted beam inside the crystal characterized by a calculable refractive index and angle of refraction. Ewald thus was able to account by a dynamical theory for the reflection, refraction, and dispersion of X-rays by perfect crystals. In 1968, reviewing his life's work on the interaction of electromagnetic waves with crystals, Ewald commented, "[My] interest always centered on the perfect crystal in which I saw the preferred material for exacting optical investigations. Herein lies a strong limitation, and an abstraction which in important aspects is contrary to nature. I am happy to see how others have been, still are, and will be, carrying on beyond my limitation."[4]

In his thesis Bethe modeled the Davisson-Germer scattering experiment by assuming the crystal to be semi-infinite to the right of the plane $z = 0$. The Hamiltonian, which governs the motion of the electron, was taken to be

$$H = -\frac{\hbar^2}{2m}\nabla^2 + V(\boldsymbol{r}) \tag{E9}$$

with

$$V(\boldsymbol{r}) = 0 \text{ for } z < 0 \tag{E10a}$$

and

$$V(\boldsymbol{r}) = \sum_n v(\boldsymbol{r} - \boldsymbol{R}_n) \text{ for } z > 0. \tag{E10b}$$

$v(\boldsymbol{r} - \boldsymbol{R}_n)$ is the potential contributed by ion n at the position $\boldsymbol{R}_n$.

The incident electron of momentum $\boldsymbol{k}_0$ was described by a plane wave $e^{i\boldsymbol{k}_0\cdot \boldsymbol{r}}$ when outside the crystal. When inside the crystal, the electron's wave function had to satisfy the Schrödinger equation:

$$-\frac{\hbar^2}{2m}\nabla^2\psi(\boldsymbol{r}) + V(\boldsymbol{r})\psi(\boldsymbol{r}) = E\psi(\boldsymbol{r}), \tag{E11}$$

with $V(\boldsymbol{r})$ the potential (E10b) it experiences inside the crystal. Furthermore, the solutions for $z < 0$ and those for $z > 0$ and their first derivatives had to be continuous across the plane $z = 0$.

The potential $V(\boldsymbol{r})$ embodies the translational symmetry of the crystal: $V(\boldsymbol{r})$ thus has the property that if $\boldsymbol{a}_1, \boldsymbol{a}_2, \boldsymbol{a}_3$ are the basis vectors that generate the lattice

$$V(\boldsymbol{r}) = V(\boldsymbol{r} + \boldsymbol{l}) \tag{E12}$$

with $\boldsymbol{l} = l_1\boldsymbol{a}_1 + l_2\boldsymbol{a}_2 + l_3\boldsymbol{a}_3$ and $l_1, l_2, l_3 = \pm 1, \pm 2, \pm 3, \ldots$

By virtue of this periodicity the potential can be represented by a Fourier series. Ewald (1916–18) had shown that a very effective representation of such a periodic potential was given by the Fourier series

$$V(\boldsymbol{r}) = \sum_{q_1,q_2,q_3=-\infty}^{+\infty} \tilde{V}_{q_1,q_2,q_3} e^{i\boldsymbol{q}\cdot\boldsymbol{r}} = \sum_{\boldsymbol{q}} \tilde{V}_{\boldsymbol{q}} e^{i\boldsymbol{q}\cdot\boldsymbol{r}} \tag{E13a}$$

where q is a vector in the reciprocal lattice,

(E13b) $$q = q_1 g_1 + q_2 g_2 + q_3 g_3$$

with q_1, q_2, q_3 integers and $g \cdot l = 2\pi \times$ integer.

If the origin of the coordinate system in (E11) shifted by l, it is clear that by virtue of (E12) $\psi(r + l)$ is also a solution of (E11). If the level E is nondegenerate, then it must be that $\psi(r + l) = c(l)\psi(r)$ with $|c|^2 = 1$. (If there is degeneracy, this can also be made the case for each wave function of the degenerate set.) Since the result of two successive displacements $l_1 + l_2$ multiplies the wave function by $c(l_1 + l_2) = c(l_1)c(l_2)$, log c is additive and must depend linearly on the displacement vector l, hence $c(l) = e^{ik \cdot l}$ so that

(E14) $$\psi(r + l) = e^{i\mathbf{k}\cdot\mathbf{l}}\psi(r)$$

The vector $\boldsymbol{k}$ is only defined up to an arbitrary vector K of the reciprocal lattice, since by the definition of the latter $e^{iK \cdot l} = 1$. To investigate the properties of the solutions for $z > 0$, Bethe sought solutions of the form

(E15) $$\psi_{z>0}(\boldsymbol{r}) = \sum_{q_1, q_2, q_3} e^{i(\boldsymbol{k}+\boldsymbol{q})\cdot \boldsymbol{r}} u_{\boldsymbol{q}}$$

and obtained the following set of equations for $u_{\mathbf{q}}$:

(E16) $$\left(E - \tilde{V}_{000} - \frac{\hbar^2 (\boldsymbol{k}+\boldsymbol{q})^2}{2m} \right) u_{\boldsymbol{q}} = \sum_{\boldsymbol{h}} \tilde{V}_{\boldsymbol{h}} u_{\boldsymbol{q}-\boldsymbol{h}}$$

In a perturbative expansion of these equations, assuming that in zeroth approximation the wave function inside the crystal is that of a free electron, $\psi = \exp(i\boldsymbol{k}\cdot\boldsymbol{r})$, $E_0 = V_{000}$. Since

$$\int e^{-i\boldsymbol{k}'\cdot \boldsymbol{r}} V(\boldsymbol{r}) e^{+i\boldsymbol{k}\cdot\boldsymbol{r}} d^3 r = \int e^{-i\boldsymbol{k}'\cdot\boldsymbol{r}} \sum_{\boldsymbol{q}} \tilde{V}_{\boldsymbol{q}} e^{i\boldsymbol{q}\cdot\boldsymbol{r}} e^{+i\boldsymbol{k}\cdot\boldsymbol{r}} d^3 r$$

$$= \sum_{\boldsymbol{q}} \tilde{V}_{\boldsymbol{q}} \int e^{-i\boldsymbol{k}'\cdot\boldsymbol{r}+i\boldsymbol{q}\cdot\boldsymbol{r}++i\boldsymbol{k}\cdot\boldsymbol{r}} d^3 r$$

(E17) $$= \begin{cases} \tilde{V}_{\boldsymbol{q}} \; if \; \boldsymbol{k}+\boldsymbol{q}-\boldsymbol{k}'=0 \\ 0 \quad \text{otherwise} \end{cases} ,$$

there are only matrix elements between states whose wave vectors, $\boldsymbol{k}$, $\boldsymbol{k}'$, are such that they differ by a vector in the reciprocal lattice: $\boldsymbol{k}' - \boldsymbol{k} = g$. Therefore, to first order the electron's wave function in the potential

$$V(\boldsymbol{r}) = \sum_{\boldsymbol{q}} \tilde{V}_{\boldsymbol{q}} e^{i\boldsymbol{q}\cdot\boldsymbol{r}}$$

is

$$\psi_1 = e^{i\boldsymbol{k}\cdot\boldsymbol{r}} + \frac{2m}{\hbar^2} \sum_{g, g\neq 0} \frac{\tilde{V}_{\boldsymbol{q}} e^{i(\boldsymbol{k}+\boldsymbol{q})\cdot\boldsymbol{r}}}{\left[\boldsymbol{k}^2 - (\boldsymbol{k}+\boldsymbol{q})^2\right]}, \tag{E18}$$

and the corresponding energy to that order is given by

$$E_1 = \frac{2m}{\hbar^2} \sum_{g, g\neq 0} \frac{\left|\tilde{V}_{\boldsymbol{q}}\right|^2}{\left[\boldsymbol{k}^2 - (\boldsymbol{k}+\boldsymbol{q})^2\right]}. \tag{E19}$$

When $\boldsymbol{k}^2 = (\boldsymbol{k} + \boldsymbol{g})^2$, the denominator vanishes and degenerate perturbation must be applied. Doing so, Bethe found that for certain ranges of the energy there are no eigenstates $\psi_{\boldsymbol{k}}$ of the system.

Now the crystal in the plane $z = 0$ has a two-dimensional periodicity such that

$$V(\boldsymbol{r}) = V(\boldsymbol{r} + l_1\boldsymbol{a}_1 + l_2\boldsymbol{a}_2). \tag{E20}$$

By an argument similar to the above Bethe inferred that the wave function for $z < 0$, representing an incoming beam of electrons with momentum $\boldsymbol{k}_0$ in the $\boldsymbol{k}_0/k_0$ direction, energy $E = \hbar^2 k_0{}^2/2m$, and a scattered beam moving in the $-z$ direction having the appropriate symmetry on the plane $z = 0$ must be of the form

$$\psi_{z<0}(\boldsymbol{r}) = e^{i\boldsymbol{k}_0\cdot\boldsymbol{r}} + \sum_{q_1, q_2} e^{i(\boldsymbol{k}_{0\|}+\boldsymbol{q})\cdot\boldsymbol{r}_\|} u_{\boldsymbol{q}}(z) \tag{E21}$$

where $\boldsymbol{k}_{0\|}$, $\boldsymbol{r}_\|$, denotes the component of $\boldsymbol{k}_0$,$\boldsymbol{r}$, parallel to the $z = 0$ surface. The $\boldsymbol{q}$'s are two-dimensional vectors with the property $\boldsymbol{q}\cdot\boldsymbol{a}_1 = 2\pi \times$ integer, $\boldsymbol{q}\cdot\boldsymbol{a}_2 = 2\pi \times$ integer solution. They are the basis vectors for the reciprocal lattice of the surface defined by $\boldsymbol{a}_1$ and $\boldsymbol{a}_2$. Now the solutions for $z \leq 0$ satisfy

$$-\frac{\hbar^2}{2m}\nabla^2\psi = E\psi \quad \text{with} \quad E = \frac{\hbar^2 k^2}{2m}, \tag{E22}$$

which translates into the following for $u_{\boldsymbol{q}}$:

$$u_{\boldsymbol{q}}(z) = \alpha_{\boldsymbol{q}} e^{-i\left(\frac{2m}{\hbar^2}E - (\boldsymbol{k}_{0\|}+\boldsymbol{q})^2\right)^{1/2} z}, \tag{E23}$$

and hence,

$$\text{(E24)} \qquad \psi_{z<0}(\boldsymbol{r}) = e^{i\boldsymbol{k}_0\cdot\boldsymbol{r}} + \sum_{q_1,q_2} e^{i(\boldsymbol{k}_{0\parallel}+\boldsymbol{q})\cdot\boldsymbol{r}_\parallel)+i\left(\frac{2m}{\hbar^2}E-(\boldsymbol{k}_{0\parallel}+\boldsymbol{q})^2\right)^{1/2}z}\alpha_{\boldsymbol{q}}.$$

The scattered beam by virtue of symmetry considerations alone is composed of a series of discrete beams, each with a different, parallel component of momentum $\left(\frac{2m}{\hbar^2}E-(\boldsymbol{k}_{0\parallel}+\boldsymbol{q})^2\right)^{1/2}$ and directions determined by $\boldsymbol{k}_{0\parallel}, \boldsymbol{q}$ and E. The determination of the $\boldsymbol{\alpha}_q$, and hence the intensities, is much more complicated and depends sensitively on the potential, the energy, and the approximation made.

APPENDIX F

The *Habilitationsschrift* Defense

EINLADUNG

zu der

am Samstag, den 3. Mai 1930,
vormittags 12 Uhr s.t.,

in der grossen Aula der Universität unter dem Vorsitze des derzeitigen Prodekans der philosophischen Fakultät II. Sektion

Herrn Professor Dr. A. Wilkens

behufs Erlangung der venia legendi an der Bayerischen Ludwig-Maximilians-Universität zu München stattfindenden

PROBEVORLESUNG

des

Dr. phil. Hans Bethe.

Habilitationsschrift:

Zur Theorie des Durchgangs schneller Korpuskularstrahlen durch Materie.

Probevorlesung:

Neuere Entwicklung der Elektronentheorie der Metalle.

Thesen:

1. Die Born'schen Abstossungskräfte bei Ionenkristallen sind nicht elektrostatischer Natur, sondern von derselben Art wie die homöopolaren Bindungskräfte (Elektronenaustausch); ausgenommen sind nur Salze mit extrem kleinem Kation (Li).
2. Ein Atom in einem Kristall hat nie eine geringere Symmetrie als seiner Lage entspricht.
3. Alle Elektronen eines Metalls, die nicht in abgeschlossenen Schalen sitzen, beteiligen sich an der Elektrizitätsleitung. Eine anderweite Angabe der „Anzahl freier Elektronen" ist physikalisch sinnlos.
4. Das mittlere elektrostatische Potential in einem Metall ist wesensverschieden von der Bildkraft der älteren Theorie; beide zusammen geben erst die Austrittsarbeit für Elektronen.
5. Aus der Intensität der Röntgeninterferenzen lässt sich der Ionisierungs-Zustand der Atome eines Kristalls nicht feststellen.
6. Die Intensität der Röntgen-Interferenzen wird durch den Comptoneffekt nicht vermindert.
7. Ein „Ramaneffekt im Röntgengebiet" existiert nur, wenn die Wellenlänge der Ramanlinie stärker von der des auffallenden Lichts abweicht als die der Comptonlinie.
8. Die Summe der Oszillatorstärken einer Spektralserie eines komplizierten Atoms ($Z > 4$) ist nicht gleich der Anzahl Elektronen im Grundniveau der Serie. Insbesondere ist für die K-Serie die Summe wesentlich kleiner als zwei, die übliche Dispersionsformel im Röntgengebiet also nicht richtig.
9. Durch Stoss schneller Elektronen werden innere Schalen gleicher Hauptquantenzahl (z. B. die Teilschalen der L-Schale) nicht gleichstark angeregt.
10. Die nichtstationäre Behandlungsweise zeitabhängiger Störungen in der Wellenmechanik ist sicherer als die stationäre.

Notes

Abbreviations

Unless otherwise indicated, all Bethe quotations are from the author's interviews with Hans Albrecht Bethe.

Bethe to Anna Bethe. Bethe made available all the letters he had written to his mother until she went into a nursing home on Long Island in 1940. They are now in Rose Bethe's possession.

HABCW I, II, & III. Interview of Hans Albrecht Bethe by Charles Weiner, 1966–1972. Niels Bohr Library and Archives, American Institute of Physics, College Park, MD. Available at http://www.aip.org/history/ohilist/4504_2.html.

HABJG I & II. Interview of Hans Albrecht Bethe by Judith Goodstein, February 17, 1982, and January 28, 1993. Available at http://resolver.caltech.edu/CaltechOH:OH_Bethe_H.

HABLH. Interview of Hans Albrecht Bethe by Lillian Hoddeson, April 29, 1981, Niels Bohr Library and Archives, American Institute of Physics, College Park, MD. Available at http://www.aip.org/history/ohilist/4505.html.

HABM I & II. The Bethe Memoirs are contained in two five-by-seven-inch leather-covered notebooks, each filled with some fifty pages of handwritten material. Bethe wrote them in the mid- and late 1990s to tell his children and grandchildren the story of his life. The notebooks are in Rose Bethe's possession.

HABTK. Interview of Hans Albrecht Bethe by Thomas Kuhn, January 17, 1964. Niels Bohr Library and Archives, American Institute of Physics, College Park, MD. Available at http://www.aip.org/history/ohilist/4503.html.

Introduction

1. Jasanoff 2007.
2. Schweber 2000.
3. Quoted in Greene 2007, 728.
4. Nye 2006; Richards 2006; Porter 2006; Terrall 2006.
5. Greene 2007, 727.
6. Ibid., 757.
7. Ibid., 756.
8. Ibid., 757.
9. One of the reasons is undoubtedly the size of the two communities. There are many more solid state physicists, some of them—like John Bardeen and Philip W. Anderson—being "off-scale" (see Hoddeson 2002; Anderson 2004). Another reason is the much narrower set of questions asked by high energy theorists, in contrast to the huge diversity and complexity of condensed matter physics phenomena to be understood.
10. They are: the computation by group theoretical methods of the energy levels of an impurity atom embedded in a crystal lattice, the quantitative estimation of the energy loss of charged particles as they travel through matter, and the energy spectrum and wave functions of a linear chain of interacting spins.
11. Bethe, Bacher, and Livingston 1986.
12. Richard Lewontin in his essays in the *New York Review of Books* has repeatedly emphasized these points (see Lewontin 2001).
13. Bethe 1934c.
14. This was already in evidence in his careful citing of references to the works of others in his two *Handbuch* articles of 1933 and in his *Review of Modern Physics* articles of 1937.
15. Knorr Cetina 1999, 1.
16. In his review of Meyer and Brenner's (1997) four-volume *German-Jewish History in Modern Times,* David Sorkin pointed out that it was German-Jewish émigré historians who wrote a good deal of the history of German Jewry after World War II, a history that Sorkin characterized as the "émigré synthesis." In that synthesis the term *assimilation* was replaced with the "neutral, social scientific categories of acculturation and integration," thus emphasizing "the multiform nature of the process"

(Sorkin 2001, 533). It is in that sense that I use the adjective "assimilated" to describe the families of Bethe's Jewish friends.

17. Strenski 1997, 6.

18. Fears that Germany might build an atomic bomb before the Allies motivated the frantic efforts on the part of the British and American physicists to make one. Thus even though by the beginning of 1945 it was clear that Germany was defeated, and various voices were being raised questioning the further efforts to produce a bomb, only one physicist left Los Alamos because of moral considerations: Joseph Rotblat. In fact, the pace of work at Los Alamos increased during 1945. Oppenheimer, with Bohr's blessing, had convinced most of the scientists that it was essential to demonstrate to the world the existence and potency of these weapons and bring them under international control.

Los Alamos was unique in its enormous concentration of first-rate scientists who constantly gave proof of what could be accomplished by working together on very circumscribed goals for what they believed was a just cause.

19. H. A. Bethe's FBI files. I am indebted to Alex Wellerstein for making these files available to me.

20. Though isolated—and perhaps because of its isolation—Los Alamos created that rare situation in the lives of individuals and communities when they feel in touch with much more than themselves. During the few years spent there, almost everyone, and in particular the physicists, felt whole. It was a merging of commitments, aspirations, and talents that had been fragmented before. Not only did the participants feel whole as individuals—their moral, intellectual, and creative passions all being channeled into the task at hand—but an atmosphere of wholeness permeated the entire enterprise, transmuting it into a kind of magic and enshrining it in the minds of those who had been there. Bethe in many ways personified the integration of the multifaceted nature of the enterprise: the theoretical and the practical, the down-to-earth and the idealistic, the rational and the passionate, the individual and the collective.

21. Bourdieu 1980.

22. See Gottfried 2006a; Drell 2006; Brown and Lee 2009.

23. Interview with Rose Bethe, April 28, 2008.

24. The launching of Sputnik by the Soviet Union in November 1957 startled and frightened the American public. Shortly thereafter President Dwight Eisenhower revitalized the role and the office of President's Science Adviser. He appointed James Killian to that post on a full-time basis and reconstituted the existing Committee into PSAC. I. I. Rabi, who was the chair of the existing committee and had acted as the President's Science Adviser on a part-time basis, had made the initial suggestion to Eisenhower "that he move the Committee in a closer relation to him and have, again, a full-time Science Adviser who would report directly to him: a person

with whom he could get along and personally liked to talk [to]" (Rabi in Golden 1986, 11).

25. Golden 1986.

26. PSAC created a panel system composed of experts who were not necessarily members. These experts addressed specific issues and reported back to PSAC. Bethe served on the panels dealing with strategic military problems, with detections of nuclear tests, and with ABM systems throughout the 1960s.

27. The interagency committee met frequently during the first three months of 1958. The representative of the Los Alamos Laboratory generally supported a test ban; the representative of the Livermore Laboratory was rather against it but agreed with most of the technical conclusions. However, he brought into the discussions a new development from Livermore, namely the possibility of continuing nuclear weapons tests underground. The representative of the State Department was strongly in favor of a test ban. The director of the Nuclear Weapons Division of the Atomic Energy Commission raised many objections, especially when the final version of the report was being discussed. The representative of the Department of Defense, General Loper, raised fewer objections during the discussion, but finally refused to sign the report on behalf of the DOD. There were several representatives of the Air Force Technical Applications Center (AFTAC), the organization that was charged with collecting radioactive debris and getting other evidence of foreign nuclear weapons tests. They were very helpful in providing information.

28. During the time that Bethe served on PSAC, and until Johnson's presidency, PSAC and the special assistant—whether Killian, Kistiakowsky, or Jerome Wiesner—worked well together. As Killian commented in 1962, "the committee brought to the Special Assistant a range of views, an objectivity and an uninhibited freedom of comment that no single science adviser could match" (Killian 1962, iv).

29. It was Ernest Lawrence, his vision and his administrative style, that set the tone at Livermore. His protégés, Herbert York, John Foster, and Harold Brown, all of whom had obtained their doctorates working at the Berkeley Radiation Laboratory, became the directors of the Livermore Laboratory and went on from there to important and influential governmental positions in Washington. Livermore from its inception was a hybrid collective of human beings embedded in conventions and personal relationships, of tools, equipment, material and technical devices, algorithms, computer codes, texts, and of deeds and achievements that gave it meaning. Its workings involved delineated groups of human beings collaborating with each other, with devices, computers, and instruments, and thereby creating new tools, new atomic and thermonuclear devices, and new groupings. These assembled groups of scientists, engineers, mathematicians, administrators, and others had cognitive properties that were radically different from the cognitive properties of any individual. Lawrence was cognizant of this and had capitalized on this insight in building the Rad Lab during

the 1930s. It was exported to Livermore, which in fact was part of the Berkeley Rad Lab during its initial stage. At Livermore, as Edwin Hutchins pointed out, Lawrence and Teller had extended their particular cognitive powers by creating an environment in which they could exercise those powers. See Hutchins 1995.

30. York 1987, 77–78.

31. Kistiakowsky became Eisenhower's science adviser after Killian had resigned to become MIT's president. See Kistiakowsky 1976.

32. See Negele 2006; Holt and Brown 2006; Brown and Lee 2009.

33. See Bethe 2006; Bahcall and Salpeter 2006; Brown 2006a,b; Brown and Lee 2006; and Brown and Lee 2009.

34. See Bernstein 1980, 2006; Ioffe 2006; Brown and Lee 2009.

35. Weber 1919, 560.

1. Growing Up

1. Schatzman 1970, Foreword. During the previous ten years or so, Bethe had been working on the nuclear many-body problem, where the same remark applies with even greater force.

2. For Ewald's work in neurobiology see Thauer 1955, Fischer 1957, Holts 1955, Pauly 1987.

3. The *Habilitation* allowed him to become a *Privatdozent,* which gave him the right to teach at a university. *Assistent* was a position equivalent to a present-day postdoctoral research associateship or a (nontenured) lectureship or assistant professorship. In German universities at the time, the professor who was the director of an institute controlled the budget and essentially all nontenured appointments, hence the use of "*Assistent* to."

4. Thauer 1955, Fischer 1957, Holts 1955.

5. Appendix A shows the Bethe family tree. Lisbeth, Albrecht's only sister, was the oldest offspring. She never married and took care of Albrecht's mother after their father died. Erich became a highly esteemed Greek scholar and a professor of Greek in Leipzig; Martin became a medical doctor in Stettin; Karol, who was two years older than Albrecht, became a not too successful merchant in Hamburg.

6. HABM I.

7. For an informative and very readable account of Richard Hertwig—and that of his equally famous brother, Oscar Hertwig—see Goldschmidt 1956, 76–109. In 1898, when he was studying medicine in Munich, Goldschmidt was a student of Richard Hertwig, who had become professor of zoology at the Ludwig Maxmilians University in 1885. Hertwig, and the way he ran his laboratory, convinced Goldschmidt to become a zoologist, and Goldschmidt returned to Munich as Hertwig's

Assistent in 1903. Hertwig's research concerned the relationship between the nucleus and the cytoplasm of a cell, and physiological studies of the development of sea urchin and frog embryos.

8. Richards 2008. Haeckel is also known for his recapitulation theory—"ontogeny recapitulates phylogeny"—which asserts that an individual organism's development, or ontogeny, parallels and summarizes its species' entire evolutionary development, or phylogeny (Gould 1977).

9. De Bont 2009, 203. Goldschmidt (1956) tells of the Hertwig brothers' meeting with Carl Vogt at the Naples station in the early 1880s. Vogt, a remarkable zoologist and comparative anatomist, was a "witty, aggressive, and even fanatical for the causes of evolution, freethinking, and German democracy of the 1848 brand." The serious and puritanical Hertwig brothers did not get along with Vogt. Vogt composed and distributed an "elaborate poem in Homeric style," which was "as clever as it is mean, . . . lampooning the Hertwigs as the two Ajaxes" (p. 84). The Hertwigs never again set foot in the Naples station and discouraged their students from going there. However, Albrecht Bethe did so after obtaining his doctorate from Richard Hertwig.

10. When Albrecht became Ewald's *Assistent,* Ewald was an *ausserordentlicher Professor* (associate professor). He later became the *ordentlicher Professor* (full professor) of physiology in Strassburg. In German universities the three main academic ranks were: *ordentlicher Professor* (full professor), *ausserordentlicher Professor* (associate professor), and *Privatdozent* (instructor). *Ordentlicher* and *ausserordentlicher* professors were salaried governmental officials. For a detailed history of the German educational system see Ringer 1969a.

11. Albrecht Bethe's *Habilitationsschrift* is entitled "On the influence of the labyrinth of the shark on its body position and swimming motions."

12. Fruton 1990, 225.

13. Alsatian is a German dialect. For the connections between Rhineland Jewry and the Jews of southwestern Germany and Alsace, see Schwarzfuchs 2003.

14. Victor Klemperer, who converted in 1903 before being inducted into the Bavarian army, reported in his autobiographical writings that all he had to do to become Protestant was to say yes, shake hands with the pastor, and vow loyalty to the Church. He was thereafter presented with a bill for 14 marks and 75 pfennigs (Elon 2002, 230).

15. The date of Anna's sister's death became etched in Hans's mother's memory. Hans and Rose Ewald were married on that same date—and Hans warned Rose never to mention the date of his marriage to his mother.

16. An *Assistent*'s salary was of the order of 100 marks a month—enough for a single person to live on—and the income from the course fees as a *Privatdozent* amounted to another 100 marks. The mark of 1900 was worth approximately four 2010 U.S. dollars.

17. The scale of dowry amounts was "common knowledge." Around 1900 a Jew-

ish university graduate in Berlin could expect "about 75,000 marks as a dowry from the bride's family—clearly a fortune" (Richarz 1997, 84). Evidently if the bridegroom was an *Assistent* and not Jewish, the dowry was much higher.

18. The crab was chosen because arthropods have relatively simple nervous systems but complex and diversified behaviors. In 1990, physiologist Ernst Florey characterized the task that Albrecht Bethe had set for himself as the " ambitious undertaking . . . [of] mapping all the neurons that constitute this 'simple nervous system' and discovering the function of each of them . . . [In this,] Bethe's program anticipated the research effort of neurophysiologists of the second half of our century: to establish the neural basis of behavior" (quoted in Shuranova and Burmistrov 2006). In their critical review of Albrecht Bethe's work on the neurophysiology of arthropods, Shuranova and Burmistrov concluded that it "was extremely important for that time, and also seems to be very interesting today. It forces modern neurophysiologists to think about some essential problems, which are little studied experimentally despite the enormous progress in the field of nerve-muscle physiology in crustaceans."

19. The general class of these chemicals is now called pheromones.

20. A. Bethe 1897a, 1897b, 1898.

21. For the strange and tragic subsequent history of Beer see http://www.shfri.net/trans/mildenberger/mildenberger.pdf. Albrecht had first met von Uexküll while staying for a short time in Kühne's institute in Heidelberg. Again, for informative, candid, and very readable personal accounts of Driesch and von Uexküll see Goldschmidt 1956, 68–71. For a more scholarly, very informative, and insightful account of the views of Driesch and von Uexküll see Harrington 1996.

22. Harrington 1996, 42. Another element that contributed to Bethe's important role in the historical origins of behaviorism is his work on the plasticity of the nervous system and, in particular, his work with Kurt Goldstein during World War I on the plasticity of the human brain. The behaviorists extrapolated Bethe and Goldstein's findings on the plasticity of the human brain, and considered human beings as an infinitely malleable black box interposed between stimulus and response. B. F. Skinner's *Walden II,* with its denial that the human brain is, at birth, hard-wired to express "human nature," is the ultimate articulation of behaviorism.

23. Baldwin 1901, 201.

24. Until 1910, when Paul Ehrlich and Sahachiro Hata's "magic bullet" Salvarsan was clinically introduced as an effective treatment against syphilis, the conventional therapy consisted in using mercury salts, which had only limited efficacy.

25. As recounted by Rose Bethe, July 26, 2009.

26. Richard Ewald was also the uncle of Paul Ewald, who later became Bethe's father-in-law.

27. Albrecht Bethe, as told to Ernst Fischer and related to me by Anne Fischer, October 25, 1991.

28. HABTK.

29. Early Greek was written right to left but later its direction changed to *boustrophedon* (with the direction of writing changing every line). *Boustrophedon* means "ox-turning," a reference to plowing.

30. As recounted by Rose Bethe, July 26, 2009.

31. To prepare to succeed at *Gymnasium,* many students attended preparatory schools (called *Vorschule*) for three years. Some of these were privately operated; others were run by the local government.

32. When Hans was eight years old he sent his mother a typewritten note with birthday wishes: "A. from me. 1. elephant son 2. elephant daughter 3. hare son/monkey son 4. hare daughter." Illustrative of his sense of humor Hans also sent her wishes from "6. pastry board and noodle roller 8. board with pots and pans like Klara. Size. 60 long and 40 wide." Letter from Bethe to Mama. Kiel, June 22, 1914.

33. When Hans's mother emigrated to the United States in 1939 she brought with her a large trunk that contained some of her personal belongings. After she died the trunk was stored in the cellar of the Bethe house on Deer Park Road in Cayuga Heights. Hans never opened the trunk, nor did Hans's wife, Rose, while Hans was alive. Hans and Rose's daughter Monica, in her yearly visit to her mother, has helped her go through all of Hans's *Nachlass* in the house. So in the summer of 2010 the trunk was finally opened. It contained the young Hans's writings and drawings from age four to age fifteen, poems that Hans's mother had written for him, and numerous photographs.

34. Monica Bethe is translating all of these materials and assembling them for publication.

35. HABCW II.

36. The Königliche University of Frankfurt-am-Main changed its name to Johann Wolfgang Goethe University in 1932.

37. For the early history of Frankfurt University, see the address of the rector on the occasion of its fiftieth anniversary (Rammelmayer 1965). Before the university opened, extended discussions had been held between the initiators of the idea of founding a *königliche Universität* in Frankfurt and state and national authorities, to guarantee that there would not be any discrimination against Jews holding full professorial positions at the university. Paul Ehrlich had been denied such a position in Berlin, his outstanding research notwithstanding. The Jewish population of Frankfurt probably gave approximately 90 percent of the 8 million marks donated toward the establishment of the university.

38. Embden had been Albrecht's colleague in Strassburg.

39. The house was still standing when Bethe visited Frankfurt after the war.

40. HABCW II.

41. Bethe-Kuhn, no date.

42. HABCW II.

43. For Max Weber's involvement in German politics during and after World War I, see his wife's biography of him (Marianne Weber 1975) and Mayer 1944 and Mitzman 1985. For Weber's correspondence with the *Frankfurter Zeitung,* see Mayer 1944, 75, 77–78. See also Ringer 2004, 74–75.

44. In their last six years at the *Gymnasium* students were taking between six and eight hours of Latin per week and, in their last four years, an additional eight hours of Greek per week. According to Karl Guggenheim, a classmate of Hans, "in all these fields we had very good teachers. On the other hand instruction in mathematics and physics was poor and the teachers were bad." Karl Guggenheim to S. S. Schweber, June 11, 1993.

45. Bethe did not recall how the diagnosis was made—only that it was not made by X-ray.

46. It is interesting to note that Bethe's chest X-rays showed scars, which caused difficulties when he applied for an immigration visa to the United States in 1934. The United States did not admit persons with tuberculosis. However, the doctors decided that the scars were indeed encapsulated and that Bethe was cured.

47. Rose Bethe, in an interview on July 26, 2009, put it thus: "Kreuznach and Odenwaldschule enabled him eventually to be happy."

48. Meccano sets consist of reusable metal strips, plates, wheels, axles and gears, motors, and so on. The strips and plates have small, regularly spaced holes, through which nuts and bolts can be inserted to connect the pieces, and, similarly, connect axles to wheels. With these pieces, one can build elaborate models and mechanical devices.

49. In fact, Germany's armies had been soundly defeated on the battlefields in the summer of 1918, and its military leadership had asked the kaiser to make peace. The armistice signed on November 11, 1918, was the recognition by Germany that it was defeated and was surrendering.

50. See Margaret MacMillan's (2002) riveting account of the various negotiations that culminated in the 1919 Treaty of Versailles. See also her article "Ending the War to End All Wars," *New York Times,* Dec. 25, 2010.

51. See Ringer 1969b.

52. In July 1914, a dollar was worth 4.2 marks; in January 1919, the exchange rate stood at 8.9 marks to the dollar; 192 marks in January 1922; 17,972 marks in January 1923; and 353,412 marks by July 1923. From August 1923 until the currency was stabilized on November 15, 1923, the exchange rate catapulted from 5 million to 4,200 billion to the dollar. The *Rentenmark,* which was introduced on November 15, was worth a thousand billion old marks, i.e., a trillion old marks! (Ringer 1969b, 63).

53. In the 1970s Bethe gave a talk on the history of the atomic bomb for which he prepared the following table outlining the hyperinflation:

Inflation numbers

10/18	10/19	5/20	4/21	1/22	7/22	10/22	11/22	12/22	1/23
1.5	5.7	12	15	45	96	430	1800	1800	2800

3/23	7/23	8/1/23	8/24/23	9/1/23	9/20/23	10/1/23
5500	40,000	260,000	1.1M	2.4M	43M	57M

10/10/23	10/20/23	11/1/23	11/5/23	11/12/23	11/20/23
700M	2.9G	31G	100G	150G	1000G

54. Even though by 1923 some of the stamps were worth a *pfennig* or less, they now have a catalog value of over 150 Euro.

55. HABTK.

56. Interview with Karl Guggenheim, Jerusalem, August 17, 1993.

57. This is corroborated by Bethe's school record. He consistently got a "sehr gut" or a 1 in mathematics (1 being the highest grade and 5 the lowest). The grade "sehr gut" was given sparingly. He consistently got a "sehr gut" or a 2 in physics, and mostly "gut" in all the other subjects, except in gymnastics and handwriting. Upon graduation he got a superlative "excellent" in mathematics.

58. Karl Guggenheim to S. S. Schweber, June 11, 1993.

59. "Schiller's plays were read more than anybody else's. Lessing is the most lovable of the three. He wrote a lovely comedy, *Mina von Baumheld,* which is very much in the spirit of English writing of the 18th century. He also wrote the very good *Nathan der Weise.* Schiller is rather bombastic. At the time we swallowed him very happily, and some of those plays by Schiller are really very good. *Wallenstein* for instance is a very good play. And in that case, he didn't even distort history. He was a historian, but he miserably distorted Mary Stewart, and the play was totally oblivious of any historical background. And Schiller miserably distorted also Don Carlos of Spain, the son of Phillip, but he correctly depicted Phillip as an unbending tyrant. But Don Carlos apparently was a very unstable and confused character, and he is made a great hero in that play."

60. Heidelberg is some ninety kilometers due south of Frankfurt. The train ride took about two hours.

61. Thus Bethe vividly recalled that in 1928, when Kurt Weil and Bertolt Brecht's *Die Dreigroschenoper* (Threepenny Opera) was first performed, "Prof. Sachs sang and played on the piano the songs in it." Bethe often referred to and quoted the lyrics of this play.

62. Paul eventually retired to England.

63. The daughter of the Levis, Hilde, became a close friend of Hans.

64. Bethe to Anna Bethe. At some early stage Bethe started numbering all the letters he wrote by the ordinal number of the particular letter, and by page numbers. For example, letter 196 was a six-page letter he wrote to his mother on 18 VII 27,

with the front side of each sheet numbered, the first one 971 and the last one 975. Thus 196/972 would correspond to the back side of the first sheet of that letter.

65. Lee 2007, 36.

66. Hans's father continued to support Anna "very handsomely" and Anna herself wrote Hans that her future was financially secure. Bethe recalled that "my father decided that if he wanted to have any money to live on, more than just by himself, he would have to have an extra income. And so he became the editor of a big encyclopedia of physiology, known as *Handbuch der normalen und pathologischen Physiologie, mit Berücksichtigung der experimentellen Pharmakologie,* and published by Springer in twenty-odd volumes, of which I have two by accident. It was very highly regarded, I think, for quite a number of years, and may still be quite good."

67. Baden-Baden is famous as a spa, located on the western foothills of the Black Forest on the banks of the Oos River, some seventy kilometers due west of Stuttgart, near the French border.

68. "Der dreitägige 'famose Schäddel' Seiner Mama, mit der ihn immer eine unsichtbare Nabelschnur verbindet." 5.VII.30.

69. Interview with Barbara and Mortimer Litt, Boston, Massachusetts, Jan. 23, 2011.

70. Kant 1998.

71. Albrecht remarried in 1928 and fathered two daughters.

72. Doris Overbeck to S. S. Schweber, Jan. 28, 1990.

73. In the spring of 1918, a much more conservative professor had been elected rector, and when the revolution came in November 1918, he asked that Albrecht take over again, because Albrecht was more in tune with the necessity of governmental change. Albrecht thus became rector again from November 1918 until the spring of 1919, at which time the Weimar government took over. From that time until 1923 Albrecht was quite active in politics. Although fairly close in political outlook to the social democrats, some of the socialists didn't like him; the communists liked him even less. He once told Hans that he had attended a meeting with the social democrats and communists in the early 1920s as one of the representatives of the democratic party. The communists at the meeting were quite "rabid," and had said "Well, once we get to the government, we'll hang everybody of the old parties, and you Professor Bethe will hang first."

74. The Democratic Party was founded a few days after the Armistice was signed on November 11, 1918. It asked the support of that part of the German population that had accepted Germany becoming a democratic republic and was opposed to restoring the monarchy. Its platform rejected socialism as the clear-cut solution of the problems Germany was facing after its defeat, and sought to amalgamate individual freedom and social responsibility. The economist Alfred Weber, the brother of Max Weber, and Adolf Wolff, the editor-in-chief of the *Berliner Tageblatt,* were the moving

spirits in its formation. Max Weber likewise became deeply involved in the affairs of the party, and the Frankfurt branch of the party nominated him to become the chairman. It is very likely that Albrecht Bethe and Max Weber got to know one another at that time (Holborn 1969, 537; Craig 1982; Marianne Weber 1975; Mayer 1944, 45). From 1919 until 1925 the Democrats usually garnered 15 to 20 percent of the seats in the Reichstag and formed governments in coalition with the social democrats and the centrist Catholic Party. In opposition were the communists and the two right-wing parties, each with approximately 5 percent of the seats.

75. At 4,634 meters, Monte Rosa is the second highest mountain in the Alps and the highest of Switzerland.

76. Merz 1965, 520–521.

77. John Theodore Merz (1965) devotes an entire chapter to the "Psycho-Physical View of Nature." In it he discusses the doctrine of psycho-physical parallelism, also called "the conscious automaton theory," which he considered the central conception in psychology as a natural science. Albrecht Bethe had studied in Freiburg, then the leading center of "psycho-physics." Merz noted that

> Psycho-physics having through Weber, Lotze, Fechner, and Wundt gradually evolved the notion of a partial parallelism of physical and psychical phenomena, the conception of a mathematical dependence or of function could be introduced between the measurable external processes and the hidden internal events which we call mental; the whole of the latter being looked upon as concomitant occurrences . . . of the more accessible though very complex phenomena of the nervous system and its centers . . . Having got hold of this partial formula, which in some cases admits even a rigorous mathematical expression [as in the case of Fechner's formula $S = K \text{Log} I$, the presumed relationship between the psychological sensation, S, and the physical intensity, I, of the stimulus triggering it], psycho-physics had no pressing need of investigating its meaning any further; . . . similar general inquiries into the origin of gravitation, of atoms, of the essence of energy or inertia having proved to be of little or no use in furthering astronomy, chemistry, thermodynamics. (Merz 1965, 517–518)

Interestingly, in a footnote, Merz indicated that the doctrine of psycho-physical parallelism was "prepared by earlier thinkers, such as Descartes, . . . , by Spinoza, and by Leibniz's doctrine of pre-established harmony." It is likely that Albrecht Bethe's studies in Berlin, Munich, Naples, and Strassburg and his subsequent research had given him a mastery of all these viewpoints and their subsequent developments.

78. Johannes Müller's analysis of neural conduction had led Du Bois–Reymond and Helmholtz to investigate sensory processing in hearing and seeing. The quantitative analysis of sensory and perceptual experiences marked the beginning of a new

disciplinary approach distinct from the natural sciences. Albrecht's approach derived from that tradition.

79. Uexküll explicitly stated that when asking questions about the function of an animal, he looked with "coolness at physiology . . . I seek only for a fitting expression in order to make the plan [what he called the *Bauplan*] of the animal *anschaulich*." For him biology was in its essence "Anschauung" (Harrington 1996, 40).

80. Harrington 1996, 40.

81. See Harrington (1996) and Ringer (1969a) for both an elaboration of these themes and for references to the subject.

82. Kurt Goldstein was a neurologist and psychologist. See Harrington 1996.

83. Kurt Goldstein joined the faculty of Frankfurt University and became the head of the Institute for Research into the Consequences of Brain Injuries, which looked after soldiers who had suffered brain injuries during the war and other patients with brain injuries or disorders (see Harrington 1996, ch. 5). I do not know how much interaction there was between Albrecht Bethe and Goldstein during the latter's stay in Frankfurt.

84. See, for example, Bethe and Fischer 1931.

85. In his interview with Thomas Kuhn, Bethe commented, "My father knew more mathematics, I think more than was usual at the time for a physiologist; he knew calculus, which was rare among physiologists at the time." On Albrecht's knowledge of thermodynamics, see Kremer 1990.

86. Cahan 1993, 1. See also Gray 2008.

87. See Kremer 1990.

88. Cahan 1993, 2.

89. Ibid., 8.

90. Helmholtz 1995.

91. Hale 1908, 3.

92. Presumably, like most other biologists, Albrecht had become aware of Mendel's work through the independent botanical experiments of de Vries, Correns, and Tschermak in 1900.

93. Bethe, A., H. A. Bethe, and Terada 1924.

94. See Richards, R. J. (2003) and Hagner (2003) for insightful accounts of the interaction of biology with medicine during the nineteenth century.

95. Fischer 1957.

96. Albrecht himself had made important contributions to staining techniques in neuroanatomy and was very familiar with the organic chemistry of dyes. It is very likely that Albrecht made Hans aware of Paul Ehrlich's synthesis of Salvarsan as a cure for trypanosome infections such as syphilis, and how Ehrlich and Hata had secured this result. The "magic bullet" Salvarsan was the first *cure* of any disease by chemicals, and Ehrlich had made his discoveries and done his research in Frankfurt! Paul Ehr-

lich's introduction of Salvarsan and his winning the Nobel Prize in Medicine in 1908 was hailed in Frankfurt. His institute, the Royal Institute of Experimental Therapy, was established in Frankfurt in 1897. See Ehrlich's 1908 Nobel Lecture (Marquardt 1951) and Lenoir 1997, ch. 7.

97. That is, it did not fit into the Mertonian model of the sociology of science, characterized by a commitment to an ethos that incorporates communalism, disinterestedness, universalism, and organized skepticism. See Lenoir 1997, Introduction, for a very perceptive and succinct overview of Robert Merton's and Joseph Ben-David's sociology of science.

98. Albrecht had written these words as a postscript to his third and concluding communication on the central nervous system of *Carcinus maenas.*

99. Thauer 1955. Albrecht was blessed in being able to enjoy doing research until the very end of his life.

100. Weber made this statement in a letter to one of the young men who had participated in the summer 1917 meeting of members of the *Freideutsche Jugend* with scholars, artists, and political writers. This meeting was underwritten by the Jena publisher and book dealer Eugen Diederrichs. Weber was one of the *Lebenpraktiker* at that meeting (Marianne Weber 1975, 596–597).

101. Goldschmidt 1956, 102. See also Goldschmidt 1960, 66–76.

102. Trude's family was not religious but observed some of the traditional dietary laws, went to synagogue on Saturdays "when there was a reason for it," and always did so on the high holidays. Trude was one of three daughters. The oldest was very gifted in art, but died in the influenza epidemic of 1918. The youngest chose to help their father in the family business.

103. Interview with Trude Fraenkel, Jerusalem, Aug. 18, 1993.

104. Neugarten 1919.

105. Trude's husband was Adolf Fraenkel, whom she had met in Freiburg in 1914. He served in the German army during the war and later came to Frankfurt to study medicine, passing his examination in 1922. Thereafter Trude and he went to live in Mainz. In 1935 they emigrated to Palestine, where he became a well-known specialist in diseases of the ears, nose, and throat.

106. Karl belonged to Blauweiss and Kadimah, the nonreligious Zionist youth organizations. See Glatzer 2009 and Gordon 2003.

107. As mentioned earlier the Physiology Institute was compartmentalized. The animal physiology course was taught by Albrecht Bethe, and the vegetative physiology course (which became physiological biochemistry) was taught by Emden. Interview with Karl Guggenheim, Jerusalem, August 17, 1993. Karl contracted tuberculosis in 1929 and was advised by his physicians not to pursue a career in clinical medicine. He then went on to do laboratory-based medical research, first in bacteriology and then, after he emigrated to Palestine in 1935, in nutrition. He established the department of nutrition at Hebrew University and became the professor of nutri-

tional sciences at Hadassah Medical School. He retired in 1974 and thereafter he wrote two books on the history of nutritional medicine.

108. Fischer 1957.

109. The rest of the letter was taken up with information concerning Anna, Hans's mother: her health, living circumstances, and his visits to her. Bethe also asked his uncle to provide him with information concerning the genealogy of the Bethes. Martin did so, and this is the basis for Appendix A.

110. HABCW II.

111. Bethe to Anna Bethe, letter 196 dated 18 VII 27.

112. Hesse 1951, 37.

113. Ibid., 38.

114. For an informative and insightful analysis of the German university system at the time see Max Weber 1919. I discuss the system in Chapter 4. Weber's essay also presents a perceptive comparison of German and American universities. Weber characterizes the German university as "plutocratic" in contrast to the "bureaucratic" American university.

115. See Lenoir 1997, Introduction, for a succinct and insightful account of Pierre Bourdieu's sociology of science and of Bourdieu's account of cultural, economic, political, and symbolic capital, their interchangeability, and how they determine the dynamics of disciplinary and institutional formation and maintenance.

2. Maturing

1. This in contrast to Albrecht Bethe. In his interview with Charles Weiner Bethe recalled, "My father was quite good with his hands, and in the year '23 he built a very elementary radio receiver so that we were among the first radio listeners in Germany." Making a crystal set radio receiver required a great deal of dexterity! It also indicates that Albrecht was keeping up with technological developments.

2. Gray 2008, 1.

3. Wachsmuth had been an *Assistent* to Helmholtz. He became the first rector of Frankfurt University and wrote its history (Wachsmuth 1929).

4. For a complete list of the courses Hans took while attending Frankfurt University, see Appendix B.

5. Graetz 1920. Leo Graetz, the son of the eminent Jewish historian Heinrich Graetz, was a distinguished experimental physicist who was a friend of Max Planck. He was the professor of experimental physics in Munich until his retirement in 1920, when he was succeeded by Wilhelm (Willy) Wien. There is a copy of *Die Atomtheorie* in Bethe's *Nachlass.* I do not know whether the copy was bought by him or whether he had found it in his father's library.

6. Incidentally, "for fun" during his first year at the university Hans took a course on double-entry bookkeeping with Prof. Dr. Kalverlam, and a summer course with

Prof. Dr. Frobenius (Organization of People and States) in which he was taught that "5000 years ago matriarchal societies were the rule."

7. Madelung did his *Habilitation* in Göttingen on the structure of crystalline solids, and later successfully calculated the binding energy of ionic crystals such as sodium chloride. In 1922 he wrote a very informative and instructive introduction to mathematical methods in theoretical physics (Madelung 1922).

8. Elsasser 1978, 15.

9. HABM.

10. Ibid.

11. Hilde Levi's autobiography is on deposit in the Niels Bohr Archives. I thank Finn Aaserud for making it available to me.

12. Erwin Strauss recalled that when he and Hans "climbed mountains together, Hans took the trouble to explain very difficult subjects to me in a way that I could understand them. I was always very interested in Physics as a boy, but I must confess that Hans's unusual gift in this field was one of the reasons which discouraged me to study sciences." After getting his *Abitur* Erwin joined a bank. On one occasion he got Hans a paid assignment at the bank, which ought to have taken a week to do. Hans completed it in less than two days. I believe Erwin's invitation to work at his bank resulted in Bethe writing a twenty-page pamphlet titled "How do I compute the yield of investments designed to yield a fixed rate of interest?" (Bethe 1929e). The pamphlet probably was distributed free of charge to those customers of the bank able to understand the simple mathematics of its exposition. Bethe's deep interest and expertise in economic matters stemmed from his having had to deal with Germany's devastating post–World War I inflation. These concerns were reflected in his taking an accounting course at Frankfurt University in the summer of 1924 and his keeping himself fully informed of economic developments throughout his life. Erwin emigrated to the United States in 1933, became an economist, and worked for the U.S. government in Washington, D.C., until his retirement. The only time they saw one another after they both left Germany was at Erwin's wedding in Alabama in 1938.

13. Hans left a carefully annotated photographic record for 1926–1927, including some of the group's later hiking and skiing activities, in three photo albums.

14. Fips Boehm and Werner Sachs later married.

15. The quotations to the end of this section are from an interview with Hilde Levi in Copenhagen, December 14 and 15, 1991.

16. Bethe to Anna Bethe, letter 95, dated October 27, 1926.

17. Bethe to Anna Bethe, letter 86/425, dated November 14, 1926.

18. Only in their homes had German Jews felt a measure of freedom within the wider hostile environment; this experience left its mark on the atmospheres of the homes of these assimilated families. During a good part of the nineteenth century the home was the space where Jews had maintained their identity, a distinctiveness that

was threatened on the one hand by assimilation and on the other, if not by the hostility of the host culture, then certainly by the segregation that it enforced.

19. Financial factors were undoubtedly involved. The dowry of Jewish women who married upper-class Germans was considerably higher than the dowry of those marrying Jewish men. This was the case with Bertha and Carl Fitting.

20. Elsasser 1978, 13.

21. Mosse 1985, 6.

22. Bruford 1959, 1975.

23. Mosse 1985, 6–7.

24. von Humboldt, quoted in Harwood 1993, 277.

25. Beyerchen 1983, 31.

26. In contrast to the Netherlands, in Great Britain and France the bourgeoisie had emerged through economic activities, and were thus identified with the operation of the market. In Germany, the educated and cultural component of the bourgeoisie became identified with the State, its bureaucracy, and its educational system. In Germany the economic bourgeoisie (the *Wirtschaftsbürgertum*) are distinguished from the educated and cultural bourgeoisie (the *Bildungsbürgertum*); the latter played a "hegemonic role in the State apparatus." Turner (1998, vi) delineated the political context of Germany after its unification by the following characteristics: "the dominance of the State, the power of the bureaucracy, the bureaucratic incorporation of the *Bildungsbürgertum,* and the political influence of the landowners" (the Junkers).

27. Sorkin 1983, 1999.

28. One of the reasons that Humboldt could conceive of being able to implement such reforms at the university level was that the German universities enjoyed a semi-autonomous corporate status. McClelland (1980) has likened the universities to the medieval craft guilds. *Lehr-* and *Lernfreiheit,* the freedom to teach and to learn, highlighted the autonomy of academic disciplines and became the twin pillars on which neohumanism was anchored.

29. See Holborn 1964, 457–484.

30. The literature on German Jews is vast, and the field is marked by sharp differences of opinion among historians. I have found the writings of Martin Meyer (1988, 2001), Meyer and Brenner (1997), and Shulamit Volkov (2001, 2006) especially valuable, and I am indebted to them for much of my presentation. In addition, David Sorkin's (2001) lengthy, informative, insightful, and constructively critical review of Meyer and Brenner's (1997) four-volume *German-Jewish History in Modern Times* was likewise particularly helpful.

31. See Mendes-Flohr and Reinhartz 1980, 101–129. Recall that Prussia had remained free of French rule. See Holborn 1964.

32. Brenner in Brenner et al. 2003, 2.

33. This by virtue of a memorandum he wrote to the chancellor of Prussia in

1809 and of Humboldt's activities at the Council of Vienna. See Holborn 1964, 436–446.

34. Humboldt's genuine commitment to Jewish emancipation is indicated by his actions. In 1814 Humboldt wrote to the minister of justice, protesting the fact that Jews were not permitted to perform autopsies and urging the minister to abolish a measure so based on prejudice. Humboldt was influential in charting the Prussian position regarding the rights of Jews at the Congresses of Vienna in 1814 and 1815. On June 14, 1815, shortly before the Congress adjourned, Humboldt wrote his wife: "I have always been favorably disposed toward [the acquisition of equal rights by the Jews of Germany]. It is . . . an idea of my youth, for Alexander and I were regarded, even when we were children, as bulwarks of Judaism."

But Humboldt's contact with Jews had greatly diminished after 1810—evidently by design. In another letter to his wife from Vienna he indicated that his efforts there on behalf of the political rights of Jews were the last embers of his devotion to Henriette Herz. "She has herself almost become a Christian. All are deserting the ancient gods."

35. Humboldt to the chancellor of Prussia, September 30, 1816 (Kohler 1918, 36). Humboldt's paper is in Kohler 1918, 72–83. See also http://www.archive.org/stream/jewishrightsatco00kohliala/jewishrightsatco00kohliala_djvu.txt.

36. This was the position that Abbé Grégoire advocated during the French Revolution. See Kaufmann 2003, 79–82.

37. Jews always constituted a small minority in Germany. In 1815 there were approximately 350,000 Jews in all the German-speaking states. See Volkov 2001.

38. See Geuss 1996 and Kaplan 1997.

39. See Sorkin 2001 and Meyer and Brenner 1997.

40. Sorkin 1983.

41. Martin Meyer's writings are the standard sources for the history of the Reform movement in both Europe and the United States. See Meyer 1988, 2001, and Meyer and Brenner, 1997.

42. Mosse 1985, 1.

43. Quoted in Elon 2002, 59.

44. Ibid., 159–161.

45. See Cresti 2003 and Ehrenfeld 2003.

46. The Constitution incorporated legislation that the North German Confederation had adopted somewhat earlier, barring discrimination on religious grounds.

47. Meyer 1988, 135.

48. The trend continued. By 1933, 55 percent of German Jewry lived in cities of over 100,000, with 32 percent living in Berlin. In 1910, the 535,200 Jews in Germany constituted 0.93 percent of the population. In 1933, the 499,700 Jews made up 0.77 percent of the population.

49. Richarz 1997, 14.

50. For a narrative of Jewish contributions to German culture during the nineteenth and twentieth centuries see Slezkine 2004.

51. Josiah Wedgwood was Charles Darwin's maternal grandfather. See, for example, Desmond and Moore 1991, 5.

52. See Lowenstein 1997.

53. Volkov 1996, 81, 83. Volkov also pointed to the difficulties faced by emancipated, *Bildungsträger* German Jews by virtue of the inner tensions and contradictions in the assertively emancipatory Enlightenment ideals encapsulated in *Bildung:* an emphasis on rationality, yet a growing realization of the role of the passions; a stress on the primacy of individualism, yet a commitment to the notion of an all-encompassing humanity; a commitment to freedom, yet loyalty to what became an absolutist state in Germany. Additionally, *Bildung* manifested another deep inner strain from its inception: the tension between its equality postulate—it was assumed that everyone could acquire it—and its elitist consequences. *Bildung* was responsible for the creation of a special *Gymnasium*- and university-educated service class recruited from the aristocracy and from the upper bourgeoisie, "all joined in culture, in love of the arts, and in social and intellectual aspirations." The possession of these attributes became a marker of social identity and a sign of distinction. Defining social division according to educational criteria became the accepted norm. *Bildung* thus played a divisive rather than an integrative role. Moreover, this process of redefinition of the criteria of social divisions also engendered a systematic "closing of the ranks," excluding everyone not part of the particular elite (Volkov 1996, 90).

54. Beyerchen (1983) notes that there has always been a deep and popular strain of anti-intellectualism, which after the French Revolution and occupation manifested itself in the deprecation by Romantic thinkers of the rational conception "of a human society based on individual liberty and their praising of an intuitive, idealized image of a national community based on personalities formed by tradition and language." See also Ringer (1969a) who stresses the Idealistic rather than the Romantic element and sees the Idealistic tradition as representing an insurgent intellectual elite challenging the dominance of an aristocracy of birth and championing the claims of intellect and merit.

55. Ringer 1969a. Mosse (1985, 4) quotes Berthold Auerbach, considered by his fellow Jews to be an authoritative voice for Judaism in the nineteenth century: "formerly the religious spirit proceeded from revelation, the present starts from *Bildung.*" The emancipated German Jews' secular religion of humanity had at its core *Bildung,* and trumpeted the ideals of rationalism and progress. Auerbach characterized it as "an inner liberation and deliverance of man, his true rebirth; not through words or customs, but through his deeds, his character, the totality of his life, the cleansing and healing of all human labor."

56. Although Jews nominally had the right to be appointed to a university professorship, in 1909–1910 fewer than 3 percent of the full professors at German universi-

ties were Jewish (Rowe 1986). Yet Jewish students constituted approximately 7 percent of the total number attending *Gymnasia.* At the time Jews made up less than 1 percent of the total population of Germany. Thus access to university teaching careers still required conversion, particularly in the humanistic faculties. Many Jews did convert, as indicated by the relatively large percentage of the professoriate—around 9 percent—who had parents, grandparents, or great-grandparents who had converted. Personal memoirs and other documents attest to the pressures exerted upon Jewish candidates to convert. By the early twentieth century forty-four holders of university chairs were Jews who had converted and only twenty-five identified themselves as Jews. These twenty-five individuals constituted 2.5 percent of the German *Ordinarii*—more than the percentage of Jews in the general population but considerably less than the percentage of Jews in the student population. At this same time Jews constituted 12 percent of the *Extraordinarii* and 14.5 percent of the *Privatdozenten* (Volkov 2006).

57. In Spain before the fourteenth century, the position of the Catholic Church was that anyone converting to Catholicism would be accepted as an equal, having the same rights as anyone born into a Catholic family (Yerushalmi 1982). However, converted Jews were discriminated against because, it was asserted, that by virtue of "blood" they remained Jews. This, occurring in the fifteenth century, was the first instance of racism based on biology. For some of the consequences of scientific racism and anti-Semitism based on biology see Gilman 1986 and Stepan and Gilman 1991.

58. Earlier in the century, pogroms occurred in various German communities in the wake of Napoleon's emancipation of the Jews in the German states he occupied. In the 1880s some political parties adopted anti-Semitism as part of their platform (Elon 2002).

59. Medical reforms instituted after the Revolution of 1848 granted physicians a legal monopoly over medical care and required the members of the profession to be trained at universities. Physicians' accreditation was contingent on their satisfying all the demands of a curriculum determined and administered by the professoriate. These reforms coincided with the transformation in Germany of the field of physiology at midcentury. In a famous manifesto, Hermann von Helmholtz, Emil Du Bois–Reymond, Ernst Brücke, and Carl Ludwig vowed to reconstitute physiology on a physico-chemical foundation. The success of their reductionist and deterministic approach gave physiology a status of equal rank with physics. As a result of their productive program of research, the field became an autonomous scientific discipline and a core element in the new curriculum for the education of German physicians. Rudolf Virchow, in particular, forcefully argued the value of physiology in securing the scientific foundations of advances in medicine. He also argued that the exposure to scientific methodology required for understanding experiments in physiology would provide valuable mental training for the physician in his medical practice (Lenoir 1997, 1998).

60. In the 1847 manifesto Helmholtz, Brücke, and Du Bois–Reymond asserted that there is no "vital" force responsible for living organisms and that research on organisms should be based only on the laws of physics and chemistry. Beginning in the late 1850s physiology replaced logic in the Prussian medical examination, creating new academic opportunities within the field of physiology (Harwood 1993). On converted Jews' Enlightenment-inspired outlook, see Ben-David 1991.

61. See Philippson 1962.

62. Stoetzler 2008, 3.

63. Concerted efforts to disenfranchise German Jews included a petition presented to Bismarck signed by over 200,000 people.

64. Stoetzler 2008, 3–4.

65. Quoted in ibid., 1.

66. Jews whose ancestry was Polish, as in the case of assimilated Jews from Breslau and Silesia, probably felt particularly vulnerable. We shall see that Peierls's father was one such Jew.

67. See Kauders 2009 for an introduction to this much-covered subject.

68. See Scholem 1976, in particular his essays on the myth of the German-Jewish dialogue (61–71).

69. Katz 1986, Volkov 2006.

70. See Rabinbach 1997.

71. The values of excellence, originality, and achievement as determinants in professorial appointments in German universities emerged in the early nineteenth century. Ben-David (1991) attributes the emergence of these values to the alleged autonomy of science and competition among universities, whereas Turner (1971, Turner, Kerwin, and Woolwine 1984) attributes it to ideological factors. By the late 1830s these institutional values were fairly well recognized, even if not always adhered to.

Helmholtz is a case in point (Cahan 1993): between 1845 and 1847 Helmholtz was stuck in Potsdam, isolated as an army physician. However, Johannes Müller and Alexander von Humboldt recognized his abilities and brought him into the academic system, first in Berlin, and then in Königsberg, and from there to a professorial appointment in Heidelberg. In Heidelberg Helmholtz, Gustav Kirchhoff, and Robert Bunsen demonstrated the potential of a system whose guiding principles were excellence, originality, and achievement. In 1871, the new Reich recognized this, and brought Helmholtz to Berlin. Helmholtz, in turn, brought to Berlin Eduard Zeller (1872), Kirchhoff (1875), Treitschke (1874), August Kundt (1888), Planck (1889), and, in a subsequent generation, Einstein (1914). They wanted those they thought the best for the "best" place. Berlin rose as Heidelberg declined. Conversely, people who were incompetent or carried unacceptable ideological baggage (for example, Eugen Dühring and Friedrich Zöllner) were driven from the system. The system was far from perfect—personal connections and other such factors did at times play a role in appointments and promotions, but Germans were able to control favoritism. Despite pervasive anti-Semitism in German academia, during the second half of the nineteenth century numerous Jews, nonpracticing Jews, or converted Jews were able to obtain positions in German universities or *Technische Hochschulen.* For statistics see Wenkel 2007. Many of these individuals worked in new or interdisciplinary fields (theoretical physics, abstract mathematics, biophysics, biochemistry, and physiology) rather than long-established fields such as anatomy and astronomy. It is worth noting that, unlike the German universities, British, French, and American universities very rarely appointed Jews or individuals with Jewish roots before the end of the century. I thank David Cahan and Jed Buchwald for very helpful remarks on this subject.

72. Thus Henri Poincaré asserted:

> Mr. Einstein is one of the most original thinkers I have ever met. In spite of his youth, he has already achieved a very honorable place among the leading savants of his age. What one has to admire above all is the facility with which he adapts himself to new concepts and knows how to draw from them every possible conclusion. He has not remained attached to classical principles, and when faced with a problem of physics he is prompt in envisaging all its possibilities. A problem which enters his mind unfolds itself in anticipation of new phenomena which may one day be verified by experiment . . . The role of mathematical physics is to ask the right questions, and experiments alone can resolve them. The future will show more and more the worth of Mr. Einstein, and the university intelligent enough to attract this young master is certain to reap great honor. (Seelig 1954, 163)

73. Meissner had a Jewish wife. After Hitler's rise to power he emigrated to the United States and obtained a position at Purdue University.

3. Becoming Bethe

1. Elkana 1981.

2. Heisenberg 1974, 47.

3. Craig 1982, 92.

4. Heisenberg claimed that Munich's emphasis on trade and tool-making "produced a different economic atmosphere. The mere scramble for profit has always counted for less in Munich" (Heisenberg 1974, 52).

5. Justus von Liebig had been invited to the University of Munich by Maximilian II in 1852, where Liebig established an important center of chemical research. Adolf von Baeyer, who in 1883 synthesized indigo, succeeded Liebig in 1875.

6. In 1912, its faculty included Adolf von Baeyer, Heinrich Wieland, and Hans Fischer (all of whom won Nobel Prizes) in chemistry; Oscar Hertwig in zoology; Paul

Groth in mineralogy and crystallography; Nobel laureate Wilhelm Röntgen in experimental physics; and Arnold Sommerfeld in theoretical physics. The humanities and social sciences were staffed by equally eminent academicians, including Heinrich Wölfflin in art history and Lujo Brentano in economics.

7. Following Sommerfeld's recent work in electron theory, Röntgen was instrumental in having Sommerfeld appointed to the post in 1905. Sommerfeld's position was initially somewhat unusual: his chair was an appointment in the university and his quarters were located there, but as part of the mathematical physics cabinet of Bavaria's scientific instrument collection. Thus it was the Bavarian state's bureaucracy rather than the university authorities that oversaw Sommerfeld's position.

8. The 1929 edition became the second volume of *Atomic Structure and Spectral Lines.*

9. Kirkpatrick 1949, 312–313.

10. Sommerfeld spent the fall semester of 1922–23 at the University of Wisconsin. During the spring of 1923 he traveled extensively on the West Coast, giving lectures on atomic and molecular structure at several universities, including Berkeley and Caltech. He again visited the United States in 1928.

11. Seth's (2010) analysis of the Sommerfeld School and Eckert's work (1993, 1999, 2003; Eckert et al., 1984, 1985; and Eckert and Märker 2003), particularly Eckert 2003, have been most helpful in shaping my views of Sommerfeld.

12. See Olesko 1991.

13. Sommerfeld and Wiechert designed a harmonic analyzer—an analog computer—that calculated the distribution of temperature over a given area as a function of depth using empirical data registered at various depths. The analyzer was built in Volkmann's institute and performed exceedingly well. Unknown to Sommerfeld and Weichert, Kelvin had built a similar analog device in Glasgow. When the Königsberg scientific society offered a prize to assess the accuracy of the predictions made by the analyzer, Sommerfeld accepted the challenge. The mathematical problem to be solved was to find a solution of the partial differential equation for heat conduction with boundary conditions that mimic the surface of the ground near the station at which the temperature measurements were recorded. Sommerfeld approximated the surface of this terrain by two intersecting planes, and reduced the problem to finding the solution of a linear differential equation on a Riemann surface of several sheets, a method which he would hone with great effectiveness and apply to the problem of optical diffraction. However, he had to withdraw his paper because of an error in the boundary conditions. He later commented that this failure had its roots in a distinctive attitude of the period, which was "clinging to mathematical generalities rather than studying the particularities of the problem and using numerical methods" (Born 1952, 266).

14. One semester of the lectures on theoretical physics Sommerfeld delivered in

Munich, the one devoted to the partial differential equations of mathematical physics, dealt with these matters. The content of these lectures became the sixth volume of his *Lectures on Theoretical Physics.*

15. Rowe 1989.

16. See Rowe 1986 and Gray 2008 for details of Klein's biography.

17. Klein's influence is readily apparent in the papers on the mathematical theory of diffraction that Sommerfeld wrote for his *Habilitationsschrift* to become a *Privatdozent* in mathematics (Sommerfeld 1968b, vol. 1). Klein had fostered Riemann's approach to the theory of complex variables and their use in physical problems. In the papers Sommerfeld submitted for his *Habilitation* he elaborated and improved the methods he had developed in his heat conduction paper. A rigorous solution for the diffraction problem was given as a complex integral over a Riemann surface with two branches. The method—which Poincaré characterized as "extremely ingenious"—was very powerful and the solution well suited for numerical calculations. These two papers by Sommerfeld have become classics in mathematical physics.

18. Sommerfeld 1968a.

19. *Bonze* evidently derived from the Japanese *bonzo,* for the most important personage in a Buddhist monastery. *Geheimrat,* when conferred to a professor by the state, was an honorific that indicated an exceptional standing in a field or fields. The title was abolished after 1918.

20. Born 1952, 277. David Rowe's (2004) beautiful article provides an insightful account of Klein's personality and influence.

21. Frank Nelson Cole, quoted in Rowe (2004, 95). Cole was a student in Göttingen during the early 1890s. He later became the secretary of the American Mathematical Society.

22. Rowe 2004.

23. Warwick 2003.

24. Peierls 1985, 23.

25. Klein had recognized that the contemporary disciplinary organization of the sciences in German universities and technical institutes prevented the creation of a forum in which the various disciplines would cooperate in tackling problems and stimulate new ideas through cross-fertilization. Klein had a vision to establish institutes in Göttingen devoted to various applied sciences, staffed by outstanding scientists with different disciplinary backgrounds. Although Friedrich Althoff, the powerful official in the Prussian Ministry of Culture in charge of the funding of education, strongly supported Klein's idea, he did not have the resources to underwrite such a project. In 1893 Klein attended the World's Columbian Exposition in Chicago, where he learned that private funds had been responsible for the recent founding of several American universities, such as the University of Chicago, Clark University, Johns Hopkins, and Stanford. He took the idea home and got monies from individ-

ual donors and various industries to eventually establish several applied science institutes in Göttingen: the Institute for Applied Electricity, with H. T. Simon as its head; the Institute for Applied Mechanics, with Ludwig Prandtl as its director; and the Institute for Geophysics, with Emil Wiechert as its director (Klein 1899; Eckert et al. 1992, 14–15).

26. This gradual enlargement in the scope of Sommerfeld's interests is reflected in the research that he carried out during this period. In 1899, he solved the problem of the propagation of electromagnetic waves along thin wires when the finite diameter and the conductance of the wire are taken into account—a problem whose solution had eluded Poincaré, J. J. Thomson, Rayleigh, and Drude. See Seth (2010) for an informative, detailed exposition of Sommerfeld's research.

27. Technologies like the gyroscope are but one example of the pre-established harmonies—this one between universities, industry, and the state—that Klein referred to in his famous speech in the summer of 1918 anticipating a German victory. Forman (1971) quotes Klein's speech in his introduction. For further details on Sommerfeld's involvement and that of his students in *Kriegphysik* during World War I, see Seth 2010.

28. See Seth (2010) for details. While at these two institutions Sommerfeld published papers on the strength of materials, vibrations in electrical generators, the hydrodynamical theory of lubrication, and ballistics. This research helped convince the engineering community of the usefulness of mathematics in the solution of their problems. Sommerfeld's powers were such that, in addition to these activities, he worked also on foundational physics problems, devoting considerable efforts to the Lorentz-Abraham theory of the electron. In Lorentz's electrodynamics electrons were assumed to be rigid spheres with a fixed electric charge density and the electromagnetic field outside these charged objects was assumed to obey Maxwell's equations in free space. The problem was to find the equations of motion for such a rigid electron under the action of its own field and an external field. J. J. Thomson had shown that, by virtue of its self-interaction, the electron acquired an increase in inertia and had suggested that possibly the entire mass of the electron might be of electromagnetic origin. In three lengthy papers Sommerfeld obtained explicit formulas for electrons moving with arbitrary accelerations using the method of Fourier transforms. The papers were written before Einstein published his first paper on relativity, and so Sommerfeld also considered the motion of electrons moving faster than the velocity of light—and obtained solutions that later were rediscovered in connection with the explanation of Cerenkov radiation, the radiation emitted by charged particles moving in a medium with a velocity greater than the velocity of light in that medium.

29. Eckert 2003, 176.

30. Sommerfeld 1968b, vol. 4, 675.

31. See Cahan (1985) for a very informative account of the institutionalization of physics in German universities during the nineteenth century.

32. Jungnickel and McCormmach 1986.

33. Sommerfeld's interest in hydrodynamics had been stimulated by the Isar Company, which had a contract to build banks to channel the Isar River, which was prone to flooding in Munich; the company had made generous grants to Sommerfeld's institute (Eckert 1999; Seth 2010).

34. Wilhelm Röntgen had accepted the professorship in 1900 and remained in Munich until his death in 1925.

35. Paul Epstein was a student in Sommerfeld's seminar from 1910 to 1914 (Epstein 1964–65).

36. Ewald 1962, 32–33.

37. Sommerfeld 1910.

38. Mehra 1975, 39.

39. See Rowe 2004.

40. Einstein to Sommerfeld, January 14, 1908, in Einstein 1995, vol. 5.

41. Einstein to Sommerfeld, September 29, 1909, in Einstein 1995, vol. 5, 134.

42. Einstein to Laub, December 31, 1909 in Einstein 1995, 146.

43. See the Bethe-Sommerfeld correspondence for 1931, Sommerfeld *Nachlass,* Deutsches Museum, Munich. They may be accessed at: http://www.deutsches-museum.de/en/archives/archive-online/.

44. Strutt 1942.

45. HABCW II.

46. Seth 2010. Seth offers valuable insights into Sommerfeld's approach to theoretical physics. He has characterized aspects of the development of theoretical physics in Germany during the first two decades of the twentieth century in terms of a dichotomy between two kinds of theoretical physics, distinguished by their methods, worldviews, discourse, and techniques: the physics of principles—with Planck, Einstein, and Bohr as its most prominent proponents; and the physics of problems—with Sommerfeld as its paradigmatic exponent (Seth 2007, 41). Seth did not take into account the role played by Klein, Hilbert, and Minkowski in these developments, and their influence on Einstein and Sommerfeld. If they are taken into account the dichotomy is not as sharp as Seth indicates. See Schweber 2008b, van Dongen 2010.

47. Seth 2010.

48. I use the term *School* to denote the cadre of students Sommerfeld trained, the problems they addressed, their particular approach to the problems they tackled, as well as the social context while they were part of Sommerfeld's seminar. See Geison and Holmes 1993.

49. Eckert 2003; quotation on p. 180.

50. Hiebert 1984, 179.

51. That it is a "harmony" relates mathematics to music: in the Pythagorean corpus, the relation is to the music of the spheres; in the early modern period, a relation between the irrationals and the musical scales. See, for example, Pesic 2005, 2010.

52. For Planck's views on these matters, and their influence on Einstein, see Schweber 2008a.

53. For Leibniz's *Monadology* see Leibniz 1965, 1967. For an accessible introduction to Leibniz's *Monadology* see Buchdahl 1969 and Rutherford's essay on Leibniz's late-period metaphysics in Jolley (1995, 124–175). Funkenstein, one of the scholars who probed Leibniz's writings and philosophy most deeply, interpreted Leibniz's "notorious 'pre-established harmony' of monads" as asserting that "without interacting, all of them are nonetheless interconnected, and each of them mirrors in its own individual way all the others" (Funkenstein 1986, 106).

54. Einstein 1954, 226–227. In that same article, alluding to Hilbert's call since the beginning of the century to axiomatize physics, Einstein noted:

> If, then, it is true that the axiomatic basis of theoretical physics cannot be extracted from experience but must be freely invented, can we ever hope to find the right way? Nay, more, has this right way any existence outside our illusions? Can we hope to be guided safely by experience at all when there exist theories (such as classical mechanics) which to a large extent do justice to experience, without getting to the root of the matter? I answer without hesitation that there is, in my opinion, a right way, and that we are capable of finding it. Our experience hitherto justifies us in believing that nature is the realization of the simplest conceivable mathematical ideas. I am convinced that we can discover by means of pure mathematical constructions the concepts and the laws connecting them with each other, which furnish the key to the understanding of natural phenomena. Experience may suggest the appropriate mathematical concepts, but they most certainly cannot be deduced from it. Experience remains, of course, the sole criterion of the physical utility of a mathematical construction. But the creative principle resides in mathematics. In a certain sense, therefore I hold it true that pure thought can grasp reality, as the ancients dreamed. (Einstein 1954, 274)

Lewis Pyenson and Leo Corry have noted that this was not a statement Einstein would have made before 1918. Einstein came to adopt this position only after his struggle with the mathematical formulation of the theory of general relativity; and it became the way he formulated the various versions of his unified theory of gravitation and electromagnetism (Pyenson 1985, Corry 2003). After his creation of the theory of general relativity, Einstein came to believe that the mathematical description of the unity of the physical world was to be in terms of local fields that are continuous functions on the space-time continuum. He came to attribute his success in formulating the theory of general relativity—his towering achievement—as stemming from the role mathematics played in the search, and in particular, the role played by the geometric constraints he imposed when choosing the gravitational field variables and the partial differential equations they were to satisfy. Fields were the fundamental entities and Einstein hoped that the empirically observed particles would emerge as localized non-singular solutions of the field equations. In the Herbert Spencer lecture Einstein delivered in Oxford in 1933, he gave a clear exposition of his views regarding the pre-established harmony between physics and mathematics. See van Dongen 2010.

55. Pyenson 1983, 1985.

56. See Holborn 1964, 152–163. Paul Forman pointed out to me that Oswald Spengler (1926, Preface to vol. 1) concluded by declaring his debt to "those to whom I owe practically everything: Goethe and Nietzsche. . . . But Goethe was, without knowing it, a disciple of Leibniz in his whole mode of thought. And, therefore, that which has . . . taken shape in my hands I am . . . proud to call a *German philosophy*." An indication of how influential a philosopher Leibniz remained well into the twentieth century, especially but not only in Germany, is that Bertrand Russell was a Leibnizian before he became an antimetaphysician, and as late as the publication of his 1945 *History of Western Philosophy,* Russell put Leibniz in a class by himself.

57. See for example, Ferreirós and Gray 2006, 383–385; Gray 2008, 277–281.

58. For example, Planck rejected the divinity of Christ, though he believed that the teachings of Christ contained "the elements of all true religions" (Heilbron 1986, 87).

59. Recall Leibniz's first article in the Leibniz-Clarke exchange: "Mr. Newton and his supporters once again have a curious opinion of God's works. According to them, God needs to reset his watch occasionally. Otherwise it would cease to work. He was not farseeing enough to equip it with a perpetual movement. God's machine on their view is even so imperfect that he is obliged to clean it every now and then by an extraordinary intervention, and even to repair it, as a watchmaker might repair his work." Samuel Clarke, Newton's mouthpiece, in his second response to Leibniz in their famous exchange noted that "The present state of the Solar System, for example, according to the laws of motion that are now established, will one day fall into confusion, and then it may be redressed or it may receive a new form." At the end of the eighteenth century, Pierre-Simon Laplace believed he had proven that the solar system was stable, and that Leibniz had been right and Newton wrong; and therefore—as he told Napoléon when presenting him with a copy of his *Exposition du Système du Monde*—God was unnecessary in his account (Laplace 1813).

60. Merz in his 1907 biography of Leibniz stressed that Leibniz had "entered on philosophical researches after he had familiarized himself with logical, legal, and mathematical studies" (Merz 1907, 137). The principle of contradiction (that two contradictory statements cannot be true at the same time) and the principle of sufficient reason (that nothing is without a reason) had their roots in logic.

61. See, for example, Planck 1915/1958, 1932.

62. Merz 1907, 137, 150.

63. See Corry 2004, 103.

64. In this connection it is interesting to note Reichenbach's comment: "When I,

on a certain occasion, asked Professor Einstein how he found his theory of relativity, he answered that he found it because he was so strongly convinced of the harmony of the universe" (Reichenbach 1949, 292).

65. Minkowski 1923, 91.

66. *Hier tritt nun das Relativitätprinzip als ein wirklisches neues physikalisches Gezetz ein.* Quoted in Corry 1997b.

67. See Darrigol 1995, 4.

68. Principle of relativity as quoted in Darrigol 1995. Poincaré's list of useful principles included the two principles of thermodynamics, the principle of action and reaction (Newton's third law), the principle of relativity, the principle of conservation of mass, and the principle of least action. All the principles, except Carnot's, could be extracted from mechanics.

69. See Reid's biographies of Hilbert and of Courant (Reid 1970, 1976, 1986) for an informative and sensitive description of the atmosphere in Göttingen from the time of Klein's arrival there until 1933. There is in particular a very insightful exploration of Hilbert's and Minkowski's friendship.

70. On Ludwig Fleck see Fehr et al. 2009, Introduction.

71. It was published as Minkowski 1915.

72. Pyenson 1977, 2. Pyenson quoted in Corry 1997b, 6n8. See also Galison 1979, 93.

73. The Lorentz transformations are the mathematical statements of the relations between the space-time coordinates of two frames of references moving with constant velocity relative to one another—such that observers in either frame, when measuring the velocity of light, will find it to be the same.

74. Believing in the possibility of mathematical formulations of specific theories that would quantitatively represent physical phenomena or processes was, of course, a prior tacit metaphysical presupposition.

75. Einstein 1954, 228.

76. Ibid. See Einstein 1995, introductions to vol. 7 and vol. 9, where the editors discuss Einstein coming to the distinction between constructive and principle theories. The eminent Einstein scholar John Stachel believes that the distinction between constructive and principle theories became less sharp in time (Stachel 2002).

77. The equivalence principle states that one cannot differentiate the results of an experiment carried out in an inertial frame in which there is a constant gravitational field from one carried out in a frame moving relative to it with a (certain) constant acceleration without any gravitational field present. The principle explains the numerical identity (in appropriate units) of the inertial and gravitational mass of an object.

78. See van Dongen 2010, ch. 1, for a concise and insightful presentation of Einstein arriving at his field equations.

79. See Utiyama 1956.

80. See Merton 1968. Robert Merton noted that eminent scientists will often get more credit for work that is more or less the same as that of a lesser-known researcher. He named the phenomenon the "Matthew effect."

81. Klein 1964.

82. The relation of statistical mechanics to thermodynamics is further complicated by probabilistic assumptions in the case of the micro canonical distribution. The sense in which statistical mechanics is a constructive theory is difficult to answer.

83. Einstein in 1921 contrasted Hilbert's way of understanding geometry to Kant's. For Kant the axioms of geometry were a priori, whereas for Hilbert the axioms did not rest on intuition and had no connection to experience. The axioms were *free creations of the human mind.* See Hiebert 1984 and van Dongen 2010 for expositions of Einstein's changing views on approaches to the formulation of physical theories.

84. See Hiebert 1984 on the Einstein-Sommerfeld friendship.

85. Einstein 1995–.

86. Sommerfeld 1968b, vol. 4, 611.

87. For Sommerfeld's views regarding "number mysticism" before the advent of quantum mechanics see Seth 2010.

88. Sommerfeld 1968b, 615.

89. Sommerfeld 1949a.

90. Pauli 1994, 64.

91. Sommerfeld 1949a.

92. See Schweber 2008a.

93. "Jenes statistische Verhalten der vielen gleichen Einzelsysteme, die keinerlei Kontakt miteinander haben (fensterlose Monaden), ohne doch anderseits kausal determiniert zu sein, ist ja in der Quanten-mechanik als letzte, nicht weiter reduzierbare Tatsache aufgefasst" (Pauli 1964, 709). See also Laurikainen 1988, Pauli 1994.

94. Bethe 1954.

95. Francis Bacon in his *Great Instauration* had spoken of the "luciferous and fructiferous" qualities of experiments.

96. Notes of Rose Bethe to S. Schweber, July 2011.

97. Quoted in Eckert and Märker 2003, 23. In this 1928 letter to his wife, knowing of Einstein's difficulties with the new quantum mechanics, Sommerfeld went on to say, "Einstein, who always looks at the whole, makes his life more difficult."

98. The English Garden is a very big park, about the size of Central Park in New York, but in a much smaller city. It is much more developed than Central Park.

99. Cassidy 1992, 100.

100. Incidentally, the same story is told with James Chadwick being replaced by Linus Pauling. In his *Memoirs* Teller tells the story of John van Vleck's encounters with Sommerfeld on the successive days of his arrival in Munich. On the first day he greeted him as Herr Sommerfeld, with barely a grunt in reply from Sommerfeld. The next day Sommerfeld responded almost audibly upon being greeted as Doktor Sommer-

feld. When addressed as Herr Professor, Sommerfeld clearly nodded and replied. And when van Vleck greeted him with "Guten Morgen, Herr Geheimrat" Sommerfeld replied with a smile: "Aber Ihr Deutsch wird jeden Tag besser." [But your German is improving every day!] (Teller 2001, 47).

101. This was in contrast to the impressions of the previous generation of Sommerfeld students. Bethe noted, "My father-in-law, Paul Ewald, took his Ph.D. with Sommerfeld in 1912, sixteen years earlier than I, and then Sommerfeld was fairly young, and was very approachable."

102. Later published as Sommerfeld 1928a,b.

103. Wali 1991, 61.

104. Somebody else gave the part on nondegenerate perturbation theory, and Bethe presented degenerate perturbation theory.

105. T. S. Kuhn, "Interview with H. A. Bethe, 01/17/1964." Archive for the History of Quantum Physics.

106. Sommerfeld, like every other professor of theoretical physics at the time, gave a three-year round of lectures that covered all of theoretical physics, beginning with Mechanics, and Mechanics of Deformable Bodies, continuing with Electrodynamics, Optics, and Thermodynamics, and Statistical Mechanics, and concluding with the Partial Differential Equations of Physics. The lectures were four hours per week, given for thirteen weeks in the winter and eleven weeks in the summer. In addition to the lectures was a weekly recitation for the discussion of problems. The lectures were published as separate books and subsequently translated into English.

107. Elsasser 1978, 34.

108. Gregor Wentzel was born in Düsseldorf, Germany, in 1898. His father, a lawyer who held a quasi-governmental position at a state bank in one of the Prussian provinces, was a man of broad intellect with many and varied interests. Wentzel's mother was likewise deeply committed to intellectual and cultural matters. Wentzel's education was relatively standard. He attended a *Gymnasium* that emphasized Latin, modern languages, the natural sciences, and mathematics. His physics teacher got him excited about astronomy and this sparked his interest in the sciences in general. He went to the University of Freiburg in 1916 with the idea of becoming an astronomer. Military service, from 1917 through 1918, interrupted his studies. After the Armistice he returned to Freiburg for a semester and a half, and in the fall of 1919 matriculated at Greifswald. Both at Freiburg and at Greifswald he studied mostly mathematics, studying physics on his own. His reading of Hermann Weyl's *Raum, Zeit und Materie* made a deep impression on him. The one physics course he did take at Greifswald was a seminar with Johannes Stark, who was assisted by Rudolf Seeliger, who had obtained his doctorate with Sommerfeld. Impressed by Wentzel's performance, Seeliger suggested he go study with Sommerfeld. Wentzel arrived in Munich in November 1920 and was admitted to Sommerfeld's seminar. A novel derivation of Peter Debye's work on the dispersion of light by permanent dipoles in a presentation

to the seminar, shortly after he arrived in Munich, got Wentzel accepted as a doctoral student. He was awarded his doctorate in 1921 with a dissertation in which he disentangled and clarified the X-ray line spectra of atoms. He stayed on in Munich, and for his *Habilitationsschrift* he analyzed multiple scattering of electrons by atoms.

It was in Munich that Wentzel first met Pauli and Heisenberg, who were fellow students in Sommerfeld's seminar. Wentzel and Pauli became close friends, a friendship that lasted until Pauli's death. In the fall of 1926 Wentzel accepted an associate professorship at the University of Leipzig, where Heisenberg was the professor of theoretical physics. In 1928 Wentzel succeeded Erwin Schrödinger at the University of Zurich. At the same time, Pauli accepted the theoretical physics professorship at the Eidgenössische Technische Hochschule, and together they made Zurich a world center of theoretical physics.

109. Bethe recalled that

> To become a theorist you had to have experimental physics as a minor, and for that you were supposed to take the advanced lab, and listen to one or two courses in experimental physics. And I have totally forgotten what experimental course I listened to, I think one, by Ruckart, who is otherwise not known, I believe.
>
> I did a number of the experiments, but failed in one of them.
>
> There was Millikan's oil drop, and then there was an experiment on electromagnetic waves using two parallel wires to generate the wave, and we then determined the wavelength. You had a sliding bridge across the wires, which somehow gave the capacitance. There were lots of electric experiments. The experiment which finally terminated my experimental career was called a dilatometer, to measure volume changes. It contained a glass plate and another glass, and it may have been shaped spherically, and you then heated the apparatus, and you observed interference when you did that, it's a tough experiment.
>
> I never built any apparatus. (HABTK)

110. Quotations in this paragraph are from HABTK.

111. See the interview of Walter Heitler by John Heilbron for Heitler's intellectual biography. Available at http://www.aip.org/history/ohilist/4662_2.html.

112. Elsasser 1978, 16.

113. It was published in 1929 in German and in 1930 in English (see Sommerfeld 1929).

114. T. S. Kuhn, "Interview with H. A. Bethe," 01/17/1964, p. 12. Heisenberg, using the Schrödinger approach, had given an explanation of the helium spectrum in 1926 by noting that the Pauli principle demanded that the wave functions of the two electrons be antisymmetric. In 1927 Heitler and London had formulated the theory of the homopolar chemical bond.

115. Peierls 1985, 23–24.

116. The ascription of classical concepts to the description of microscopic entities, their properties, and their measurements relies on the experimental context in which the microscopic entities are described. The description of a particle in terms of its position is complementary to the description in terms of its momentum, in the sense that the manifestations of either position or momentum depend on mutually exclusive measurements. Moreover, the information obtained through these experiments exhausts the possible objective knowledge of the kinematics of the entity. For Bohr's interpretation of quantum mechanics see Jan Faye's entry, "The Copenhagen Interpretation of Quantum Mechanics," in the Stanford Encyclopedia of Philosophy at http://plato.stanford.edu/entries/qm-copenhagen/#4.

117. Heisenberg describes the uncertainty principle: "The more precisely the position is determined [at a given time], the less precisely the momentum is known at this instant, and conversely" (Heisenberg 1925). The principle applies to other pairs of noncommuting variables. Sommerfeld undoubtedly would have read Heisenberg's paper when it was published and it surely would have struck him, if for no other reason than that the microscope is a central feature of the argument and that Heisenberg had nearly failed his doctoral examination because he could not answer Willy Wien's questions regarding the optics of a microscope.

118. Sommerfeld had hoped that the "Proclamation" would be circulated only among German scientists and would not be made public. But that hope was dashed when the *Frankfurter Zeitung* published it on April 28, 1915. For a thorough presentation of the policies regarding the annexation of the territories in the East and the West that the military victories would bring under German control see Holborn 1969.

119. Heilbron 1986, 72; Wolff 2003. The first of the three points in the "Proclamation" asked that when citing the literature "the English should no longer receive greater consideration" than German scientists "as had often happened in the past." But it was often understood as an appeal not to cite.

120. After the abdication of Kaiser Wilhelm and the signing of the armistice that ended World War I in early November 1918, Einstein became very active politically. Addressing a crowd of over a thousand at a meeting of the *Bund Neues Vaterland* in mid-November 1918—at which time, revolutionary "Soviets" were being established in several German cities—Einstein called himself "an old democrat" and exhorted his listeners "to stand guard, lest the old class tyranny of the right be replaced by a class tyranny of the left." On learning that Einstein believed in the principles set forth in the constitution of the Weimar republic—tenets that characterized "the new era"—and that Einstein wanted to work to help ensure their realization, Sommerfeld wrote Einstein a letter in mid-December 1918 expressing his consternation and apprehension, and depicted the new era as "miserable and stupid" (quoted in Einstein 1995, 7: xxiii).

121. Heilbron 1986, 32.

122. Einstein was the only "German" scientist invited to the 1924 Solvay Congress: because of his stature, and because he was considered Swiss. Upon receiving the invitation Einstein wrote Lorentz, the organizer of the conference, that Sommerfeld had urged him not to attend because no German physicist had been invited. Einstein concurred: one should not bring politics into scientific issues and individual scientists are not responsible for the acts of their governments. He declined the invitation and asked not to be invited to future meetings. The lack of contact between French and German scientists until 1927 was detrimental to both.

123. See Bacciagaluppi and Valentini 2010.

124. Sommerfeld 1954, 86.

125. See Hoddeson and Baym 1980, 8–10.

126. Ibid., 13–15.

127. See Casimir 1960, 139.

128. At the beginning of the century, Sommerfeld had worked actively on the Drude-Lorentz theory of metals and had periodically given lecture courses on the theory of solids. For an assessment of Sommerfeld's contribution to the developments of solid state theory after the advent of quantum mechanics see Hoddeson, Baym, and Eckert 1992, particularly 157–160.

129. Sommerfeld's research became the main subject of his research seminar. Its participants included Peierls, Eckart, Houston, and Pauling. The research also became the content of a lecture course during the summer of 1927.

130. Metals that are good electrical conductors are also good thermal conductors. Gustav Wiedemann and Rudolph Franz discovered the phenomenological fact that, to a good approximation, the ratio of the thermal conductivity to the electrical conductivity of a metal varies linearly with temperature, with the proportionality constant roughly the same for all metals. The electrical conductivity σ is defined in terms of the current j_E produced by an electric field E, $j_E = \sigma E$. In the Drude model $j_E = nev$, with n the number of electrons per unit volume, e the electron charge, and v the (average) velocity they acquire in the mean free time τ between collisions with an ion. Their mean free path l is $l = v\tau$. In the Drude model $\sigma = ne^2/m$. Similarly, the thermal conductivity κ_Q is defined by $j_Q = -\kappa_Q \mathbf{grad}\ T$, with T the temperature. Sommerfeld could show that in the Drude model with the conduction electrons obeying Fermi-Dirac statistics

$$\frac{\kappa}{\sigma T} = \frac{\pi^2 k_B^2}{3e^2}$$

with k_B Boltzman's constant. For the Sommerfeld calculations using Fermi-Dirac statistics see Hoddeson and Baym 1980, 15–17, and Sommerfeld and Bethe 1933.

131. Jammer 1966, 247. X-rays with wavelengths comparable to interatomic distances in crystals made possible the investigation the atomic structure of crystals. The

periodic arrangement of the atoms in a crystal was the essential feature responsible for diffraction effects. De Broglie was suggesting that electrons such that their wavelength was of the order of 10^{-8} cm would be suitable to explore their wave property through diffraction effects by crystals. This required that their energy $E = h^2/2m\lambda^2$ be of the order of 10 to 40 eV.

132. Regarding Compton's experiments, see Stuewer 1975. Quotation in Klein 1964.

133. Klein 1964, 32.

134. Hilbert had developed a severe case of anemia. See Reid 1986.

135. Davisson and Kunsman 1923.

136. All quotations in this paragraph are from Elsasser 1978, 60.

137. See Elsasser 1925 for Elsasser's paper. See also Elsasser 1978 and Gehrenbeck 1974, 182–187.

138. Born 1927.

139. Davisson and Germer 1927.

140. The position of any atom in the crystal is thus specified by the vector $l_1\mathbf{a}_1 + l_2\mathbf{a}_2 + l_3\mathbf{a}_3$ with l_1,l_2,l_3 integers. See, for example, James 1930 for a contemporary introductory text to X-ray crystallography.

141. The diffraction pattern in the transmission of X-rays through thin foils of crystalline solids had been observed in Sommerfeld's institute by Friedrich and Knipping, on the heels of von Laue's suggestion that X-rays might be diffracted by crystals. See Forman 1969.

142. For an account of Davisson and Germer's experiment see Gehrenbeck 1974, 1978. For their biographies, see Gehrenbeck 1981.

143. Bethe 1927.

144. Actually Patterson recounted that immediately after studying the Davisson-Germer paper he came up with "an incorrect interpretation of why the diffraction angle did not check with the lattice spacing. 'I interpreted this as a change in lattice spacing near the surface of the crystal whereas Bethe correctly explained it a few weeks later in terms of a refractive index effect.'" Patterson quoted in Ewald 1962, 612.

145. Bethe 1981. Though aware of Patterson's paper when he wrote his paper, Bethe evidently had not been aware of Eckart's and Fritz Zwicky's papers. There are no references to them in it.

146. Willy Wien was the editor of the *Handbuch der Experimentalphysik.* He felt strongly that a new edition of the *Handbuch der Physik* edited by Hans Geiger and Karl Scheel had too strong a theoretical bias. The *Handbuch der Experimentalphysik* was to rectify that situation.

147. In an epilogue Ott (1928) mentioned Davisson and Germer's experiment, cited Bethe's first *Naturwissenschaften* paper and gave him sole credit for interpreting their findings by attributing an index of refraction to electron waves.

148. Pauli had taken Ewald's course in the winter semester, 1919–20 (Enz 2002, 53).

149. The assignment of the indices of the order of the maxima of the reflection was different from that of the first *Naturwissenschaften* paper.

150. See Ewald 1916–18.

151. In Bethe's (1933a) *Handbuch* article on solid state physics where he discusses electron diffraction, he considers a *finite* crystal, and calculates not only angles of reflection but also the intensity of the reflected and transmitted beams.

152. Bethe did some calculations of intensities assuming $v_n(r - R_n)$ is Coulomb-like.

153. See Pendry 1974 and Duke 1974.

154. See the preface by G. P. Thompson to Beeching 1946; and Laue 1948.

155. See for example Pendry 1974 and Duke 1974.

156. HABCW.

157. "Votum von Arnold Sommerfeld zür Annahme der Dissertation von Bethe (Theorie der Elektronenbeugung im Festkörper)." Arnold Sommerfeld to University of Munich, July 24, 1928. Sommerfeld Papers, Deutsches Museum (Archiv NL 89, 030, Mappe Gutachten). http://www.lrz.de/~Sommerfeld/KurzFass/05941.html.

158. Doris Overbeck to S. S. Schweber, January 28, 1990.

159. See Bethe's letters to Peierls during Bethe's convalescence in Lee 2007, 21–30.

160. Peierls 1985, 5.

161. Ibid., 11.

162. Ludwig Fulda was a popular poet, playwright, and translator during the Wilhelminian Empire, and later during the Weimar Republic. He fought against government censorship, particularly as one of the founders of the Goethe federation in Berlin, and organized the first German performances of Henrik Ibsen's *Ghosts* in Augsburg—officially not public performances, in order to avoid censorship. In the fall of 1914 after World War I had broken out Fulda drafted the "Appeal of the 93" professors, defending the actions of the German army in the war, particularly in Belgium. On his seventieth birthday in the summer 1932, Fulda was awarded the Goethe medal for science and art by President Paul von Hindenburg. On May 5, 1933, the medal was rescinded on the grounds that Fulda was a Jew. Fulda died at the age of seventy-six in 1939 in Berlin, where he was waiting for a residence permit for the United States, where his son Karl Hermann had immigrated in 1936.

163. Peierls 1985, 8.

164. Ibid., 16.

165. Fritz Haber was born in Breslau, Germany, where his parents were prominent members of the Jewish community. He and Rudi's father, who was also born in Breslau, knew one another from childhood. As a chemist at the University of Karlsruhe before World War I, Haber and Carl Bosch developed the Haber process, the catalytic synthesis, at high temperature and high pressure, of ammonia from hydrogen

and atmospheric nitrogen. For this Haber won the Nobel Prize in 1918. During World War I while serving in the German army, Haber developed poison gases to be used against enemy troops. Haber's wife fervently opposed this work and committed suicide, very likely in response to his having personally overseen the first successful use of chlorine at the Second Battle of Ypres in April 1915. She shot herself early in the morning on May 15, 1915. That same morning, Haber left for the eastern front to oversee poison gas release against the Russians.

166. See Bethe's letter to Peierls concerning the visit in Lee 2007, 19–20.

167. My account of Teller growing up is based on his *Memoirs* (Teller 2001).

168. Deák 2009. The following small list of some Hungarian Jews or Hungarians whose parents had been Jewish makes the point: Szilard, von Neumann, Teller, Theodore von Kármán, George Pólya, Michael Polanyi, Karl Polanyi, Georg Lukács, Robert Capa, Arthur Koestler, Karl Mannheim, Eugene Ormandy, Georg Solti, Joseph Szigeti, George Szell, Fritz Reiner, André Kertész, Ferenc Molnar, Alexander Korda, Brassaï, Marcel Breuer, László Moholy-Nagy.

169. The government and the press did not take as firm a stand as Deák (2009) suggests. In 1888 at the age of twenty-eight Gustav Mahler was appointed the musical director of the Budapest Opera, at the time a "lackluster company." During his first year there he produced and conducted a brilliant season that prevented bankruptcy. Yet right-wing Hungarian newspapers "rarely missed an opportunity to assail his Jewishness along with his musical innovations." He was accused in parliament of running the opera as a "Jewish–bourgeois preserve." These persistent attacks wore down Mahler's resistance and he resigned his post at the end of his third year in Budapest (Grunfeld 1979, 45).

170. Teller 2001, 6, 7.

171. Ibid., 32.

172. Ibid., 33.

173. Ibid., 15.

174. The Schütz-Harkány family had been Jewish and had converted.

175. Teller 2001, 37.

176. Ibid., 39.

177. For a detailed account of Teller's study in Karlsruhe, Munich, and Leipzig see Hargittai 2010.

178. Teller 2001, 48.

179. Ibid., 49.

4. Beyond the Doctorate

1. Poincaré 1952, 141.

2. Hole theory is discussed in the section "Some Metaphysics," below.

3. Although Peierls's letters to Bethe before the early 1930s have been lost, after

his departure from Munich in the summer of 1928, Peierls in his letters from Leipzig very probably kept Bethe informed about what Heisenberg and his students were working on and what was going on in Bohr's institute in Copenhagen.

4. These were very hard times in German universities, so Hund did not have a regular position for an *Assistent,* but had to supply funding from several sources. The salary was quite poor, and so Bethe chose to remain in the Frankfurt position. "It would probably have been better for my development as a physicist to have gone to Hund, who was a far more productive physicist than Madelung" (HABCW II).

5. Lee 2007, 54–55.

6. That Vera was of Jewish descent did not affect Albrecht's professional life after 1933, as he himself had no Jewish roots. He retired from his Frankfurt professorship in 1936 and was allowed to continue working in his laboratory.

7. HABCW II.

8. Ibid.

9. Lee 2007, 48.

10. Elsasser 1978, 115.

11. See Dirac 1928. See Bethe's letter to Peierls, written as he was studying Dirac's papers shortly after they had been published in the *Proceedings of the Royal Society of London*; see Lee 2007, 34.

12. Lee 2007, 49.

13. Madelung did suggest that Bethe look at an experiment that Pohl had performed to try to disprove the attribution of wave properties to electrons. Pohl had constructed a lattice of wires which he charged alternatively positively and negatively and passed a beam of electrons through it. Some of the electrons were deflected to the right, some to the left. In fact, he got two maxima, one on the right side, and one on the left side. Pohl wanted to show that one did not need waves or quantum mechanics to explain these maxima. But because the wires were 2 to 3 mm apart Madelung believed this could not have anything to do with the de Broglie wavelength of the low energy electrons being transmitted through the lattice. In a paper submitted to the *Zeitschrift für Physik* entitled "The passage of electrons through the electric field of a lattice," Bethe showed that with the Pohl arrangement "one would indeed obtain two maxima—even though there was no wavelength, no diffraction. It is just the property of the electric fields between these wires" (HABLH, 16–17).

14. Recall that Richard Ewald, with whom Albrecht had done his *Habilitation* in Strassburg, was Paul Ewald's uncle.

15. Erwin Richard Fues had obtained his doctorate with Sommerfeld in 1920 and his *Habilitation* with Ewald in 1924. He thereafter became a *Privatdozent* at the Stuttgart Technische Hochschule, and Ewald's *Assistent.* He was away from Stuttgart from the fall of 1927 until the spring of 1929 because he had obtained a Rockefeller Foundation Education Board fellowship and had gone to Zurich to work with Schrödinger and thereafter to Copenhagen to work with Bohr.

16. If H is the Hamiltonian describing a physical system, define the functional

$$I[\chi]=\frac{\int \chi^* H \chi d\tau}{\int \chi^* \chi d\tau}.$$

If ψ is a solution of $(H - E)\psi = 0$, then $I[\psi] = E$. Moreover $I[\psi]$ is stationary under an infinitesimal change $\psi \to \psi + \delta\psi$ (provided $\delta\psi$ satisfies appropriate boundary conditions), i.e., $\delta I[\psi] = 0$. If the eigenfunctions ψ_r of H satisfying $H\psi_r = E_r\psi_r$ form a complete set, then an arbitrary χ can be expanded in terms of the ψ_r and one readily verifies that $I[\chi] \geq E_1$ where E_1 denotes the lowest eigenvalue of H. Thus an upper bound to the ground-state energy E_1 is obtained by using a suitable trial function in I. This is the basis for all the variational methods mentioned in the text. See Moiseiwitsch 1966, 163–170.

17. Hartree 1928.

18. Ibid. Hartree's method was immediately generalized by Fock, who approximated the n-particle electronic wave function by a determinant constructed with the n single particle wave function, thus automatically satisfying the Pauli principle.

19. See Bethe's "one- and two-electron" article (Bethe 1933a, 368) and Mott and Jones 1936, 44–46.

20. Hartree 1957.

21. Bethe 1930a.

22. Bethe 1930b. The article can be found in Bethe's (1997) *Selected Works.* Using a variational function that Hylleraas had introduced in the helium problem of the form $(1 + \alpha u + \beta t^2) \exp(-\gamma s)$ with $s = r_1 + r_2$, $t = r_1 - r_2$, and $u = |r_2 - r_1| = r_{12}$, and three parameters, α, β, γ, Bethe found that the resulting Rayleigh-Ritz upper bound on the energy lies below the binding energy of the hydrogen atom. The result depended on the presence of the term αu in the wave function, which by its dependence on the inter-electronic distance took into account correlations between the electrons. Subsequent variational calculations employing a large number of parameters obtained the value of 0.75 . . . eV for the binding energy of H^-, with an accuracy of many, many decimals. Bethe's paper was the point of departure for numerous investigations of the stability of three- and four-particle systems interacting by pure Coulomb forces as a function of the masses and charges of the particles. See Martin 1998, or CERN-TH/98–23, quant-ph/9801047.

23. See Chandrasekhar 1944.

24. Weisskopf, quoted in Stefan 1998, 50.

25. Ewald 1929.

26. Fritz London had been Ewald's *Assistent* during the academic year 1926. Ewald knew firsthand Heitler and London's theory of the covalent bonding of the hydrogen molecule. Some of London's work had been out while he was in Stuttgart. See Gavroglu 1995.

27. The story regarding Ewald not knowing quantum mechanics when Bethe came to Stuttgart began to be repeated in Bethe's interviews and became remembered as actually having been the case. This from a man endowed with an exceptional memory, who in his recollections took great pains to give proper credit to everyone concerned! Once again this makes evident—if indications are necessary—the tricks our memory plays on us. Bethe in his interviews with me did express regret for not having thanked Ewald.

28. Conant 1970, 71.

29. Incidentally, except for Maria Goeppert, a *Doktorandin* of Born in Göttingen, all the young theorists were men.

30. Not even to Sommerfeld. Actually, Sommerfeld did not offer much help to Bethe, and only read Bethe's Ph.D. manuscript after it had been completed.

31. Dong-Wong Kim informed me that the frequency of indicating the help received from other researchers in the Cavendish was low during Maxwell's (1874–1879) and Rayleigh's (1879–1884) regimes but increased under J. J. Thomson's directorship. After 1895 authors acknowledged their friends' advice, discussions, calculations, and so on, and expressed gratitude with a note at the end of their papers. See Kim 2002, 113. Expressing thanks became the norm in Manchester and at the Cavendish under Rutherford. Being Rutherford's son-in-law, and having his office in the Cavendish, Fowler adopted these norms and imparted them to his students.

32. Pais 1991.

33. Bethe 1981, 75. Rupp's findings could not be reproduced by other experimenters. It was subsequently discovered that many of Rupp's experiments were "faked." Rupp "withdrew them explicitly in a short note in the *Zeitschrift für Physik* 95 (1935), complete with a note by his psychiatrist that 'psychogenic dream like-states [had] entered his research'" (Bethe 1981, 75).

34. The classical theory had been worked out in great detail by Born and his co-workers, by Madelung, Ewald, and others (see Born 1923).

35. According to the Pauli principle the electronic wave function must be antisymmetric under the exchange of the variables describing the electrons. A wave function of two electrons having the same position and the same spin would therefore vanish.

36. Bethe 1929b, 4.

37. What is striking in Bethe's approach to the problem of cohesion of crystalline solids is that—using modern parlance—he was aware of the notion of broken symmetry and was explicit in stating that symmetry determines dynamics, that is, determines the equations of motion to be used.

38. Mermin and Ashcroft 2006. An English translation of Bethe's (1930b) article can be found in Bethe's (1997) *Selected Works.* In contrast to the assumption in his thesis that the electrons involved in electron diffraction by crystals were "quasi-free," Bethe in this article was concerned with the description of the electronic states of the

atoms in the crystal. Bethe assumed the "quasi-free" atoms' electronic states to closely resemble those of the free atom but perturbed by the periodic electric field produced by the other atoms in the crystal. The levels—which for the free atom possess the degeneracies implied by invariance under the full rotation group—are split when the atom's environment has the reduced symmetry of its crystalline environment. Bethe determined the splitting for each of the point groups describing the crystalline environment as well as the degeneracies that remained. Thus in a field of cubic symmetry Bethe found that a level with integral angular momentum J is split into levels which have threefold, twofold, or no degeneracy; whereas levels with half-odd integral J are split into levels with twofold or fourfold degeneracy.

39. Bethe et al. 1989, 11.

40. Mott and Jones 1936.

41. Bethe 1930c. He returned to these problems when at Cornell and with a student wrote a paper on the absorption spectra of rare earth salts (Bethe and Spedding 1937).

42. Herman Mark was born in Vienna. He had fought in the Austrian ski troops during World War I on both the Russian and Italian fronts. He escaped from an Italian prison camp by pretending to be an Englishman. Before becoming the head of the I. G. Farben research center in Mannheim he had taught at the Technical University of Karlsruhe. Paul Ewald was there for the academic year 1926–27. Edward Teller, who was just beginning his university studies at the time, studied with both Mark and Ewald that year. Mark later accepted a position at the University of Berlin and became a staff member of the Kaiser-Wilhelm Institute in Dahlem. He went on to become the "father" of polymer physics.

43. Felix Bloch, Nevill Mott, and Rudolf Peierls were the other young theorists invited by Debye.

44. Debye 1930.

45. The story varied as it was repeated. In the 1990s it was "Bethe is my student. I need him. I want him back immediately." Bethe told Charles Weiner in 1969 that Sommerfeld had written Ewald a letter which stated: "I need Bethe and what are you doing with him? You have no right to have him" (HABCW II).

46. The Bethe-Peierls correspondence is located in the Bodleian Library at the University of Oxford and has been translated and published by Lee (2007).

47. Herman Seligmann was the youngest brother of Bethe's grandmother Olo. Hans never knew him as he died very young. He was survived by his wife, Jeanette, and their two daughters. They lived in Munich "in modest circumstances." All three "were killed by the Nazis" (HABM).

48. The *Notgemeinschaft der Deutschen Wissenschaft* (Emergency Association of German Science) was founded in the fall of 1920 on the initiative of Fritz Haber, Max Planck, and Ernst von Harnack in order to raise funds to support the research of German scientists. It was successful in obtaining funds from the German federal gov-

ernment and from American foundations. Planck briefly headed the *Notgemeinschaft* until Schmidt-Ott was installed as president. See Heilbron 1986, 90–92. Regarding Hans's fellowship, one mark at that time was equivalent to approximately two dollars now.

49. Born 1926a,b,c.

50. It is interesting to note that Born in a letter to Bohr on December 18, 1926, told him of Wentzel's derivation in first Born approximation of Rutherford's formula for Coulomb scattering and added "I have myself succeeded in systematically working through the collision formulae for neutral hydrogen atoms. One obtains there a formula which embraces at the same time the collision of α rays and of electrons. In the same way one can calculate the excitation probability and the scattering of the excited electrons, on which I am presently engaged" (quoted in Bohr 1987, 208). Born therefore gave Elsasser a problem that he felt could be successfully resolved in a reasonable amount of time.

51. Elsasser 1978, 71–72. Elsasser concluded his paper with the following acknowledgment: "I cordially thank Prof. Born for suggesting this work and his constant encouraging sympathy during its course, furthermore I owe Mr. J. R. Oppenheimer the best thanks for many valuable suggestions."

52. See Thomson 1904.

53. See Rutherford, Chadwick, and Ellis 1930, chs. 3, 5, and 14, for a review of the work done on the subject before 1930.

54. See Niels Bohr's (1987) *Collected Works,* and Jens Thorsen's highly informative introduction. See also Williams 1945.

55. Bohr 1987, 3. The range of the fission fragments when uranium is disintegrated by neutron absorption played an important role in their identification, and much theoretical work was done on the subject during World War II.

56. Bethe et al. 1989, 4, 5.

57. Bethe referred to Born for his formulation of scattering theory. It is also clear that he had read Dirac's (1927b) *Zeitschrift für Physik* paper in which a general theory of collisions in nonrelativistic quantum mechanics is formulated in momentum space.

58. See Bethe and Jackiw 1968, ch. 11, where the details of the general approach are given.

59. In fact, even explicit knowledge of the potential function governing the interactions is not needed. See Bethe and Jackiw 1968, ch. 11, for an exposition of the Bethe method in atomic and molecular physics.

60. Bethe's evaluation of the logarithms in the expression for the energy loss in the passage of fast charged particles through atomic hydrogen proved to be very useful in connection with his famous Lamb shift calculation in 1947 (Schweber 2011).

61. Details of Bethe's calculation can be found in Bethe and Ashkin 1953, Bethe 1964, and more comprehensively in Bethe and Jackiw 1968, ch. 17.

62. Bohr in his 1913 paper had derived an expression for the energy loss per unit

length of a charged particle of mass M, charge Ze, velocity V, traversing matter whose density of electrons is n. The atomic electrons were assumed bound by a force whose distance dependence is assumed linear, hence each electron in the atom could be characterized by a certain frequency ν. Bohr had determined

$$b_{\max} = \frac{V}{<\nu>\sqrt{1-\frac{V^2}{c^2}}},$$

with $<\nu>$ an appropriate average frequency for electrons in the absorbing material, and $b_{\min} = \frac{Ze^2}{mV^2}$.

63. Bethe 1930b.

64. Bethe et al. 1989, 6.

65. Born 1975, 234.

66. Bethe 1933a.

67. Bethe-Sommerfeld correspondence, Sommerfeld *Nachlass,* Deutsches Museum, Munich, at http://deutsches-museum.de/en/archives/archive-online/. See Eckert and Märker 2003, Lee 2007, Brown and Lee 2009.

68. The literal translation is "the well-being of my soul."

69. That is, more than nuclear physics. The Kapitza Club was an informal gathering of physicists working in the Cavendish Laboratory in Cambridge University. It was originally organized in 1922 by Peter Kapitza, then a young Russian physicist who was a favorite of Rutherford. He invented new methods for producing extremely strong magnetic fields and investigated the properties of materials under those conditions. The Kapitza Club met once a week after dinner in its members' college rooms. Kapitza's intention in forming the club was to provide a forum for unconstrained discussion of developments in physics free of the reserve that characterized laboratory discussions. It also took on the characteristics of a theoretical physics colloquium. Membership of the club was by invitation, and in the early years members were almost all, apart from Kapitza, theorists. When Bethe visited Cambridge there were some twenty-five members.

70. Bethe-Sommerfeld correspondence, Bethe to Sommerfeld, November 2, 1930.

71. For biographical material on Blackett see Nye 2004 and Hore 2003.

72. Blackett 1933, 67.

73. In his lectures Dirac was following Dirac 1930a. The story goes that Dirac did so because he had carefully thought about how to say things in his book, and did not know how to improve its presentation. Bethe quotation is Bethe-Sommerfeld correspondence, Bethe to Sommerfeld, November 2, 1930.

74. Bethe-Sommerfeld correspondence, Bethe to Sommerfeld, November 2, 1930.

75. The following is an amusing story—though undoubtedly apocryphal—that

Bethe told his son Henry when he was very young, regarding his knowledge of English in 1930. To go to England Bethe took a ferry across the channel. On the ferry he picked up an English newspaper. One of the columns carried a headline "Manslaughter gets 7 years in jail" which he read as "Man's laughter gets 7 years in jail." When in Cambridge he therefore always had a very serious face and didn't laugh when he heard a joke. When asked why, he explained that according to the newspaper a man was given seven years in prison for laughing!

76. Bethe interview with Schweber, December 9, 1996.

77. See Hendry's highly informative introductions to the various sections of his *Cambridge Physics in the Thirties* (Hendry 1984).

78. HABCW I.

79. The fine structure constant, $2\pi e^2/\hbar c$, is the dimensionless constant that can be constructed out of *e*, the charge of the electron; $\hbar$, Planck's constant; and *c*, the velocity of light. Eddington had published a paper in 1928 in which he derived a relativistic equation for two electrons. According to his theory, based on Dirac's spin matrices, the fine structure constant had to have the value 1/136. Eddington presented this work at the Royal Society. Rutherford objected vehemently, indicating that the empirical value of the fine structure constant was nearer 1/137. Eddington replied saying, "Well, perhaps, it should be 136 from the matrices and one for the teapot." In a later paper Eddington did derive the value 1/137 for the fine structure constant (Mott 1984, 130).

80. The first meeting of the club was held on January 6, 1901, at which time it was agreed that the club be called the $\nabla^2 V$ Club and that the "object of this Club be the discussion of Questions in Mathematical Physics." It usually met in the rooms of the president of the club. During Bethe's stay in Cambridge Dirac was the president, and it met in his rooms in St. John's College. The records of the meetings are available in the Archives of the History of Quantum Physics. The record for the meeting of November 11, 1930, states Mssrs. Bethe, Cairns, Feather, and Riezler "were entertained as guests."

81. The London Club met monthly either in London or in Cambridge to discuss recent advances in physics. It had been organized by Blackett so that physicists from Bristol, London, and other southern England cities could attend. The brief entry for the meeting on April 21, 1931, records that Bethe talked on "'Passage of fast particles through matter.' Signed P M S Blackett." The minutes of the Kapitza Club are available in the Archives for the History of Quantum Physics (AHQP). See Kuhn 1967.

82. See the extensive letters on the subject from Bethe to Peierls in Lee 2007, 64–83.

83. See Lee 2007, 87.

84. Bethe 1932.

85. The fullest exposition of the theory is to be found in the lengthy article Bethe and Ashkin 1953.

86. See Mott 1987 and Bohr 1987, 204–216, where some of Mott's research is discussed.

87. Mott was born on September 30, 1905, and died August 8, 1996 (see Davis 1998).

88. Mott 1930a.

89. Mott 1928. Wentzel in November 1926 had succeeded in deriving Rutherford's cross-section for the scattering of a charged particle by a Coulomb field by applying Born's approximation method to lowest order. It is a specific property of the $1/r$ potential that the quantum mechanical and classical cross sections are the same (see Williams 1945).

90. Mott 1929b.

91. Mott 1929c.

92. See Mott 1987, 24–29.

93. Ibid., 25.

94. For biographical material on Guido Beck, see the transcript of the interview of John Heilbron with Guido Beck, April 22, 1967 (American Institute of Physics), Niels Bohr Library and Archives (http://www.aip.org/history/ohilist/4500.html), and Darrigol 1992.

95. Bethe, interview with Schweber, December 10, 1996.

96. Lee 2007, 86.

97. Ibid., 86.

98. HABCW I.

99. HABJG I.

100. HABM II.

101. Ibid.

102. Bethe in Segrè 1970, 59–60.

103. Ibid., 60.

104. Bethe and Bethe 2002, 27.

105. Now known as the Fermi-Thomas model of the atom.

106. Bethe and Bethe 2002, 28. He also called his *ψarum* his "Jewel Box."

107. In the next-to-last paragraph of the paper, Bethe indicated that in a future paper the method would be extended to two- and three-dimensional lattices "and its physical implications for cohesion, ferromagnetism, and electrical conductivity will be derived" (Bethe 1931b).

108. Ludloff 1931.

109. See Gaudin 1983, 1–15; Bethe 1931b. The English version of the paper is included in Bethe 1997, 155–184.

110. After reading Bethe's article, Pauli wrote Heisenberg: "As a pure mathematical *tour de force* (*Kraftleistung*) I much admired the work of Bethe, in that he did correctly what Ludloff had done wrong" (Pauli 1985, 94).

111. Giamarchi 2004, 37.

112. Batchelor 2007. See also Batchelor 2006, Sutherland 2004.

113. Brown and Lee 2006, 8.

114. Bethe's second stay in Rome was from March through May 1932.

115. In 1932, Placzek went to Copenhagen and worked with Bohr until 1938, with periods of fellowships or visiting professorships at the universities of Kharkov, Jerusalem, and Paris. In 1938 he rejoined Bethe in Cornell.

116. Placzek 1934.

117. HABCW II.

118. The *Festschrift* is Marshak 1966.

119. Bethe and Salpeter 1957.

120. Bacher and Weisskopf 1966, 2.

121. Bethe 1981, 73–74.

122. Bacher and Weisskopf 1966, 2.

123. Mott 1987, 48.

124. The interview continued, with Mermin noting that he thought that things have changed somewhat. "These days the student's name tends to come first, if at all possible, if it's a collaboration." Bethe then commented that this was to "advertise" the student. Excerpts from the transcript of a half-hour videotape recorded on February 25, 2003, in which David Mermin interviewed the ninety-seven-year-old Bethe about the early days of solid state physics and the role he played in its development (Bethe and Mermin 2004).

125. HABCW I.

126. See Gavroglu 1995.

127. Dirac 1927a.

128. Dirac 1927b.

129. In addition, between 1934 and 1935 Pauli proved that the requirements that a quantum field theory conform to the requirements of special relativity, that is, that it be Lorentz invariant, that the energy of the field system have a lower bound, and that field observables at two points that cannot be connected by light signals from one to the other commute, determined that integer spin particles obey Bose-Einstein statistics, and that particles with odd half-integer spin obey Fermi-Dirac statistics.

130. This was a consequence of the assumption that the interaction between fields are of the form $J(x)A(x)$ (where $J(x)$ is an observable referring to one of the field systems, e.g., the matter field's electromagnetic current, and $A(x)$ to an observable of a second field system). The term in the Hamiltonian describing the interaction of the electromagnetic field, $A_\mu(x)$—now conceived as a noncommuting operator—with the current operator of the matter field, $j^\mu(x)$—similarly conceived as an non-commuting operator—is (as in the classical description) of the form $j^\mu(x)A_\mu(x)$. When the (quantized) electromagnetic potential $A_\mu(x)$ is expanded in a Fourier series, the expansion coefficients are operators that can be interpreted as creation and annihilation operators for photons. The interaction is thus responsible for the creation and annihilation of photons.

131. For an informative, insightful, and accessible assessment of Dirac's contributions to the development of the quantum theory and quantum field theory, see Kragh 1990.

132. Dirac 1930a, Introduction. In 1977 Dirac recalled: "I was very well aware that there was an enormous mass difference between the proton and the electron but I thought that in some way the Coulomb force between the electrons in the sea might lead to the appearance of a different rest mass for the proton" (Dirac 1977, 20).

The idea of the hole theory was suggested to Dirac by the chemical theory of valency, according to which electrons in an atom form closed shells that do not contribute at all to the valency. An electron outside closed shells contributes to the atom's valence, as can electrons in an incomplete shell or through a hole in a closed shell. "One could apply the same idea to the negative energy states and assume that normally all the negative energy states are filled up with electrons, in the same way in which the closed shells in the chemical atom are filled up" (Dirac 1977, 50). In 1929–1930, while writing his book on quantum mechanics, Dirac was also working on the description of the "chemical atom."

133. Dirac 1977, 20.

134. In Bristol Dirac went on to say, "There being no holes we can call protons, we must assume that protons are independent particles. The proton will now also have negative energy states, which we must again assume to be all occupied. The independence of the electron and proton according to this view allow us to give them any masses we please, and further there will be no mutual annihilation of electron and protons." But the difficulties had been gotten over "only at the expense of the unitary theory of the nature of electrons and protons." However Dirac was not yet ready to give up his unitary view for he added: "At the present it seems too early to decide what the ultimate theory of the proton will be. One would like, if possible, to preserve the connection between the protons and electrons, in spite of the difficulties it leads to" (Dirac 1930a, 605–606).

135. Dirac 1930b.

136. Dirac 1984, 61.

137. Dirac 1977, 21.

138. See Wigner's comment on Pauli and the neutrino in Stuewer 1979. Dirac was of course pleased with the experimental verification of his prediction of antielectrons. But Bohr at first did not believe in Anderson's discovery, not even after Blackett found clear evidence for electron-positron pair production in cloud chamber photographs of cosmic ray showers. When the evidence became more and more convincing, one of Bohr's final remarks was: "even if all this turns out to be true, of one thing I am certain: that it has nothing to do with Dirac's theory of holes!" Pauli likewise objected to Dirac's new version of his hole theory. Pauli wrote Dirac in April 1933: "I do not believe in your perception of 'holes,' even if the existence of the 'antielectron' is proved."

139. Fermi 1926.

140. Thus the dynamics of an electron was described in hole theory by the Dirac equation with the negative energy states assumed all to be occupied. How to treat the negative energy states in the case of two interacting electrons was not answered until they got a field theoretic description.

141. See Enz 2002 for a presentation of Pauli and Heisenberg's formulation of QED.

142. Heitler's (1936) *The Quantum Theory of Radiation* was based on Fermi's formulation of QED in the *Reviews of Modern Physics* article. Heitler's book was the point of departure in 1939 for Feynman addressing quantum electrodynamics in his Lagrangian formulation: Fermi's article was the source of Weisskopf's insight that the Lamb shift could be interpreted as the effect of the zero-point energy vacuum fluctuations of the electromagnetic field on the motion of the electron in the hydrogen atom. In addition, Fermi also showed how to handle the gauge conditions in QED when the electromagnetic field is described by vector and scalar potentials.

143. Fermi 1927.

144. Amaldi and Fermi 1936, 305.

145. Fermi 1932; see also Schweber 2002.

146. Bethe 1930a. Wentzel had already indicated these properties in the scattering theory course that Bethe had taken with Wentzel in Munich in 1926; see the references to Wentzel's articles on the subject in Bethe 1930a. In 1930 Sommerfeld, with his student Schur, had been calculating detailed properties of the photoelectric effect. This was the stimulus for Bethe giving a time-dependent description of the process.

147. Bethe 1955.

148. The confusion which reigned before the discovery of the neutron is perhaps best captured in the remarks made by Bohr in his 1930 Faraday lecture:

> Still, just as the account of those aspects of atomic constitution essential for the explanation of the ordinary physical and chemical properties of matter implies a renunciation of the classical ideals of causality, the features of atomic stability, still deeper-lying, responsible for the existence and the properties of atomic nuclei, may force us to renounce the very idea of energy balance. I shall not enter further into such speculations and their possible bearing on the much debated question of the source of stellar energy. I have touched upon them here mainly to emphasize that in atomic theory, notwithstanding the recent progress, we must be still be prepared for new surprises.

See also Gamow 1931.

149. See Peierls 1986.

150. Pauli, letter to the "Dear radioactive ladies and gentlemen," December 4, 1930, in Pauli 1985. Pauli thought of his neutrinos as also explaining the "false" statistics of the N^{14} and Li^{6} nuclei and stated that they "differ[ed] from light quanta in

that they do not travel with the speed of light: The mass of the neutr[ino] must be of the same order of magnitude as the electron mass and, in any case, not larger than 0.01 proton mass." Indicative of the conservatism of the theorists is that Pauli labeled his suggestion of the existence of a new particle as "desperate" and did not publish his proposal. He did present it formally to the 1933 Solvay Conference, which is where Fermi first heard of Pauli's hypothesis. Mehra 1975.

151. Segrè 1979, 49–50.

152. Fermi 1934.

153. See Kragh 1992 and Roqué 1992. Møller made use of correspondence arguments—still in the vogue in Copenhagen at the time—to derive the scattering matrix element for the transition from the initial two-electron state to the final state after the scattering; Breit made heavy use of the Heisenberg-Pauli formalism, the consistency of which was questionable and the clarity of which was not always readily discernable. There was also the question of the dependence of the result on the particular gauge that is adopted. The aim of Bethe and Fermi was to indicate the relation between the two approaches, and more important, discuss how perturbation theory could be used to generate transparent results.

154. Bethe and Bethe 2002, 29.

155. "Statistics" refers to whether the particles are bosons, in which case there is no restriction as to the number of particles that can occupy a given quantum state, or fermions, in which case only one particle can occupy a given quantum state. A (one-particle) quantum state can be characterized either by the position and the spin of the particle or by its momentum and its spin. Thus no two identical fermions can be at the same position if they have the same spin.

156. Recall Laplace's demiurge "would embrace in the same formula the movements of the greatest bodies of the universe and those of the lightest atom" (Laplace 1951, 4).

157. It is constants of nature—Planck's constant, $= 6.626 \times 10^{-27}$ erg s, the velocity of light, $c = 3 \times 10^{10}$ cm/s, the masses of "elementary" entities, such as that of the electron $m_e = 9.1 \times 10^{-28}$ g, and that of the proton, $m_p = 1.7 \times 10^{-24}$ g—that demarcate domains. Other domains of nature are delineated by physical constants (usually with the dimension of a length), but in these cases the constants are usually characteristics of the *particular* experimental or physical situation. For example, in the case of fluid flow on a macroscopic scale, the density, viscosity, and temperature of the fluid and the size of the channel in which the flow takes place—all adjustable parameters—determine the domain and the valid description (i.e., whether by hydrodynamical equations or by Boltzmann-like equations). Every system is characterized by a characteristic mass, M, a characteristic length, L, and a characteristic time, T. A quantum mechanical description is necessary if ML^2/T is the order of h in magnitude.

158. In the acknowledgments of his paper, Smith thanked Bethe for "suggesting the problem and for helpful discussions in connection with it" (Smith 1932).

159. After having earned his doctorate in physics in 1906 in Erlangen, Geiger went to Manchester University and worked with Rutherford. In 1909, together with Ernest Marsden, he performed the famous Geiger-Marsden experiment, in which α-particles were scattered by the gold atoms in a thin gold foil. Their findings led Rutherford to create his nuclear model of the atom. With Marsden, Geiger developed the (charged particle) counter that now bears his name.

160. W. Henneberg used the Born approximation to calculate the cross section for ionization of electrons in the K shell of atoms by slow protons and α-particles and found good agreement with experiments. In the 1930s Bethe was not aware that Henneberg was "1/4 Jewish." Henneberg died in 1940 on the Russian front. See Bethe and Ashkin 1953, 170.

161. Tübingen is some fifty kilometers due south of Stuttgart.

162. For an overview of the political developments in Germany from 1848 to 1945 see Holborn 1969.

163. Gustav Stresemann had formed a coalition government in September 1923 in the wake of the riots and severe hyperinflation that had been triggered by the French and Belgian occupation of the Ruhr.

164. Indicative of what he knew of German newspapers in the mid-1990s, Bethe commented: "[As compared to the *Frankfurter Allgemeine Zeitung*] one of the Hamburg papers may be still better. One is called *Die Zeit* and the other *Die Welt*. *Die Zeit* is a daily and the other is a weekly. The daily is quite conservative. Now the Munich *Süddeutsche Zeitung* seems to be the most readable of the major papers."

165. In a letter to Peierls in June 1928, Bethe indicated that he "voted for the social democrats knowing that the democratic party wouldn't play an important role in the Bavarian parliament." The Social Democrats did make significant gains both in the national and in the local Bavarian elections but did not obtain a majority in either parliament (Lee 2007, 37).

166. In 1936 Geiger accepted the physics professorship at the Berlin Technische Hochschule and also became the editor of the *Zeitschrift für Physik*. Together with Heisenberg and Max Wien he took a stand against the Nazis' attempt to eliminate the teaching of the theories of relativity and of quantum mechanics, which Stark and others had labeled "Jewish Physics." See Beyerchen 1979 and the entry on Geiger in the *Dictionary of Scientific Biography*.

167. See Beyerchen (1983, 38–41) for possible reasons that the Nazi government encouraged dismissed personnel to leave Germany, rather than confining them.

168. A German mark in 1933 was equivalent to roughly two dollars in 2000.

169. In January 1919 Weber complemented that lecture with an equally famous one, "Politics as a Vocation."

170. Although Weber had in mind both natural and social scientists, I will limit the discussion to the natural scientists.

171. Lassman and Velody 1989, 167–168.

172. Ibid., 134.

173. "Nothing is worthy of man as man unless he can pursue it with passionate devotion" (ibid., 135).

174. "We know of no artist who has ever done anything but serve his work and only his work" (ibid., 137).

175. Kahler 1989, 38.

176. Lassman and Velody 1989, 137.

177. Helmholtz surely looms large in Weber's essay. Helmholtz 1995, 82.

178. At the celebration of Victor Weisskopf's eightieth birthday at the American Academy of Arts and Sciences in Cambridge, Massachusetts.

5. England, 1933–1935

1. See http://www.lrz.de/~Sommerfeld/PersDat/00384.html for Sommerfeld's correspondence with William Henry Bragg and http://www.lrz.de/~Sommerfeld/PersDat/00820.html for Sommerfeld's correspondence with William Lawrence Bragg.

2. See Ewald's account of Bragg's work in Ewald 1962, 57–73, 82–83. Also Bernal 1962 and Andrade 1962.

3. Phillips 1979, 98.

4. Ibid., 105.

5. Ibid., 105–106.

6. See Bethe and Hildebrandt 1988, 167–168, for an account of the conference.

7. HABCW.

8. The Council had been founded by William Beveridge, Rutherford, and A. V. Hill a few weeks after the racial laws had been enacted in Germany. The Council raised funds to assist scholars who had been forced to flee the Nazi regime, and also acted as an information center, putting affected academics in touch with institutions that could best help them.

9. Lee 2007, 108.

10. Eckert and Märker, 2003, 2:196. Rudi and Genia spoke Russian with one another.

11. The Bethe-Peierls correspondence has been beautifully edited by Sabine Lee (2007). Both the original German and an English translation are included in the book. Brief biographical information about persons mentioned and references to articles mentioned in the letters are included in footnotes.

12. Lee 2007, 34.

13. Ibid., 477.

14. Letter dated October 27, 1986, in Lee 2007, 474–475.

15. Letter dated September 9, 1995, in Lee 2007, 492–493.

16. In 1930 James had written a book on X-ray crystallography that went through five editions, the latest appearing in 1961.

17. Letter in Lee 2007, 490.

18. Bethe established that in general the order of the arrangements of two different kinds of atoms A and B on a face-centered cubic lattice is a function of

$$x = \exp(-V_B/kT)$$

where k is the Boltzmann constant, T the absolute temperature, and

$$V_B = 1/2(V_{AA} + V_{BB} - 2V_{AB}),$$

with V_{AA} the interaction potential between two nearest A atoms, V_{BB} that between two B atoms and V_{AB} that between an A and a B atom. To arrive at his conclusion required Bethe to analyze the energies of the many possible configurations. The analogy would be to prove that

$$\frac{n!}{r!(n-r)!}$$

is the total number of arrangements of n coins into r heads, and $n - r$ tails irrespective of order, by exhaustively counting the possibilities for a number of small cases.

19. See Mott and Jones (1936) and Nix and Shockley (1938) for early presentations of Bethe's theory of order-disorder transitions in alloys; Muto and Takagi (1955) for a later one. They include Kirkwood's extension of Bethe's analysis and present a detailed exposition of Bethe's generative contribution and how subsequent work derived from it.

20. The Ising model is a mathematical model of ferromagnetism that assumes the elementary entities to be spins located in a lattice (of 1, 2, 3, or d dimensions) at positions labeled by i and described by variables s_i that can be in one of two states, usually taken to be $+1$ and -1. Each spin interacts only with its nearest neighbors. The energy of the Ising model is defined to be

$$E = -\sum_{i=j} J_{ij} s_i s_j.$$

For each pair, if $J_{ij} > 0$, the interaction is called ferromagnetic, and if $J_{ij} < 0$, the interaction is called antiferromagnetic. See Brush 1967.

21. If in the Ising model the J_{ij} are the same for all pairs equal to J, the ground state is the state in which all the spins are up—in other words, all the $s_i = +1$ for all i. Bethe had noted that the order is destroyed by a single mismatch in which two neighboring sites have opposite spins. This costs an energy $\Delta E = 2J$, and because the mismatch can occur any of N places it contributes an entropy $\Delta S = \log N$. The change in free energy in the creation of a mismatch is thus $\Delta E = 2J - T \log N$,

which is less than zero for N large enough. Therefore, there cannot be an ordered state for $T \neq 0$ for a large system. Peierls applied that reasoning to the two-dimensional Ising model and, by calculating energy changes and entropy contribution for different configurations, proved that an ordered state could exist in two and three dimensions for $T \neq 0$ and suitable J.

22. Naturally occurring samples of a chemical element consist of several isotopes. The nuclei of two isotopic atoms have the same charge but different masses. Since the nuclear electric charge alone determines the chemical properties of the atom, two isotopes will have the same chemical behavior. The atomic mass of every isotope is nearly an integer.

23. Gordin 2004.

24. In the early 1960s a similar situation existed for the then known subnuclear "elementary particles." A pattern, called the eightfold way, was discerned by Murray Gell-Mann and by Yuval Ne'eman and, as had been the case with Mendeleev's table, gaps were noted. New particles were predicted and found. The further elucidation by Gell-Mann and by George Zweig of the eightfold-way patterning for the baryons and the mesons was based on the properties of the (assumed) constituents of the objects being classified. These constituents were called quarks by Gell-Mann and the appellation has stuck. This was the analogue of Bohr's theory of the chemical elements of the early 1920s. And as in the case of the chemical elements, the more "foundational" explanation for the regularities had to await the formulation of quantum chromodynamics.

25. Of all the electromagnetic properties of the nucleus, its electric charge, Ze, is by far the most important for atomic physics. The finite size of the nucleus, its magnetic moment, and so on all give rise to small corrections.

26. Bethe and Bacher 1936, 83.

27. There is one important and crucial exception before 1932: in 1928, on the basis of the new quantum mechanics, Gamow and, independently, Condon and Gurney, formulated a model to explain the emission of α-particles by heavy nuclei. Gamow's explanation was important because it implied that just as α-particles could "leak" through the potential barrier from the inside of the nucleus to the outside, similarly, charged particles could be transmitted through the barrier from outside to inside the nucleus. This insight led Cockcroft and Walton to design the first accelerator.

28. Hiebert 1988, 76. See also Hendry 1984; Hughes 1998, 2002b.

29. Stuewer 1979; see also Hendry 1984; Hughes 1998, 2002b; Mladjenović 1998.

30. Chadwick 1932. See also the recollections of Norman Feather, Chadwick, and Philip Dee of the discovery of the neutron in Hendry 1984, 31–48.

31. Heisenberg 1932.

32. See Wigner 1979.

33. See Wigner 1933. The analogy with the forces of homopolar binding proved useful. The strong homopolar bond between two hydrogen atoms is the result of the confinement of their two electrons in the space between the two nuclei, this when the electrons are in a singlet spin state. Under those circumstances, a third hydrogen atom cannot be readily attached to the H_2 molecule because the Pauli principle makes it impossible to pack a third electron between the nuclear charges. A drop of liquid hydrogen thus has an energy approximately proportional to the number of hydrogen molecules, and therefore proportional to the number of atoms present. Heisenberg in his papers on nuclear forces and nuclei conceived of the neutron and the proton as "molecules" in order to explain saturation. See Carson 1996.

34. See McMillan 1979 and Livingston 1969 for a brief history of particle accelerators. See also Cockcroft's recollections in Hendry 1984, 74–81.

35. At the time all the Cambridge physicists were calling the deuteron *diplon*; people elsewhere were calling it *deutron.* Rutherford or Chadwick commented that "'deutron' is an impossible name, because deutron and neutron when you have a cold sound just the same," and so it became the deuteron.

36. Wigner 1933.

37. See David Brink's article, "The story of Majorana's exchange force," based on the lecture he delivered at the International Conference on Ettore Majorana's Legacy and the Physics of the 21st Century in Commemoration of the Centennial of Majorana's Birth, Catania, Italy, 5–6 Oct 2006, p. 2. Available at http://cdsweb.cern.ch/record/1115598.

38. The list is taken from the paper on the structure of the deuteron that Bethe and Peierls (1934) wrote after their visit to the Cavendish and their discussion with Chadwick and Goldhaber. The separation of the center-of-mass motion yields the following relative motion equations

$$
\begin{aligned}
\left(\frac{\hbar^2}{M}\nabla^2 + E\right)\psi(r,s,\sigma) &= V(r)\psi(r,s,\sigma) \\
&= (-1)^l V(r)\psi(r,s,\sigma) \\
&= (-1)^{l+S+1} V(r)\psi(r,s,\sigma).
\end{aligned}
$$

The Majorana interaction replaces r by $-r$, and is equivalent to an ordinary potential according to whether the angular momentum of the relative motion is even or odd. For a system of two particles, the spin function is symmetric for the triplet state, that is, if $S = 1$, or antisymmetric for $S = 0$, hence the $(-1)^{S+1}$ factor. The potential in Eq.(3) was proposed by Heisenberg. It exchanges both position and spin, hence the factor $(-1)^{S+1+l}$ above. Because one cannot have a wave function that is symmetrical in all four particles without violating the Pauli principle, Heisenberg forces cannot account for the binding energy of ^{4}He. (They lead to saturation already for the deuteron.) An exchange potential that is a combination of Heisenberg and Majorana forces leads to saturation because of the alteration in attraction and repulsion according to the *l* value and yields ^{4}He as a saturated configuration.

39. See Goldhaber 1979 for a detailed exposition of the experiment.

40. The results depended on what was taken to be the mass of the neutron.

41. Bethe and Peierls 1934.

42. The Columbia experiments had yielded a value close to 30×10^{-24} cm^2, whereas Fermi and his group in Rome had obtained a value of around 80×10^{-24} cm^2. Fermi explained soon thereafter that the value depended on a chemical binding effect of protons in molecules.

43. Bethe 1979, 14.

44. On February 25, 1934, Bethe wrote Sommerfeld, "Peierls and I have also thought about some things concerning the neutrino, the neutral particle with spin and approximately the mass of an electron, that Pauli invented about 3 years ago so as to put into order the statistics of nuclei and the continuous spectrum" (Bethe-Sommerfeld correspondence).

45. Bethe and Peierls 1934.

46. Cowan and Reines and colleagues, using antineutrinos produced in the Savannah River reactor, did observe such a reaction and found for the cross section the value $(11 \pm 2.6)\ 10^{-44}$ cm^2.

47. I have not given references to these works. The calculations can be found in chapters 3 and 4 of Heitler 1936, together with the references to the literature. For a succinct and insightful history of cosmic rays until the mid-1930s, see Rasetti 1936, ch. 7.

48. Heitler 1936, 281.

49. Heitler 1936, 160–161.

50. No electromagnetic radiation will be emitted if the impact parameter, *b,* is greater that the atomic radius, r_a; when $b > r_a$ the atomic electrons shield the charged projectile particle from the Coulomb field of the nucleus.

51. See Cassidy 1992 and Galison 1979 for a detailed account of these events. See also Heitler 1936 for the contemporaneous report of the issues and calculations.

52. Bethe's letter had begun with, "I don't know—but in every letter I need to apologize for its lateness. In this case I have an especially bad conscience: first of all you are awaiting an answer, and secondly I was made especially glad by your letter after Christmas. Most of all, that you are again quite well, and have completely overcome the depression of last summer!"

53. Published as Bethe and Heitler 1934; Bethe 1934a.

54. "Progress in Physics" 1935, 49–50.

55. Ibid., 2.

56. Ibid., 16.

57. Evidently Bethe and Peierls had investigated what later became known as the

Gamow-Teller version of the Fermi β-decay theory, which involves derivatives of the operators in the interaction terms of the theory. See Bethe's comments to Gamow's lecture. Ibid., 66. See also his comments on page 166 regarding the stability of ^{3}He.

It is interesting to note that once again Bethe's and Elsasser's paths crossed. Elsasser, like Bethe, had to leave Germany in 1933. He obtained a temporary job in Paris with Joliot and began working in nuclear physics. In 1934 he published several important papers in the *Journal de Physique* that established—on the basis of special nuclear stability—what were later called *magic nuclei* with 2, 8, 20, 50, and 82 neutrons or protons. He commented on these at the conference. See Elsasser 1935.

58. The following presentations were made at the session on cosmic radiation: C. D. Anderson and S. H. Neddermeyer, "Fundamental Processes in the Absorption of Cosmic-Ray Electrons and Photons"; P. Auger and L. Leprince-Ringuet, "Some Measurements of Cosmic Radiation at High Altitudes"; P. M. S. Blackett, "The Absorption of Cosmic Rays"; I. S. Bowen, R. A. Millikan, and H. V. Neher, "A Very High Altitude Survey of the Effects of Latitude upon Cosmic-Ray Intensities and an Attempt at a General Interpretation of Cosmic-Ray Phenomena"; A. H. Compton and R. D. Bennett, "A Study of Cosmic-Ray Bursts at Different Altitudes"; G. Hoffman, "The Connection between Cosmic Radiation and Atomic Disintegration"; B. Rossi, "Some Results Arising from the Study of Cosmic Rays"; G. Bernadini, H. A. Bethe, C. W. Gilbert, V. Posejpal, and D. Skobeltzyn, "Discussion on Cosmic Radiation."

59. HABCW.

60. While in Munich, Smith did some research on Hartree-Fock methods for excited states of atoms and published a paper on the subject in which he thanked Bethe in the acknowledgments (Smith 1932).

61. Bethe in a letter of June 9, 1934, to Sommerfeld commented, "Bragg has returned from America, where for 1/2 year he was a visiting professor (at Cornell). He found that the American mentality has become much more European under the influence of the crisis—skeptical, and the dollar no longer the only measure of value. After America, he (Bragg) is physically and otherwise very venturesome."

62. Farrand papers, Cornell University Archives. The letter continues:

> A motion favoring his appointment for the coming year with the distinct understanding that such an appointment was to be looked upon as a trial period, was passed with all voting favorably except two. Prof. Richtmyer who in the discussion had expressed disapproval of such a move had asked to be excused to meet another engagement shortly before this motion was made and put, and Prof. Howe did not vote after having explained that he would like to refrain from voting if in so doing it would not be construed that he was opposed to such an appointment of Dr. Bethe. In the discussion, arguments strongly fa-

voring Dr. Bethe were presented by many of those present including both younger and older men in the department.

63. The position had been occupied by Walter Heitler.

64. Margaret Morrison has stressed this point about models. See Morrison 1999, Morgan and Morrison 1999.

6. Hilde Levi

1. The information in this chapter regarding Hilde Levi is based on my interviews with her in Copenhagen on December 14 and 15, 1991. The quoted material is from that interview. She also taped an autobiography, which was transcribed and deposited in the Bohr Archive (hereafter *Autobiography*). I thank Finn Aaserud, the director of the Niels Bohr Archive, for permission to consult Hilde Levi's *Autobiography.*

2. See Noren (1976) for a photographic history of the Wallachs.

3. Albert Hesse was the editor of the *Chemisches Zentralblatt* from 1902 until 1923. He devoted most of his life to the systematic coordination and abstracting of pure and applied chemical literature. He was not Jewish.

4. When he was in Munich, Bethe visited the Wallachs at their summer house on several occasions.

5. Elisabeth Schumann was an outstanding soprano. When Hilde Levi knew her, Schumann was singing in Mozart operas with both the Vienna and the London (Covent Garden) opera companies, as well as in Wagner operas at the Salzburg Festival. In 1937 Schumann emigrated to the United States and thereafter taught at the Curtis Institute of Music in Philadelphia.

6. Hilde Levi to S. S. Schweber, January 9, 1992.

7. Ibid.

8. Information regarding the Fischers and insights on Frankfurt Jewry were given to me in a memorable interview with Anne Fischer in Richmond, VA, on October 25, 1991.

9. Brandt and Hight 1987, 1.

10. The collection included many works by Ernst Ludwig Kirchner, Emil Nolde, Wassily Kandinsky, and August Macke.

11. Brandt and Hight 1987, 7.

12. Nelson produced a great quantity of work (collected in nine volumes of *Gesammelte Schriften*) in a tragically short life. He was an insomniac who worked day and night, and succumbed to pneumonia at the age of forty-five.

13. See Reid (1970) for Nelson's interaction with Hilbert and with Courant. For Nelson's influence on Karl Popper see Hacohen 2000.

14. Nelson was a neo-Marxist and an active member of the Internationalische Kampfbund.

15. Interview with Anne Fischer, Richmond, VA, October 25, 1991. Anne Fischer was a remarkable woman. For a sketch of her life see her interview with Marilyn Scott in the *Alumnus Journal of the Virginia Commonwealth University* for Reunion 1992.

16. The Fischers emigrated to the United States, settling in Richmond, Virginia. Ernst became the chairman of the physiology department at the Medical College of Virginia, and Anne became a social worker. When Anne died in 2008 at the age of 105, the collection was sold to the Virginia Museum of Fine Arts.

17. Unless otherwise specified, all the quotes are from notes taken during an interview with Hilde Levi on December 14, 1991.

18. Letter 81, November 9, 1926, Hans Bethe to his parents.

19. Levi, *Autobiography,* 1–2.

20. Hans met Margie after Richard Willstätter had resigned from the University of Munich. Richard Willstätter was born in Karlsruhe in 1872. His family was Jewish. At age eighteen he entered the University of Munich to study chemistry and became a student of Adolf von Baeyer. He received his doctorate in 1894 with a thesis on the structure of cocaine. He continued research into the structure of other plant alkaloids and synthesized several of them, including chlorophyll. In 1902 he accepted the position of *Professor extraordinarius* in Munich but left for Zürich in 1905 to become professor at the ETH. In 1912 he became professor of chemistry at the University of Berlin and the director of the Kaiser Wilhelm Institute for Chemistry. In 1915 he won the Nobel Prize for Chemistry "for his researches on plant pigments, especially chlorophyll." Willstätter returned to Munich in 1916 as the successor of von Baeyer. During the 1920s he investigated the mechanisms of enzyme reactions and helped establish the fact that enzymes were chemical substances.

His Nobel biography indicates that in 1924 Willstätter's career came to "a tragic end when, as a gesture against increasing antisemitism, he announced his retirement. Expressions of confidence by the Faculty, by his students and by the Minister failed to shake the fifty-three-year-old scientist in his decision to resign. He lived on in retirement in Munich . . . Dazzling offers both at home and abroad were alike rejected by him." In 1938 with the help of one of his former students he escaped to Switzerland. He spent the last three years of his life in Muralto, a small village near Locarno, writing his autobiography, *Aus meinem Leben.* It was translated into English as *From My Life* in 1965. Willstätter died of a heart attack in 1942. See Willstätter 1965; Robinson 1953.

21. Levi, *Autobiography,* 2. In Berlin she also attended advanced lectures in physics given by Max Bodenstein and Peter Pringsheim at the Friedrich-Wilhelm Universität, which after 1949 became known as Humboldt University of Berlin.

22. Levi arrived in Copenhagen on April 24, 1934, sixteen days after Franck. See Aaserud 1990, 122.

23. Hilde Levi to S. S. Schweber, January 9, 1992.

24. Loschka J. Michel to S. S. Schweber, January 14, 1992.

25. Hilde Levi to S. S. Schweber, January 9, 1992.

26. Ibid.

27. Ibid.

28. For an insightful, informative, detailed description of the Bohr Institute, the experimental and theoretical activities carried out in it from its inception until World War II, as well as sensitive portraits of Bohr, Hevesy, Franck, and an account of their mutual friendship, see Aaserud 1990. Aaserud also describes Hilde Levi's association with Franck (Aaserud 1990, 132–137) and with Hevesy (Aaserud 1990, 137–146).

29. Hilde Levi to S. S. Schweber, January 9, 1992.

30. Aaserud 1990, 221. In her interview with Charles Weiner on October 28, 1971, Levi stated that she accepted Bohr's offer because of "the way he [Bohr] made it so interesting, it was not sort of a matter of obedience, it was just being very quickly convinced that it was frightfully interesting and you should do that." Quoted in Aaserud 1990, 221. Her scientific and technical work as Franck's and Hevesy's assistant, as well as her later scientific investigations, are described in her interview with Charles Weiner on October 28, 1971, which is on deposit at the Niels Bohr Library, American Institute of Physics, College Park, MD.

31. Levi, *Autobiography.*

32. Quoted in Reid 1986, 366. Born, Franck, and Courant, the directors of the theoretical physics, experimental physics, and mathematics institutes, respectively, were all affected by the dismissal orders. The three were close friends and discussed their possible courses of action with one another. See Reid 1976, 355–376, for their reactions. Courant's reaction to Einstein's resignation of his Prussian citizenship is worth noting. In a letter to Franck written the day after Einstein's move had been made public in the newspapers Courant wrote: "Even though Einstein does not consider himself a German he has received so many benefits from Germany that it is no more than his duty to help dispel the disturbance he has caused . . . What hurts me particularly is that the renewed wave of anti-Semitism is . . . directed indiscriminately against every person of Jewish ancestry, no matter how truly German he may feel within himself, no matter how he and his family have bled during the war and how much he has contributed to the general community. I cannot believe that such injustice can prevail much longer—in particular, since it depends so much on the leaders, especially Hitler, whose last speech made a quite positive impression on me" (Reid 1976, 364–365). Courant's impression of Hitler changed very rapidly thereafter.

33. Levi, *Autobiography,* 6.

34. Ibid.

35. Ibid., 9.

36. Copies of the letters from James Franck to Hilde Levi can be found in the James Franck *Nachlass* in the University of Chicago Archives. The entire correspondence has been deposited in the Niels Bohr Archive. In one of the letters Franck identified with Heinrich Heine and his exile, using Heine's words, "Ich hatte einst ein schönes Vaterland" (I once had a beautiful homeland), to describe his own feelings.

37. Hertha Sponer received her doctorate in physics in 1920 in Göttingen with Peter Debye. She then went for a year to the University of Tübingen as an assistant to James Franck. She returned to Göttingen in 1921, completed her *Habilitation* with Debye, and became the first woman to become a *Privatdozent* in physics there. Sponer published extensively on her experimental work in atomic and molecular physics and on the interpretation of her findings by quantum mechanics. By 1932 she had become an associate professor of physics in Göttingen. In 1933, after being dismissed, she went to teach at the University of Oslo as a visiting professor, and in 1936 accepted an appointment at Duke University as a professor of physics. She remained there until her death in 1968. In 1946 she married James Franck. See http://www.phy.duke.edu/history/DistinguishedFaculty/HerthaSponer/.

38. See Bethe's interview with Judith Goodstein in 1982 (HABJG I, 11) regarding his views concerning theory and experiments in nuclear physics before the discovery of the neutron by Chadwick in 1932.

39. Bohr to Bethe, June 10, 1932. Bohr Correspondence, Niels Bohr Archives.

40. Bethe 1979, 20.

41. Bethe to Bohr, September 28, 1934.

42. Bethe to Bohr, November 5, 1936 (in German).

43. Bohr to Bethe, November 23, 1934.

44. George Washington University and the Carnegie Institution sponsored these Conferences on Theoretical Physics annually, and Gamow, Teller, and Tuve were their organizers. The theme of the 1937 conference was "Elementary Particles." In attendance were J. Bartlett, H. Bethe, F. Bloch, N. Bohr, G. Breit, A. Crane, C. Critchfield, J. Franck, W. Furry, G. Gamow, L. Hafstad, K. Herzfeld, F. Kalckar, M. Plesset, I. Rabi, L. H. Thomas, E. Teller, M. Tuve, J. A. Wheeler, and E. P. Wigner (Harper et al. 1997, 17).

45. Aage Bohr to S. S. Schweber, February 14, 1992.

46. Finn Aaserud both in his book on the Bohr Institute (Aaserud 1990) and in his brief biographical memoir of Hilde Levi told of her career. In the acknowledgments of his book he captured Hilde's feisty character: "When I moved to Copenhagen in 1980, I soon came to know Hilde Levi . . . She was now working on a biography of George Hevesy; with whom she had worked as an assistant from the mid 1930s until the Second World War. Never hesitating to provide unrestrained criti-

cism of my work, she soon became, in effect, my on-site adviser and close friend" (Aaserud 1990, ix).

7. Cornell University

1. Cornell had been a partner of Samuel Morse in laying the first lines between Baltimore and Washington. He subsequently established a vast network that was bought by Western Union.

2. Becker 1943; on Cornell, see p. 67, on White, p. 68.

3. Ibid., 123.

4. Note that no mention is made of religion.

5. Bishop 1962, 41–42.

6. Becker 1943, 66. White was the youngest senator, and chaired the committee on education; Cornell was chairman of the committee on agriculture.

7. Rogers 1942, 46–51.

8. For White's career after his presidency and his writing his famous and influential *A History of the Warfare of Science with Theology in Christendom* (1896) see White's *Autobiography* (1905) and Moore 1979.

9. Johns Hopkins University, chartered in 1874, was the second explicitly nondenominational institution of higher learning in the United States. John Coit Gilman, who became its first president in 1875, noted when offered the presidency that "The trustees whom he [Johns Hopkins] selected are responsible neither to ecclesiastical nor legislative supervision; but simply to their own conviction of duty and the enlightened judgment of their fellow men."

10. Nor did White's forthright advocacy of reason in *Warfare of Science* (White 1896) help matters.

11. Felix Adler was the founder of the Ethical Culture movement. For details of his life see Kraut 1979, Friess 1981, and Schweber 2000.

12. Kraut 1979, 96.

13. Adler's university lectures, according to an editorial in the Cornell *Review,* were "calculated to develop in young minds, at least, strongly rationalistic views" (Hofstadter and Metzger 1955, 355). Influenced by the writings of the evolutionary anthropologist E. B. Tylor, Adler in one of his lectures indicated that some of the central doctrines of Christianity—such as the idea of the immaculate conception—could be found in other religions as well. He also pointed to the similarities between some of the statements of Buddha and those of Jesus. As a result he was bitterly attacked in the local press. See Rogers 1942, 70–71.

14. In an open letter sent to Cornell's undergraduates and alumni after Adler's three-year appointment had expired, Andrew White noted that the lectures, though they "revealed much rare knowledge and great ability in . . . presentation," had drawn

criticism from "sundry denominational newspapers—the organ of various sectarian colleges." He strongly defended the appointment but implied that Adler had voluntarily withdrawn (Open letter of May 4, 1877, Andrew White Papers, Cornell Archives). The real reason for not renewing his contract—the faintheartedness of the administration—was revealed in a letter the vice-president of Cornell wrote Adler toward the end of his appointment:

> You are surprised at my silence about your chances of returning here. . . My charge is the University . . . Truth and Liberty will take care of themselves, their triumph is secure, but an institution may be injured irreparably, and no such great injury can befall it as a collision with either Truth or Liberty. Should one who guides it expose it to such a chance? . . . True wisdom, it seems to me, forbids my bringing on a contest where victory would not be of the greatest importance, but where defeat would be lasting injury. This is the reason why I said nothing to you about returning on a new engagement. Had I the power I would make such an engagement, for I believe that your lectures here did nothing but good. (Hofstadter and Metzger, 1955, p. 340)

15. Kraut 1979, 104.

16. The Morrill Land Grant College Act of 1862 specified that "the leading object" of new land grant colleges "shall be, without excluding other scientific and classical studies . . . to teach such branches of learning as are related to agriculture and the mechanics arts . . . in order to promote the liberal and practical education of the industrial classes in the several pursuits and professions of life" (Becker 1943, 34 and ch. 1).

17. For the case of Harvard University during the first third of the present century see Conant 1970, 76 ff.; see also DeVane 1965.

18. Paul Hartman has written an informative account of physics at Cornell (Hartman 1993). See also "Seventy Years of Physics at Cornell" written in 1958 by "Retiring Professors Howe and Grantham" (27 pp. Mimeographed notes). Both of these publications are available from the physics department of Cornell University.

19. Shorter-term visitors included Aston, Sommerfeld, Ehrenfest, Kramers, and G. P. Thomson.

20. The course culminated with a discussion of the Bohr atom.

21. Kennard was born in 1885 in Columbus, Ohio, and received a bachelor of arts degree from Pomona College in 1907. He was awarded a Rhodes Scholarship in 1908 with which he went to Oxford and earned a bachelor of science degree in 1911. Thereafter he enrolled in Cornell's graduate school and obtained a doctorate in physics in 1913. Except for a brief period from 1914 to 1916 as an instructor at the University of Minnesota, he spent most of his academic career (1912 to 1946) at Cornell. He was appointed professor of physics at Cornell in 1926. Kennard published his *Kinetic Theory of Gases* in 1938. After Richtmyer's death, Kennard edited the third

(1942), fourth (1947), and fifth (1955) editions of Richtmyer's *Introduction to Modern Physics.* During World War II he served as a scientist at the David W. Taylor Model Basin working on problems in fluids and hydrodynamics, and stayed there thereafter.

22. K. T. Compton came in 1927, W. F. G. Swann in 1928, A. H. Compton in 1929; E. C. Kemble visited Cornell during the summer 1930 and W. V. Houston during the summer of 1932.

23. Lloyd P. Smith came to Cornell in 1926 as a Coffin Fellow and stayed on as an instructor while earning his doctorate.

24. While in Munich, Smith did some research on Hartree-Fock methods and published a paper on the subject (Smith 1932).

25. Hartman 1993, 182.

26. Bethe, in Bernstein 1980, 44.

27. Ibid., 468–479.

28. But by the mid-1930s vigorous theoretical centers existed at MIT, Harvard, Princeton, Michigan, Wisconsin, Caltech, and Berkeley, staffed principally by American-born theorists.

29. Stuewer 1984.

30. Konopinski to Samuel Goudsmit, February 20, 1937, SG papers.

31. The Washington Conferences 1935–1942:

Date	Topic	Attendance
19–21 Apr 1935	Nuclear Physics	35
27–29 Apr 1936	Molecular Physics	60
15–20 Feb 1937	Elementary Particles	26
21–23 Mar 1938	Stellar Energy	34
26–28 Jan 1939	Low Temperature	53
21–23 Mar 1940	Interior of the Earth	56
22–24 May 1941	Elementary Particles	33
23–25 Apr 1942	Stellar Evolution & Cosmology	25

The data come from material presented by Karl Hufbauer at the meeting of the History of Science Society in 1980. The attendance figures refer to those invited members and "those present informally."

32. Bethe to John A. Fleming, February 10, 1937.

33. Livingston Farrand graduated from Princeton in 1888 and went on to obtain a medical degree from Columbia University in 1891. He became interested in psychology and anthropology and, after joining Franz Boas on expeditions to the Pacific Northwest, he became a professor of anthropology at Columbia in 1903, eventually assuming the chair of the department. He subsequently became president of the University of Colorado and accepted the presidency of Cornell in 1921. One of the major developments of Cornell during his presidency was the establishment of Cornell Medical School in New York City, which he organized.

34. Farrand, address on May 27, 1923, on the occasion of the celebration of the 100th anniversary of the founding of Trinity College. Farrand Papers, Olin Library, Cornell University.

35. Carl Becker to Farrand, January 25, 1929. Farrand Papers, Box 2, Cornell University Archives.

36. The committee included Robert Hutchins, Robert Millikan, Harlow Shapley, Robert Sproul, and Oswald Veblen. As of March 1937 grants had been made to seventy-five scholars. Each grant was of the order of $2,000 and allowed the scholar to stay for one year at a specified university (Farrand Papers, Box 14, Folder 11).

37. Speech at the 150th anniversary of the American Philosophical Society, April 2, 1930. Farrand Papers.

38. Ibid., 12–13.

39. Ibid., 24.

40. Ibid., 34.

41. Day gave some other quotations of Mussolini: "War alone brings to its highest tension all human energy and puts the stamp of nobility on the people who have the courage to lead it," and "Believe, obey, fight."

42. Day 1941, 34–37.

43. Day 1937, 29.

44. At Day's inauguration Conant also emphasized that "Liberty and the life of the true university are closely linked." Democracy and the independence of universities "have marched forward and retreated side by side" (Conant 1937, 14).

45. After World War II Cornell established a comprehensive course of study of contemporary Russian civilization, at the time a pioneering enterprise that aroused criticism in conservative circles. Day responded: "It is part of the respect we owe to our youth to deny it no knowledge that will enable it to bear, as it will bear resolutely and willingly, and in the enduring tradition of freedom, the weight of the world that is descending on its shoulders" (*New York Times,* March 24, 1951, 13).

46. Day 1937, 32–34.

47. Hartman 1993, pp. 169–213.

48. Richtmyer "had 94 papers and 60 meetings abstracts to his credit, mostly in X-rays" (Hartman 1993, Schrödinger 1982, 168).

49. In the early 1930s Gibbs had established that there was a fine structure in the H_α line of hydrogen, the transition from the $n = 3$ to the $n = 2$ levels. His findings were in rough agreement with the predictions of the Dirac equation for an electron in a Coulomb field, but seemed to indicate that the $2p_{1/2}$ and $2s_{1/2}$ levels were not degenerate—what later became known as the Lamb shift. However, he could not resolve the splitting and give a dependable value for it.

50. During the spring semester of 1935 Bethe taught two graduate courses: Physics 475, Quantum Theory of Spectra and Radiation; Physics 476, Quantum Mechanics of Solids. The following year he taught a yearlong course for seniors, Physics

200, Introduction to Theoretical Physics, and a graduate yearlong lecture course, Physics 415, Special Topics in Physics. Interestingly, during the academic year 1966–67, he taught two hours of "informal study" in the Physics 200 course that Kennard was teaching that year.

51. At the time there were only three ranks in the department: instructor, assistant professor, and full professor. Promotion to professorship meant tenure.

52. Robert Bacher, after-dinner speech, Los Alamos, November 19, 1986.

53. In the paper Schwinger had generalized the Dirac-Fock-Podolsky (Dirac, Fock, and Podolsky 1932) many-time formulation of quantum electrodynamics by describing the electrons by second-quantized field operators. What Schwinger would later call the interaction representation (Schwinger 1948) was introduced for all the field operators (electromagnetic as well as the matter field operators) and the interaction between charged particles calculated to order e^2. Although Schwinger never published the paper, the manuscript was typed, and it is a testament to his impressive maturity in physics at age sixteen.

54. Bethe to Rabi, July 10, 1935. Rabi Papers, Box 1, Folder 8: Bethe. Collections of the Manuscript Division, Library of Congress.

55. For Condon's biography see Morse 1976; for DuBridge's biography see Greenstein 1997; for more about Lark-Horovitz see http://www.physics.purdue.edu/prizes_awards/horovitz.shtml; for an obituary of Tate see Pegram 1950; for Loomis see Conant 2002; and for van Vleck see http://nobelprize.org/nobel_prizes/physics/laureates/1977/vleck-autobio.html.

56. The following material is taken from Bacher's interview with Charles Weiner at the California Institute of Technology, June 30, 1966. Available at http://www.aip.org/history/ohilist/27978.html.

57. In his interview with Weiner, Marshak related that as a graduate student at Columbia and later at Cornell he had heard rumors that Bethe's "Jewishness" had been discussed when his appointment was being considered at Cornell in 1934. Rabi had repeatedly tried to obtain an appointment for Bethe at Columbia after 1935, but was turned down.

58. Gamow 1937b, vii–viii.

59. The experiment resulted in a publication by Weekes, Livingston, and Bethe in 1936.

60. "Nuclear Physics at Cornell." Report to the Board of Trustees, 1947.

61. In 1940–41, when the design for a second larger cyclotron was being discussed, Bethe worked out a theory of the fringing fields and wrote a manuscript entitled "Elementare Theorie des Fringing Effects." See also van Vleck to Bethe, December 10, 1941. Bethe Papers, Box 2.

62. Wigner 1955, 5–6.

63. For a biography of Kirkwood see Rice and Stillinger 1999.

64. A joint paper was published later (Bethe and Kirkwood 1939).

65. Bethe-Sommerfeld correspondence, Bethe to Sommerfeld, August 1, 1936.

66. Bethe was very much involved in the deliberations about and the design of the planned new cyclotron. In November 1937, Bethe and Rose sent a letter to the *Physical Review* entitled "On the Maximum Energy Obtainable from a Cyclotron." They pointed out the serious difficulties they believed would be encountered in accelerating charged particles to higher energies than in existing cyclotrons, because the relativistic variation of mass would destroy "either resonance or focusing." The letter got Bethe into a serious controversy with Lawrence and others at Berkeley. The exchanges and resolution of the conflict are detailed in Heilbron and Seidel 1989. In June 1941 the Board of Trustees of Cornell University accepted President Day's recommendation and appropriated the necessary funds for the construction of the cyclotron and the building that was to house it. But the project was delayed "in view of the impossibility of securing the necessary materials for this highly important item of apparatus so long as the defense program is on."

67. Symposium on Photoelectricity and Thermionics, July 4–6, 1935. Speakers included local speakers as well as S. Dushman and I. Langmuir from General Electric, and J. A. Becker, C. J. Davisson, H. E. Ives, and F. C. Nix from the Bell Telephone Laboratory. Others were the Symposium on Nuclear Physics, July 2–4, 1936; and the Symposium on the Structure of Metallic Phases, July 1–3, 1937.

8. The Happy Thirties

1. Stuewer 1979, vii. In May 1967 and May 1969 as well the American Institute of Physics had held workshops to explore the history of nuclear physics (Weiner 1972).

2. Stuewer 1979, xi–xii.

3. Ibid., 160.

4. Ibid., 11.

5. See Strömgren 1937 and 1938. Regarding Weizsäcker it is interesting to note that he, too, like Bethe and Gamow, wrote a monograph on nuclear physics in 1937 (Weizsäcker 1937). For his research on nucleosynthesis, see Weizsäcker 1938, and chap. 12, sect. 5, of Chandrasekhar 1939.

6. I have not provided references to the large number of experimental papers on neutron, proton, and α-particle induced nuclear reactions that are peripherally referred to in this chapter, for it is not intended as a history of nuclear physics. The relevant references can be found in the Bethe Bible and in Rasetti 1936, chapter 6. I have based my version of the history of nuclear reaction principally on Rasetti 1936, which is a remarkably lucid and informed contemporaneous textbook on nuclear physics and the history of its development. I have also found Amaldi 1986 very helpful.

7. Bethe kept the notes and calculations he made in connection with his research at Cornell, and they have been deposited in the Cornell University Archives. A study

of these notes—usually on 8.5 × 11 inch paper—reveals Bethe's uncanny ability to do integrations in the complex plane, to approximate integrals, and to get insights into differential equations. The notes include page after page of numerical approximations, with numerous tables of numbers. These notes indicate that Bethe made mistakes (parts of the calculations are crossed out) and the pages are not all neat. But the overall impression is that of a tank moving forward, leveling all obstacles and building the bridges that are necessary to get to the end point.

8. Regarding the appropriateness of the model, Bethe remarked in his paper (Bethe 1935g): "It is not likely that the approximation made in this paper, i.e., taking the nucleus as a rigid body and representing it by a potential field acting on the neutron is really adequate . . . Anyway, it is the only practical approximation in many cases." Perrin and Elsasser's (1935) paper was based on similar assumptions.

9. Bohr 1936. For particularly insightful expositions of the history of these developments see Peierls 1940, Wigner 1955, Friedman and Weisskopf 1955, Peierls 1986, and Mottelson 2008.

10. Quoted in Mottelson 2008, 121.

11. Ibid.

12. Frisch evidently constructed the model for Bohr.

13. Bethe 1979, 23.

14. The time dependence of the wave function describing an excited state originally in the state n at time $t = 0$ is given by

$$e^{-i(E_n - \frac{i\Gamma_n}{2})t}$$

so that at time t the probability of finding the nucleus in the state n is proportional to $e^{-\Gamma t}$. When the energy is sufficiently high, a particle j can be emitted with different energies, corresponding to different states of the residual nucleus. The partial widths are then subdivided further to take this into account.

15. Polanyi and Wigner 1925.

16. Bethe and Placzek 1937.

17. Thus in the Bethe Bible, Bethe stressed the importance of the papers by Elsasser (1933) and Guggenheimer (1934) on the nuclear shell model. In his Minnesota address, "The Happy Thirties," Bethe made a special point to refer to these papers.

18. Bethe 1936a.

19. See Gamow 1931 for the first discussion of nuclei modeled in analogy with liquid drops.

20. For the history of the liquid drop model of the nucleus see Stuewer 1997 and the references therein.

21. The model was elaborated after the war by Barschall and by Feshbach, Porter, and Weisskopf in their optical model of nuclear scattering.

22. These contributions of Bethe include his calculations of the level density of the compound nucleus and his contributions to the introduction of thermodynamic

concepts into nuclear physics, which were introduced into a larger field of prewar experimental findings by Bohr, Kalckar, Oppenheimer, Peierls, Placzek, Weisskopf, and Wigner, among others.

23. All the quotations are from the interview of Robert Marshak by Charles Weiner on June 15, 1970. Niels Bohr Library and Archives, American Institute of Physics, College Park, MD. Available at http://www.aip.org/history/ohilist//4760_1.html.

24. The mathematics team won the city-wide competition the year Marshak was captain.

25. The Pulitzer Scholarship entitled the recipient to free tuition at Columbia College for four years, which in those days was $400 a year, and an additional cash award of $250 a year.

26. Interview of Robert Marshak by Charles Weiner, June 15, 1970.

27. Ibid.

28. Ibid.

29. In his interview with Weiner, Marshak added: "I heard sort of vaguely that it was extremely difficult to get an assistantship in the Columbia physics department, and in particular I think I heard that it was extremely difficult for Jewish students to obtain assistantships. I didn't feel that the fact I only was awarded a tuition grant and nothing more for the first year was in any way reflecting a possible discriminatory practice, because I realized that I had just made the changeover and that was the best I could expect and was grateful for the tuition waiver."

30. Interview of Robert Marshak by Charles Weiner, June 15, 1970.

31. Newton was one of the first to speculate about the consequences of his laws for cosmic evolution. In 1692 he wrote Richard Bentley that he had been thinking about a cosmogony accounting for the evolution of stellar objects and that he had worried about the stability of the world at large. Because, according to the law of universal gravitation, every massive object attracts every other massive object, Newton asked, what is to prevent the collapse of the cosmos? He told Bentley that he had convinced himself that if space were finite, gravitation being universal and attractive, a (nonrotating) uniform distribution of matter would be unstable and would "convene" into a single cluster. However, he had concluded that if space were infinite, it would be possible to understand the clumping of matter into stellar objects: "if the matter was evenly disposed throughout an infinite space . . . some of it would convene into one mass and some into another, so as to make an infinite number of great masses, scattered great distances from one another throughout all that infinite space. And thus might the Sun and the fixed Stars be formed, supposing the matter were of a lucid nature."

32. On the eighteenth century, see, for example, Greene 1959. It is often forgotten that both Kant and Laplace believed that stars are entirely similar to the sun and that their genesis is due to processes identical to those that were responsible for the

formation of the sun. Kant already asserted that: "Millions and whole myriads of millions of centuries will flow on, during which always new worlds and systems of worlds will be formed after each other in the distant regions away from the center of Nature" (Kant 1755). For an illuminating account of cosmogenetic theorizing until the end of the nineteenth century see Poincaré 1913.

33. Hale 1908, p. 5.

34. In 1859, the very year that Darwin published his *Origin of Species,* Kirchhoff succeeded in determining the chemical composition of the sun using a spectroscope.

35. See, for example, Henry Norris Russell's Halley lecture of 1933 (Russell 1933) for a history of these developments.

36. Hale 1908, 3.

37. At the beginning of the twentieth century Robert Stawell Ball commented:

> A question has sometimes been asked as to the most important discovery in astronomy which has been made in the century just closed. If, by the most important discovery, we mean that which has most widely extended our knowledge of the Universe, I do not think there need be much hesitation in stating the answer. It seems to me beyond a doubt that the most astonishing discovery of the last century in regard to the heavenly bodies is that which has revealed the elementary substances of which the orbs of heaven are composed. This is the more interesting and instructive because it has taught us that the materials of the sun, of the stars, and of the nebulae are essentially the elements of which our own earth is formed , and with which chemists had already become well acquainted" (Ball 1902, 269).

38. Thomson's contributions can be found in Thomson 1882–1911, 5:141–144, 205–230. The modern version of the argument goes as follows: The total energy of a star in hydrostatic equilibrium under its own gravitation is, by the virial theorem, equal to half its potential energy. From dimensional considerations the potential energy must be proportional to GM^2/R, hence:

$$E = q\frac{GM^2}{R}$$

where q is a numerical constant of order 1 depending on the density distribution in the star. As the star contracts, energy will be released at the rate

$$-\frac{dE}{dt} = q\frac{GM^2}{R^2}\frac{dR}{dt}.$$

This rate on the contraction hypothesis is identified with the luminosity L of the star. Conversely, an observed luminosity L implies a rate of contraction

$$-\frac{1}{R}\frac{dR}{dt} = -\frac{LR}{qGM^2}.$$

Inserting the value of 2 ergs/gr sec for the luminosity, and 1.985×10^{33} gr and 6.951×10^{10} cm for the mass and radius of the sun, one obtains for (1/R) (dR/dt) approximately 5×10^{-8} year^{-1}, which implies that the sun must change very substantially in a time of the order of 2×10^{7} years.

39. Burchfield 1990.

40. Kelvin estimated the age of the earth by assuming its entire mass was initially molten and calculating how long it would take to reach the present temperature gradient inside the earth. For the conductivity of the materials involved he used the conductivity of the surface rocks.

41. Burchfield 1990.

42. Quoted in Wilson 1983, 206. In 1929 Rutherford carefully measured uranium-lead and thorium-lead ratios in radioactive terrestrial rocks—and determined a minimum age of the earth at between 1 and 2 billion years.

43. See Hufbauer 1981 for the history of astronomers taking up the energy generation problem in stars.

44. Eddington 1926, 291.

45. See Philip and DeVorkin 1977 and DeVorkin and Kenat 1983.

46. Hufbauer 1981.

47. Eddington 1926, 301.

48. The history of stellar evolution until 1920 is insightfully presented in Strömgren 1951a and DeVorkin 1984. Thomson defined convective equilibrium in the following terms: "Any fluid under the influence of gravity is said to be in convective equilibrium if the density and the temperature are so distributed throughout the whole fluid mass that the surfaces of equal density and of equal temperature remain unchanged when currents are produced in it by any influence so that changes of pressure due to the inertia of motions are negligible." Thomson 1882–1911, 5:254–283. He continued: "the essence of convective equilibrium is that if a small spherical or cubic portion of the fluid in any position, P, is ideally enclosed in a sheath impermeable to heat and expanded or contracted to the density of the fluid at any other place P′, its temperature will be altered, by the expansion or contraction, from the temperature which it had at P to the actual temperature of the fluid at P′."

49. A polytropic change of a system is a quasistatic change of state carried out in such a way that the specific heat remains constant (at some prescribed value) during the entire process: $dQ/dT = c =$ constant. Thus an *adiabatic process* is a polytropic process of zero specific heat, and an *isothermal process* is one of infinite heat capacity.

50. Eddington's (1926) classic *The Internal Constitution of the Stars* gives a beautiful account of these models. An informative and concise exposition is given by Cowling (1966) in his Royal Astronomical Society presidential lecture.

51. See Strömgren 1951b, DeVorkin 1984, and Philip and DeVorkin 1977. Other startling facts were the mass-luminosity law and the properties of variable stars.

52. See DeVorkin and Kenat 1983, Strömgren 1951a,b, Gingerich 1995, and Hufbauer 2000 for a historical overview of these developments.

53. A sensitive method for estimating surface temperatures of stars depends on the study of the relative intensities of the numerous dark thin lines, called Fraunhofer lines, that are embedded in the continuous spectra of stars. These dark lines are due to the selective absorption of light by the stellar atmosphere. The relative absorbing power of different atoms depends on the temperature. The appearance of this absorption line pattern from star to star permits an estimate of their surface temperature from the character of the spectrum. Astronomers divided the observed range of stellar atmosphere into ten groups, called by different letters: O, B, A, F, G, K, M, R, N, and Meghnad Saha formulated the theory that gives the relation between the temperature of the absorbing gas and the character of the absorption spectrum.

54. Regarding opacity, the number of interactions a particle undergoes in traversing a distance dx in a target of density n is $n\sigma dx$, which defines the effective cross section σ for absorption or scattering. The mean free path l of the particle in the target is defined so that $n\sigma l = 1$, i.e., there is one interaction in traversing the distance l. If the traversed matter consists of a variety of absorbers and scatterers, each with its density n_i and cross section σ_i, the mean free path will be

$$l = \frac{1}{\sum_i n_i \sigma_i} = \frac{1}{\rho\kappa}.$$

The fact that all particle densities are proportional to the mass density ρ has been used to define the opacity κ.

55. The history of stellar structure and evolution from 1924 to 1938 is reviewed in detail in Hufbauer 2000.

56. Eddington 1926, 15.

57. See Chandrasekhar 1939.

58. Russell 1939.

59. Hufbauer 1981.

60. Gamow 1931.

61. My biographical account of Gamow is based on Gamow 1970, Reines 1972, and Harper et al. 1997. See also Karl Hall's impressive doctoral dissertation (1999).

62. One of his students at the school was Lev Bronstein, who later became better known as Leon Trotsky. Bronstein did not think him a very good teacher and organized a petition to have him dismissed.

63. Friedmann was born on June 4, 1888, in St. Petersburg. He entered *Gymnasium* in 1897 and the university in the fall of 1906. His teachers there were Vladimir Steklov and Paul Ehrenfest. While a student at the university he organized a "mathematical academy" that included his close friend Ya. D. Tamarkin, as well as M. F. Petelin, V. I. Smirnov, Yakov A. Shohat, and A. S. Besicovich, who all became well-

known mathematicians. In 1913 he got a position at the Physical Observatory of St. Petersburg. His first assignment was processing the observations obtained from mete-reographs flown from kites, which introduced him to meteorology. His research contributions were so outstanding that he was sent to the Geophysical Institute in Leipzig to study theoretical meteorology with Vilhelm Bjerknes. There he studied the problem of weather prediction and wrote a paper with Theodor Hesselberg, "The order of magnitude of meteorological elements in the space and time derivatives," which has become a classic. In it they assessed the relative magnitude of the various terms in the hydrodynamical equations of an incompressible fluid. During World War I Friedmann became involved in aerodynamics and studied the motion of a projectile released from a moving aircraft, that is, methods for dropping bombs from airplanes. He also taught aerodynamics in Kiev. In 1918, after the revolution, Friedmann was given a professorial appointment at Perm University, where he helped set up the Perm Physico-Mathematical Society. He there wrote an outstanding text, *The Hydrodynamics of Incompressible Fluids,* and initiated the statistical study of fully developed turbulence. In 1920 he returned to Petrograd as senior physicist in charge of the mathematical department of the Geophysical Observatory. Among the young researchers he invited to join him there was V. A. Fock. He also gave courses at the university on general relativity, and these resulted in his important contributions to relativistic cosmology. In 1925 Friedmann was appointed the director of Main Geophysical Observatory. He died in August 1925 of typhoid fever, which he contracted on his way home to Petrograd following an accident in the balloon he was flying to make meteorological observations in the stratosphere.

64. Reines 1972, 289.

65. Edward U. Condon and Ronald W. Gurney gave the same explanation at the same time. See Gurney and Condon 1929.

66. Houtermans was born on January 22, 1903, near the then-German Baltic port of Danzig. He was raised by his mother in Vienna. His mother was a gifted intellectual with wide-ranging interests in the humanities and the sciences. She was the first woman to obtain a doctorate in chemistry at the University of Vienna.

67. Casimir 2010.

68. Shortly after Gamow's paper appeared, von Laue and Kudar independently suggested that the light elements might be formed through the inverse process of alpha decay but, like Houtermans, found values much too low for the probability per unit time for the process to occur even under stellar conditions. Regarding available lab energy, α-particles from the radioactivity of thorium, radium, and other heavy elements were of course available. It is with such 2 MeV alpha particles that Rutherford in 1919 had observed the reaction ${}_2He^4 + {}_7N^{14} \rightarrow {}_8O^{17} + {}_1H^1$.

69. Atkinson and Houtermans assumed that the collision cross section is given by πr_0^2, where r_0 is the collision radius.

70. Somewhat smaller periods are obtained if the collision cross section is assumed to be $2\pi\lambda^2$ where λ is the de Broglie wavelength of the particle.

71. This was changed to the more sedate title, "On the Question of the Possibility of the Synthesis of Elements in Stars," by the editor of the *Zeitschrift für Physik.*

72. Gamow 1935.

73. Rutherford published his lectures the following year in a little book entitled *The Newer Alchemy* (Rutherford 1937).

74. Rutherford 1937, 67.

75. In 1929 Rutherford made use of the relative abundance of ^{238}U and ^{235}U and their natural radioactive decay into isotopes of lead to determine the age of these elements, that is, the time at which they were produced, assuming that both of these uranium isotopes were formed at the same time. He had found this time to be 3,000 million years.

76. Gamow 1937a, 234.

77. Gamow 1938, 198.

78. In the early 1990s Karl Hufbauer interviewed Bethe to obtain his account of the events that led to his solution of the stellar energy problem. Hufbauer extensively researched the conference and the contributions of the participants, which he wrote up in a paper, "Hans Bethe's breakthrough solution of the stellar-energy problem," which was never published (Hufbauer 2000). It is the most detailed and reliable account of this important episode. I thank him for his permission to quote from it. See also Rebsdorf 2007; Rebsdorf's article presents a detailed biographical portrait of Bengt Strömgren, a major contributor to the conference, as well as an informative account of the conference itself. The date of the conference had been set to accommodate Strömgren, who was returning to Denmark. See Rebsdorf 2007.

79. Gamow and Teller 1938.

80. Strömgren 1937. See Rebsdorf 2007, and the interview of Strömgren by Gordon Baym and Lillian Hoddeson in May 1967, available at http://www.aip.org/history/ohilist/5070_1.html.

81. Williams et al. 1937.

82. Interview of S. Chandrasekhar by S. Weart on May 17, 1977. Niels Bohr Library and Archives, American Institute of Physics, College Park, MD. Available at http://www.aip.org/history/ohilist/4551_1.html.

83. Chandrasekhar, Gamow, and Tuve 1938.

84. Bethe erred. Astronomers had indicated that the abundance of nitrogen in the sun was 14 percent by mass, but it is actually less than half a percent. The ^{14}N (p, γ) reaction—the slowest one—limits the rate of the CNO cycle. Bethe had too large an abundance, so that the overall rate was too fast. Also, he had a slightly higher central temperature, 18 million degrees, as compared to the presently accepted one of 15 million.

85. For all nuclei lighter than carbon the reaction with protons generates an α-particle with the recovery of the original nucleus. Nuclei heavier than fluorine lead to radiative capture of the proton and destroy the original nucleus.

86. Hufbauer 2000.

87. *Kristallnacht* did convince them, and they emigrated in the spring of 1939.

88. The A. Cressy Morrison Astronomical Prize of $500 was established in 1938. It was to be awarded to the author of the paper judged by the Council of the New York Academy of Sciences to be "the most meritorious contribution on the subject of solar and stellar energy."

89. Hufbauer 2000.

90. In Samuelsson and Sohlman 1998.

91. This was a new effect, and the calculations to determine the opacity were complicated, involving the Fermi-Dirac distribution at finite temperature and requiring knowledge of the equation of matter in a white dwarf at very high temperatures and pressures. Bethe helped Marshak overcome all the difficulties he encountered, and these investigations eventually resulted in a joint paper (Marshak and Bethe 1940).

92. Interview of Marshak by Weiner, June 15, 1970.

93. For a biography of Weisskopf, see Jackson and Gottfried 2003.

94. Weisskopf 1991, 5.

95. Ibid.

96. For insightful accounts of fin-de-siècle Vienna and in particular the Jewish setting see the biography of Popper by Hacohen (2000). See also Janik and Toulmin 1973 and Coen 2007.

97. Weisskopf and Wigner 1930.

98. See Weisskopf 1991 and the interview of Weisskopf by Kuhn and Heilbron, American Institute of Physics, 1965, for more details on Weisskopf's scientific peregrinations during the 1930s. Interview available at http://www.aip.org/history/ohilist/4944.html.

99. Interview of Victor Weisskopf by Charles Weiner and Gloria Lubkin, September 22, 1966. Niels Bohr Library and Archives, American Institute of Physics, College Park, MD. Available at http://www.aip.org/history/ohilist/4944.html and http://www.aip.org/history/ohilist/4945_1.html.

100. Fermi et al. 1941.

101. Bethe (1940b) argued against calling the Yukawa-Anderson-Street particle a "heavy electron," and for linguistic reasons agreed with Bhabha that it ought to be called *meson.*

102. For further details concerning the history of the field theoretic interpretation of nuclear forces see Brown and Rechenberg 1996.

103. Cassidy 1981, Galison 1983, Rossi 1983, Brown and Rechenberg 1991.

104. Kemmer 1938, 9.

105. Bethe 1939b, 1940a, 1940b.

106. For the strong criticism of Bethe's theory by Pauli and Heisenberg see Brown and Rechenberg 1996, 219–221.

107. Heitler (1940) and Bhabha (1941) independently indicated how this could

come about. They observed that the cross section for the scattering of a neutral scalar meson on a nucleon would be quite small, because the direct and crossed Born terms very nearly cancel. For a charged meson the cross section is large, because only one of the Born terms exists. They pointed out that if in addition to the nucleons there existed low-lying excited states of the nucleon—"isobars"—with charge $+2e$ and $1e$, then there would be two Born terms that could nearly cancel, and thus produce a small cross section as in the neutral case. Somewhat earlier, Heisenberg (1939) had noted a different mechanism. He calculated the scattering of a neutral vector meson coupled to the spin of a nucleon and found that the large Born approximation result was greatly suppressed by "reaction effects." He observed that the self-field of the nucleon, that is, that part of the meson field which is attached to the nucleon and carried with it, was responsible for an increased inertia of the spin motion and resulted in a cross section of the order of a^2 where a is the assumed radius of the nucleon.

The next important step was taken by Wentzel, who in 1940 showed that the simplest nontrivial static model, that of a charged scalar meson field coupled to a fixed scatterer, could be solved quantum mechanically in the limit of a large coupling constant. He found that the solution exhibited isobars; however, these isobars did not result in a particularly small cross section because their excitation energy was not small enough. Oppenheimer and Schwinger (1941) extended Wentzel's strong coupling calculations to the case of neutral pseudoscalar mesons. They found that Heisenberg's (1939) classical result for the scattering was reproduced, and noted that the solution gave rise to isobars that had been overlooked by Heisenberg and reduced the cross section for meson-nucleon scattering by the Heitler-Bhabha mechanism. Stimulated by Wentzel's pioneering work on strong coupling theory, Schwinger gave a more extensive treatment of the strong coupling limit of the charged scalar mesotron field. He observed that the large coupling should serve to bind mesotrons in stationary states around the nucleon and give rise to isobaric states, which he exhibited by using a series of canonical transformations (Schwinger 1970).

108. Oppenheimer 1941, 39.

109. Ibid.

110. Ibid., 50.

111. See Pauli's "Meson Theory," the set of lectures he gave at the MIT Radiation Laboratory toward the end of the war, and Pauli 1946 for a summary of these activities.

9. Rose Ewald Bethe

1. See Philippson 1962.

2. Rose Bethe, interviewed by S. S. Schweber, December 15, 2006. Unless otherwise noted all the quotations in the present chapter are from this interview.

3. For the atmosphere of Göttingen at that time see Reid 1970, 1976.

4. Born was then a *Privatdozent* and von Kármán an assistant in Ludwig Prandtl's applied mechanics institute.

5. Bethe and Hildebrandt, 1988, 136. Bethe wrote the extensive first part of the biographical memoir dealing with Paul Peter Ewald's personal life. Ella, Paul Peter's wife, was the source of much of the information in it.

6. Ibid.

7. Quoted in Reid 1970, 108.

8. Bethe and Hildebrandt 1988, 141.

9. Ibid.

10. Reid 1970, 131. Reid indicated that if Hilbert felt a draft in a restaurant or at a concert he would borrow a fur from one of the ladies present.

11. Rose referred to Hans Fallada's "Bei uns zu Heim" for a description of the "formal" dinners when guest speakers were invited to the house.

12. The Adlers eventually emigrated to Brazil. The mother had obtained a position in the Yellow Fever Institute in Rio de Janeiro. The father and daughter joined her shortly thereafter.

13. It was Paul Peter Ewald who, in 1911, while working on the refraction light in its passage through matter, gave von Laue the idea that X-rays could be used to analyze crystal structure by their X-ray diffraction patterns.

14. Ella was a distant relative of Mrs. Courant.

15. See Reid 1970, 132, 140–141. See also the oral interview of Dr. Alfred Landé and Mrs. Landé by Charles Weiner, October 3, 1973. Available at http://www.aip.org/history/ohilist/4727.html.

16. See Reid 1976, 1986.

17. Hans corroborated this. In a letter to his mother dated May 8, 1937, Hans wrote her about the forthcoming trip: "I would like to be together with the Tellers: They are both so nice, and are among the few people with whom I can really enjoy nature, who are never boring and never not congenial. We like each other ever more, and Mici gave me this compliment in Washington last Monday: 'You are the nicest guest.'"

18. In 1935 Teller had accepted a professorship at George Washington University in Washington, D.C., and joined George Gamow. A condition of Gamow's acceptance of a professorship at George Washington was that another theoretician would be appointed at a professorial rank and that the university would sponsor an annual theoretical physics conference like the ones Bohr organized in Copenhagen.

19. Ewald had resigned his post in Stuttgart and had obtained a position in Belfast in 1939.

20. Bethe to Anna Bethe, July 23, 1938, aboard the SS *Bremen,* the ship that took him to Europe.

21. Weisskopf 1991, 19.

22. See Josephson 1987.

23. Weisskopf 1991, 108.

24. Bethe to Anna Bethe, November 30, 1938.

25. Bethe to Peierls, August 22, 1938. Lee 2007, 235.

26. Hans's mother was born in Strasbourg, which, according to American law, was considered French although at the time of her birth it was part of Germany. Thus she came under the French quota, which was empty, rather than the German quota, which was full for three years to come. As a consequence she was able to emigrate in June of 1939.

27. Bethe to Anna Bethe, May 11, 1939.

28. Rose remembered that he told her, "Your mother-in-law need[s] a lot of care and you are not doing a good job."

29. This word is identified with Nazi rhetoric, so there is a touch of sarcasm in Bethe using it. It might be rendered in English as "national comrade."

30. Arendt 1978, quoted in Bell 2002.

Conclusion

1. Rabi, in Bromley and Hughes, 1970, 184.

2. And Hughes corroborated his contention by narrating the history of the role of theoretical physics at the Cavendish under the directorship of J. J. Thomson, a former Tripos Wrangler, and under Rutherford, who believed that "while experimental facts came first," "the more attractive clothing they can be wrapped in the better." In 1920 Rutherford had feared that Einstein's popularity after Eddington's corroboration of the validity of general relativity would "ruin many scientific men in drawing them away from the field of experiment to the broad road of metaphysical conceptions" (Hughes 1998, 343–344).

3. Rutherford got the University of Cambridge to establish a lectureship in theoretical physics, telling the university that "with a colleague on the mathematical side he would be much happier about the future of physics in this University because the development of physics in the last decade has tended to bring the theoretical and experimental worker in closer contact to their mutual advantage, and I feel it is of vital importance to the future of Physics in general in Cambridge that everything possible should be done to promote this close association on which future progress so much depends." Quoted in Hughes 1998, 349. See also Hughes 2002b, 2005.

4. Weisskopf continued, "Some of them could get 'in'; Williams for example, was not known by anybody but he had such a wonderful, buoyant personality that he was accepted right away. Others did not have this luck, and it was not always dependent on ability in physics, although it was, of course, to some extent." Oral interview with Dr. Victor Weisskopf by Charles Weiner and Gloria Lubkin, MIT, September 22, 1966. Niels Bohr Library and Archives, Center for History of Physics, College Park, MD. Available at http://www.aip.org/history/ohilist/4945_1.html.

5. Nietzsche 1983, 199–230. Quoted in Pletsch 1991.

6. For an overview of Bethe's approach to physics that includes the post–World War II period see G. Brown 2006a,b.

7. See Morrison 1999, Morgan and Morrison 1999.

8. Bazerman 1988.

9. See Smith 2002.

10. See Brown and Lee 2009.

11. Hiring activities at Los Alamos were intense. Feynman accepted an offer from Cornell in 1944, and Morrison did the same a little later. Several experimentalists were also added to the faculty of the department. By 1945 Bethe was holding faculty meetings of the Cornell contingent at Los Alamos. The plans that Harvard formulated for the expansion of its science and engineering departments, as well as the records of the committee that formulated them, can be found among the papers of Harlow Shapley, Harvard University Archives. Those of Princeton can be found in the Departmental files deposited in the Mudd Archives, Princeton University. Bethe to Gibbs, November 20, 1944. Bethe Papers, Cornell University.

12. Bacher to Gibbs, November 4, 1944. Bethe Papers.

13. Bacher to Gibbs, December 6, 1944. Bethe Papers.

14. The proposal went on to detail the administrative structure for the running of the department and research laboratory, and for the appointment of faculty, graduate students, and teaching and research assistants. See Bethe Papers. The document also indicates that each "part" had an initial budget of roughly $200,000 provided by the university.

15. Day to Bacher, September 16, 1946, Bethe Papers. In October 1946 the Office of Naval Research informed Day that it "will undertake to execute a contract" for the construction of the accelerator for the nuclear studies laboratory. The Laboratory of Nuclear Studies was dedicated in October 1948, by which time the construction of a betatron was well under way. On October 29, 1945, the physics faculty adopted the following resolution:

> We, the members of the staff, conscious of the unusual opportunity furnished to this department by the University's willingness to finance a new and very extensive research project, and aware of the difficulties inherent in such an arrangement, do hereby put ourselves on record as pledging our individual and collective support to the men named and the plans as outlined, declaring our several intentions to deal frankly and objectively with whatever problems may arise—all to the purpose that we may better ensure the prestige of the University, the unity of the group, and the satisfaction of every individual in the group.

16. The Pocono Conference was held from March 30 to April 2, 1948; the third, Oldstone, was held April 11–14, 1949. See Schweber 1994, ch. 4.

17. The ravages of inflation are illustrated by the fact that the average expenditure

to accommodate the twenty-five attendees for a night at the comfortable and elegant inns where the conferences took place was $200.

18. Kramers at the time was visiting the Institute for Advanced Study, after having finished chairing the UN Scientific and Technological Committee on Nuclear Energy and teaching at Columbia.

19. MacInnes recorded in his diary that "it was immediately evident that Oppenheimer was the moving spirit of the affair." Darrow, in his diary, gave a revealing account of Oppenheimer:

> As the conference went on the ascendency of Oppenheimer became more evident—the analysis (often caustic) of nearly every argument, that magnificent English never marred by hesitation or groping for words (I never heard "catharsis" used in a discourse on [physics], or the clever word "mesoniferous" which is probably O's invention), the dry humor, the perpetually-recurring comment that one idea or another (incl. some of his own) was certainly wrong, and the respect with which he was heard. Next most impressive was Bethe, who on two or three occasions bore out his reputation for hard & thorough work, as in analyzing data on cosmic rays variously obtained. (Diary entry for June 3, 1947. Darrow Papers. Niels Bohr Library, AIP.)

20. For Bethe's post–World War II contributions to quantum electrodynamics see Dyson 2006.

21. Bethe (1947) did acknowledge the comments of Weisskopf and Schwinger, who first claimed that hole theory must be used to obtain convergence in this problem, that a hole theoretic calculation of energy level differences would be finite. Their insight justified the high energy cut-off Bethe introduced in his calculation.

22. In his talk at Shelter Island II in 1983 commemorating Shelter Island I, Lamb at the historical session made the following remarks:

> Kramers was there that year. (2) When I heard Kramers talk at Shelter Island I it seemed to me that he mainly said that we should be applying the methods used by Lorentz for the classical electron. I could not see that he indicated how such a program could be carried out, and hence did not derive great inspiration from his talk . . .*** Unlike Weisskopf, who was working on these problems actively before this wonderful time, Kroll and I were only inspired when we found out how Bethe had subtracted the self-energy of a free electron.
>
> ***Editors' note: Bethe makes the following interjection: "He did inspire me."

23. Dresden 1987.

24. The importance of Bethe's calculation is apparent from Weisskopf's reaction: after studying the manuscript he received on June 11, Weisskopf wrote to Bethe that he was

quite enthusiastic about the result. It is a very nice way to estimate the effect and it is most encouraging that it comes out just right. I am very pleased to see that Schwinger's and my approach seems to be the right one after all. Your way of calculating is just an unrelativistic estimate of our effect, as far as I can see.

I am all the more pleased about the result since I tried myself unsuccessfully to estimate the order of magnitude of our expression. I was unable to do this but I got more and more convinced that the method was sound . . . I would like to talk it over with you especially the "korrespondenz Deutung" of the effect.

But he added:

I do not quite agree with your treatment of the history of the problem in your note. That the $2S_{1/2}-P_{1/2}$ split has something to do with radiation theory and hole theory was proposed by Schwinger and myself for quite some time. We did not do too much about it until shortly before the conference. We then proposed to split an infinite mass term from other terms and get a finite term shift, just as I demonstrated it at the conference. Isn't that exactly what you are doing? Your great and everlasting deed is your bright idea to treat this at first unrelativistically. "Es möchte doch schon sein" if this were indicated in some footnote or otherwise.
See you soon I hope
Yours
Vicki

Bethe put such a footnote into his paper, and answered Weisskopf a few days after receiving Weisskopf's letter. Weisskopf soon acknowledged that Bethe's "abstract" was "*harmloser*" (much more harmless) than he initially thought and agreed, "Let's forget about patent claims."

25. I owe the notion of a crucial calculation to my colleague and friend Howard Schnitzer at Brandeis.

26. Bethe 1947.

27. See, for example, Polkinghorne 1989.

28. Also, whereas the Shelter Island and Pocono conferences looked upon quantum electrodynamics (QED) as a self-contained discipline, the Rochester conferences saw "particle physics" come into its own, with QED as one of its subfields—albeit one with a privileged, paradigmatic position.

Appendix C

1. Quotation from Pauli 1964, 216. I have based my brief presentation of Heisenberg's and Dirac's approach to quantum mechanics in 1925 on Aitchison,

MacManus, and Snyder 2004 and Gottfried 2011. For a fuller presentation see Mehra and Rechenberg 1982–86, Darrigol 1992, Kragh 1979, 1990, 1999, Farmelo 2009, Pais et al. 1998. The whole history of the invention of quantum mechanics is under review at the Max Planck Institute for the History of Science under Jurgen Renn's guidance.

2. The opening sentence of Heisenberg's (1925) paper stated: "In this paper it will be attempted to secure foundations for a quantum theoretical mechanics which is exclusively based on relations between quantities which in principle are observable."

3. Born 1978, 217.

4. Born and Jordan 1925a.

5. Born and Jordan 1925b.

6. Born, Heisenberg, and Jordan 1925.

7. See Przibram 1967.

8. The thesis is reprinted in Ludwig 1968, 73–93.

9. Gregor Wentzel, interviewed by T. S. Kuhn, February 3–5, 1964, Chicago, for the Archives for the History of Quantum Physics (from notes; not tape recorded).

10. Wentzel 1964.

11. The quotation is from Bloch 1976, 23. On November 3, 1925, Schrödinger wrote to Einstein that he had read "with the greatest interest the ingenious thesis of Louis de Broglie" (Moore 1989, 192).

12. Moore 1989, 196.

13. Wien to Schrödinger, February 6, 1926. Mehra and Rechenberg 1987, 534.

14. Quoted in Mehra and Rechenberg 1987, 537.

15. Sommerfeld to Pauli, February 3, 1926. Quoted in Mehra and Rechenberg 1987, 617.

16. Schrödinger, "Quantisierung als Eigenvertproblem," reprinted in Schrödinger 1982 and Ludwig 1968, 94–105.

17. ψ should be real, finite everywhere, single valued, continuous, and twice differentiable.

18. Initially Schrödinger considered the wave equation obtained for the matter wave field of an electron in the electric field of a proton with the dynamics being described relativistically. He assumed that

$$(*) \qquad E = h\nu = \frac{m_0 c^2}{\sqrt{1-\left(\frac{v^2}{c^2}\right)}} - \frac{e^2}{r},$$

with the boundary condition that ψ be single valued, continuous, twice differentiable, and vanish at infinity. Eq.(C26) with v given by (*) gave the wrong energy terms for the bound states, and thus did not yield the Balmer formula. Schrödinger then analyzed the nonrelativistic description.

19. The lecture took place on July 21, 1926.

20. Heisenberg 1974, 72.

21. Ibid., 73.

22. Bethe probably attended the lecture and witnessed the sharp exchanges. He evidently was unaware at the time of the importance of the altercations and in his interview with me could not confirm being present at the lecture.

23. Born 1968, 35.

24. For an insightful account of the genesis of Born's probabilistic interpretation of the ψ function see Pais 1986, 258–261.

25. Pais 1986, 260.

26. Pais 1986, 261.

27. These two papers are Born 1926b,c.

28. Wigner reported that Einstein objected to the interpretation of Schrödinger waves that obey Schrödinger's equation as a guiding field. He realized that if the guiding field is taken as propagating in three-dimensional space, then in a collision between two particles where both only follow the guiding field, energy and momentum will not be conserved, because one of the particles will not know how the other follows the guiding field. To conserve energy and momentum the guiding field for the two particles must propagate in the six-dimensional configuration space of the two particles—and Einstein did not conceive of this possibility. "The field guides not the particles but the configuration. Schrödinger's wonderful accomplishment was that he introduced the guiding field, not for particles but for the configuration" (Wigner 1979, 166).

Appendix E

1. See Sigurdsson's insightful review of Cruickshank et al. (1992) for the background of Ewald's 1916–17 papers (Sigurdsson 1996).

2. If damping is included, this formula gives fairly good values for the index of refraction as a function of frequency once the meaning of the n_i is further clarified.

3. At the macroscopic level this is accounted for by ascribing to the crystal a dielectric constant. It was Lorentz who first made the connection between the microscopic theory and Maxwell's macroscopic theory.

4. Quoted in Sigurdsson 1996, 87.

References

A full list of Bethe's publications can be found in *Nuclear Physics* A 762 (2005):13–49; in Bethe 1997, 585–605; and in Brown and Lee 2006, 283–315; references to Bethe's work here are listed in chronological order.

Hans A. Bethe's Publications

Bethe, A., H. A. Bethe, and Y. Terada. 1924. "Versuche zur Theorie der Dialyse" [Experiments Concerning the Theory of Dialysis]. *Zeitschrift für Physikalische Chemie* 112:250–269.

Bethe, H. A. 1927. "Über die Streuung von Elektronen an Kristallen" [On the Scattering of Electrons by Crystals]. *Die Naturwissenschaften* 15:786–788.

——— 1928. "Theorie der Beugung von Elektronen an Kristallen" [Theory of the Diffraction of Electrons by Crystals]. *Annalen der Physik* 87:55–129.

——— 1929a. "Berechnung der Elektronenaffinität des Wasserstoffs" [Calculation of Electronic Affinity of Hydrogen]. *Zeitschrift für Physik* 57:815–821. English translation in Bethe 1997, 73–76.

——— 1929b. "Termaufspaltung in Kristallen" [Splitting of Terms in Crystals]. *Annalen der Physik* 3:133–208. English translation in Bethe 1997, 1–72.

——— 1929c. "Über den Durchgang von Kathodenstrahlen durch gitterförmige elektrische FeldeRoyal" [Passage of Cathode Rays through Electric Fields Formed by Grids]. *Zeitschrift für Physik* 54:703–710.

——— 1929d. "Vergleich der Elektronenverteilung im Heliumgrundzustand nach verschiedenen Methoden" [Comparison of the Distribution of Electrons in the Helium Ground State as Calculated by Different Methods]. *Zeitschrift für Physik* 55:431–436.

——— 1929e. "Wie errechne Ich den Ertrag meiner festverzinslichen Pappiere" [How do I compute the yield of investments designed to yield a fixed rate of interest?]. Frankfurt am Main: Frankfurter Societäts-Druckerei G.m.b.H. Abtailung Buchverlag.

——— 1930a. "Über die nichtstationäre Behandlung des Photoeffekts" [Non-Stationary Treatment of the Photoelectric Effect]. *Annalen der Physik* 4:443–449.

——— 1930b. "Zur Theorie des Durchgangs schneller Korpuskularstrahlen durch Materie" [Theory of Passage of Fast Corpuscular Rays through MatteRoyal]. *Annalen der Physik* 5:325–400. English translation in Bethe 1997, 77–154.

——— 1930c. "Zur Theorie des Zeemaneffektes an den Salzen der seltenen Erden" [Theory of the Zeeman Effect in the Salts of Rare Earth]. *Zeitschrift für Physik* 60:218–233.

——— 1931a. "Change of Resistance in Magnetic Fields." *Nature* 127:336–337.

——— 1931b. "Zur Theorie der Metalle. I. Eigenwerte und Eigenfunktionen der linearen Atomkette" [Theory of Metals. Part I. Eigenvalues and Eigenfunctions of the Linear Atomic Chain]. *Zeitschrift für Physik* 71:205–226. English translation in Bethe 1997, 155–184.

Bethe, H. A., G. Beck, and W. Riezler. 1931. "On the Quantum Theory of the Temperature of Absolute Zero." *Die Naturwissenschaften* 19:39. Also reprinted in Bethe 1997, 185.

Bethe, H. A. 1932. "Bremsformel für Elektronen relativistischer Geschwindigkeit" [Scattering of Electrons of Relativistic Velocity]. *Zeitschrift für Physik* 76:293–299.

Bethe, H. A., and E. Fermi. 1932. "Über die Wechselwirkung von zwei Elektronen" [On the Interaction of Two Electrons]. *Zeitschrift für Physik* 77:296–306.

Bethe, H. A. 1933a. "Quantenmechanik der Ein und Zwei-Elektronenprobleme" [Quantum Mechanics of One- and Two-Electron Problems]. In H. Geiger and K. Scheel, *Handbuch der Physik,* 24, pt. 1: *Quantentheorie,* ed. A. Smekal, 273–560. Berlin: Julius Springer.

——— 1933b. "Theorie des Ferromagnetismus" [Theory of Ferromagnetism]. In P. J. W. Debye, ed., *Magnetismus,* 74–81. Leipzig: Hirzel.

Sommerfeld, A., and Bethe, H. A. 1933. "Elektronentheorie der Metalle." In H. Geiger and K. Scheel, *Handbuch der Physik* 24, pt. 2: *Aufbau der zusammenhängenden Materie,* ed. A. Smekal. Berlin: Julius Springer.

Bethe, H. A., and H. Fröhlich. 1933. "Magnetische Wechselwirkung der Metallelektronen. Zur Kritik der Theorie der Supraleitung von Frenkel" [Magnetic Interaction of Metallic Electrons. About Criticism of Frenkel's Theory of Superconductivity]. *Zeitschrift für Physik* 85:389–397.

Bethe, H. A. 1934a. "The Influence of Screening on the Creation and Stopping of Electrons." *Proceedings of the Cambridge Philosophical Society* 30:524–539.

——— 1934b. "Quantitative Berechnung der Eigenfunktion von Metallelektronen" [Quantitative Calculation of the Eigenfunction of Electrons in Metals]. *Helvetica Physica Acta* 7, supp. 2:18–23.

——— 1934c. "Zur Kritik der Theorie der Supraleitung von R. L. Schachenmeier" [On Criticism of R. Schachenmeier's Theory of Superconductivity]. *Zeitschrift für Physik* 90:674–679.

Bethe, H. A., and W. Heitler. 1934. "On the Stopping of Fast Particles and on the Creation of Positive Electrons." *Proceedings of the Royal Society of London A* 146:83–112. Reprinted in Bethe 1997, 187–218.

Bethe, H. A., and R. Peierls. 1934. "The Neutrino." *Nature* 133:532–533, 689–690. Reprinted in Bethe 1997, 219.

Compton, A. H., and H. A. Bethe. 1934. "Composition of Cosmic Rays." *Nature* 134:734–735.

Bethe, H. A. 1935a. "Ionization Power of a Neutrino with Magnetic Moment." *Proceedings of the Cambridge Philosophical Society* 31:108–115.

——— 1935b. "Masses of Light Atoms from Transmutation Data." *Physical Review* 47:633–634.

——— 1935c. "Memorandum on Cosmic Rays." *Carnegie Institution of Washington Yearbook* 34:333–335.

——— 1935d. "On the Annihilation Radiation of Positrons." *Proceedings of the Royal Society of London A* 150:129–141.

——— 1935e. "Photoelectric Disintegration of the Diplon." Papers and discussions, International Conference on Physics, 1934, London and Cambridge, 93–94. London: Physical Society.

——— 1935f. "Statistical Theory of Superlattices." *Proceedings of the Royal Society of London A* 150:552–575. Reprinted in Bethe 1997, 245–270.

——— 1935g. "Theory of Disintegration of Nuclei by Neutrons." *Physical Review* 47:747–759.

Bethe, H. A., and R. Peierls. 1935a. "Quantum Theory of the Diplon." *Proceedings of the Royal Society of London A* 148:146–156. Reprinted in Bethe 1997, 223–234.

——— 1935b. "The Scattering of Neutrons by Protons." *Proceedings of the Royal Society of London A* 149:176–183. Reprinted in Bethe 1997, 235–244.

Bethe, H. A. 1936a. "An Attempt to Calculate the Number of Energy Levels of a Heavy Nucleus." *Physical Review* 50:332–341.

——— 1936b. "Nuclear Radius and Many-Body Problem." *Physical Review* 50:977–979.

Bethe, H. A., and R. F. Bacher. 1936. "Nuclear Physics. Part A: Stationary States of Nuclei." *Reviews of Modern Physics* 8:82–229.

Weekes, D. F., M. S. Livingston, and H. A. Bethe. 1936. "A Method for the Determination of the Selective Absorption Regions of Slow Neutrons." *Physical Review* 49:471–473.

Bethe, H. A. 1937. "Nuclear Physics. Part B: Nuclear Dynamics, Theoretical." *Reviews of Modern Physics* 9:69–244.

Bethe, H. A., and G. Placzek. 1937. "Resonance Effects in Nuclear Processes." *Physical Review* 51:450–484.

Bethe, H. A., and M. E. Rose. 1937a. "Kinetic Energy of Nuclei in the Hartree Model." *Physical Review* 51:283–285.

——— 1937b. "The Maximum Energy Obtainable from the Cyclotron." *Physical Review* 52:1254–1255.

Bethe, H. A., and F. H. Spedding. 1937. "The Absorption Spectrum of $Tm_2(SO_4)_3{:}8H_2O$." *Physical Review* 52:454–455.

Hoffman, J. G., M. S. Livingston, and H. A. Bethe. 1937. "Some Direct Evidence on the Magnetic Moment of the Neutron." *Physical Review* 51:214–215.

Livingston, M. S., and H. A. Bethe. 1937. "Nuclear Physics. Part C: Nuclear Dynamics, Experimental." *Review of Modern Physics* 9:245–390.

Rose, M. E., and H. A. Bethe. 1937. "Nuclear Spins and Magnetic Moments in the Hartree Model." *Physical Review* 51:205–213.

Bethe, H. A. 1938a. "The Binding Energy of the Deuteron." *Physical Review* 53:313–314.

——— 1938b. "Coulomb Energy of Light Nuclei." *Physical Review* 54:436–439.

——— 1938c. "Magnetic Moment of Li^7 in the Alpha-Particle Model." *Physical Review* 53:842.

——— 1938d. "A Method for Treating Large Perturbations." *Physical Review* 54:955–967.

——— 1938e. "The Oppenheimer–Phillips Process." *Physical Review* 53:39–50.

——— 1938f. "Order and Disorder in Alloys." *Journal of Applied Physics* 9:244–251.

Bethe, H. A., and C. L. Critchfield. 1938. "The Formation of Deuterons by Proton Combination." *Physical Review* 54:248–254. Reprinted in Bethe 1997, 247–354.

Bethe, H. A., M. E. Rose, and L. P. Smith. 1938. "The Multiple Scattering of Electrons." *Proceedings of the American Philosophical Society* 78:373–383.

Konopinski, E. J., and H. A. Bethe. 1938. "The Theory of Excitation Functions on the Basis of the Many-Body Model." *Physical Review* 54:130–138.

Bethe, H. A. 1939a. "Energy Production in Stars." *Physical Review* 55:434–456. Reprinted in Bethe 1997, 355–378.

——— 1939b. "The Meson Theory of Nuclear Forces." *Physical Review* 55:1261–1263.

Bethe, H. A., and W. J. Henderson. 1939. "Evidence for Incorrect Assignment of the Supposed Si^{27} Radioactivity of 6.7-Minute Half-Life." *Physical Review* 56:1060–1061.

Bethe, H. A., and J. G. Kirkwood. 1939. "Critical Behavior of Solid Solutions in the Order–Disorder Transformation." *Journal of Chemical Physics* 7:578–582.

Bethe, H. A., and R. Marshak. 1939. "The Physics of Stellar Interiors and Stellar Evolution." *Reports on Progress in Physics* 6:1–15.

Bethe, H. A., F. Hoyle, and R. Peierls. 1939. "Interpretation of Beta-Disintegration Data." *Nature* 143:200–201.

Rose, M. E., and H. A. Bethe. 1939. "On the Absence of Polarization in Electron Scattering." *Physical Review* 55:277–289.

Bethe, H. A. 1940a. "A Continuum Theory of the Compound Nucleus." *Physical Review* 57:1125–1144.

——— 1940b. "The Meson Theory of Nuclear Forces. Part 1: General Theory." *Physical Review* 57:260–272.

——— 1940c. "The Meson Theory of Nuclear Forces. Part 2: Theory of the Deuteron." *Physical Review* 57:390–413.

——— 1940d. "Recent Evidence on the Nuclear Reactions in the Carbon Cycle." *Astrophysical Journal* 92:118–121.

Bethe, H. A., S. A. Korff, and G. Placzek. 1940. "On the Interpretation of Neutron Measurements in Cosmic Radiation." *Physical Review* 57:573–587.

Bethe, H. A., and L. W. Nordheim. 1940. "On the Theory of Meson Decay." *Physical Review* 57:998–1006.

Holloway, M. G., and H. A. Bethe. 1940. "Cross Section of the Reaction $N^{15}(p,\alpha)$ C^{12}." *Physical Review* 57:747.

Marshak, R., and H. A. Bethe. 1940. "The Generalized Thomas–Fermi Method as Applied to Stars." *Astrophysical Journal* 91:329–343.

Blanch, G., A. N. Lowan, R. Marshak, and H. A. Bethe. 1941. "The Internal Temperature-Density Distribution of the Sun." *Astrophysical Journal* 94:37–45.

Bethe, H. A. 1947. "The Electromagnetic Shift of Energy Levels." *Physical Review* 72:339–341. Reprinted in Bethe 1997, 297–400.

Bethe, H., and J. Ashkin. 1953. *Passage of Radiations through Matter.* In E. Segrè, ed., *Experimental Nuclear Physics,* vol. 1, 166–357. New York: John Wiley and Sons.

Bethe, H. A. 1954. "Review of R. Courant and D. Hilbert, *Methods of Mathematical Physics.*" *Science* 119(3080):75–76.

——— 1955. "Remarks at the Memorial Symposium in the Honor of Enrico Fermi." *Reviews of Modern Physics* 27(3):253.

Bethe, H. A., and E. E. Salpeter. 1957. *Quantum Mechanics of One- and Two-Electron Atoms.* New York: Academic Press.

Bethe, H. A. 1958. "Review of Junk's *Brighter than a Thousand Suns.*" *Bulletin of the Atomic Scientists* 12:426–428.

——— 1964. *Intermediate Quantum Mechanics.* Notes by R. W. Jackiw. New York: W. A. Benjamin.

——— 1967/1997. "Nobel Lecture." In Bethe 1997, 379–396.

——— 1968. "Energy Production in Stars." *Physics Today* 20(9):36–44.

Bethe, H. A., and R. W. Jackiw. 1968. *Intermediate Quantum Mechanics,* 2nd ed. New York: W. A. Benjamin.

Bethe, H. A. 1979. "The Happy Thirties." In Stuewer 1979, 9–31.

——— 1980. "Recollections of Solid State Theory, 1926–1933." In N. F. Mott, ed., *The Beginnings of Solid State Physics: A Symposium,* 49–51. London: Royal Society.

——— 1981. "Reminiscences on the Early Days of Electron Diffraction." In Goodman 1981.

Bethe, H. A., R. F. Bacher, and M. S. Livingston. 1986. *Basic Bethe: Seminal Articles on Nuclear Physics.* Introduction by R. H. Stuewer. New York: American Institute of Physics and Tomash Publishers.

Bethe, H. A., and G. Hildebrandt. 1988. "Peter Paul Ewald 1888–1985." *Biographical Memoirs of Fellows of the Royal Society* 34:135–176.

Bethe, H. A., P. A. M. Dirac, W. Heisenberg, E. P. Wigner, O. Klein, and E. M. Lifshitz. 1989. *From a Life of Physics.* Singapore: World Scientific.

Bethe, H. A. 1997. *Selected Works of Hans A. Bethe. With Commentary.* Singapore: World Scientific.

Bethe, H. A., and H. Bethe. 2002. "Enrico Fermi in Rome." *Physics Today* 55:28–29.

Bethe, H. A., and N. D. Mermin. 2004. "A Conversation about Solid State Physics." *Physics Today* 57/6:53–56.

Bethe, H. A. 2006. "My Life in Astrophysics." In Brown and Lee 2006, 27–44.

General Bibliography

Aaserud, F. 1990. *Redirecting Science: Niels Bohr, Philanthropy, and the Rise of Nuclear Physics.* Cambridge: Cambridge University Press.

Aitchison, I. J. R., D. A. MacManus, and T. M. Snyder. 2004. "Understanding Heisenberg's 'Magical' Paper of July 1925: A New Look at the Calculational Details." *American Journal of Physics* 72(11):1370–1379.

Amaldi, E. 1986. "A Few Flashes on Hans Bethe's Contributions during the Thirties." In Molinari and Ricci 1986, 1–16.

Amaldi, E., and E. Fermi. 1936. "On the Absorption and Diffusion of Slow Neutrons." *Physical Review* 50:899–928.

Anderson, P. W. 2004. *A Career in Theoretical Physics,* 2nd ed. Hackensack, NJ: World Scientific.

Andrade, E. N. da C. 1962. "In memoriam. William Henry Bragg." In Ewald 1962, 308–327.

Arendt, H. 1978. *The Jew as Pariah: Jewish Identity and Politics in the Modern Age,* ed. R. H. Feldman. New York: Grove.

——— 2000. "'What Remains? The Language Remains': A Conversation with

Günter Gaus." In P. Baehr, ed., *The Portable Hannah Arendt,* 3–24. New York: Penguin.

Atkinson, R. d'E. 1931. "Atomic Synthesis and Stellar Energy." *Astrophysical Journal* 73:250–347.

Atkinson, R. d'E., and F. G. Houtermans. 1929a. "Transmutation of Light Elements in Stars." *Nature* 123:567–582.

——— 1929b. "Zur Frage der Aufbaumöglicheit der Elemente in Sternen." *Zeitschrift für Physik* 54:656–665.

Bacciagaluppi, G., and A. Valentini. 2010. *Quantum Theory at the Crossroads.* Cambridge: Cambridge University Press.

Bacher, R. F. 1932. *Atomic Energy States, as Derived from the Analyses of Optical Spectra.* New York: McGraw-Hill.

Bacher, R. F., and V. Weisskopf. 1966. In Marshak 1966.

Bacon, F. 1980. *The Great Instauration.* Edited by J. Weinberger. Arlington Heights, Ill.: AHM Pub. Corp.

Bacon, G. E. 1966. *X-ray and Neutron Diffraction.* Oxford and New York: Pergamon Press.

Bahcall, J. N., and E. E. Salpeter. 2006. "Solar Energy Generation and Solar Neutrinos." In Brown and Lee 2006, 147–156.

Baldwin, J. M., ed. 1901–05. *Dictionary of Philosophy and Psychology.* New York: Macmillan.

Ball, R. S. 1902. *The Earth's Beginning.* New York: D. Appleton.

Barschall, H. H. 1952. "Regularities in the Total Cross Sections for Fast Neutrons." *Physical Review* 86:431.

Batchelor, M. T. 2006. "Bethe Ansatz." *Encyclopedia of Mathematical Physics,* 253–257. Boston: Elsevier.

——— 2007. "The Bethe Ansatz after 75 Years." *Physics Today* 60:36–40.

Bazerman, C. 1988. *Shaping Written Knowledge.* Madison, WI: University of Wisconsin Press.

Becker, C. 1943. *Cornell University: Founders and the Founding.* Ithaca: Cornell University Press.

Beeching, R. 1946. *Electron Diffraction,* 2nd ed. London: Methuen.

Bell, K. 2002. "Auden, Exile and Community." In S. Ouditt, ed., *Displaced Persons: Conditions of Exile in European Culture.* Aldershot, UK: Ashgate.

Beller, M. 1999. *Quantum Dialogue: The Making of a Revolution.* Chicago: University of Chicago Press.

Ben-David, J. 1991. *Scientific Growth: Essays on the Social Organization and Ethos of Science,* ed. and with an intro. by G. Freudenthal. Berkeley: University of California Press.

Berger, B., ed. 1989. *Shapers of Their Own Lives.* Berkeley: University of California Press.

Bernal, J. D. 1962. In Ewald 1962, 374–383.

Bernstein, J. 1980. *Hans Bethe, Prophet of Energy.* New York: Basic Books.

——— 2006. "Hans Bethe at the *New Yorker.*" In Brown and Lee 2006, 111–114.

Bethe, A. 1897a. "Das Nervensystem von *Carcinus Maenas*: Ein anatomisch-physiologischer Versuch." *Archiv für mikroskopische Anatomie und Entwicklungsmechanik* 50:460–546.

——— 1897b. "Vergleichende Untersuchungen über die Funktionen des Centralnervensystems der Arthropoden."

——— 1898. "Dürfen wir den Ameisen und Bienen psychologische Qualitäten zuschreiben?" *Pflugers Archiv für die gesamte Physiologie des Menschen und der Tiere* 70:15–99.

Bethe, A., and E. Fischer. 1931. "Die Anpassungsfähigkeit (Plastizität) des Nervenssystem." *Handbuch der normalen und pathologischen Physiolgie.* Bg. 15/2: 1045–1130 and 1175–1220.

Bethe-Kuhn, A. n.d. *Das Neugierige Sternlein. Ein Märchenspiel in 6 Bildern.* Musik von Hans Hermann. Berlin: Drei-Masten Verlag.

Beyerchen, A. D. 1979. *Scientists under Hitler: Politics and the Physics Community in the Third Reich.* New Haven: Yale University Press.

——— 1983. "Anti-Intellectualism and the Cultural Decapitation of Germany under the Nazis." In Jackman and Borden 1983, 29–45.

Bishop, M. 1962. *A History of Cornell.* Ithaca: Cornell University Press.

Bjerge, T., and C. H. Wescott. 1935. "On the Slowing Down of Neutrons in Various Substances Containing Hydrogen." *Proceedings of the Royal Society A* 150:709–728.

Blackett, P. M. S. 1933. "The Craft of Experimental Physics." In H. Wright, ed., *University Studies, Cambridge 1933.* London: Nicholson and Watson.

Bloch, F. 1976. "Reminiscences of Heisenberg and the Early Days of Quantum Mechanics." *Physics Today* 29(12):23–27.

Bohr, N. 1936. "Nuclear Capture and Nuclear Constitution." *Nature* 137:344–348.

——— 1984. *Collected Works,* vol. 5: *The Emergence of Quantum Mechanics (Mainly 1924–1926),* ed. Klaus Stozenburg. Amsterdam: North-Holland.

——— 1987. *Collected Works,* vol. 8: *The Penetration of Charged Particles through Matter (1912–1954),* ed. Jens Thorsen. Amsterdam: North-Holland.

Born, M. 1920. *Der Aufbau der Materie; drei Aufsätze über moderne Atomistik und Elektronentheorie.* Berlin: J. Springer.

——— 1923. *Atomtheorie des festen Zustandes (Dynamik der Kristallgitter)* 2. Aufl. Berlin: B.G. Teubner.

——— 1926a. "Quantummechanik der Stossvorgänge." *Zeitschrift für Physik* 38:803–827. English translation in Ludwig 1968 as "Quantum Mechanics."

——— 1926b. "Zur Quantummechanik der Stossvorgänge." *Zeitschrift für Physik* 37:863–867.

——— 1926c. "Zur Wellenmechanik der Stossvorgänge." *Göttingen Nachrichten,* 146–160.

——— 1927. "Physical Aspects of Quantum Mechanics." *Nature* 119:354–357.

——— 1933. *Optik; ein Lehrbuch der elektromagnetischen Lichttheorie.* Berlin: Springer.

——— 1952. "Arnold Johannes Wilhelm Sommerfeld 1868–1951." *Obituary Notices of Fellows of the Royal Society* 8(21):275–296.

——— 1968. *My Life and My Views.* New York: Charles Scribner's Sons.

——— 1978. *My Life. Recollections of a Nobel Laureate.* London: Taylor and Francis.

Born, M., and A. Einstein. 1971. *The Born-Einstein Letters,* trans. Irene Born. London: Macmillan.

Born, M., W. Heisenberg, and P. Jordan. 1925. "Zur Quantenmechanik II." *Zeitschrift für Physik* 35:557–615. English translation in van der Waerden 1968.

Born, M., and P. Jordan. 1925a. "Zur Quantenmechanik." *Zeitschrift für Physik* 34:858–888. English translation in van der Waerden 1968.

——— 1925b. "Zur Quantentheorie aperiodischer Vorgänge." *Zeitschrift für Physik* 33:479–505.

Born, M., and R. Oppenheimer. 1927. "Zur Quantentheorie der Moleküle." *Annalen der Physik* 84:457–484.

Born, M., and E. Wolf. 1959. *Principles of Optics; Electromagnetic Theory of Propagation, Interference, and Diffraction of Light,* with contributions by A. B. Bhatia and others. London: Pergamon.

——— 1964. *Principles of Optics,* 2nd rev. ed. New York: Pergamon.

Bourdieu, P. 1980. *The Logic of Practice.* Stanford, CA: Stanford University Press.

Brandt, F. R., and E. H. Hight. 1987. *German Expressionist Art: The Ludwig and Rosy Fischer Collection.* Seattle: Distributed by the University of Washington Press.

Breit, G., and E. Wigner. 1936. "Capture of Slow Neutrons." *Physical Review* 49:519–531.

Brenner, M., V. Caron, and U. R. Kaufmann. 2003. *Jewish Emancipation Reconsidered: The French and German Models.* Tübingen: Mohr Siebeck.

Bromley, A., and V. W. Hughes, eds. 1970. *Facets of Physics.* New York: Academic Press.

Brown, G. E. 2006a. "Hans Bethe and Astrophysical Theory." In Brown and Lee 2006, 175–184.

——— 2006b. "Hans Bethe and His Physics." In Brown and Lee 2006, 1–26.

Brown, G. E., and C. H. Lee. 2006. *Hans Bethe and His Physics.* Singapore: World Scientific.

Brown, G. E., and S. Lee. 2009. "Hans Bethe 1906–2005. A Biographical Memoir." Washington, D.C.: National Academy of Sciences.

Brown, L. M. 1985. "How Yukawa Arrived at the Meson Theory." *Progress of Theoretical Physics,* supp. 85:13–19.

——— 2002. "The Compton Effect as One Path to QED." *Studies in History and Philosophy of Science Part B: Studies in History and Philosophy of Modern Physics* 33(2):211–249.

Brown, L. M., M. Dresden, L. Hoddeson, and M. Riordan, eds. 1995. *The Standard Model.* Cambridge: Cambridge University Press.

Brown, L. M., and H. Rechenberg. 1988. "Nuclear Structure and Beta Decay (1932–1933)." *American Journal of Physics* 56/11:982–988.

——— 1991. "Quantum Field Theories, Nuclear Forces, and the Cosmic Rays (1934–1938)." *American Journal of Physics* 59:595–605.

——— 1996. *The Origin of the Concept of Nuclear Forces.* Philadelphia: Institute of Physics.

Bruford, W. H. 1959. "The Idea of Bildung in Wilhelm von Humboldt's Letters." In *The Era of Goethe: Essays Presented to James Boyd,* 17–46. Oxford: Basil Blackwell.

——— 1975. *The German Tradition of Self-Cultivation.* Cambridge: Cambridge University Press.

Bruner, J. 1992. *Acts of Meaning.* Cambridge, MA: Harvard University Press.

Brush, S. G. 1967. "History of the Lenz-Ising Model." *Reviews of Modern Physics* 39:883–893.

Buchdahl, G. 1969. *Metaphysics and the Philosophy of Science; the Classical Origins: Descartes to Kant.* Cambridge, MA: MIT Press.

Burchfield, J. D. 1990. *Lord Kelvin and the Age of the Earth,* with a new afterword. Chicago: University of Chicago Press.

Cahan, D., ed. 1985. "The Institutional Revolution in German Physics, 1865–1914." *Historical Studies in the Physical Sciences* 15(2):1–66.

———, ed. 1993. *Hermann von Helmholtz and the Foundations of Nineteenth-Century Science.* Berkeley: University of California Press.

———, ed. 2003. *From Natural Philosophy to the Sciences.* Chicago: University of Chicago Press.

Cao, T. Y. 1997. *Conceptual Developments of 20th Century Field Theories.* Cambridge: Cambridge University Press.

——— 2010. *From Current Algebra to Quantum Chromodynamics: A Case for Structural Realism.* Cambridge: Cambridge University Press.

Carson, C. 1996. "The Peculiar Notion of Exchange Forces. I. Origins in Quantum Mechanics, 1926–1928. Studies in History and Philosophy of Modern Physics 27/1:23–45.

Casimir, H. B. G. 1960. "Pauli and the Theory of the Solid State." In Fierz and Weisskopf 1960, 137–139.

——— 2010. *Haphazard Reality: Half a Century of Science.* Amsterdam: Amsterdam University Press.

Cassidy, D. 1981. "Cosmic Ray Showers, High Energy Physics, and Quantum Field

Theories: Programmatic Interactions in the 1930s." *Historical Studies in the Physical Sciences* 12:1–39.

——— 1992. *Uncertainty: The Life and Science of Werner Heisenberg.* New York: Freeman.

Chadwick, J. 1932. "The Existence of a Neutron." *Proceedings of the Royal Society of London A* 136:692–708.

Chadwick, J., and M. Golhaber. 1934. "A Nuclear Photo-effect: Disintegration of the Diplon by γ-rays." *Nature* 134:237–238.

——— 1935. "The Nuclear Photoelectric Effect." *Proceedings of the Royal Society of London A* 151:479–493.

Chandrasekhar, S. 1939. *An Introduction to the Study of Stellar Structure.* Chicago: University of Chicago Press.

——— 1944. "The Negative Ions of Hydrogen and Oxygen in Stellar Atmospheres." *Reviews of Modern Physics* 16(3–4):301–306.

——— 1951. "The Structure, the Composition and the Source of Energy in the Stars." In Hynek 1951, 598–674.

Chandrasekhar, S., G. Gamow, and M. A. Tuve. 1938. "The Problem of Stellar Energy." *Nature* 141(3578):982.

Charpa, U., and U. Deichmann. 2007. *Jews and Sciences in German Contexts.* Tübingen: Mohr Siebeck.

Coen, D. 2007. *Vienna in the Age of Uncertainty: Science, Liberalism, and Private Life.* Chicago: University of Chicago Press.

Conant, J. B. 1937. "The Role of the Endowed University in American Higher Education." Proceedings and Addresses at the Inauguration of Edmund Ezra Day, October 8, 1937. Ithaca: Printed for the University.

——— 1970. *My Several Lives: Memoirs of a Social Inventor.* New York: Harper and Row.

——— 2002. *Tuxedo Park.* New York: Simon and Schuster.

Corry, L. 1997a. "David Hilbert and the Axiomization of Physics." *Archive for the History of the Exact Sciences* 51:83–198.

——— 1997b. "Hermann Minkowski and the Postulate of Relativity." *Archive for the History of the Exact Sciences* 51:273–314.

——— 1998. "The Influence of David Hilbert and of Hermann Minkowski on Einstein's Views over the Interrelation between Physics and Mathematics." *Endeavour* 22(3):97–99.

——— 2003. *Modern Algebra and the Rise of Mathematical Structures,* 2nd rev. ed. Basel: Birkhäuser.

——— 2004. *David Hilbert and the Axiomatization of Physics.* Dordrecht: Kluwer Academic.

——— 2006. "Axiomatics, Empiricism, and *Anschauung* in Hilbert's Conception of Geometry: Between Arithmetic and General Relativity." In Ferreirós and Gray 2006, 133–156.

Courant, R., and D. Hilbert. 1924. *Methoden der matematische Physik.* Berlin: Springer.

Cowling, T. G. 1966. "The Development of the Theory of Stellar Structure." *Quarterly Journal of the Royal Astronomical Society* 7:121–137.

Craig, G. A. 1982. *The Germans.* New York: New American Library.

Cresti, C. 2003. "*Kultur* and *Civilisation* in the Franco-Prussian War." In Brenner, Caron, and Kaufmann 2003, 92–109.

Cruickshank, D. W. J., H. J. Juretschke, and N. Kato, eds. 1992. *P. P. Ewald and His Dynamical Theory of X-ray Diffraction: A Memorial Volume for Paul P. Ewald, 23 January 1988–22 August 1985.* Oxford: Oxford University Press.

Dalitz, R., and R. Peierls, eds. 1997. *Selected Scientific Papers of Sir Rudolf Peierls.* Singapore: World Scientific.

Darrigol, O. 1992. *From c-Numbers to q-Numbers: The Classical Analogy in the History of Quantum Theory.* Berkeley: University of California Press.

——— 1995. "Henri Poincaré's Criticism of *Fin de Siècle* Electrodynamics." *Studies in History and Philosophy of Modern Physics* 26(1):1–44.

Davis, E. A., ed. 1998. *Nevill Mott: Reminiscences and Appreciations.* London: Taylor & Francis.

Davisson, C., and L. Germer. 1927. "The Scattering of Electrons by a Single Crystal of Nickel." *Nature* 119:558–560.

Davisson, C., and C. H. Kunsman. 1923. "The Scattering of Low Speed Electrons by Platinum and Magnesium." *Physical Review* 22: 242–258.

Day, E. E. 1937. "Inaugural Address." Proceedings and Addresses at the Inauguration of Edmund Ezra Day, October 8, 1837. Ithaca: Printed for the University.

——— 1941. *The Defense of Freedom.* Ithaca: Cornell University Press.

Deák, I. 2009. "Heroes from Hungary." *New York Review of Books* 56(18) 24.

De Bont, R. 2009. "Between the Laboratory and the Deep Blue Sea: Space Issues in the Marine Stations at Naples and Wimereux." *Social Studies of Science* 39(2):199–227.

Debye, P., ed. 1928. *Probleme der modernen Physik; Arnold Sommerfeld zum 60. Geburtstage gewidmet von seinen Schülern.* Leipzig: Verlag von S. Hirzel.

Debye, P. 1930. *Elektroneninterferenzen.* Leipzig: Verlag von S. Hirzel.

Desmond, A., and Moore, J. 1991. *Darwin: The Life of a Tormented Evolutionist.* New York: Norton.

DeVane, W. C. 1965. *Higher Education in Twentieth Century America.* Cambridge, MA: Harvard University Press.

DeVorkin, D. 1984. "Stellar Evolution and the Origin of the Hertzsprung-Russell Diagram." In O. Gingerich, ed., *Astrophysics and Twentieth-century Astronomy to 1950,* pt. A, vol. 4: *The General History of Astronomy,* ed. M. Hoskin, 90–108. Cambridge: Cambridge University Press.

DeVorkin, D., and R. Kenat. 1983. "Quantum Physics and the Stars (II): Henri Norris Russell and the Abundances of the Elements in the Atmospheres of the Sun and Stars." *Journal of the History of Astronomy* 14:180–222.

Dirac, P. A. M. 1927a. "The Quantum Theory of the Emission and Absorption of Radiation." *Proceedings of the Royal Society of London A* 114:243–265.

——— 1927b. "Über die Quantenmechanik der Stossvorgänge." *Zeitschrift für Physik* 44:585–595.

——— 1928. "The Quantum Theory of the Electron." *Proceedings of the Royal Society (London) A* 117:610-624.

——— 1930a. "A Theory of Electrons and Protons." *Proceedings of the Royal Society (London) A* 117:610–624.

——— 1930b. "The Proton." *Nature* 26:605–606.

——— 1977. "Recollection of an Exciting Era." In C. Weiner, ed. *History of Twentieth Century Physics.* Proceedings of the International School of Physics "Enrico Fermi," Course 57, 109–146. New York: Academic Press.

——— 1984. "Blackett and the Positron." In Hendry 1984, 61–62.

Dirac, P., V. A. Fock, and B. Podolsky. 1932. "On Quantum Electrodynamics." *Physikalische Zeitschrift der Sowjetunion* 2:468–479.

Drell, S. 2006. "Shaping Public Policy." In Brown and Lee 2006, 251–262.

Dresden, M. 1987. *H. A. Kramers: Between Tradition and Revolution.* Berlin: Springer.

Duke, C. B. 1974. "The Characterization of Solid Surfaces"; "Elastic Low–Energy Electron Diffraction"; and "Surface Crystallography," all in F. O. Goodman, ed., *Dynamic Aspects of Surface Physics,* 52–210. Proceeedings of the International School of Physics "Enrico Fermi," Course 58. Bologna: Editrice Compositori.

Dunning, J. R., et al. 1935. "Interactions of Neutrons with Matter." *Physical Review* 48:265–280.

Dupré, J. S., and S. A. Lakoff. 1962. *Science and the Nation: Policy and Politics.* Englewood Cliffs, N.J.: Prentice-Hall.

Dyson, F. J. 2006. "Hans Bethe and Quantum Electrodynamics." In Brown and Lee 2006, 157–164.

Eckert, M. 1987. "Propaganda in Science: Sommerfeld and the Spread of the Electron Theory of Metals." *Historical Studies in the Physical and Biological Sciences* 18:191–233.

——— 1993. *Die Atomphysiker: eine Geschichte der theoretischen Physik am Beispiel der Sommerfeldschule.* Braunschweig: Vieweg.

——— 1999. "Mathematics, Experiments and Experimental Physics: The Early Days of the Sommerfeld School." *Physics in Perspective* 1:238–252.

——— 2003. "The Practical Theorist: Sommerfeld." *Philosophia Scientiæ* 7(2):167–188.

Eckert, M., and K. Märker, eds. 2003. *Arnold Sommerfeld: Wissenschaftlicher Briefwechsel,* vol. 2: *1919–1951.* Berlin: GNT-Verlag; Munich: Deutsches Museum.

Eckert, M., W. Pricha, H. Schubert, and G. Torkar. 1985. *Geheimrat Sommerfeld-Theoretischer Physiker. Eine Dokumentation aus seinem Nachlass.* Munich: Deutsches Museum.

Eckert, M., et al. 1984. *Geheimrat Sommerfeld, theoretischer Physiker: eine Dokumentation aus seinem Nachlass.* Munich: Deutsches Museum.

Eckert, M., et al. 1992. "The Roots of Solid-State Physics before Quantum Mechanics." In Hoddeson, Braun, et al. 1992, 3–87.

Eddington, A. S. 1926. *The Internal Constitution of the Stars.* Cambridge: Cambridge University Press.

Ehrenfeld, J. 2003. "Citizenship and Acculturation." In Brenner, Caron, and Kaufmann 2003, 155–167.

Einstein, A. [1933]. "The Method of Theoretical Physics." Herbert Spencer Lecture, Oxford. In A. Einstein, 1934, *Mein Weltbild.* Amsterdam: Querido Verlag.

——— 1949. "Remarks Concerning the Essays Brought Together in This Cooperative Volume." In Schilpp 1949, 665–688.

——— 1954. *Ideas and Opinions.* New York: Bonanza Books.

——— 1973. *Ideas and Opinions,* ed. Carl Seelig. London: Souvenir Press.

——— 1995–. *The Collected Papers of Albert Einstein.* Princeton: Princeton University Press.

Vol. 5: *The Swiss Years. Correspondence, 1902–1914,* trans. Anna Beck.

Vol. 7: *The Berlin Years: Writings, 1918–1921,* ed. Michael Janssen et al., 2002.

Vol. 9: *The Berlin Years: Correspondence, January 1919–April 1920,* ed. Diana Kormos Buchwald et al., 2004.

Elkana, Y. 1981. "A Programmatic Attempt at an Anthropology of Knowledge." In E. Mendelsohn and Y. Elkana, *Sciences and Cultures: Anthropological and Historical Studies of the Sciences,* 1–68. Dordrecht: Reidel.

Elon, A. 2002. *The Pity of It All: A Portrait of the German-Jewish Epoch, 1743–1933.* New York: Henry Holt.

Elsasser, W. M. 1925. "Bemerkungen zur Quantenmechanik freier Elektronen." *Naturwissenschaften* 13:711.

——— 1927. "Zur Theorie der Stossprozesse bei Wasserstoff." *Zeitschrift für Physik* 45:522–538.

——— 1933. "Sur le Principe de Pauli dans les Noyaux." *Le Journal de Physique et le Radium* 4:253–256; 5 (1934):389–397, 635–639.

——— 1935. "La Structure des Noyaux Atomiques, Complexes." *Annales de l'Institut Henri Poincaré* 5(3):223–262.

——— 1978. *Memoirs of a Physicist in the Atomic Age.* New York: Science History Publications.

Enz, C. 2002. *No Time to Be Brief: A Scientific Biography of Wolfgang Pauli.* Oxford: Oxford University Press.

Epstein, P. 1964–65. Interview with Paul S. Epstein by Alice Epstein. Caltech Archives 2004, available at http://oralhistories.library.caltech.edu/73/.

Ewald, P. P. 1916–18. "Zur Begründung der Kristalloprik," I, II, III. *Annalen der Physik* 49 (1916):1–38, 117–141; 54 (1918):519–556.

——— 1929. "Some Modern Developments of Wave Mechanics and Their Bearing on the Understanding of Crystal Structure." *Transactions of the Faraday Society* (London) 67:402–409.

——— 1962. *Fifty Years of X-ray Diffraction.* Published for the International Union of Crystallography. Utrecht: Oosthoek's Uitgeversmaatschappij.

——— 1968. "Personal Reminiscences." *Acta Crystallographica* A24:1–3. Reprinted in Cruickshank et al. 1992, 125–127.

——— 1969. "The Myth of Myths: Comments on P. Forman's Paper 'The Discovery of the Diffraction of X-rays by Crystals.'" *Archives for History of the Exact Sciences* 6:72–81.

Farmelo, G. 2009. *The Strangest Man: The Hidden Life of Paul Dirac, Mystic of the Atom.* New York: Basic Books.

Fehr, J., N. Jas, and I. Lowy, eds. 2009. *Penser avec Fleck: Investigating a Life Studying Life Sciences.* Zurich: Collegium Helveticum Heft 7.

Feist, R. 2004. "Metaphysics, Mathematics, and Pre-established Harmony." In W. Sweet, ed., *Approaches to Metaphysics,* 75–92. Dordrecht: Kluwer Academic.

Feldman, G. D. 1993. *The Great Disorder: Politics, Economics, and Society in the German Inflation, 1914–1924.* New York: Oxford University Press.

Fermi, E. 1926. "Argomenti pro e conto la ipotesi dei quanti di luce." *Nuovo Cimento* 3:47–54.

——— 1927. "Sul mechanismo dell' emissione nella meccanica ondulatoria." *Rend. Lincei* 5:795–800.

——— 1932. "Quantum Theory of Radiation." *Reviews of Modern Physics* 4:87–132.

——— 1934. "Versuch einer Theorie der β-Strahlen. I." *Zeitschrift für Physik* 88:161–171.

——— 1962. *Collected Works,* vol. 1: *Italy 1921–1938.* Chicago: University of Chicago Press.

Fermi, E., and E. Amaldi. 1935. "Sull'assorbimento dei Neutroni Lenti." *Ricerca Scientifica* 6:334–347.

Fermi, E., G. Breit, I. I. Rabi, E. P. Wigner, J. R. Oppenheimer, and J. H. van Vleck. 1941. *Nuclear Physics.* Philadelphia: University of Pennsylvania Press.

Ferreirós, J., and Gray, J. J. 2006. *The Architecture of Modern Mathematics: Essays in History and Philosophy.* New York: Oxford University Press.

Feshbach, H., C. E. Porter, and V. F. Weisskopf. 1954. "Model for Nuclear Reactions with Neutrons." *Physical Review* 96:448–464.

Fierz, M., and V. Weisskopf, eds. 1960. *Theoretical Physics in the Twentieth Century: A Memorial Volume to Wolfgang Pauli.* New York: Interscience.

Fischer, E. 1957. "Albrecht Bethe." *Ergebnisse der Physiologie Biologischen Chemie und Experimentallen Pharmakologie* 49:1–22.

Fishbane, M., and J. Glatzer Wechsler, eds. 1997. *The Memoirs of Nahum Glatzer.* Cincinnati: Hebrew Union College.

Forman, P. 1969. "The Discovery of the Diffraction of X-rays by Crystals: A Critique of the Myths." *Archives for History of the Exact Sciences* 6:38–71.

——— 1971. "Weimar Culture, Causality and Quantum Theory, 1918–1927: Adaptation of German Physicists and Mathematicians to a Hostile Intellectual Environment." *Historical Studies in the Physical Sciences* 3:1–115.

——— 1978. "The Reception of an Acausal Quantum Mechanics in Germany and Britain." In S. H. Mauskopf, ed., *The Reception of Unconventional Science,* 11–50. AAAS Selected Symposium 25. Boulder: Westview Press for AAAS.

——— 1987. "Review of Basic Bethe: Seminal Articles on Nuclear Physics, 1936–1937." *Isis* 78(3):293.

Forman, P., and A. Hermann. 1981. "Sommerfeld, Arnold." *Dictionary of Scientific Biography,* C. C. Gillespie, ed. in chief. New York: Scribner's Sons.

Forman, P., and J. Sánchez-Ron, eds. 1996. *National Military Establishments and the Advancement of Science and Technology.* Boston: Kluwer Academic.

Friedman, F. L., and V. Weisskopf. 1955. "The Compound Nucleus." In W. Pauli, ed., *Niels Bohr and the Development of Physics.* London: Pergamon Press.

Friedrich, W., P. Knipping, and M. Laue. 1912. "Interferenz-Erscheinungen bei Rötgenstrahlen." *Bayerische Akad. d. Wiss. zu München, Sitzungsber. math.-phys. Kl.,* 303–322. Reprinted in *Annalen der Physik* 41 (1913):971–988. English translation in Bacon 1966, 89–108.

Friess, H. L. 1981. *Felix Adler and Ethical Culture: Memories and Studies.* New York: Columbia University Press.

Fröhlich, H., W. Heitler, and N. Kemmer. 1938. "On the Nuclear Forces and the Magnetic Moments of the Neutron and Proton." *Proceedings of the Royal Society of London A* 166:154.

Fruton, J. 1990. *Contrasts in Scientific Styles: Research Groups in the Chemical and Biochemical Sciences.* Memoirs of the American Philosophical Society 191. Philadelphia: American Philosophical Society.

Funkenstein, A. 1986. *Theology and the Scientific Imagination from the Middle Ages to the Seventeenth Century.* Princeton: Princeton University Press.

Galison, P. 1979. "Minkowski's Space-Time: From Visual Thinking to the Absolute World." *Historical Studies in the Physical Sciences* 10:85–121.

——— 1983. "The Discovery of the Muon and the Failed Revolution against Quantum Electrodynamics." *Centaurus* 26:262–316.

Gamow, G. 1928. "Zur Quantentheorie der des Atomkerns." *Zeitschrift für Physik* 51:204–212.

——— 1931. *Constitution of Atomic Nuclei and Radioactivity.* Oxford: Clarendon Press.

——— 1935. Nuclear Transformation and the Origin of the Chemical Elements. *Ohio Journal of Science* 35:406–413.

——— 1937a. *Structure of Atomic Nuclei and Nuclear Transformations.* Oxford: Clarendon Press.

——— 1937b. "Nuclear Energy Sources and Stellar Evolution." *Physical Review* 53:595–604.

——— 1938. "L'évolution des étoiles du point de vue de la physique moderne." *Annales de l'Institut Henri Poincaré* 8:193–211.

——— 1947. *Atomic Energy in Cosmic and Human Life: Fifty Years of Radioactivity.* Cambridge: Cambridge University Press.

——— 1956. In A. Beers, ed., *Vistas in Astronomy,* vol. 2. New York: Pergamon.

——— 1970. *My World Line: An Informal Biography.* New York: Viking.

Gamow, G., and F. G. Houtermans. 1929. "Zur Quantenmechanik des Radioactiven Kerns." *Zeitschrift für Physik* 52:496–509.

Gamow, G., and E. Teller. 1938. "The Rate of Selective Thermonuclear Reactions." *Physical Review* 53(7):608–609.

Garwin, R., and von Hippel, F. 2006. "In Memoriam: Hans Bethe." In Brown and Lee 2006, 273–278.

Gaudin, M. 1983. *La Fonction d'Onde de Bethe.* Paris: Masson.

Gavroglu, K. 1995. *Fritz London: A Scientific Biography.* Cambridge: Cambridge University Press.

Gehrenbeck, R. K. 1974. "C. J. Davisson, L. H. Germer, and the Discovery of Electron Diffraction." Ph.D. diss., University of Minnesota, 1974.

——— 1978. "Electron Diffraction: Fifty Years Ago." *Physics Today* 31(1):34–41.

——— 1981. "Davisson and Germer." In Goodman 1981, 12–27.

Geison, G. L., and F. L. Holmes. 1993. "Research Schools: Historical Reappraisals." *Osiris* 2d series. Volume 8.

Geuss, R. 1996. "Kultur, Bildung, Geist." *History and Theory* 35(2):151–164.

Giamarchi, T. 2004. *Quantum physics in one dimension.* New York: Oxford University Press.

Gilman, S. L. 1986. *Jewish Self-hatred: Anti-Semitism and the Hidden Language of the Jews.* Baltimore: Johns Hopkins University Press.

Gilman, S., and J. Zipes. 1997. *Yale Companion to Jewish Writing and Thought in German Culture 1096–1996.* New Haven: Yale University Press.

Gingerich, O. 1995. "Report on the Progress in Stellar Evolution to 1950." In P. C. van der Kruit and G. Gilmore, *Stellar Populations.* Dordrecht: Kluwer Academic Press.

Glatzer, N. 2009. *Essays in Jewish Thought.* Tuscaloosa: University of Alabama Press.

Glazer, N. 1972. *American Judaism,* 2nd ed. Chicago: University of Chicago Press.

Golden, W. T. 1986. "President's Scientific Advisory Board Revisited." *Science, Technology, and Human Values* 11(2):5–19.

Goldhaber, 1979. "The Nuclear Photoeffect." In Stuewer 1979, 81–110.

Goldschmidt, R. 1956. *Portraits from Memory: Recollections of a Zoologist.* Seattle: University of Washington Press.

——— 1960. *In and Out of the Ivory Tower: The Autobiography of Richard B. Goldschmidt.* Seattle: University of Washington Press.

Goodman, P., ed. 1981. *Fifty Years of Electron Diffraction.* Dordrecht: D. Reidel, for the International Union of Crystallography.

Gordin, M. D. 2004. *A Well-Ordered Thing: Dmitrii Mendeleev and the Shadow of the Periodic Table.* New York: Basic Books.

Gordon, P. E. 2003. *Rosenzweig and Heidegger: Between Judaism and German Philosophy.* Berkeley: University of California Press.

Gottfried, K. 2006a. "Hans Bethe." In Brown and Lee 2006, 125–130.

——— 2006b. "Obituary: Hans Bethe." In Brown and Lee 2006, 279–282.

——— 2011. "P. A. M. Dirac and the Discovery of Quantum Mechanics." *American Journal of Physics* 79(3):261–268.

Gould, S. J. 1977. *Ontogeny and Phylogeny.* Cambridge, MA: Belknap Press of Harvard University Press.

Graetz, L. 1920. *Die Atomtheorie in ihrer neusten Entwicklung. Sechs Vortrege.* Stuttgart: J. Engelhorns.

Gratton, L. 1963a. "Stellar Evolution. Introductory Lecture." In Gratton 1963b, 1–10.

———, ed. 1963b. *Star Evolution.* Proceedings of the International School of Physics "Enrico Fermi," Course 28. New York: Academic Press.

Gray, J. 2008. *Plato's Ghost: The Modernist Transformation of Mathematics.* Princeton: Princeton University Press.

Greene, B. 2006. "The Universe on a String." *New York Times,* Op-Ed page, October 20.

Greene, J. C. 1959. *The Death of Adam: Evolution and Its Impact on Western Thought.* Ames: Iowa State University Press.

Greene, M. 2007. "Writing Scientific Biography." *Journal of the History of Biology* 40:727–759.

Greenspan, N. T. 2005. *The End of the Certain World: The Life and Science of Max Born.* New York: Basic Books.

Greenstein, J. L. 1997. "Lee Alvin DuBridge." *Biographical Memoirs of the National Academy of Sciences* 72:88–113.

Grunfeld, F. W. 1979. *Prophets without Honour: A Background to Freud, Kafka, Einstein and Their World.* New York: McGraw-Hill.

Guggenheimer, K. 1934. "Remarques sur la Constitution des Noyaux Atomiques." *Le Journal de Physique et le Radium* 5:253–256, 475–485.

Gurney, R. W., and E. U. Condon. 1929. "Wave Mechanics and Radioactive Disintegration." *Physical Review* 33:127–140.

Hacohen, M. H. 2000. *Karl Popper: The Formative Years, 1902–1945: Politics and Philosophy in Interwar Vienna.* Cambridge: Cambridge University Press.

Hagner, M. 2003. "Scientific Medicine." In Cahan 2003, 49–87.

Hale, G. E. 1908. *The Study of Stellar Evolution.* Chicago: University of Chicago Press.

Hall, K. 1999. "Purely Practical Revolutionaries: A History of Stalinist Theoretical Physics." Ph.D. diss., Department of the History of Science, Harvard University.

Hargittai, I. 2010. *Judging Edward Teller.* Amherst, NY: Prometheus Press.

Harper, E., W. C. Parke, and G. D. Anderson. 1997. *The George Gamov Symposium.* San Francisco: Astronomical Society of the Pacific.

Harrington, A. 1996. *Reenchanted Science: Holism in German Culture from Wilhelm II to Hitler.* Princeton: Princeton University Press.

Hartman, P. 1993. *The Cornell Physics Department: Recollections and a History of Sorts.* Ithaca: s.n.

Hartree, D. R. 1928. "The Wave Mechanics of an Atom with a non-Coulomb Central Field." Parts I, II, and III. *Proceedings of the Cambridge Philosophical Society* 24:89; 24:111; 24:426.

——— 1957. *The Calculation of Atomic Structures.* New York: Wiley.

Harwood, J. 1987. "National Styles in Science: Genetics in Germany in the United States between the World Wars." *Isis* 78:390–414.

——— 1993. *Styles of Scientific Thought: The German Genetics Community, 1900–1933.* Chicago: University of Chicago Press.

Heilbron, J. L. 1986. *The Dilemmas of an Upright Man.* Berkeley: University of California Press.

Heilbron, J. L., and R. Seidel. 1989. *Lawrence and His Laboratory: A History of the Berkeley Laboratory,* vol. 1. Berkeley: University of California Press.

Heisenberg, W. 1925. "Über quantentheoretische Umdeutung kinematischer und mechanischer Beziehungen" [Quantum-Theoretical Re-interpretation of Kinematic and Mechanical Relations]. *Zeitschrift für Physik* 33:879–893. English translation in van der Waerden 1968.

——— 1932. "Über den Bau der Atomkerne. I." *Zeitschrift für Physik* 77:1–11.

——— 1974. *Across the Frontiers,* trans. Peter Heath. New York: Harper & Row.

Heitler, W. 1936. *The Quantum Theory of Radiation.* Oxford: Clarendon Press.

Helmholtz, H. von. 1995. *Science and Culture: Popular and Philosophical Essays,* ed. and with an intro. by David Cahan. Chicago: University of Chicago Press.

Hendry, J., ed. 1984. *Cambridge Physics in the Thirties.* Bristol: Adam Hilger.

Hershberg, J. 1993. *James B. Conant.* New York: Harper and Row.

Hesse, H. 1951. *Siddhartha.* New York: New Directions.

Hiebert, E. 1984. "Einstein's Image of Himself as Philosopher of Science." In E. Mendelsohn, ed., *Transformation and Tradition in the Sciences: Essays in Honor of I. Bernard Cohen,* 175–190. Cambridge: Cambridge University Press.

——— 1988. "The Role of Experiment and Theory in the Development of Nuclear Physics in the Early 1930s." In D. Batens and J. P. van Bendegem, eds., *Theory and Experiment: Recent Insights and New Perspectives on Their Relation,* 55–76. Dordrecht: D. Reidel.

Hoddeson, L. H. 2002. *True Genius: The Life and Science of John Bardeen: The Only Winner of Two Nobel Prizes in Physics.* Washington, D.C.: Joseph Henry Press.

Hoddeson, L. H., and G. Baym. 1980. "The Development of the Quantum Mechanical Electron Theory of Metals: 1900–28." In Mott 1980, 8–23.

Hoddeson, L. H., G. Baym, and M. Eckert. 1992. "The Development of the Quantum Mechanical Electron Theory of Metals: 1926–33." In Hoddeson, Braun, et al. 1992, 88–181.

Hoddeson, L. H., E. Braun, J. Teichmann, and S. Weart. 1992. *Out of the Crystal Maze: Chapters from the History of Solid-State Physics.* New York: Oxford University Press.

Hofstadter, R., and Metzger, W. P. 1955. *The Development of Academic Freedom in the United States.* New York: Columbia University Press.

Holborn, H. 1964. *A History of Modern Germany, 1648–1840.* Princeton: Princeton University Press.

——— 1969. *A History of Modern Germany, 1840–1945.* New York: A. A. Knopf.

Holt, J., and G. E. Brown. 2006. "Hans Bethe and the Nuclear Many-Body Problem." In Brown and Lee 2006, 201–238.

Holton, G. 1974. "Striking Gold in Science: Fermi's Group and the Recapture of Italy's Place in Physics." *Minerva* 12:59–198.

——— 1978. *The Scientific Imagination: Case Studies.* Cambridge: Cambridge University Press.

——— 1983. "The Migration of the Physicists to the United States." In Jackman and Borden 1983, 169–188.

Holts, R. 1955. "Albrecht Bethe." *Münchener Medizinische Wochenschrift* 97/21 (Seile 707–708):1–6.

Hore, P., ed. 2003. *Patrick Blackett: Sailor, Scientist and Socialist.* London: Frank Cass.

Hoyle, F. 1994. *Home Is Where the Winds Blows: Chapter from a Cosmologist's Life.* Mill Valley, CA: University Science Books.

Hufbauer, K. 1981. "Astronomers Take Up the Stellar-Energy Problem." *Historical Studies in the Physical Sciences* 11(2):277–303.

——— 2000. "Hans Bethe's Breakthrough Solution of the Stellar-Energy Problem." Unpublished manuscript.

Hughes, J. 1998. "Modernists with a Vengeance: Changing Cultures of Theory in Nuclear Science, 1920–1930." *Studies in History and Philosophy of Science. Part B: Studies in History and Philosophy of Modern Physics* 29:339–367.

——— 2002a. "Craftsmanship and Social Service: W. H. Bragg and the Modern Royal Institution." In F. James, ed., *"The Common Purposes of Life": Essays on the History of the Royal Institution of Great Britain, 1799–1999,* 225–247. Aldershot, UK: Ashgate.

——— 2002b. "Radioactivity and Nuclear Physics." In M. J. Nye, ed., *The Cambridge History of Science,* vol. 5: *The Modern Physical and Mathematical Sciences,* 350–374. Cambridge: Cambridge University Press.

——— 2005. "Redefining the Context: Oxford and the Wider World of British Physics, 1900–1940." In R. Fox and G. Gooday, eds. *Physics in Oxford 1839–1939: Laboratory, Learning and College Life.* Oxford: Oxford University Press.

Hutchins, E. 1995. *Cognition in the Wild.* Cambridge, MA: MIT Press.

Hynek, J. A., ed. 1951. *Astrophysics: A Topical Symposium Commemorating the Fiftieth Anniversary of the Yerkes Observatory and a Half Century of Progress in Astrophysics.* New York: McGraw-Hill.

Ioffe, B. 2006. "Hans Bethe and the Global Energy Problem." In Brown and Lee 2006, 263–272.

Jackman, J. C., and C. M. Borden, eds. 1983. *The Muses Flee Hitler: Cultural Transfer and Adaptation 1930–1945.* Washington, D.C.: Smithsonian Institution Press.

Jackson, D., and K. Gottfried. 2003. "Victor Frederick Weisskopf 1908–2002." *Biographical Memoirs of the National Academy of Sciences* 84:1–27.

James, R. W. 1930. *X-ray Crystallography.* New York: Dutton.

James, W. 1902. *Talks to Teachers on Psychology: And to Students on Some of Life's Ideals.* New York: Henry Holt.

Jammer, M. 1966. *The Conceptual Development of Quantum Mechanics.* New York: McGraw-Hill.

Janik, A., and S. Toulmin. 1973. *Wittgenstein's Vienna.* New York: Simon and Schuster.

Jasanoff, S. 2007. "A Splintered Function: Fate, Faith, and the Father of the Atomic Bomb." *Metascience* 17(3):51–356.

Jolley, N. 1995. *The Cambridge Companion to Leibniz.* Cambridge: Cambridge University Press.

Josephson, P. R. 1987. "Early Years of Soviet Nuclear Physics." *Bulletin of Atomic Scientists* 43(10):16–28.

Jungnickel, C., and R. McCormmach. 1986. *The Intellectual Mastery of Nature: Theoretical Physics from Ohm to Einstein,* 2 vols. Chicago: University of Chicago Press.

Kafka, F. 1974. *I Am a Memory Come Alive: Autobiographical Writings by Franz Kafka,* ed. N. Glatzer. New York: Schocken Books.

Kahler, E. von. 1989. "For Science—Against the Intellectuals among Its Despisers." In Lassman and Velody 1989a.

Kant, I. 1755. *Allgemeine Naturgeschichte und Theorie des Himmels.* Königsberg: Johann Friederich Petersen.

——— 1933. *Immanuel Kant's Critique of Pure Reason,* trans. N. K. Smith. London: Macmillan.

——— 1998. *Groundwork of the Metaphysics of Morals,* trans. and ed. M. Gregor, with an intro. by C. M. Korsgaard. Cambridge: Cambridge University Press.

Kaplan, M. 1997. "1812. The German Romance with *Bildung* Begins." In Gilman and Zipes 1997, 124–129.

———, ed. 2005. *Jewish Daily Life in Germany: 1618–1945*. Oxford: Oxford University Press.

Katz, J. 1986. *Jewish Emancipation and Self-Emancipation*. Philadelphia: Jewish Publication Society.

Kauders, A. 2009. "Weimar Jewry." In A. McElligott, ed., *Weimar Germany*. Oxford: Oxford University Press.

Kaufmann, U. R. 2003. In Brenner, Caron, and Kaufmann 2003.

Kemmer, N. 1938. "Quantum Theory of Einstein-Bose Particles and Nuclear Interactions." *Proceedings of the Royal Society of London A* 166:127–153.

——— 1939. "The Particle Aspect of Meson Theory." *Proceedings of the Royal Society of London A* 173:91–116.

Kevles, D. 1978. *The Physicists*. New York: A. Knopf.

Killian, J. 1962. "Introduction." In J. S. Dupré, A. Sanford, and S. A. Lakoff, *Science and the Nation: Policy and Politics*. Englewood Cliffs, NJ: Prentice-Hall.

Kim, D. W. 2002. *Leadership and Creativity: A History of the Cavendish Laboratory, 1871–1919*. Dordrecht: Kluwer.

Kirkpatrick, P. 1949. "Address Presenting the 1948 Oersted Medal to Arnold Sommerfeld." *American Journal of Physics* 17:312–314.

Kistiakowsky, G. B. 1976. *A Scientist at the White House*. With an intro. by Charles S. Maier. Cambridge, MA: Harvard University Press.

Klein, F. 1899. "Über die Neueinrichtungen für Elektroteknik und allgemeine technische Physik an der Universität Göttingen." *Physikalische Zeitschrift* 1:143–145.

Klein, M. 1964. "Einstein and the Wave-Particle Duality." *The Natural Philosopher* 3:1–49.

Knorr Cetina, Karin. 1999. *Epistemic Cultures: How the Sciences Make Knowledge*. Cambridge, MA: Harvard University Press.

Kohler, M. J. 1918. *Jewish Rights at the Congresses of Vienna (1814–1815) and Aix-la-Chapelle (1818)*. New York: American Jewish Committee.

Kragh, H. 1979. "On the History of Early Wave Mechanics." *Centaurus* 26:154–197.

——— 1990. *Dirac: A Scientific Biography*. Cambridge: Cambridge University Press.

——— 1992. "Relativistic Collisions: The Work of Christian Møller in the Early 1930s." *Archive for History of Exact Sciences* 43:299–328.

——— 1999. *Quantum Generations: A History of Physics in the Twentieth Century*. Princeton: Princeton University Press.

Kramers, H. 1927. "La diffusion de la lumière par les atomes." *Atti. Cong. Internationa Fisica Como* [Transactions of Volta Centenary Congress] 2:545–557.

Kraut, B. 1979. *From Reform Judaism to Ethical Culture: the Religious Evolution of Felix Adler*. Cincinnati: Hebrew Union College Press.

Kremer, R. 1990. *The Thermodynamics of Life and Experimental Physiology*. New York: Garland.

Krieger, M. H. 1992. *Doing Physics: How Physicists Take Hold of the World.* Bloomington: Indiana University Press.

Kriplovich, I. B. 1992. "The Eventful Life of Fritz Houtermans." *Physics Today* 45(7):29–37.

Kuhn, T. S. 1962. *The Structure of Scientific Revolutions.* Chicago: University of Chicago Press.

——— 1967. *Sources for the History of Quantum Physics.* Philadelphia: American Philosophical Society.

Lamb, W. E., and L. Schiff. 1938. "On the Electromagnetic Properties of Nuclear Systems." *Physical Review* 53:651.

Laplace, P. S. 1813. *Exposition du Système du Monde.* Paris: Mme ve Courcier.

——— 1951. *A Philosophical Essay on Probabilities.* New York: Dover.

Lassman, P., and E. Velody. 1989a. *Max Weber's "Science as a Vocation,"* London: Unwin Hyman.

Lassman, P., and E. Velody, with H. Martins. 1989. "Max Weber on Science, Disenchantment and the Search for Meaning." In Lassman and Velody, 1989a, 159–204.

Latour, B. 2004. "Why Has Critique Run Out of Steam? From Matters of Fact to Matters of Concern." *Critical Inquiry* 30(2):225–248.

Laue, M. von. 1931. "The Diffraction of an Electron-Wave at a Single Layer of Atoms." *Physical Review* 37:53–59.

——— 1948. *Materiewellen und ihre Interferenzen.* Leipzig: Akademische Verlagsgesellschaft Geest und Portig.

Laurikainen, K. V. 1988. *Beyond the Atom: The Philosophical Thought of Wolfgang Pauli.* New York: Springer-Verlag.

Lee, S. 2007. *The Bethe-Peierls Correspondence.* Singapore: World Scientific.

Leibniz, G. W. 1965. *Monadology and Other Philosophical Essays,* trans. P. Schrecker and A. M. Schrecker. Indianapolis: Bobbs-Merrill.

——— 1967. *The Leibniz-Arnauld Correspondence,* trans. H. T. Mason, intro. by C. H. R. Parkinson. Manchester: Manchester University Press.

Lenoir, T. 1997. *Instituting Science: The Cultural Production of Scientific Disciplines.* Stanford: Stanford University Press.

——— 1998. "Revolution from Above: The Role of the State in Creating the German Research System, 1810–1910." *American Economic Review* 88(2):22–27.

Lewontin, R. C. 2001. *The Triple Helix: Gene, Organism, and Environment.* Cambridge, MA: Harvard University Press.

Livingston, M. S. 1969. *Particle Accelerators: A Brief History.* Cambridge, MA: Harvard University Press.

Lorentz, H. A. 1916. *The Theory of Electrons and Its Applications to the Phenomena of Light and Radiant Heat,* 2nd ed. New York: G. E. Stechert.

———, ed. 1923. *The Principle of Relativity: A Collection of Original Memoirs on the Special and General Theory of Relativity, by H. A. Lorentz, A. Einstein, H.*

Minkowski and H. Weyl, with notes by A. Sommerfeld, tr. by W. Perrett and G. B. Jeffery. New York: Dodd.
Lowenstein, S. M. 1997. "Jewish Participation in German Culture." In Meyer and Brenner 1997, 305–335.
Ludloff, H. 1931. "Zur Frage der Nullpunktentropie des festen Körper von Standpunkt der Quantenstatistik." *Zeitschrift der Physik* 68(7–8):433–492.
Ludwig, G., ed. 1968. *Wave Mechanics.* New York: Pergamon.
Lurier, J. 1980. "The Physics Department at Cornell between World Wars." B.A. thesis, Department of History, Cornell University.
MacMillan, M. 2002. *Paris 1919: Six Months that Changed the World.* New York: Random House.
Madelung, E. 1922. *Die mathematischen Hilfsmittel des Physikers.* Berlin: J. Springer.
Magueijo, J. 2009. *A Brilliant Darkness: The Extraordinary Life of Ettore Majorana.* New York: Basic Books.
Mahaux, C., and H. A. Weidenmuller. 1979. "Recent Developments in Compound-Nucleus Theory." *Annual Review of Nuclear and Particle Science* 29:1–31.
Majorana, E. 1933. "Über die Kerntheorie." *Zeitschrift für Physik* 82:137–145.
Malament, D. B., ed. 2002. *Reading Natural Philosophy: Essays in the History and Philosophy of Science and Mathematics to Honor Howard Stein on His 70th Birthday.* La Salle, IL: Open Court.
Marquardt, M. 1951. *Paul Ehrlich.* New York: Henry Schuman.
Marshak, R. E. 1939. "Contributions to the Theory of the Internal Constitution of Stars." Ph.D. diss., Physics Department, Cornell University.
——— 1966. *Perspectives on Modern Physics: Essays in Honor of Hans A. Bethe.* New York: Interscience Publishers.
Marshak, R., and Bethe, H. A. 1939. "Physics of Stellar Interiors and Stellar Evolution." *Reports on Progress in Physics* 6:1–15.
Martin, A. 1998. "Stability of Three- and Four-body Coulomb Systems." *Ukrainskij fizičeskij žurnal* 43(6–7):680–687.
Mayer, J. P. 1944. *Max Weber and German Politics.* London: Faber and Faber. Repr. London: Routledge, 1998.
McClelland, C. 1980. *State, Society, and University in Germany, 1700–1914.* New York: Cambridge University Press.
McMillan, E. 1979. "Early History of Particle Accelerators." In Stuewer 1979, 111–156.
Mehra, J. 1975. *The Solvay Conferences on Physics: Aspects of Development of Physics since 1911.* Dordrecht: D. Reidel.
Mehra, J., and H. Rechenberg. 1982–86. *The Historical Development of Quantum Theory,* 6 vols. New York: Springer.
——— 1987. *Erwin Schrödinger and the Rise of Wave Mechanics.* New York: Springer-Verlag.

Mendes-Flohr, P., and J. Reinhartz. 1980. *The Jew in the Modern World: A Documentary History.* New York: Oxford University Press.
Mermin, N. D., and N. W. Ashcroft. 2006. "Hans Bethe's Contribution to Solid-State Physics." In Brown and Lee 2006, 189–200.
Merton, R. K. 1968. "The Matthew Effect in Science." *Science* 159 (3810):56–63.
——— 1988. "The Matthew Effect in Science, II: Cumulative Advantage and the Symbolism of Intellectual Property." *Isis* 79:606–623.
Merz, J. T. 1907. *Leibniz.* Edinburgh: William Blackwood and Sons.
——— 1965. *The History of European Scientific Thought in the Nineteenth Century,* 2 vols. New York: Dover. Previously published as J. T. Merz. 1896–1914. *A History of European Thought in the Nineteenth Century,* 4 vols. Edinburgh: W. Blackwood.
Meyer, M. A. 1988. *Response to Modernity: A History of the Reform Movement in Judaism.* New York: Oxford University Press.
——— 2001. *Judaism within Modernity: Essays on Jewish History and Religion.* Detroit: Wayne State University Press.
Meyer, M. A., and M. Brenner, eds. 1997. *German-Jewish History in Modern Times,* vol. 3: *Integration in Dispute 1871–1918.* New York: Columbia University Press.
Mildenberger, F. 2006. "The Beer/Bethe/Uexküll Paper (1899) and Misinterpretations Surrounding 'Vitalistic Behaviorism.'" *History and Philosophy of the Life Sciences* 28(2):175–189.
Minkowski, H. 1915. "Das Relativitätprinzip [1907]." *Annalen der Physik* 47:927–938. Quoted in Corry 1997b.
——— 1923. "Space and Time." In Lorentz 1923, 75–91.
Mitzman, A. 1985. *The Iron Cage: A Historical Interpretation of Max Weber.* New Brunswick, NJ: Transaction Books.
Mladjenović, M. 1998. *The Defining Years in Nuclear Physics, 1932–1960s.* Philadelphia: Institute of Physics.
Moiseiwitsch, B. L. 1966. *Variational Principles.* New York: Interscience.
Molinari, A., and R. A. Ricci, eds. 1986. *From Nuclei to Stars.* A meeting in nuclear physics and astrophysics exploring the path opened by Bethe. Course XCI of the International School of Physics "Enrico Fermi." Amsterdam: North Holland.
Moon, P. B. 1992. "The (London) Physics Club, 1928–1953." *Notes and Records of the Royal Society of London* 46(1):171–174.
Moore, J. R. 1979. *The Post-Darwinain Controversies.* Cambridge: Cambridge University Press.
Moore, W. J. 1989. *Schrödinger, Life and Thought.* Cambridge: Cambridge University Press.
Morgan, M., and M. Morrison, eds. 1999. *Models as Mediators: Essays on the Philosophy of the Natural and Social Sciences.* Cambridge: Cambridge University Press.
Morrison, M. 1998. "Modelling Nature: Between Physics and the Physical World." *Philosophia Naturalis* 38:65–85.

——— 1999. "Models as Autonomous Agents." In Morgan and Morrison 1999, 38–65.

Morrison, M., and M. Morgan. 1999. "Models as Mediating Instruments." In Morgan and Morrison 1999, 10–37.

Morse, P. N. 1976. "Edward Uhler Condon." *Biographical Memoirs of the National Academy of Sciences* 48:124–151.

Mosse, G. L. 1985. *German Jews beyond Judaism.* Bloomington: Indiana University Press.

Mott, N. F. 1928. "The Solution of the Wave Equation for the Scattering of Particles by a Coulombian Center of Force." *Proceedings of the Royal Society of London A* 118:542–549.

——— 1929a. "The Scattering of Fast Electrons by Atomic Nuclei." *Proceedings of the Royal Society of London A* 124:425–442.

——— 1929b. "The Exclusion Principles and Aperiodic Systems." *Proceedings of the Royal Society of London A* 125:222–230.

——— 1929c. "The Wave Mechanics of α-Ray Tracks." *Proceedings of the Royal Society of London A* 126:79–84.

——— 1930a. *An Outline of Wave Mechanics.* Cambridge: Cambridge University Press.

——— 1930b. "The Collision between Two Electrons." *Proceedings of the Royal Society of London A* 126:259–267.

——— 1980. *The Beginnings of Solid State Physics: A Symposium.* Organized by Sir Nevill Mott, April 30–May 2, 1979. London: Royal Society.

——— 1984. "Theory and Experiment at the Cavendish circa 1932." In Hendry 1984, 125–132.

——— 1987. *A Life in Science.* London: Taylor and Francis.

Mott, N. F., and H. Jones. 1936. *The Theory of the Properties of Metals and Alloys.* Oxford: Clarendon Press.

Mott, N. F., and H. S. S. Massey. 1933. *The Theory of Atomic Collisions.* Oxford: Clarendon Press.

Mottelson, B. 1986. "The Study of the Nucleus as a Theme in Contemporary Physics." In J. de Boer, E. Dal, and O. Ulfbeck, eds., *The Lesson of Quantum Theory,* 79–98. Amsterdam: North-Holland, for the Royal Danish Academy of Sciences and Letters.

Mottelson, B. R. 2008. "Niels Bohr and the Development of Concepts in Nuclear Physics." Nishina Memorial Lectures. *Lecture Notes in Physics* 746:115–136. Berlin: Springer.

Muto, T., and Y. Takagi. 1955. "The Theory of Order-Disorder Transitions in Alloys." In F. Seitz and D. Turnbull, eds., *Solid State Physics,* vol. 1. New York: Academic Press.

Neddermeyer, S. H., and C. D. Anderson. 1937. "Note on the Nature of Cosmic Ray Particles." *Physical Review* 51:884–886.

Negele, J. W. 2006. "Hans Bethe and the Theory of Nuclear Matter." In Brown and Lee 2006, 165–174.

Nernst, W., and A. Schönflies. 1919. *Einführung in die mathematische Behandlung der Naturwissenschaften: Kurzgefasstes Lehrbuch der Differential- und Integralrechnung mit besonderer Berücksichtigung der Chemie.* Munich: R. Oldenbourg.

Neugarten, T. 1919. "Die Einfluss der H-Ionenkonzentration und die Phosphosäure auf Erregbarkeit und Leistungfähigkeit der Musklen." *Pflügers Archiv für gesamte Physiologie* 175(1–2):94–108.

Nietzsche, F. 1983. *Untimely Meditations.* Translated by R. J. Hollingdale with an introduction by J. P. Stern. Cambridge: Cambridge University Press.

Nix, F., and W. Shockley. 1938. "Order-Disorder Transitions in Alloys." *Reviews of Modern Physics* 10:1–71.

Noren, C. 1976. *The Camera of My Family.* New York: Knopf.

Nye, M. J. 2004. *Blackett: Physics, War, and Politics in the Twentieth Century.* Cambridge, MA: Harvard University Press.

——— 2006. "Scientific Biography: History by Another Means." *Isis* 97:322–329.

Olesko, K. 1991. *Physics as a Calling: Discipline and Practice in the Königsberg Seminar for Physics.* Ithaca: Cornell University Press.

Oppenheimer, J. R. 1941. "The Mesotron and the Quantum Theory of Fields." Pages 38–50 in Fermi et al. 1941.

Oppenheimer, J. R., and J. Schwinger. 1941. "On the Interaction of Mesotrons and Nuclei." *Physical Review* 16:150–152.

Oppenheimer, J. R., and R. Serber. 1937. "A Note on the Nature of Cosmic Ray Particles." *Physical Review* 51:1113.

Ott, H. 1928. *Strukturbestimmung mit Röntgeninterferenzen.* In W. Wien and F. Harms, eds., *Handbuch der Experimentalphysik,* vol. 7, part 2, 1–322.

Pais, A. 1982. "Max Born's Statistical Interpretation of Quantum Mechanics." *Science* 218:1193–98.

——— 1986. *Inward Bound: Of Matter and Forces in the Physical World.* Oxford: Clarendon Press.

——— 1991. *Niels Bohr's Times in Physics, Philosophy, and Polity.* Oxford: Clarendon Press.

Pais, A., M. Jacob, D. I. Olive, and M. F. Atiyah. 1998. *Paul Dirac: The Man and His Work.* Cambridge: Cambridge University Press.

Patterson, A. L. 1962. "Experiences in Crystallography—1924 to Date." In Ewald 1962, 612–622.

Pauli, W. 1946. *Meson Theory of Nuclear Forces.* New York: Interscience.

——— 1964. *Collected Works,* ed. R. Kronig and V. F. Weisskopf. New York: Interscience.

——— 1985. *Wissenschaftlicher Briefwechsel mit Bohr, Einstein, Heisenberg u.a. Band II: 1930–1939.* Edited by K. von Meyenn. Berlin: Springer Verlag.

——— 1994. *Writings on Physics and Philosophy,* ed. C. P. Enz and K. von Meyenn, trans. R. Schlapp. Berlin: Springer.

Pauly, P. J. 1987. *Controlling Life: Jacques Loeb and the Engineering Ideal in Biology.* New York: Oxford University Press.

Pegram, G. P. 1950. "John Torrence Tate 1889–1950." *Physics Today* 3:6–7.

Peierls, R. 1940. "The Bohr Theory of Nuclear Reactions." *Reports in Progress of Physics* 7:87–106. Reprinted in Dalitz and Peierls 1997, 283–302.

——— 1955. *Quantum Theory of Solids.* Oxford: Clarendon Press.

——— 1960. "Quantum Theory of Solids." In Fierz and Weisskopf 1960, 140–160.

——— 1980. "Recollections of Early Solid State Physics." In Mott 1980, 28–38.

——— 1984. "Reminiscences of Cambridge in the Thirties." In Hendry 1984, 195–200.

——— 1985. *Bird of Passage.* Princeton: Princeton University Press.

——— 1986. "Introduction." In N. Bohr, *Collected Works,* vol. 9: *Nuclear Physics (1929–1952),* ed. R. Peierls, 1–54. Amsterdam: North-Holland.

Pendry, J. B. 1974. *Low Energy Electron Diffraction: The Theory and Its Application to the Determination of Surface Structure.* New York: Academic Press.

Perrin, F. 1935. "Mécanisme de la Capture des Neutrons Lents par les Noyaux Legers." *Comptes Rendus de l'Académie des Sciences* 200:1749–1752.

Perrin, F., and W. M. Elsasser. 1935. "Théorie de la Capture Sélective des Neutrons par Certain Noyaux." *Comptes Rendus de l'Académie des Sciences* 200:450–452.

Pesic, P. 2005. "Earthly Music and Cosmic Harmony." *Journal of 17th Century Music* 11(1).

——— 2010. "Hearing the Irrational." *Isis* 101(3):501–530.

Philip, A. G., and D. DeVorkin, eds. 1977. *In Memory of Henri Norris Russell.* Albany: Dudley Observatory.

Philippson, J. 1962. "The Philippsons, a German-Jewish Family 1775–1933." Publications of the Leo Baeck Institute, vol. 7, 95–118.

Philippson, L. 1855. *The Development of the Religious Idea in Judaism, Christianity and Mahomedanism: Considered in Twelve Lectures on the History and Purport of Judaism.* Delivered in Magdeburg, 1847, by Ludwig Philippsohn; trans. and annotated by A. M. Goldsmid. London: Longman, Brown, Green and Longmans.

Phillips, D. 1979. "William Lawrence Bragg." *Biographical Memoirs of Fellows of the Royal Society* 25:74–143. Reprinted in J. M. Thomas and D. Phillips, eds. 1990. *Selections and Reflections: The Legacy of Sir Lawrence Bragg,* 1–70. Northwood: Science Reviews.

Placzek, G. 1934. "Rayleigh-Streuung und Raman-Effekt." In Erich Marx, ed., *Handbuch der Radiologie* 6(2):209–374. Leipzig: Akademische Verlagsgesellschaft.

Planck, M. 1915. *Eight Lectures on Theoretical Physics Delivered at Columbia University in 1909,* trans. A. P. Wills. New York: Columbia University Press.

——— 1915/1958. "Das Prinzip der kleinsten Wirkung." In M. Planck, *Vorträge und Reden,* 91–101. Braunschweig: Fried, Vieweg und Sohn.

——— 1932. *Der Kausalbegriff in der Physik.* Leipzig: J. A. Barth.

Pletsch, C. 1991. *Young Nietzsche. Becoming a Genius.* New York: The Free Press.

Poincaré, H. 1913. *Leçons sur les Hypothèses Cosmogoniques. Rédigées par Henri Vergne,* 2nd ed. Paris: A. Hermann et fils.

——— 1952. *Science and Hypothesis.* With a preface by J. Larmor. New York: Dover.

Polanyi, M., and E. Wigner. 1925. "Bildung und Zerfall von Molekülen." *Zeitschrift für Physik* 33:429–434.

Polkinghorne, J. C. 1989. *Rochester Roundabout: The Story of High Energy Physics.* New York: W. H. Freeman.

Porter, T. M. 2006. "Is the Life of a Scientist a Scientific Unit?" *Isis* 97:314–321.

"Progress in Physics. International Conference on Physics London 1934." 1935. *Nature* 136:49–50.

Przibram, K., ed. 1967. *Letters on Wave Mechanics: Schrödinger, Planck, Einstein, Lorentz,* trans. and with an intro. by M. J. Klein. New York: Philosophical Library.

Pyenson, L. 1977. "Hermann Minkowski and Einstein's Special Theory of Relativity." *Archive for History of Exact Sciences* 17:71–95.

——— 1983. *Neohumanism and the Persistence of Pure Mathematics in Wilhelmian Germany.* Philadelphia: American Philosophical Society.

——— 1985. *The Young Einstein: The Advent of Relativity.* Bristol: Adam Hilger.

Rabi, I. I. 1963. "Science in the Satisfaction of Human Aspiration." In *The Scientific Endeavor: Centennial Celebration of the National Academy of Sciences.* New York: Rockefeller Institute Press.

Rabinbach, A. 1997. *In the Shadow of the Catastrophe: German Intellectuals between Apocalypse and Enlightenment.* Berkeley: University of California Press.

Radder, H. 1983. "Kramers and the Forman Thesis." *History of Science* 21:165–182.

Rammelmayer, A. 1965. "Festrede des Rektor. Fünfzig Jahre Univerität Frankfurt am Main." In *Frankfurter Universitätreden,* vol. 33, 20–44. Ansprachen, Ehrungen und Gluckwünsche bei der Jubilaumfeier, Johann Wolfgang Goethe-Universität. Frankfurt am Main: Vittorio Klostermann.

Rasetti, F. 1936. *Elements of Nuclear Physics.* New York: Prentice-Hall.

Rebsdorf, S. O. 2007. "Bengt Strömgren: Interstellar Glow, Helium Content, and Solar Life Supply, 1932–1940." *Centaurus* 49:56–79.

Reichenbach, H. 1949. "Philosophical Significance of Relativity." In Schilpp 1949.

Reid, C. 1970. *Hilbert.* Berlin: Springer.

——— 1976. *Courant in Göttingen and New York: The Story of an Improbable Mathematician.* New York: Springer.

——— 1986. *Hilbert/Courant.* Berlin: Springer.

Reines, F., ed. 1972. *Cosmology, Fusion and Other Matters: George Gamow Memorial Volume.* Boulder: Colorado Associated University Press.

Rice, S. A., and F. H. Stillinger. 1999. "John Gamble Kirkwood." *Biographical Memoirs of the National Academy of Sciences* 77.

Richards, J. L. 2006. "Introduction: Fragmented Lives." *Isis* 97:302–306.

Richards, R. J. 2003. "Biology." In Cahan 2003, 16–48.

——— 2008. *The Tragic Sense of Life: Ernst Haeckel and the Struggle over Evolutionary Thought.* Chicago: University of Chicago Press.

Richarz, M. 1997. "Jewish Women in the Family and Public Sphere." In Meyer and Brenner 1997.

Richtmyer, F. K. 1928. *Introduction to Modern Physics,* 1st ed. New York: McGraw-Hill.

Ringer, F. K. 1969a. *The Decline of the German Mandarins: The German Academic Community, 1890–1933.* Cambridge, MA: Harvard University Press.

———, ed. 1969b. *The German Inflation of 1923.* New York: Oxford University Press.

——— 1979. *Education and Society in Modern Europe.* Bloomington: Indiana University Press.

——— 2004. *Max Weber: An Intellectual Biography.* Chicago: University of Chicago Press.

Robinson, R. 1953. "Richard Willstätter. 1872–1942." *Obituary Notices of Fellows of the Royal Society* 8(22):609–634.

Rogers, W. P. 1942. *Andrew D. White and the Modern University.* Ithaca: Cornell University Press.

Roqué, X. 1992. "Møller Scattering: A Neglected Application of Early Quantum Electrodynamics." *Archive for History of Exact Sciences* 44:197–264.

Rosenfeld, L. 1972. "Nuclear Reminiscences." In Reines 1972, 289–290.

Rossi, B. 1983. "The Decay of 'Mesotrons' (1939–1943): Experimental Particle Physics in the Age of Innocence." In L. M. Brown and L. Hoddeson, eds., *Pions to Quarks.* Cambridge: Cambridge University Press, 183–205.

Rowe, D. 1986. "'Jewish Mathematics' at Göttingen in the Era of Felix Klein." *Isis* 77(3):422–449.

——— 1989. "Klein, Hilbert and the Göttingen Mathematical Tradition." *Osiris,* 2d ser. 5:186–213.

——— 2004. "Making Mathematics in an Oral Culture: Göttingen in the Era of Klein and Hilbert." *Science in Context* 17(1–2):85–129.

Rubin, H. 1995. "Walter M. Elsasser 1904–1991." *Biographical Memoirs of the National Academy of Sciences* 1995:102–165.

Russell, B. 1945. *A History of Western Philosophy.* New York: Simon and Schuster.

Russell, H. N. 1933. *The Composition of the Stars.* Oxford: Clarendon Press.

——— 1939. "Stellar Energy and Bethe's Carbon Cycle." *Proceedings of the American Philosophical Society* 81:295–307.

Rutherford, E. 1937. *The Newer Alchemy.* New York: Macmillan.

Rutherford, E., J. Chadwick, and C. D. Ellis. 1930. *Radiations from Radioactive Substances.* Cambridge: Cambridge University Press.

Samuelsson, B., and M. Sohlman, eds. 1998. *Physics, 1963–1970.* River Edge, NJ: World Scientific.

Schatzman, E. 1970. *Physics and Astrophysics.* Lectures Given in the Academic Training Program. CERN 70–31. Geneva.

Schilpp, P. A., ed. 1949. *Albert Einstein: Philosopher-Scientist.* The Library of Living Philosophers, vol. 7. Evanston, IL: The Library of Living Philosophers.

Scholem, G. 1976. *On Jews and Judaism in Crisis,* ed. W. J. Dannhauser. New York: Schocken Books.

Schrödinger, E. 1926. "Quantisierung als Eigenvertproblem." *Annalen der Physik* 79:361–376; 79:489–527; 80:437–490; 81:109–139. All four papers are included in Schrödinger 1982.

——— 1982. *Collected Papers on Wave Mechanics, Together with His Four Lectures on Wave Mechanics.* New York: Chelsea.

Schwarzfuchs, S. 2003. "Alsace and Southern Germany: The Creation of a Border." In Brenner, Caron, and Kaufmann 2003.

Schweber, S. S. 1986. "The Empiricist Temper Regnant: Theoretical Physics in the United States." *Historical Studies in the Physical Sciences* 17(1):55–98.

——— 1988. "The Mutual Embrace of Science and the Military." In *Yearbook of the Sociology of Science 1988,* ed. E. Mendelsohn, M. Roe Smith, and Peter Weingart. Dordrecht: Kluwer Academic Publishers.

——— 1989a. "Molecular Beam Experiments, the Lamb Shift, and the Relation between Experiment and Theory." *American Journal of Physics* 57(4):299–304.

——— 1989b. "Wentzel, Gregor." *Dictionary of Scientific Biography,* C. C. Gillespie, ed. in chief. New York: Scribner's Sons.

——— 1994. *QED and the Men Who Made It: Dyson, Feynman, Schwinger and Tomonaga.* Princeton: Princeton University Press.

——— 2000. *In the Shadow of the Bomb: Bethe and Oppenheimer and the Moral Responsibility of the Scientist.* Princeton: Princeton University Press.

——— 2002. "Fermi's Quantum Electrodynamics." *Physics Today* 55(6):31–37.

——— 2006. "The Happy Thirties." In Brown and Lee 2006, 131–146.

——— 2008a. *Einstein and Oppenheimer: The Meaning of Genius.* Cambridge, MA: Harvard University Press.

——— 2008b. "Sommerfeld's Seminar and the Causality Principle." *Physics in Perspective* 10:1–41.

——— 2011. "Shelter Island Revisited." 2011 Pais Prize Lecture. APS April 2011 Meeting. Anaheim, CA.

Schwinger, J. 1948. "Quantum Electrodynamics. I. A Covariant Formulation." *Physical Review* 74:1439–1461.

——— 1970. "Charged Mesotron Field." In G. Wentzel, *Quanta: Essays in Theoretical Physics Dedicated to Gregor Wentzel,* ed. P. Freund, C. J. Goebel, and Y. Nambu, 101–138. Chicago: University of Chicago Press.

Seelig, C. 1954. *Eine Dokumentarische Biographie.* Zurich: Europa.

Segrè, E. 1970. *Enrico Fermi: Physicist.* Chicago: University of Chicago Press.

——— 1979. "Nuclear Physics in Rome." In Stuewer 1979, 35–63.

Seitz, F. 1994. *On the Frontier: My Life in Science.* New York: American Institute of Physics.

Seth, S. 2007. "Crisis and the Construction of Modern Physics." *British Journal of the History of Science* 40(1):25–51.

——— 2010. *Crafting the Quantum: Arnold Sommerfeld and the Practice of Theory, 1890–1926.* Cambridge, MA: MIT Press.

Shuranova, Z., and Y. Burmistrov. 2006. "Albrecht Bethe's Contribution to Crustacean Neuroscience: Hundred Years Ago and Today." *Crustaceana* 79(1):99–122.

Sigurdsson, S. 1991. "Hermann Weyl, Mathematics and Physics, 1900–1927." Ph.D. diss., Harvard University.

——— 1996. "Sublime and Worldly Crystals." *Annals of Science* 53:85–88.

Slezkine, Y. 2004. *The Jewish Century.* Princeton: Princeton University Press.

Smith, G. E. 2002. "From the Phenomenon of the Ellipse to an Inverse-Square Force: Why Not?" In Malament 2002, 31–70.

Smith, L. P. 1932. "Calculation of the Quantum Defect for Highly Excited S States of Para- and Orthohelium." *Physical Review* 42:176–181.

Sommerfeld, A. 1910. "Vierdimensionale Vektorenalgebra." *Annalen der Physik* 32:749–776; 33:649–698. Reprinted in Sommerfeld 1968b, vol. 2, 189 and 217.

——— 1923. "'Notes' to H. Minkowski's 'Space and Time.'" In H. A. Lorentz, A. Einstein, H. Minkowski, and H. Weyl, *The Principle of Relativity.* New York: Methuen Books.

——— 1927. *Three Lectures on Atomic Physics.* New York: E. P. Dutton.

——— 1928a. "Zur Elektronentheorie der Metalle auf Grund der Fermischen Statistik. I. Teil: Allgemeines, Strömungs- und Austrittsvorgänge." *Zeitschrift für Physik* 47(1–2):1–32.

——— 1928b. "Zur Elektronentheorie der Metalle auf Grund der Fermischen Statistik. II. Teil: Thermo-elektrische, galvano-magnetische und thermomagnetische Vorgänge." *Zeitschrift für Physik* 47(1–2):43–60.

——— 1929. *Atombau und Spektrallinien: Wellenmechanischer Ergänzungsband.* Braunschweig: F. Vieweg. Translated by H. L. Brose as *Wave-Mechanics: Supplementary Volume to Atomic Structure and Spectral Lines* (New York: E. P. Dutton, 1930).

——— 1936. "Wege zu physicalischen Erkentniss." *Scientia* 51:181–187. Reprinted in Sommerfeld 1968b, vol. 4, 608–614.

——— 1949a. *Partial Differential Equations in Physics,* trans. Ernst G. Straus. New York: Academic Press.

——— 1949b. "Some Reminiscences of My Teaching Career." *American Journal of Physics* 17:315.

——— 1954. *Optics,* trans. Otto Laporte and Peter A. Moldauer. New York: Academic Press.

——— 1968a. "Autobiographical Skizze." In Sommerfeld 1968b.

——— 1968b. *Gesammelte Schriften,* 4 vols. Hrsg. im Auftrag und mit Unterstützung der Bayerischen Akademie der Wissenschaften von F. Sauter. Braunschweig: Freid. Vieweg und Sohn.

——— 2000. *Wissenschaftlicher Briefwechsel,* vol. 1: *1892–1918,* ed. Michael Eckert and Karl Märker. Diepholz: GNT-Verlag.

Sommerfeld, A., and G. Schur. 1930. "Über den Photoeffekt in der K-Schale der Atome, insbesondere über die Voreilung der Photoelektronen." *Annalen der Physik* 4:409–432. Reprinted in Sommerfeld 1968b, vol. 4, 136–159.

Sorkin, D. 1983. "Wilhelm Von Humboldt: The Theory and Practice of Self Formation (Bildung) 1791–1810." *Journal of the History of Ideas* 44:55–73.

——— 1999. *The Transformation of German Jewry, 1780–1840.* Detroit: Wayne State University Press.

——— 2001. "The Émigré Synthesis: German-Jewish History in Modern Times." *Central European History* 34(4):531–559.

Speiser, A. 1923. *Die Theorie der Gruppen von Endlicher Ordnung, mit Anwendungen auf Algebraische Zahlen und Gleichungen sowie auf die Kristallographie.* Berlin: J. Springer.

Spengler, O. 1926. *The Decline of the West.* Authorized translation with notes by Charles Francis Atkinson. 2 vols. New York: A. A. Knopf.

Stachel, J. 2002. *Einstein from 'B' to 'Z.'* Boston: Birkhäusen.

Staley, R. 2008. *Einstein's Generation: The Origins of the Relativity Revolution.* Chicago: University of Chicago Press.

Stefan, V., ed. 1998. *Physics and Society: Essays in Honor of Victor Weisskopf.* New York: Springer.

Stein, H. 1994. "Some reflections on the structure of our knowledge in Physics." In D. Prawitz, B. Skyrms, and D. Westerståhl, eds., *Logic, Methodology, and Philosophy of Science* 9. Proceedings of the Ninth International Congress of Logic, Methodology, and Philosophy of Science, 633–655. New York: Elsevier.

Stepan, N. L., and S. L. Gilman. 1991. "Appropriating the Idioms of Science: The Rejection of Scientific Racism." In D. LaCapra, ed., *The Bounds of Race: Perspectives on Hegemony and Resistance,* 72–103. Ithaca: Cornell University Press.

Stern, F. 1999. *Einstein's German World.* Princeton: Princeton University Press.

Stoetzler, M. 2008. *The State, the Nation, and the Jews: Liberalism and the Antisemitism Dispute in Bismarck's Germany.* Lincoln: University of Nebraska Press.

Street, J. C., and E. C. Stevenson. 1937. "New Evidence for the Existence of a Particle of Mass Intermediate between the Proton and the Electron." *Physical Review* 52:1003–1004.

Strenski, I. 1997. *Durkheim and the Jews of France.* Chicago: University of Chicago Press.

Strömgren, B. 1937. "Die Theorie des Sterninnern und die Entwicklung der Sterne." *Ergebnisse der exakten Naturwissenschaften* 16:465–534.

——— 1938. "On the Helium and Hydrogen Content of the Interior of the Stars." *Astrophysical Journal* 87:520–534.

——— 1951a. "On the Development of Astrophysics during the Last Half Century." In Hynek 1951, 1–11.

——— 1951b. "The Growth of Our Knowledge of the Physics of the Stars." In Hynek 1951, 172–258.

Strutt, R. J. (Lord Rayleigh). 1942. *The Life of Sir J. J. Thomson, O. M., Sometime Master of Trinity College, Cambridge.* Cambridge: Cambridge University Press.

Stuewer, R. 1975. *The Compton Effect: Turning Point in Physics.* New York: Science History Publications.

Stuewer, R., ed. 1979. *Nuclear Physics in Retrospect: Proceedings of a Symposium on the 1930s.* Minneapolis: University of Minnesota Press.

Stuewer, R. 1984. "Nuclear Physicists in a New World: The Émigrés of the 1930s in America." *Berichte zum Wissenschafigeschicht* 7:32–40.

——— 1997. "Gamow, Alpha Decay, and the Liquid-Drop Model of the Nucleus." In Harper et al. 1997, 30–43.

Sutherland, B. 2004. *Beautiful Models: 70 Years of Exactly Solved Quantum Many-Body Problems.* River Edge, NJ: World Scientific.

Teller, E., with Judith Shoolery. 2001. *Memoirs: A Twentieth-Century Journey in Science and Politics.* Cambridge, MA: Perseus.

Terrall, M. 2006. "Biography as Cultural History of Science." *Isis* 97:306–313.

Thauer, R. 1955. "Albrecht Bethe." *Archiv für die gesammte Physiologie* 261:i–xiv.

Thomson, J. J. 1904. *Conduction of Electricity through Gases.* Cambridge: Cambridge University Press.

Thomson, W. 1882–1911. *Mathematical and Physical Papers,* 6 vols. Cambridge: Cambridge University Press.

Thorpe, C. 2006. *Oppenheimer: The Tragic Intellect.* Chicago: University of Chicago Press.

Tillman, J. R., and B. P. Moon. 1935. "Selective Absorption of Slow Neutrons." *Nature* 136:66–67.

Tropp, E. A., V. A. Frankel, and A. D. Chernin. 1993. *Alexander A. Friedmann: The Man Who Made the Universe Expand.* Cambridge: Cambridge University Press.

Turner, R. S. 1971. "The Growth of Professorial Research in Prussia, 1818 to 1848: Causes and Context." *Historical Studies in the Physical Sciences* 3:137–182.

——— 1998. "Introduction." In Mayer 1944.

Turner, R. S., E. Kerwin, and D. Woolwine. 1984. "Careers and Creativity in Nineteenth-Century Physiology: Zloczower Redux." *Isis* 75:523–529.

Tuve, M. A., N. P. Heydenburg, and L. R. Hafstad. 1936. "The Scattering of Protons by Protons." *Physical Review* 50:806–825.

Utiyama, R. 1956. "Invariant Theoretical Interpretation of Interaction." *Physical Review* 101:1597.

van der Waerden, B. L., ed. 1968. *Sources of Quantum Mechanics*. New York: Dover.

van Dongen, J. 2010. *Einstein's Unification*. Cambridge: Cambridge University Press.

Van Rahden, T. 2005. "Jews and the Ambivalences of Civil Society in Germany, 1800–1933: Assessment and Reassessment." *Journal of Modern History* 77(4):1024–1047.

van Vleck, J. H. 1932. *The Theory of Electric and Magnetic Susceptibilities*. Oxford: Oxford University Press.

Volkov, S. 1996. "The Ambivalence of *Bildung:* Jews and Other Germans." In K. L. Berghahn, ed., *The German-Jewish Dialogue Reconsidered: A Symposium in Honor of George L. Mosse,* 81–98. New York: Peter Lang.

——— 2001. "Jewish Scientists in Imperial Germany. Part I and II." *Aleph. Historical Studies in Science and Judaism* 1: 215–281.

——— 2006. *Germans, Jews, and Antisemites: Trials in Emancipation*. Cambridge: Cambridge University Press.

von Holtz, E. 1955. "Albrecht Bethe." *Die Naturwissenschaften* 42:99–101.

von Laue, M. 1931. "The Diffraction of an Electron-Wave at a Single Layer of Atoms." *Physical Review* 37:53–59.

Wachsmuth, R. 1929. *Die Gründung der Universität Frankfurt*. Frankfurt am Main: Englert und Schlosser.

Wali, K. C. 1991. *Chandra: A Biography of S. Chandrasekhar*. Chicago: University of Chicago Press.

Warwick, A. 2003. *Masters of Theory: Cambridge and the Rise of Mathematical Physics*. Chicago: University of Chicago Press.

Weber, Marianne. 1975. *Max Weber: A Biography,* trans. and ed. Harry Zohn. New York: John Wiley and Sons.

Weber, Max. 1919. "Science as Vocation" [Wissenschaft als Beruf]. Republished in *From Max Weber: Essays in Sociology,* trans. and ed. H. H. Gerth and H. Wright Mills, 129–156. Oxford: Oxford University Press.

Weiner, C., ed. 1972. *Exploring the History of Nuclear Physics; Proceedings of the American Institute of Physics-American Academy of Arts and Sciences Conferences on the History of Nuclear Physics, 1967 and 1969,* ed. C. Weiner, assisted by E. Hart. New York: American Institute of Physics.

Weisskopf, V. 1991. *The Joy of Insight: Passions of a Physicist*. New York: Basic Books.

Weisskopf, V., and E. Wigner. 1930. "Berechnung der naturlichen Linienbreite auf Grund der Diracschen Lichttheorie." *Zeitschrift für Physik* 63:54–73.

Weizsäcker, C. F. 1937. *Die Atomkerne: Grundlage und Anwendugen ihrer Theorie*. Leipzig: Akademische Verlagsgesellschaft.

——— 1938. "Die Elementumwandlungen im Innern der Sterne." *Physikalische Zeitschrift* 38:176–191.

Wenkel, S. 2007. "Jewish Scientists in German-Speaking Academia: An Overview." In Charpa and Deichmann 2007, 265–284.

Wentzel, G. 1940. "Zum Problem des statischen Mesonfeldes." *Helvetica Physica Acta* 13:269–308.

——— 1943. *Einführung in die Quantentheorie der Wellenfelder.* Vienna: F. Deuticke.

——— 1964. Transcript of an interview by Thomas S. Kuhn. February 3–5. American Institute of Physics, College Park, MD.

Weyl, H. 1923. *The Principle of Relativity.* New York: Methuen.

White, A. D. 1876. *The Warfare of Science.* New York: D. Appleton.

——— 1896. *A History of the Warfare of Science with Theology in Christendom,* 2 vols. London: Macmillan.

——— 1905. *Autobiography of Andrew Dickson White,* 2 vols. New York: Century.

Wiechert, E. 1903. "Theorie der automatischen Seiemographen." *Abhand. König. Gesellschaft Göttingen.* Math. Phys. Klasse II:1–124.

Wien, W., and F. Harms, eds. 1928. *Handbuch der Experimentalphysik,* vol. 7, pt. 2. Leipzig: Akademische Velagsgesellschaft.

Wigner, E. P. 1933. "On the Mass Defect of Helium." *Physical Review* 43:252–257.

——— 1955. "On the Development of the Compound Nucleus Model." *American Journal of Physics* 23:371–380.

——— 1979. "The Neutron: The Impact of Its Discovery and Its Uses." In Stuewer 1979, 157–178.

——— 1992. *The Recollections of E. P. Wigner as Told to Andrew Szanton.* New York: Plenum Press.

Williams, E. J. 1945. "Application of Ordinary Space-Time Concepts in Collision Problems and Relation of Classical Theory to Born's Approximation." *Reviews of Modern Physics* 17(2–3):217–226.

Williams, J. H., W. G. Shepherd, and R. O. Haxby. 1937. "Evidence for the Instability of ^{5}He." *Physical Review* 51:888–889.

Willstätter, R. 1965. *From My Life.* New York: Benjamin.

Wilson, D. 1983. *Rutherford, Simple Genius.* London: Hodder and Stoughton.

Wolff, S. L. 2003. "Physicists in the 'Krieg der Geister': Wilhelm Wien's 'Proclamation.'" *Historical Studies in the Physical and Biological Sciences* 33(2):337–368.

Yang, C. N., and M. L. Ge. 2006. "Bethe's Hypothesis." In Brown and Lee 2006, 185–188.

Yerushalmi, Y. H. 1982. "Assimilation and Anti-Semitism: The Iberian and the German Models." Leo Baeck Memorial Lecture 26. New York: Leo Baeck Institute.

York, H. F. 1987. *Making Weapons, Talking Peace: A Physicist's Odyssey from Hiroshima to Geneva.* New York: Basic Books.

Yukawa, H. 1935. "On the Interaction of Elementary Particles. I." *Proceedings of the Physico-Mathematical Society of Japan* 17:48–57.

Yukawa, H., S. Sakata, M. Kobayashi, and M. Taketani. 1938. "On the Interaction of Elementary Particles. IV." *Proceedings of the Physico-Mathematical Society of Japan* 20:720–745.

Yukawa, H., S. Sakata, and M. Taketani. 1938. "On the Interaction of Elementary Particles. III." *Proceedings of the Physico-Mathematical Society of Japan* 20:319–340.

Acknowledgments

This book has had a long gestation period. When Hans Bethe asked me to write his biography in the fall of 1988 he had in mind a scientific biography. My extended interviews with him in the early 1990s altered the project. It became clear to me that I wanted to write something more than an intellectual biography that would also encompass his activities in influencing post–World War II national and international nuclear policy. What emerged from my protracted efforts to fulfill both his and my aspirations was my book *In the Shadow of the Bomb: Bethe and Oppenheimer and the Moral Responsibility of the Scientist* (Princeton University Press, 2000) and the present volume.

To be trusted with his personal reminiscences and with his assessment of his life, and entrusted with narrating the story of his life to a wider audience, has been an exhilarating privilege. I hope that what comes through in the biography is my deep admiration and respect for him. I came to love this man. Yet I also have tried very hard to be a responsible historian in describing his life.

My biography of him could not have been written without Rose Bethe, Hans Bethe's partner in life for over sixty years. I hope that the reader will come to see that the book is also very much her biography. I cannot adequately thank her. I have dedicated the book to her as a token of my appreciation and indebtedness. Similarly, Hans and Rose's children, Henry and Monica, have been most generous in forthrightly sharing their memories and insights. I am deeply indebted to them.

When I first started doing the research for the book and interviewing people I was still involved with my history of quantum electrodynamics in the 1947–1952 period. I therefore concentrated on Bethe's work from 1939 on. There was also a special emphasis on Los Alamos because I believed that that experience had shaped what he did afterward. The scourge of time has taken away many of the people I then interviewed, among them: Robert Bacher, Francis Low, Robert Marshak, Rudolf Peierls, Marshall Rosenbluth, Ed Salpeter, Ted Taylor, Edward Teller, Robert R. Wilson, and Victor Weisskopf. Needless to say their recollections of their interactions with Bethe and their perceptions and perceptiveness were of great value.

In this connection, I would like to point to the remarkable interviews of physicists that Charles Weiner conducted in the 1960s and 1970s for the Center for the History of Physics at the American Institute of Physics and its Niels Bohr Library and Archives. His interviews of Hans Bethe, Richard Feynman, Robert Marshak, Victor Weisskopf, and others have been extremely important for my work, past and present.

Interviewing people who knew the social setting of Frankfurt when Bethe grew up there allowed me to get to know some remarkable people. Two such women whose insights and forthrightness were invaluable in writing the present volume stand out: Anne Fischer of Richmond, Virginia, whose husband was an *Assistent* of Bethe's father at Frankfurt University and who was part of the social circle of the parents of the young people who were Hans's friends; and Hilde Levi of Copenhagen, whose relation to Hans Bethe is narrated in Chapter 6. I might add that a detailed social history of women who had been raised in assimilated German Jewish families at the beginning of the twentieth century, had obtained a university education, and who after the advent of Hitler had emigrated from Germany would be most welcome.

Michel Baranger, Laurie Brown, Jed Buchwald, David Cahan, Morton Camac, Tian Yu Cao, Cathryn Carson, Stanley Cohen, Dale Corson, Freeman Dyson, Ken Ford, Paul Forman, Evelyn Fox Keller, Vounia Gissis, Sorel Gottfried, Karl Hall, Anne Harrington, David Hecht, John Heilbron, Steve Heims, Arne Hessenbruch, Erwin Hiebert, Gerald Holton, Roman Jackiw, Sheila Jasanoff, Mikhel Johnson, David Kaiser, Barbara and Mortimer Litt, Priscilla McMillan, Everett Mendelsohn, John Negele, John Rigden, Fritz Rohrlich, Howard Schnitzer, Suman Seth, Skuli Sigurdsson, Roger Stuewer, Shulamit Volkow, Alex Wellenstein, and Dusha Weisskopf all contributed in one way or another. I thank them. I would like to especially mention Paul Hartman and Karl Hufbauer. When I first began, Paul made available to me both his history of the Cornell Physics Department as well as the notebooks of the courses he took with Bethe as a graduate student in the late 1930s. Karl Hufbauer shared with me his research on Bethe's solution of the stellar energy problem, much of which, regrettably, has remained unpublished. I have tried to emulate his helpfulness and collegiality and to meet the standards of scholarship that he has set in his writings. I am very much indebted to him.

David Cassidy read the entire manuscript and made very helpful comments and suggestions. So did Kurt Gottfried and much more. Kurt was a colleague and a close friend of Bethe for over forty years. He graciously shared with me his knowledge of him, of physics, of the physics community, and of Cornell and in addition gave the hospitality of his home. I am deeply indebted to him and to his wife. Similarly, my friendship with Paul Forman made the biography a better book: I have tried to meet the standards he has set for himself as a historian.

Books such as the present one do not get written without the assistance of librarians. I would like to acknowledge Finn Aaserud's help both as a historian and as the director of the Niels Bohr Library. I cannot effectively thank Elaine Engst, the director, and David Corson, the curator of the history of science collections, of the Division of Rare and Manuscript Collections of the Kroch Library at Cornell for the assistance they gave me in my perusal of countless number of boxes of Bethe's *Nachlass,* some of them only very recently catalogued. Similarly I would like to thank Paul Miner, the administrative head of the Cornell department of physics, who went through all the departmental records and made available to me the departmental letters concerning Bethe in the 1930s. I would also like to thank Roger Meade at the Los Alamos Archives, and the librarians at the MIT and Harvard Archives for their help.

Certainly, one of the enabling conditions that made it possible to write this book was access to the amazing resources contained in the Harvard Libraries. I thank the Harvard department of the History of Science for this and for the many other benefits stemming from being an Associate in it.

The encouragement I received from Michael Fisher at Harvard University Press and his insistence to publish at the earliest possible date were very important and very meaningful. I am very grateful to him. Lauren Esdaile's efficient and knowledgable initial handling of the manuscript was very helpful. Anne McGuire carefully went over the bibliography and helped make it reliable and readily accessible. Nicholas Koenig prepared a comprehensive index. The book is a much better one for the care and thoroughness with which Katherine Brick edited the manuscript. I cannot adequately sing her praises nor satisfactorily thank her.

Needless to say, I take full responsibility for the mistakes that remain in the printed version of the book.

Finally, I would like to acknowledge the presence of Snait Gissis in my life and in this book. There is very little in the book that has not benefitted from her probing questions, her constructive criticisms, her erudition, her good judgment, and her editorial acumen. And her companionship and love have given me the best years of my life.

Index